Birkhäuser

Paolo Maria Mariano • Luciano Galano

Fundamentals of the Mechanics of Solids

Birkhäuser

Paolo Maria Mariano
DICeA
University of Florence
Firenze, Italy

Luciano Galano
DICeA
University of Florence
Firenze, Italy

Eserciziario di Meccanica delle Strutture
Original Italian edition published by © Edizioni CompoMat, Configni, 2011

ISBN 978-1-4939-4718-8 ISBN 978-1-4939-3133-0 (eBook)
DOI 10.1007/978-1-4939-3133-0

Mathematics Subject Classification (2010): 7401, 7402, 74A, 74B, 74K10

Springer New York Heidelberg Dordrecht London

Softcover re-print of the Hardcover 1st edition 2015

Printed on acid-free paper

Springer Science+Business Media LLC New York is part of Springer Science+Business Media (www.springer.com)

Preface

This book emerged from an "experiment in didactics" that has been under development at the University of Florence (Italy) since the academic year 2006/2007: teaching the mechanics of solids to undergraduate students in mechanical engineering in a deductive way from a few first principles, including essential elements of the description of finite-strain behavior and paying attention to the role of invariance properties under changes in observers.

We do not claim originality for this program and are also conscious of the advantages of other approaches: ours is just a description of the origin of a choice dictated by personal taste and history. It has been also motivated by the consciousness that a merely descriptive style can risk reducing the treatment to a list of special examples or formulas with unexplained origin, while an inductive approach could give, even indirectly, prominence to reasoning by analogy, blurring in some way the logical structure of the theory.

Since its inception, about 170 have taken the 84-hour course each year. All students had previous training in analysis and geometry, including basic elements of linear algebra. They were also trained in rational mechanics of mass points and rigid bodies, which is taught in a course of the same length. Their mathematical background had been enlarged during the course by requisite notions from tensor algebra and analysis. We collect the pertinent material in an appendix, where we clarify further the notation adopted in the text. One of us (PMM) developed the "experiment" varying the course every year according to student response and the changes in his perception. The other (LG) began later to transfer appropriate portions of the spirit of the course to analogous courses in other fields of engineering.

There are several introductory textbooks on the mechanics of solids. Some of them follow a strictly deductive program rather than being primarily descriptive or having an inductive approach. We have written this book by following our personal taste, with the goal of organizing the subject matter in a way that prepares the readers for further study, being conscious that the development of mechanics could require even modifications of the first principles. Beyond the technical aspects, our conviction is that the subject must be presented in a critical way, without giving the reader the impression that it has been constructed as an immutable structure crystallized once and for all. In fact, it seems to us that a merely dogmatic approach to mechanics does not contribute to the possibility of deeply investigating the foundational aspects of the subject. In contrast, the attention to foundational aspects is the primary tool for constructing new models, even new theories: families of interconnected models. The interest for the analysis of the theoretical foundations is not a mere interest for the formal structure of the theories; rather, it has to be stimulated in the students in university courses, even though perhaps only a few of them will be involved in research activities after the completion of their education. To us, even those who will work as professional engineers can take meaningful advantage of this type of program so that they might eventually have the flexibility to learn and (perhaps) manage new models and techniques, those that might be developed to satisfy future technological needs or, above all, for giving us a better knowledge of nature. Moreover, an attitude that favors the comprehension and analysis of the foundational aspects of mechanical theories encourages one to search for the physical meaning of every formal step we do, on the basis of our analytical, geometric, and/or computational skills.

In this spirit, we begin with the definition of bodies and deformation, recovering the kinematics of the rigid ones as a special case. In this way, we establish a link with the basic courses in rational mechanics of mass points and rigid bodies, showing how the subject matter we present is a natural continuation of the previous topics. We distinguish between the space in which we select the reference point for a body and the one in which we record shapes that we consider deformed. The second space is what we consider the physical one, the first being just a "room" used for comparing lengths, areas, volumes, with their prototypical counterparts that we declare to be undeformed. This unusual distinction allows us to clarify some statements concerning changes in observers and related invariance properties.

We distinguish also between material and spatial metrics, each defined in the pertinent space. Then finite-strain measures emerge from the comparison between one metric and the pullback of the other in the space where we decide to compare the two. Small-strain deformation tensors arise from the linearization process. This is the topic of Chapter 1.

Chapter 2 deals with the definition of observers and a class of their possible changes, those determined by rotating and translating frames (i.e., coordinate systems) in the ambient physical space. We call these changes in observers *classical*. We suggest options for them, all pertaining to the way in which we alter frames in space, indeed, irrespectively of the type of body considered; in fact, the class of changes in observers is not to be confused with the class of admissible motions for a body, although the two classes intersect.

In Chapter 3, we tackle the representation of bulk and contact actions in terms of the power they develop. We write just the external power on a generic part of the body and require its invariance under classes of isometric changes in observers. The integral balances of forces and couples emerge as a result. Then they are used to derive the action–reaction principle, the existence of the stress tensor, the balance equation in Eulerian and Lagrangian descriptions, the expression of the internal (or inner) power in both representations. The approach follows the spirit of a 1963 proposal by Walter Noll.[1]

Chapter 4 deals with constitutive issues. We discuss the way of restricting a priori the set of possible constitutive structures on the basis of the second law of thermodynamics—here presented as a mechanical dissipation inequality—and on requirements of objectivity. Our attention is essentially focused on nonlinear and linearized elasticity. We discuss also the notion of material isomorphism. Incidentally, when we foresee changes in observers in the reference (material) space, the requirement that the observers record the same material forces the change in observer itself to preserve the volume, according to the definition of material diffeomorphism, irrespectively of the type of body under scrutiny. Such classes of changes in observers become crucial in the description of material mutations, a topic not treated here, since it goes beyond the scope of this book.

In Chapter 5, we discuss variational principles in linearized elasticity. Among them, the Hellinger–Prange–Reissner and Hu–Washizu principles are additional to the material constituting the course mentioned repeatedly above. The chapter includes also Kirchhoff's uniqueness theorem, and the Navier and Beltrami–Donati–Michell equations. The latter equations are essential tools for the analyses developed in the subsequent chapter. We end the chapter with some remarks on two-dimensional equilibrium problems.

Chapter 6 deals with the de Saint-Venant problem: the statics of a linear elastic slender cylinder, free of weight, loaded just on its bases. There are two ways of discussing such a problem: in terms of displacements or stresses. We follow the second approach and are indebted to the 1984 treatise in Italian on the matter by Riccardo Baldacci.[2] The chapter ends with a proof of the basic Toupin's theorem on the de Saint-Venant principle.

Chapter 7 includes a description of some yield criteria and a discussion of their role in the representation of the material behavior. There are several criteria, introduced for various reasons, not all of the same importance. Our choice is to include in this book just the classical ones, and nothing more.

In one aspect, Chapter 8 is separate from the program followed in the course mentioned above. The chapter includes director-based models of rods, a term used here in a broad sense for rods themselves, beams, shafts, columns, etc. Their description is a revisitation in terms of invariance of the external power under changes in observers—the view followed for the three-dimensional continuum—of

[1]Noll W. (1963), La Mécanique classique, basée sur une axiome d'objectivité, pp. 47–56 of *La Méthode Axiomatique dans les Mécaniques Classiques et Nouvelles* (Colloque International, Paris, 1959), Gauthier-Villars, Paris.

[2]Baldacci R. (1984), *Scienza delle costruzioni*, vol. I, UTET, Torino.

a 1985 proposal by Juan Carlos Simo.[3] In the chapter, we include both the finite-strain and linearized treatments; the course that we taught involved just the latter one.

Chapter 9 is an overview of some bifurcation phenomena. Attention is essentially focused on the Euler rod.

This book can be used variously for a course in the mechanics of solids, with the instructor selecting some parts and neglecting others. Ours is just a proposal.

In ending this work, we have to express our gratitude to the Birkhäuser team for their help and in particular for their understanding of the reasons for our falling behind the original schedule. Among others, we mention and thank Allen Mann and Christopher Tominich for the care they have taken in following our work during different portions of its development. Also, we thank the copyeditor, David Kramer, for his work.

Firenze, Italy

Paolo Maria Mariano
Luciano Galano

April 2015

[3] Simo J. C. (1989), A finite-strain beam formulation. The three-dimensional dynamic problem. Part I, *Comp. Meth. Appl. Mech. Eng.* **49**, 55–70.

Contents

List of Figures

List of Tables

Chapter 1
Bodies, Deformations, and Strain Measures

1.1 Representation of Bodies

We have a body—a piece of the phenomenological world—in hand, and we want to evaluate its changes in shape, under the action of external agencies, with respect to a shape taken as a reference. In imagining a program for such an analysis, the first step is to specify *exactly* what we mean by the word *shape*. The observation of condensed matter at various spatial scales often reveals intricate molecular and/or atomic architectures. This observation in itself furnishes just a partial view of the material structure of a body, depending on the type of instrument used and its resolution. The representation of the material's morphology (the geometry of a body at various spatial scales) that we may propose—indeed construct—from time to time is then a partial result with respect to the complex reality. The minimal essential information that we need to make assertions about the material morphology is the specification of a region $\mathcal{B}$ that the body under investigation might in principle occupy in space, the Euclidean point space. Once such a choice has been made, further information on the geometry at finer spatial scales can be added. We could, in other words, construct a sort of hierarchy going deeper and deeper in the evaluation of details that we want to include in the representation of the body's morphology, which is what we call its *shape*.

For the moment and the purposes of this book, here we limit the hierarchy to the (let us say) zeroth degree. Hence, from now on, unless otherwise stated, we shall identify the shape of a body with the region $\mathcal{B}$ of the physical space occupied by it in certain circumstances. Among infinitely many possibilities, restrictions on the choices of $\mathcal{B}$ are imposed by what we intuitively consider bodies in our daily experience and the technical requirement that we use tools of mathematical analysis such as the divergence theorem.

According to these restrictions, an appropriate choice is to consider $\mathcal{B}$ to be a subset of the Euclidean point space $\mathcal{E}^m$, $m = 1, 2, 3$, which is *regularly open*—it is,

P.M. Mariano, L. Galano, *Fundamentals of the Mechanics of Solids*,
DOI 10.1007/978-1-4939-3133-0_1

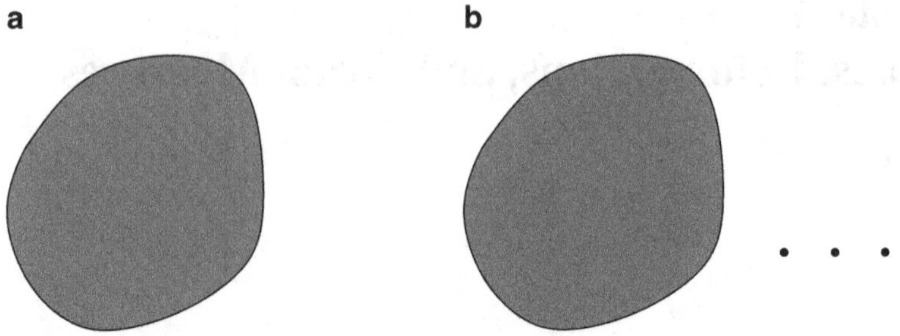

Fig. 1.1 The region indicated in panel (*a*) is an appropriate choice for $\mathcal{B}$. What is indicated in panel (*b*) is not appropriate (the aspect certifying the inappropriateness is the presence of the three dots on the right, which render the set not regularly open)

therefore, by definition, an open set that coincides with the interior of its closure[1]—and is endowed with a surface-like boundary, oriented by the (outward) unit normal n to within a finite number of corners and edges (Fig. 1.1).

From now on we fix $m = 3$, unless otherwise stated. The choice is made to help the intuition of the reader in visualizing physical concepts and the interpretation of them. Over $\mathcal{E}^3$, the translation space $\mathcal{V}^3$ is naturally defined: it is the space of vectors generated by differences of points. A basis—that is, a set of linearly independent vectors spanning $\mathcal{V}^3$, say e_1, e_2, e_3—is available. A generic point in $\mathcal{B}$ is indicated by x. Once we take a specific point in $\mathcal{E}^3$ and give it the role of the origin of a coordinate system, we transform $\mathcal{V}^3$ into $\mathbb{R}^3$. In the frame selected, x is identified by a triplet of numbers, namely x^1, x^2, x^3, when (as decided previously) the ambient space is three-dimensional; otherwise, the list is enlarged or reduced according to the dimension of the host space. In the subsequent pages, subsets of $\mathcal{B}$ with nonzero volume and the same geometric properties of $\mathcal{B}$ itself will be called *parts*. The same terminology will be used for subsets of images of $\mathcal{B}$ under a special class of maps describing the change of place of a body.

1.2 Deformations

We fix a possible location $\mathcal{B}$ for a body and take it as a *reference* to define in what sense another macroscopic shape can be considered deformed.

[1]The closure of a set $\mathcal{A}$ of a given space (which is itself a set) is the union of $\mathcal{A}$ with the set of its accumulation points, the elements not belonging to $\mathcal{A}$ but having elements of $\mathcal{A}$ in every neighborhood. The notion of open set foresees the preliminary assignment of a topology in the space where $\mathcal{A}$ is selected, namely a collection of sets (that we call *open*) containing the empty set, the whole space, and such that the union of every family of open sets is in the class, as is the intersection of every collection of finitely many open sets.

It is not necessary that the body under scrutiny occupy $\mathcal{B}$. It is sufficient that it could do so. The location $\mathcal{B}$ is a geometric setting in which we can compare lengths, areas, and volumes.

We select $\mathcal{B}$ *not* in the physical space (or better, the space that we consider so), but rather in an isomorphic copy of it. Hence we distinguish the space $\mathcal{E}^3$ (or $\mathbb{R}^3$ if you wish) where we choose $\mathcal{B}$ from the one, namely $\tilde{\mathcal{E}}^3$ (or $\tilde{\mathbb{R}}^3$), containing all the other places that we compare with the reference one—the isomorphism between the two is the natural identification $i : \mathcal{E}^3 \longmapsto \tilde{\mathcal{E}}^3$ (or $i_{\mathbb{R}} : \mathbb{R}^3 \longmapsto \tilde{\mathbb{R}}^3$)—or it can be an orientation-preserving isometry (a rotation and a translation). Other reasons justifying such a point of view will be clear later when we discuss the notion of observers and their related changes.

Once we assign coordinate systems to $\mathcal{E}^3$ and $\tilde{\mathcal{E}}^3$, we require that they have the same orientation. In the translational space $\tilde{\mathcal{V}}^3$, a basis is indicated by $\tilde{e}_1, \tilde{e}_2, \tilde{e}_3$, and a generic point in $\tilde{\mathcal{E}}^3$ is indicated by y, with components y^1, y^2, y^3.

From $\mathcal{B}$, we determine other places in $\tilde{\mathcal{E}}^3$ by means of maps

$$x \longmapsto y := \tilde{y}(x) \in \tilde{\mathcal{E}}^3, \qquad x \in \mathcal{B},$$

which we presume are (*i*) one-to-one, (*ii*) at least piecewise differentiable, and (*iii*) orientation-preserving. The last requirement implies that the generic $\tilde{y}$ maps every oriented volume in $\mathcal{B}$ onto a volume (remember the requirement of bijectivity) with the same orientation.

The image of $\mathcal{B}$ under $\tilde{y}$ (which we call the *actual shape* or *current place* of the body and indicate by $\mathcal{B}_a := \tilde{y}(\mathcal{B})$) has the same geometric properties of $\mathcal{B}$ specified above, thanks to the structural properties presumed for $\tilde{y}$.

The map $\tilde{y}$ is what we commonly call a **deformation**. Maps from $\mathbb{R}^3$ to $\tilde{\mathbb{R}}^3$ that are C^1 and preserve the orientation of $\mathbb{R}^3$ satisfy requirements (*i*) through (*iii*), but they are not the only possible deformations. In fact, the theoretical structures that we present here may give rise to boundary value problems that admit solutions satisfying the assumptions (*i*) through (*iii*) only in some weak sense. Hence, spaces larger than that of continuous and continuously differentiable point-valued functions have to be considered. The choice of such spaces is addressed by the physical features of the phenomenon that we are analyzing. We do not go into details here, for this essay has just an introductory character. However, we mention the issue just to warm the reader to be suspicious of phrases maintaining that continuum mechanics develops only by "selecting a smooth diffeomorphism defined over a compact open set of the ambient physical space and performing calculations." Such a point of view is in fact simplistic, as is clearly suggested by experience, culture, and nontrivial analysis of specific cases to those who are pushed by intellectual curiosity to go beyond introductory essays, even just a bit.

1.3 The Deformation Gradient

The spatial derivative of a deformation evaluated at a point x in $\mathcal{B}$ is denoted by F, namely

$$F := D\tilde{y}(x),$$

for all x in $\mathcal{B}$ where $\tilde{y}$ is differentiable.

Consider a smooth curve $s \longmapsto \bar{x}(s) \in \mathcal{B}$ with $s \in (-1, 1)$ and evaluate at a certain point, say $x = \bar{x}(0)$, its tangent

$$\mathsf{t} := \left. \frac{d\bar{x}(s)}{ds} \right|_{s=0}.$$

The image of the curve $s \longmapsto \bar{x}(s)$ over the current place $\mathcal{B}_a$ is determined by the deformation $\tilde{y}$ and is given by the map $s \longmapsto \tilde{y}(\bar{x}(s)) \in \mathcal{B}_a$. Its tangent at $y = \tilde{y}(\bar{x}(0))$ is given by

$$\mathsf{t}_a := \left. \frac{d\tilde{y}(\bar{x}(s))}{ds} \right|_{s=0},$$

so that by chain rule, we get

$$\mathsf{t}_a = D\tilde{y}(x)\mathsf{t} = F\mathsf{t}.$$

Then F maps linearly tangent vectors to $\mathcal{B}$ to tangent vectors to $\mathcal{B}_a$ (Fig. 1.2). In short, we write[2] $F \in \mathrm{Hom}(T_x\mathcal{B}, T_{\tilde{y}(x)}\mathcal{B}_a)$, which is tantamount to expressing the inclusion $F \in \mathrm{Hom}(\mathbb{R}^3, \tilde{\mathbb{R}}^3)$. Consider three linearly independent vectors at $x \in \mathcal{B}$. They span $\mathbb{R}^3$: linear combinations of them cover the whole space, so they are a basis. Write e_1, e_2, e_3 for them and e^1, e^2, e^3 for the dual basis,[3] the one in the space dual to $\mathbb{R}^3$, i.e., the space containing all possible linear forms over[4] $\mathbb{R}^3$, indicated by $\mathbb{R}^{3*}$. Recall that the elements of the dual of a linear space are called *covectors*. We can make an analogous choice at $y = \tilde{y}(x)$. We indicate by $\tilde{e}_1, \tilde{e}_2, \tilde{e}_3$ the basis vectors chosen at y, and by $\tilde{e}^1, \tilde{e}^2, \tilde{e}^3$ the relevant dual basis.

With respect to the (dual) basis e^1, e^2, e^3 in a neighborhood of $x \in \mathcal{B}$ and the basis $\tilde{e}_1, \tilde{e}_2, \tilde{e}_3$ at $y = \tilde{y}(x)$ in $\mathcal{B}_a$, F is given by

---

[2] $\mathrm{Hom}(A, B)$ indicates the space of linear maps from A to B. Here $T_x\mathcal{B}$ indicates the linear space of vectors tangent to *all* smooth curves on $\mathcal{B}$ crossing x. It is called the *tangent space* to $\mathcal{B}$ at x. Analogously, $T_{\tilde{y}(x)}\mathcal{B}_a$ is the tangent space to $\mathcal{B}_a$ at $y = \tilde{y}(x)$.

[3] e^i is defined by the condition $e^i \cdot e_j = e^i(e_j) = \delta^i_j$, with δ^i_j the Kronecker symbol, which is 1 for $i = j$ and 0 for $i \neq j$.

[4] Recall that there is a natural isomorphism between $\mathbb{R}^3$ and its dual counterpart, and it is defined by the metric.

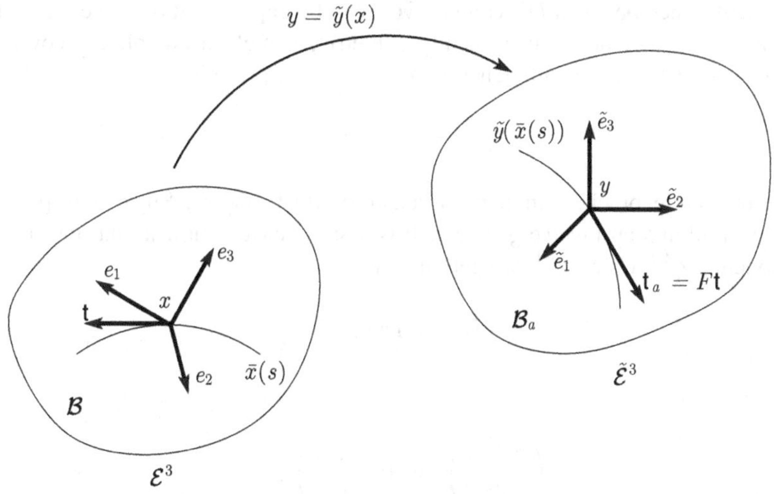

Fig. 1.2 Deformation of $\mathcal{B}$

$$F = \frac{\partial \tilde{y}^i(x)}{\partial x^A} \tilde{e}_i \otimes e^A,$$

where (and from now on) uppercase indices refer to coordinates in $\mathcal{B}$, while lowercase indices indicate coordinates in the actual shape $\mathcal{B}_a$, and we assume Einstein's convention to leave understood summation over repeated indices. The iAth component of F is just

$$F^i_A = \frac{\partial \tilde{y}^i(x)}{\partial x^A},$$

with a contravariant index[5] i and a covariant index[6] A.

Commonly, F is called a **deformation gradient**, and we follow here that tradition, although the definition of F that we have written at the beginning of this section involves the derivative $D\tilde{y}(x)$ of the map $\tilde{y}$, rather than the gradient $\nabla \tilde{y}(x)$, which has, in the bases considered previously, the form

$$\nabla \tilde{y}(x) = \left(\frac{\partial \tilde{y}^i(x)}{\partial x} \right)^A \tilde{e}_i \otimes e_A.$$

[5]When we change the basis $\tilde{e}_1$, $\tilde{e}_2$, $\tilde{e}_3$ into another basis, the ith component of F is altered as a vector.

[6]When we change the dual basis e^1, e^2, e^3 into another one, the Ath component of F is altered as a covector (the derivative of a function with respect to x, indeed).

The difference between $D\tilde{y}(x)$ and $\nabla\tilde{y}(x)$ can be appreciated once we recall that the basis e_1, e_2, e_3 determines locally a metric, namely a completely covariant positive definite tensor g with generic component g_{AB} given by

$$g_{AB} = \langle e_A, e_B \rangle_{\mathbb{R}^3},$$

where the angle brackets indicate a scalar product[7] in $\mathbb{R}^3$. Since g is positive definite, it admits an inverse g^{-1}, which is a symmetric contravariant tensor with components[8] g^{AB}. Using g^{-1}, we get, in fact,

$$\nabla\tilde{y}(x) = D\tilde{y}(x)g^{-1},$$

which is

$$\left(\frac{\partial \tilde{y}^i(x)}{\partial x}\right)^A = \left(\frac{\partial \tilde{y}^i(x)}{\partial x^B}\right) g^{BA}.$$

In this case we say that we have *raised the index A by using the metric*. When g is the identity second-rank tensor, I–this means that e_1, e_2, e_3 are orthonormal—the iAth component of $\nabla\tilde{y}(x)$ and the corresponding component of $D\tilde{y}(x)$ differ just by the nature of the Ath component, which is contravariant in one case, so that formally, we have[9]

$$\nabla\tilde{y}(x) \in \mathrm{Hom}(T_x^*\mathcal{B}, T_{\tilde{y}(x)}\mathcal{B}_a),$$

and covariant in the other case.

1.4 Formal Adjoint F^* and Transpose F^{T} of F

For a vector $a \in \mathbb{R}^3$ and covector $b \in \tilde{\mathbb{R}}^{3*}$, the **formal adjoint** F^* of F is defined by the relation

$$b \cdot Fa = F^* b \cdot a,$$

where here, the dot in $b \cdot Fa$ indicates the natural product between a vector Fa and a covector b. In other words, F takes the vector a in $\mathbb{R}^3$ and pushes it forward in $\tilde{\mathbb{R}}^3$.

---

[7] We can assume that the scalar product is defined with respect to another frame in $\mathbb{R}^3$ that we consider orthogonal.

[8] Of course we can consider the dual basis e^1, e^2, e^3 as obtained from e_1, e_2, e_3, by the action of g^{-1}, so that $e^A = g^{AB} e_B$.

[9] In other words, $\nabla\tilde{y}(x)$ maps covectors at x in $\mathcal{B}$ into vectors at $y = \tilde{y}(x)$ in the actual shape $\mathcal{B}_a$.

Then we take a linear form over $\tilde{\mathbb{R}}^3$, namely an element b of $\tilde{\mathbb{R}}^{3*}$, a covector, and the value that it assumes over the vector Fa, that is, $b \cdot Fa = b(Fa)$. Conversely, in the product $F^*b \cdot a$, we take a covector b in $\tilde{\mathbb{R}}^{3*}$, pull it back into $\mathbb{R}^{3*}$, and evaluate over $a \in \mathbb{R}^3$ the value of the covector F^*b, namely $F^*b \cdot a = F^*b(a)$.

For vectors $a \in \mathbb{R}^3$ and $\bar{a} \in \tilde{\mathbb{R}}^3$, the **transpose** F^{T} of F is defined by the relation

$$\langle Fa, \bar{a} \rangle_{\tilde{\mathbb{R}}^3} = \left\langle a, F^{\mathrm{T}}\bar{a} \right\rangle_{\mathbb{R}^3},$$

where in this case, the angle brackets on the left-hand side of the equality indicate the scalar product in $\tilde{\mathbb{R}}^3$, while the brackets on the right-hand side represent, as above, the scalar product in $\mathbb{R}^3$. In other words, F pushes forward the vector $a \in \mathbb{R}^3$ into $\tilde{\mathbb{R}}^3$, where the vector $Fa \in \tilde{\mathbb{R}}^3$ is multiplied by the vector $\bar{a} \in \tilde{\mathbb{R}}^3$ using the scalar product in $\tilde{\mathbb{R}}^3$. Conversely, F^{T} pulls back the vector $\bar{a} \in \tilde{\mathbb{R}}^3$ into $\mathbb{R}^3$, where its image, namely the vector $F^{\mathrm{T}}\bar{a}$, is multiplied by a, using the scalar product in $\mathbb{R}^3$. The two scalar product structures, the one in $\mathbb{R}^3$ and the other in $\tilde{\mathbb{R}}^3$, may in principle be different, meaning that they can be associated with different metrics.

The link between F^* and F^{T} is established by the metric g introduced previously and the companion metric $\tilde{g}$ in $\tilde{\mathbb{R}}^3$ with components determined by the basis $\tilde{e}_1, \tilde{e}_2, \tilde{e}_3$, so that $\tilde{g}_{ij} = \left\langle \tilde{e}_i, \tilde{e}_j \right\rangle_{\tilde{\mathbb{R}}^3}$.

Proposition 1. $F^T = g^{-1}F^*\tilde{g}$.

Proof. First we recall that for $a \in \mathbb{R}^3$ and $c \in \tilde{\mathbb{R}}^3$,

$$\langle Fa, c \rangle_{\tilde{\mathbb{R}}^3} = Fa \cdot \tilde{g}c = \tilde{g}c(Fa).$$

In other words, to evaluate the scalar product between the two vectors Fa and c, we can change one vector into its corresponding covector $c^\flat := \tilde{g}c$ (an element of $\mathbb{R}^{3*}$), and we compute the value of such a covector over the other vector. In this case we say that we have *lowered the index of c by using the metric*. Roughly speaking, when $\tilde{g}$ is just the second-rank identity tensor, i.e., $\tilde{g}_{ij} = \delta_{ij}$, with the right-hand side the Kronecker symbol, namely

$$\delta_{ij} = \begin{cases} 1 & \text{if} \quad i = j \\ 0 & \text{if} \quad i \neq j \end{cases},$$

we have $\langle Fa, c \rangle_{\tilde{\mathbb{R}}^3} = Fa \cdot \delta c = Fa \cdot c^\flat = c^{\mathrm{T}} Fa = F^i_A a^A c^\flat_i$. The summation over repeated indices is understood, as usual. When the metric $\tilde{g}$ does not coincide with the identity, which means that the basis $\tilde{e}_1, \tilde{e}_2, \tilde{e}_3$ is not orthogonal—in the usual jargon we say in this case that $\tilde{g}$ is not *flat*—the sum $F^i_A a^A c^\flat_i$ is affected by the coefficients $\tilde{g}_{ij}$, for we have $c^\flat_i = \tilde{g}_{ij} c^j$. Moreover, with a in $\mathbb{R}^3$ and b in $\mathbb{R}^{3*}$, we can write

$$b \cdot a = \left\langle g^{-1}b, a \right\rangle_{\mathbb{R}^3} = \left\langle a, g^{-1}b \right\rangle_{\mathbb{R}^3}.$$

In other words, when we have to evaluate b over a, i.e., when we want to compute $b(a)$, we can transform the covector b in the corresponding vector by means of the inverse metric—the covector b becomes the vector $g^{-1}b$—and we multiply $g^{-1}b$ by a using the scalar product $\langle\ ,\ \rangle_{\mathbb{R}^3}$. The natural operation is to calculate $b \cdot a = b(a)$, because we have just to compute the value of a linear function over $\mathbb{R}^3$ at the vector a. However, our previous remarks on the use of the inverse metric allow us to write

$$\left\langle a, F^{\mathrm{T}}c \right\rangle_{\mathbb{R}^3} = \langle Fa, c \rangle_{\tilde{\mathbb{R}}^3} = Fa \cdot \tilde{g}c = a \cdot F^* \tilde{g}c = \left\langle a, g^{-1}F^*\tilde{g}c \right\rangle_{\mathbb{R}^3} .$$

The previous relations prove then the initial statement. □

In components we get $(F^{\mathrm{T}})_i^A = g^{AB} F_B^{*j} \tilde{g}_{ji}$. For the scalar product we have written $\mathbb{R}^3$ or $\tilde{\mathbb{R}}^3$ as indices to the angle brackets to specify unambiguously the space where the scalar product is considered, underlining in this way the role of the different metrics, which are crucial in proving Proposition 1.

Warning: from now on, for the sake of conciseness, we shall at times use the dot for both a product of a covector with a vector and the scalar product between two vectors, leaving understood the action of the pertinent metric when necessary. We shall taking care to specify every time the meaning in the specific context.

1.5 Homogeneous Deformations and Rigid Changes of Place

A deformation of the form

$$x \longmapsto y := \tilde{y}(x) = w + F(x - x_0), \qquad x \in \mathcal{B},$$

with w an arbitrary vector in $\tilde{\mathbb{R}}^3$, x_0 an arbitrary point in $\mathcal{E}^3$, and F *constant* in space, is called **homogeneous**. An example of a homogeneous deformation is the inflation of a balloon.

A special and significant case of homogeneous deformation occurs when F is in $SO(3)$—let us write R for F in this case.[10] The deformation

$$x \longmapsto y := \tilde{y}(x) = w + R(x - x_0), \qquad x \in \mathcal{B}, \tag{1.1}$$

is an orientation preserving **isometry**, a *rigid change of place*. To recognize the rigid nature of the deformation induced by w and R, consider two points x_1 and x_2 in $\mathcal{B}$, and take their images under (1.1), calling them y_1 and y_2, respectively. By evaluating the distance between y_1 and y_2, namely $|y_1 - y_2|^2$, and using (1.1), we get

[10]We recall that $R \in SO(3)$ means that $R \in \mathrm{Hom}(\mathbb{R}^3, \mathbb{R}^3)$, $\det R = +1$, $R^{-1} = R^{\mathrm{T}}$.

$$|y_1 - y_2|^2 = \langle (y_1 - y_2), (y_1 - y_2) \rangle_{\tilde{\mathbb{R}}^3} = \langle R(x_1 - x_2), R(x_1 - x_2) \rangle_{\tilde{\mathbb{R}}^3}$$
$$= \left\langle (x_1 - x_2), R^{\mathrm{T}} R(x_1 - x_2) \right\rangle_{\mathbb{R}^3} = \langle (x_1 - x_2), (x_1 - x_2) \rangle_{\mathbb{R}^3} = |x_1 - x_2|^2 ,$$

so (1.1) is nothing more than an orientation-preserving isometry, a rigid change of place.

We then define a body to be **rigid** when it admits as deformations *only* orientation-preserving isometries.

1.6 Linearized Rigid Changes of Place

Consider parameter-dependent families of isometries defined by maps $\alpha \longmapsto w_\alpha \in \tilde{\mathbb{R}}^3$ and $\alpha \longmapsto R_\alpha \in SO(3)$, with $\alpha \in (-1, 1)$ and R_0 coinciding with the identity, which are differentiable with respect to the parameter. Define y_α by

$$x \longmapsto y_\alpha := \tilde{y}_\alpha(x) = w_\alpha + R_\alpha(x - x_0).$$

We call the map $x \longmapsto y$ a **linearized rigid deformation** (or small rigid change of place) when

$$y = R_\alpha^{\mathrm{T}} \left. \frac{d\tilde{y}_\alpha(x)}{d\alpha} \right|_{\alpha=0} .$$

By computing the derivative, we get

$$y = R_\alpha^{\mathrm{T}} \left. \frac{dw_\alpha}{d\alpha} \right|_{\alpha=0} + R_\alpha^{\mathrm{T}} \left. \frac{dR_\alpha}{d\alpha} \right|_{\alpha=0} (x - x_0).$$

Since

$$R_\alpha^{\mathrm{T}} R_\alpha = I,$$

we obtain

$$0 = \frac{d}{d\alpha}(R_\alpha^{\mathrm{T}} R_\alpha) = R_\alpha^{\mathrm{T}} \frac{dR_\alpha}{d\alpha} + \frac{dR_\alpha^{\mathrm{T}}}{d\alpha} R_\alpha = R_\alpha^{\mathrm{T}} \frac{dR_\alpha}{d\alpha} + \left(R_\alpha^{\mathrm{T}} \frac{dR_\alpha}{d\alpha} \right)^{\mathrm{T}} ,$$

which implies

$$R_\alpha^{\mathrm{T}} \frac{dR_\alpha}{d\alpha} = -\left(R_\alpha^{\mathrm{T}} \frac{dR_\alpha}{d\alpha} \right)^{\mathrm{T}} .$$

In other words, the second-rank tensor

$$W := R_\alpha^{\mathrm{T}} \frac{dR_\alpha}{d\alpha}\bigg|_{\alpha=0}$$

is skew-symmetric. By writing $\bar{c}$ for the vector $R_\alpha^{\mathrm{T}} \frac{dw_\alpha}{d_\alpha}\bigg|_{\alpha=0}$, we find that the generic linearized rigid deformation is given by

$$y = \bar{c} + W(x - x_0)$$

with $\bar{c} \in \tilde{\mathbb{R}}^3$ and $W \in \mathrm{Skw}(\mathbb{R}^3, \tilde{\mathbb{R}}^3)$.

Define now the vector field

$$x \longmapsto u := \tilde{u}(x) = y - i(x) \in \tilde{\mathbb{R}}^3,$$

where $i : \mathcal{E}^3 \longrightarrow \tilde{\mathcal{E}}^3$ is the isomorphism (the identification, here) between $\mathcal{E}^3$ and $\tilde{\mathcal{E}}^3$. We call u a **displacement**. In particular we call u a **linearized rigid displacement** when it has the form

$$u = c + W(x - x_0), \tag{1.2}$$

with c a vector in $\tilde{\mathcal{E}}^3$ given by the difference $\bar{c} - i(x)$. Since W is a skew-symmetric tensor over the three-dimensional space, by definition there exists a three-dimensional vector q such that for any other $a \in \mathbb{R}^3$,

$$Wa = q \times a,$$

so that (1.2) can be written as

$$u = c + q \times (x - x_0).$$

1.7 Kinematic Constraints on Rigid Bodies

We have already defined bodies admitting only isometries as possible deformations to be *rigid*. Each such body has only *six* degrees of freedom in three-dimensional Euclidean space: the components of c and q, in other words, the placement of a point (three parameters), and the orientation (the three Euler angles).

The definition of a rigid body is then based on a restriction on realizable deformations for the body. Further restrictions can be imposed. They constitute what we call **kinematic constraints**. Here we refer to those expressed by scalar-valued maps f, depending on the displacement, and call then *holonomic*. We distinguish two cases:

$$f(u) \leq 0,$$

which we call a *unilateral* constraint, and

$$f(u) = 0,$$

the so-called *bilateral* constraint. Specific cases of bilateral constraints appear in the next section.

1.8 Kinematics of a 1-Dimensional Rigid Body

Consider a one-dimensional rigid body. It occupies in the reference place a portion of a piecewise smooth line in $\mathcal{E}^3$. We assume that the body is subjected only to linearized rigid displacements.

The placement of the body is in reference to an orthogonal frame defined by an origin O and three linearly independent unit vectors e_1, e_2, e_3, a frame indicated by $\{O, e_1, e_2, e_3\}$. A generic point P of the body has coordinates x_P^1, x_P^2, x_P^3. The relevant position vector is $x_P = (P - O)$.

As we have already seen, an arbitrary linearized rigid displacement is defined by two vectors: the translation c and the rotation q (Fig. 1.3), namely

$$u(P) = u_P = c + q \times x_P$$

with

$$c = c^1 e_1 + c^2 e_2 + c^3 e_3, \qquad q = \omega^1 e_1 + \omega^2 e_2 + \omega^3 e_3.$$

Fig. 1.3 Rigid body in $\mathcal{E}^3$. Examples of translational and rotational constraints

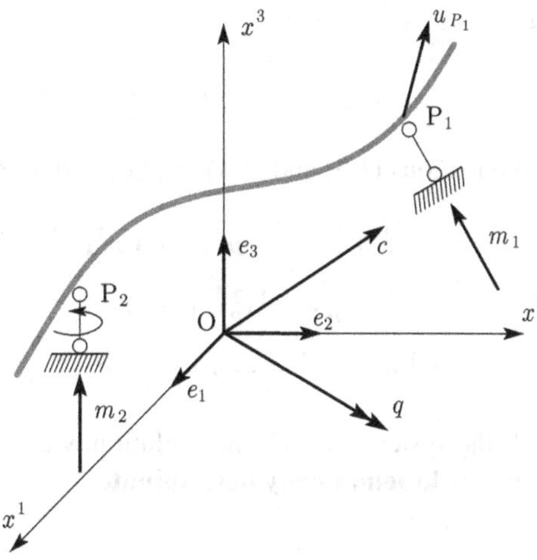

For a rigid body, every bilateral punctual holonomic constraint is a combination of two types of single constraints: (*i*) displacement u_P of a point P along a direction (simple translational constraint) and (*ii*) rotation about a direction prescribed (simple rotational constraint). Let m be such a direction (Figure 1.3 shows two possible choices of m corresponding to (*i*) and (*ii*)). In the first case (*i*), we write formally

$$u_P \cdot m = (c + q \times x_P) \cdot m = \bar{\delta}_P, \tag{1.3}$$

while in the second case (*ii*),

$$q \cdot m = \bar{\vartheta}, \tag{1.4}$$

with $\bar{\delta}_P$ and $\bar{\vartheta}$ two scalars that can be zero (in this case, the constraint is called *ideal*). The simple translational constraint at P along the direction m is made by a physical device imposed on P. Such a device is represented by a *simple pendulum* having the axis parallel to m (Fig. 1.3, point P_1). The simple rotational constraint about a direction m is exerted by a device applied to a point. We call it a *simple torsional pendulum* (point P_2 in Figure 1.3). A combination of two or more simple translational or rotational constraints at the same point generates what is called a *multiple constraint*. Its *multiplicity* is the number of simple constraints composing it.

For a rigid body, the set of equations of types (1.3) and (1.4), specifying all bilateral, holonomic constraints imposed on it, determines an algebraic system

$$\boldsymbol{A}\boldsymbol{p} = \bar{\boldsymbol{p}}, \tag{1.5}$$

where $\boldsymbol{p} = \left(c^1, c^2, c^3, \omega^1, \omega^2, \omega^3\right)^{\mathrm{T}}$, and $\bar{\boldsymbol{p}}$ is the vector of the prescribed translations and rotations. Here $\boldsymbol{A}$ is the matrix of coefficients (those related to the specific frame of reference considered), having dimensions $n_v \times 6$, with n_v the number of simple constraints. Equation (1.5) determines the kinematic state. In the selected frame, we have

$$m = m^1 e_1 + m^2 e_2 + m^3 e_3,$$

and equations (1.3) and (1.4) can be written in components as

$$\begin{aligned} & m^1 c^1 + m^2 c^2 + m^3 c^3 + \left(m^3 x_P^2 - m^2 x_P^3\right) \omega^1 + \left(m^1 x_P^3 - m^3 x_P^1\right) \omega^2 \\ & + \left(m^2 x_P^1 - m^1 x_P^2\right) \omega^3 = \bar{\delta}_P, \\ & m^1 \omega^1 + m^2 \omega^2 + m^3 \omega^3 = \bar{\vartheta}. \end{aligned}$$

- If the system has only one solution, we say that the rigid body under consideration is **kinematically determinate**.

- If the system has no solutions (this case is possible only if $\bar{p} \neq 0$), we affirm that the body is **kinematically impossible**—it is not possible to satisfy all the prescribed displacements that the constraints would impose.
- If the system has infinitely many solutions, we say that the rigid body is **kinematically indeterminate**—the number and disposition of constraints are insufficient to fix the body in space.

  Denote by $\hat{l}$ the dimension of $\ker(A)$. It is called the **lability** of the constrained body described by (1.5). Here $\hat{l}$ indicates the number of degrees of freedom that the constrained body can exploit in order to move. For a single rigid body, $\hat{l} = 6 - r_A$, with r_A the rank of A.

All these remarks apply also to systems of rigid bodies linked to one another.

A significant case occurs when for a single rigid body we have exactly six constraints ($n_v = 6$, A is square). If $\det A \neq 0$, the algebraic system (1.5) admits only one solution once $\bar{p}$ is assigned. In that case, the rigid body considered is always kinematically determinate, and we refer to this circumstance by affirming that it is **kinematically isodeterminate**. When $\bar{p} = 0$ and $\det A \neq 0$, the solution is $p = 0$, so that the number and disposition of the six constraints are necessary and sufficient to fix the body in space.

When $n_v < 6$, the body is not always kinematically determinate. If $n_v > 6$ and $\text{rank}(A) = 6$, we affirm that the rigid body is **kinematically hyperdeterminate** or **impossible**, depending on $\bar{p}$: there are too many constraints, and at least six of them are linearly independent. However, if $\text{rank}(A) < 6$, the rigid body is once again kinematically indeterminate or impossible, depending on $\bar{p}$: some constraints are ineffective, since they repeat conditions already imposed.

Consider a one-dimensional rigid body in a plane in three-dimensional space undergoing planar linearized rigid displacements. Only three degrees of freedom are involved. If the body is in the plane spanned by the vectors e_1, e_2 (axes x^1, x^2, Fig. 1.4), the translation and rotation vectors defining an arbitrary rigid displacement are respectively

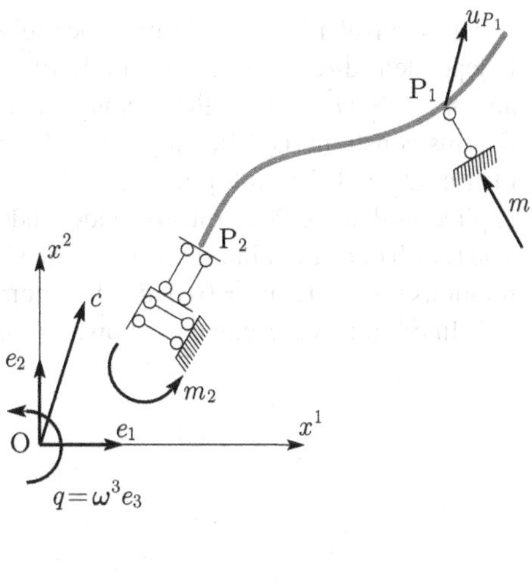

Fig. 1.4 A one-dimensional rigid body in the plane x^1x^2. Examples of translational and rotational constraints

$$c = c^1 e_1 + c^2 e_2, \qquad q = \omega^3 e_3.$$

Consider a unit vector $m = m^1 e_1 + m^2 e_2$. The constraints (1.3) and (1.4) can be expressed explicitly as

$$m^1 c^1 + m^2 c^2 + \left(m^2 x_P^1 - m^1 x_P^2\right) \omega^3 = \bar{\delta}_P \tag{1.6}$$

$$\omega^3 = \bar{\vartheta}, \tag{1.7}$$

where x_P^1 and x_P^2 are the coordinates of the point P where the constraint is applied. In this case, $\boldsymbol{A}$ is an $n_v \times 3$ matrix, and $\boldsymbol{p}$ has components c^1, c^2, and ω^3.

The kinematic classification of rigid bodies in $\mathcal{E}^2$ is the same as in the three-dimensional case. The constraints are represented by specific graphic symbols; the most commonly used in $\mathcal{E}^2$ are shown in Figure 1.5. Analogous symbols hold in $\mathcal{E}^3$ with the meaning appropriate to the different dimension of the ambient space.

- The first constraint represented in Figure 1.5 is the *simple pendulum* (or *trolley* or *bogie* or *carriage* if you wish). It obstructs the translation of the point P along the x^2 direction ($m^2 = 1$, $m^1 = 0$) and is holonomic, bilateral, and ideal. Its multiplicity is 1. The same representation holds if the body is in $\mathcal{E}^3$.
- The second constraint in Figure 1.5 is again the *simple pendulum*. It imposes on the point P that it translate along the x^2 direction of the prescribed value $\bar{\delta}_P$ ($m^2 = 1$, $m^1 = 0$) and that it be holonomic and bilateral, and that it produce a structural failure of amount $\bar{\delta}_P$. Its multiplicity is equal to 1. The displacement involved in the previous definition is *with respect to* a reference place, or ground. The same representation holds if the rigid body is in $\mathcal{E}^3$.
- The third constraint in Figure 1.5 is called a *hinge*. It eliminates the possibility for P to translate in any direction

$$u_P = 0,$$

  and is equivalent to two simple pendulums applied in P along two linearly independent directions (for example $m^1 = 1$, $m^2 = 0$ for the first pendulum and $m^1 = 0$, $m^2 = 1$ for the second). Its multiplicity is 2 (double constraint). In $\mathcal{E}^3$, this constraint is called a *spherical hinge*; the same graphical representation of it holds, but the multiplicity is 3 (it is equivalent to three simple pendulums applied in P along three linearly independent directions).
- The fourth constraint in Figure 1.5 is the *simple torsional pendulum*. It eliminates rotations ($m^3 = 1$, $\omega^3 = 0$) and is holonomic, bilateral, and ideal. Its multiplicity is 1. In $\mathcal{E}^3$, it is represented as shown in Figure 1.3.

Name	Multiplicity	Dofs constrained	Symbols
1. Pendulum	Simple	$u_P^2 = 0$	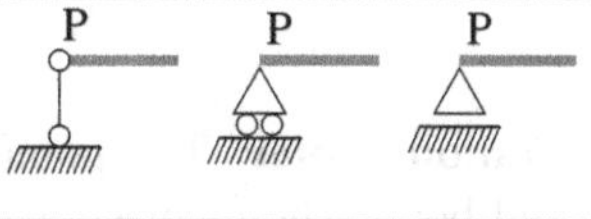
2. Sinking pendulum	Simple	$u_P^2 = -\bar{\delta}_P$	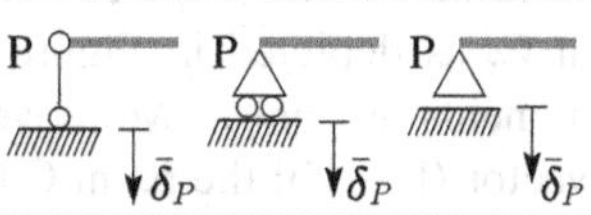
3. Hinge	Double	$u_P^1 = 0$, $u_P^2 = 0$	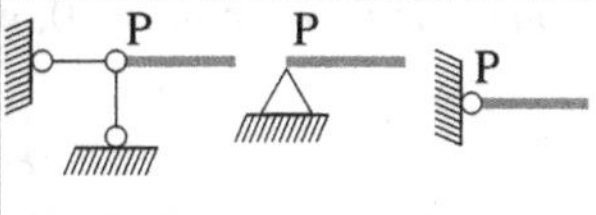
4. Torsional pendulum	Simple	$\omega^3 = 0$	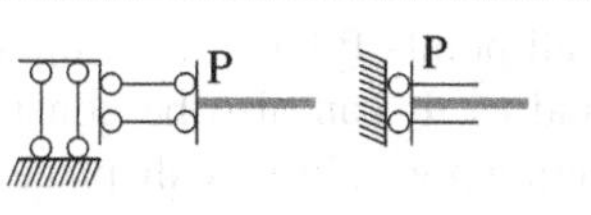
5. Glyph	Double	$u_P^2 = 0$, $\omega^3 = 0$	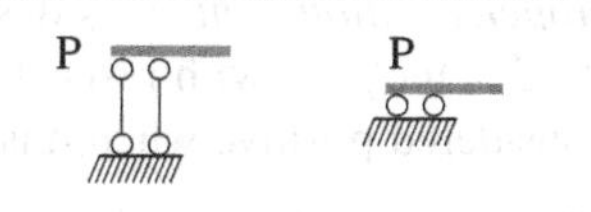
6. Joint	Triple	$u_P^1 = 0$, $u_P^2 = 0$, $\omega^3 = 0$	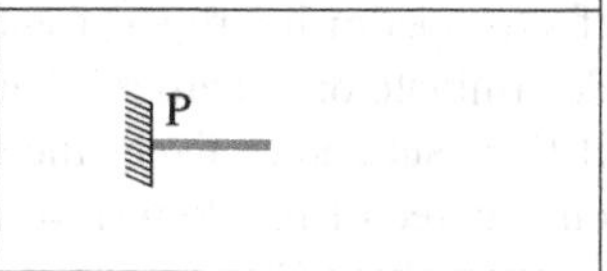

Fig. 1.5 Graphic symbols used to represent different constraints in $\mathcal{E}^2$

- The fifth constraint in Figure 1.5 is the *glyph*, a combination of a translational pendulum ($m^2 = 1$, $m^1 = 0$) and a torsional pendulum ($m^3 = 1$, $\omega^3 = 0$). It allows just translations along the x^1 direction. Its multiplicity is 2.
- The sixth constraint in Figure 1.5 is the *joint*, a combination of a hinge and a simple torsional pendulum at the same point P. Here $u_P = 0$ for every point of the body (no linearized rigid displacement is possible). Its multiplicity is 3 (triple constraint). In $\mathcal{E}^3$, this constraint has the same representation, but the multiplicity is 6, since it is equivalent to a spherical hinge applied at P plus three torsional pendulums having three linearly independent directions.

Graphics help us to represent arbitrary linearized rigid displacements, defined by translation and rotation vectors c and q, of a rigid body in a plane embedded in $\mathcal{E}^3$; a generic point P with coordinates x_P^1 and x_P^2 is subjected to a displacement of components (in the selected frame)

$$u_P^1 = c^1 - \omega^3 x_P^2, \qquad u_P^2 = c^2 + \omega^3 x_P^1.$$

If we suppose that $\omega_3 \neq 0$, we call the point C defined by the coordinates

$$x_C^1 = -\frac{c^2}{\omega^3}, \qquad x_C^2 = \frac{c^1}{\omega^3},$$

the **rotation center**. The point C is at rest and does not necessarily belong to the rigid body. The displacement compatible with the constraints is a *rotation* ω^3 about C, as depicted in Figure 1.6. The diagrams representing the displacement components are linear. Moreover, the displacement of every point P is normal to the vector (P – C); the point C has finite distance from O, and it is called a *proper rotation center*. When $\omega^3 = 0$, we obtain

$$u_P^1 = c^1, \qquad u_P^2 = c^2,$$

and all points P have the same displacement, a rigid *translation* with components c^1 and c^2. It can also be considered a rotation about a point C at infinity in the direction normal to the displacement, a point indicated by $(\mathrm{C})_\infty$ and considered an *improper rotation center*, as described in Figure 1.7. The direction with improper point C is inclined with respect to the x^1-axis of the considered frame by an angle α, considered positive when it is counterclockwise, such that $\tan\alpha = -c^1/c^2$.

- If a (proper or improper) rotation center does not exist, the body is kinematically determinate or hyperdeterminate, and $\hat{l} = 0$.
- If C exists and is fixed, the body is kinematically indeterminate and has only one degree of freedom ($\hat{l} = 1$); the linearized rigid displacement is an arbitrary rotation about C or a translation.
- If every point of a straight line can play the role of rotation center, the rigid body is kinematically indeterminate and has two degrees of freedom ($\hat{l} = 2$); all its possible displacements are given by the linear combination of two arbitrary independent displacements, each one obtained by fixing arbitrarily two different positions of C on the line.
- If every point in the plane can play the role of rotation center, the rigid body is kinematically indeterminate and has three degrees of freedom ($\hat{l} = 3$) in the same plane.

A constraint limits the possible places of the rotation center C of a rigid body. Figure 1.8 reports some prominent cases. In the plane, a simple constraint restricts the possible placement of C to a straight line; a double constraint (the combination of two simple constraints) determines uniquely the position of C; a triple constraint states that C does not exist. The kinematic analysis of a rigid body in the plane reduces to finding the rotation center. Less simple is the analysis in three dimensions, since a rotation axis is always available but not necessarily a rotation center.

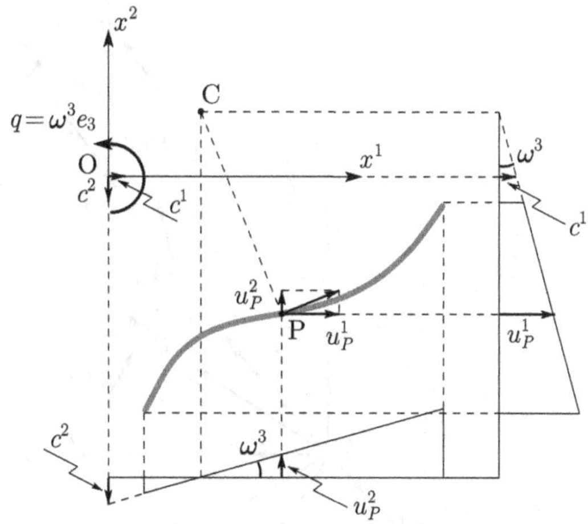

Fig. 1.6 Linearized displacements of a one-dimensional rigid body in $\mathcal{E}^2$ when $\omega^3 \neq 0$

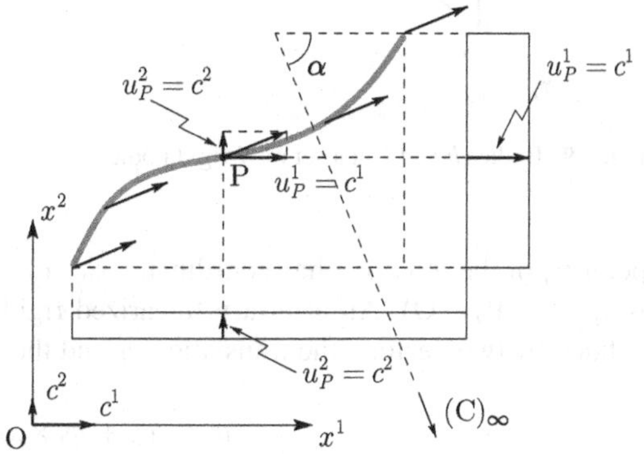

Fig. 1.7 Translation ($\omega^3 = 0$) of a one-dimensional rigid body in $\mathcal{E}^2$

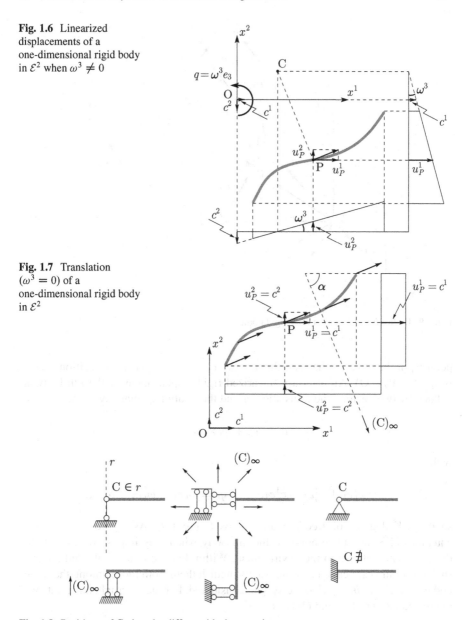

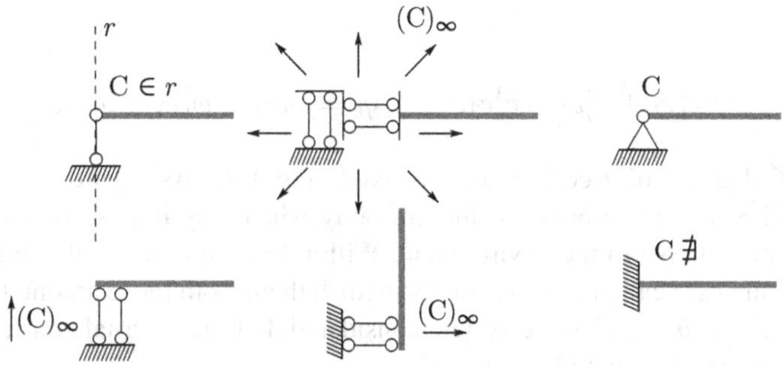

Fig. 1.8 Positions of C given by different ideal constraints

1.9 Kinematics of a System of 1-Dimensional Rigid Bodies

The description of the kinematics of a single rigid body admits a natural generalization to a system of *K* rigid bodies. We refer their position to a frame defined by an origin O and three linearly independent orthogonal unit vectors e_1, e_2, e_3. A generic

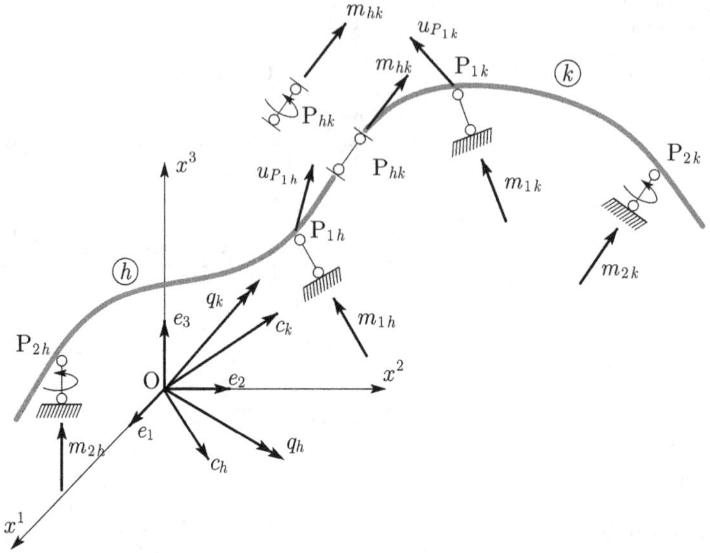

Fig. 1.9 Bodies h and k in a set of K rigid bodies in $\mathcal{E}^3$

point P_h of the hth body has coordinates $x^1_{P_h}, x^2_{P_h}, x^3_{P_h}$. The pertinent position vector is $x_{P_h} = (P_h - O)$. An arbitrary linearized rigid displacement of the hth body is defined by two vectors: the translation c_h and the rotation q_h, namely

$$u_h = c_h + q_h \times x_{P_h}$$

with

$$c_h = c^1_h e_1 + c^2_h e_2 + c^3_h e_3, \qquad q_h = \omega^1_h e_1 + \omega^2_h e_2 + \omega^3_h e_3,$$

so that $6K$ degrees of freedom are involved (Fig. 1.9). As regards constraints, equations (1.3) and (1.4) apply to the hth body when they impose restrictions to displacements relative to the environment. With reference to a set of rigid bodies, such constraints are called *external* to distinguish them from those among the same bodies, which are *internal* to the system considered. For the external constraints we rewrite equations (1.3) and (1.4) as

$$u_{P_h} \cdot m_h = (c_h + q_h \times x_{P_h}) \cdot m_h = \bar{\delta}_{P_h}, \tag{1.8}$$

$$q_h \cdot m_h = \bar{\vartheta}_h. \tag{1.9}$$

In components, evaluated with respect to an orthogonal frame, they read

$$m_h^1 c_h^1 + m_h^2 c_h^2 + m_h^3 c_h^3 + \left(m_h^3 x_{P_h}^2 - m_h^2 x_{P_h}^3\right)\omega_h^1 + \left(m_h^1 x_{P_h}^3 - m_h^3 x_{P_h}^1\right)\omega_h^2$$
$$+ \left(m_h^2 x_{P_h}^1 - m_h^1 x_{P_h}^2\right)\omega_h^3 = \bar{\delta}_{P_h},$$
$$m_h^1 \omega_h^1 + m_h^2 \omega_h^2 + m_h^3 \omega_h^3 = \bar{\vartheta}_h.$$

Let us assume that every body in the system admits at least one simple constraint linking it to another body. Such a constraint, in fact an internal one, limits relative displacements. In this case, equations (1.8) and (1.9) change into

$$(u_{P_h} - u_{P_k}) \cdot m_{hk} = \bar{\delta}_{P_{hk}}, \tag{1.10}$$

$$(q_h - q_k) \cdot m_{hk} = \bar{\vartheta}_{hk}, \tag{1.11}$$

with the indices h and k referring to the hth and kth bodies connected at the point P_{hk}, where there is a simple constraint. Here $\bar{\delta}_{P_{hk}}$ and $\bar{\vartheta}_{hk}$ are the values of possible relative displacements and/or rotations imposed. The assignment of constraints with multiplicity greater than 1 is analogous to the case of a single body.

If we apply n_v simple constraints to a set of K bodies, the relevant set of equations of type (1.8), (1.9), (1.10), (1.11) constitutes a system of n_v algebraic equations with $6K$ unknowns, again say

$$\boldsymbol{A}\boldsymbol{p} = \bar{\boldsymbol{p}},$$

with $\boldsymbol{p} = \left(c_1^1, c_1^2, c_1^3, \omega_1^1, \omega_1^2, \omega_1^3, \ldots, c_K^1, c_K^2, c_K^3, \omega_K^1, \omega_K^2, \omega_K^3\right)^{\mathrm{T}}$ and $\boldsymbol{A}$ an $n_v \times 6K$ matrix in this case. Hence, systems of rigid bodies can be classified into *determinate*, *isodeterminate*, *indeterminate*, *hyperdeterminate impossible* families, exactly as we have done for the single rigid body. Reduction to the two-dimensional case (Fig. 1.10) will be treated explicitly in the exercises discussed in Section 1.11.

1.10 The Flat-Link Chain Method

As we have already pointed out, the kinematic analysis of a single rigid body reduces to the determination of the rotation center. A generalization to a system of K bodies is called the *flat-link chain method*. For the sake of simplicity, we present it first in the case $K = 2$: two rigid bodies constrained by each other. If they have a linearized rigid displacement, we can evaluate the center of rotation C_1 of the body indexed by 1 and the center of rotation C_2 of the body 2 with respect to the environment. These points are the so-called *absolute rotation centers*. Since the graphs of the displacements of the two bodies are linear, they intersect in one point, called the *relative rotation center*, denoted by C_{12}, since it represents the rotation center of one body in the relative displacement with respect to the other. The point C_{12} can be proper or improper (the latter case occurs when $\omega_1^3 = \omega_2^3$). As with external constraints, internal constraints give analogous information about the position of C_{12} (Fig. 1.8).

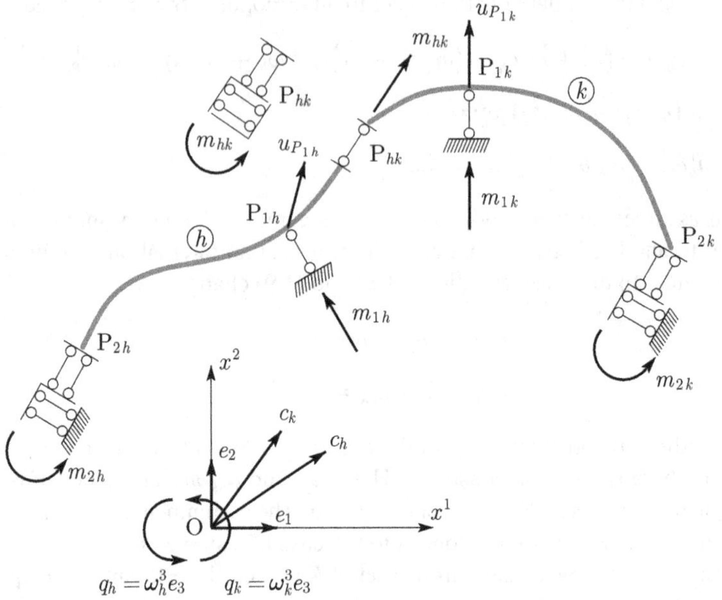

Fig. 1.10 Bodies *h* and *k* in a set of *K* rigid bodies in $\mathcal{E}^2$

Theorem 2 (First theorem on flat-link chains). *Given two rigid bodies* 1 *and* 2 *constrained by each other, if they undergo linearized rigid displacements, the points* C_1, C_2, C_{12} *are aligned.*

Formally, we write

$$C_1, C_2, C_{12} \checkmark . \tag{1.12}$$

Figure 1.11 shows two typical cases. In the first case, we have $\omega_1^3 \neq \omega_2^3$, so that C_{12} is a proper point and the two bodies have the same displacement at C_{12}. In the second case, $\omega_1^3 = \omega_2^3$; C_{12} is an improper rotation center, and the displacement diagrams of the two bodies are parallel straight lines. Even in the second case, equation (1.12) holds. The flat-link chain method consists in finding C_1, C_2, and C_{12} associated with the constraints and the equation (1.12). Prominent cases are listed below:

- C_1 and C_2 do not exist $\Leftrightarrow$ the bodies 1 and 2 are at rest ($\hat{l} = 0$).
- C_1 does not exist and C_2 is uniquely determined $\Leftrightarrow$ body 1 is fixed and 2 rotates around C_2 ($\hat{l} = 1$).
- C_1, C_2 and C_{12} are uniquely determined and are distinct points on a straight line $\Rightarrow$ the bodies 1 and 2 rotate around their own centers as in Figure 1.11 ($\hat{l} = 1$).
- C_1, C_2 and C_{12} are uniquely determined and $C_1 \equiv C_{12} \neq C_2 \Rightarrow$ body 2 is fixed and body 1 rotates around C_1 ($\hat{l} = 1$).

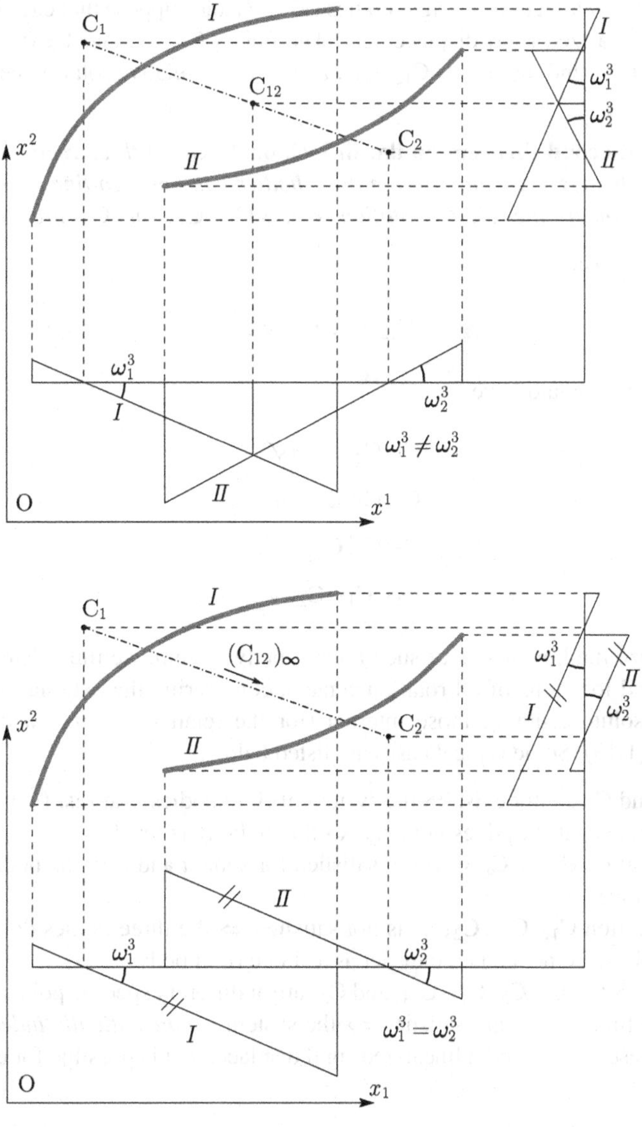

Fig. 1.11 Position of the center C_{12} in two significant cases

- C_1, C_2 and C_{12} are uniquely determined and $C_1 \equiv C_2 \neq C_{12} \Rightarrow$ the two bodies move together as a single rigid body ($\hat{l} = 1$).
- C_1, C_2, and C_{12} are uniquely determined and $C_1 \equiv C_2 \equiv C_{12} \Rightarrow$ the two bodies rotate independently around the respective centers ($\hat{l} = 2$).
- C_1, C_2, and C_{12} are not uniquely determined and equation (1.12) holds for different positions of the centers $\Rightarrow$ the system of two bodies satisfies $\hat{l} \geq 2$.

Consider now a set of three rigid bodies ($K = 3$) and suppose that each body may undergo a linearized rigid displacement. The points C_1, C_2, and C_3 (the absolute rotation centers) and the points C_{12}, C_{13}, and C_{23} (the relative rotation centers) are well defined.

Theorem 3 (Second theorem on flat-link chains). *Given three reciprocally connected rigid bodies 1, 2, and 3, if the three bodies can undergo linearized relative rigid displacements, the relative rotation centers C_{12}, C_{13}, and C_{23} are aligned.*

Formally, we write

$$C_{12}, C_{13}, C_{23} \checkmark .$$

In this case, we should have

$$\begin{aligned} &C_1, C_2, C_{12} \quad \checkmark , \\ &C_1, C_3, C_{13} \quad \checkmark , \\ &C_2, C_3, C_{23} \quad \checkmark , \\ &C_{12}, C_{13}, C_{23} \checkmark . \end{aligned} \tag{1.13}$$

To analyze the kinematics of such a three-body system, we must determine the existence and locations of all rotation centers, considering the external constraints (for the absolute centers), those internal (for the relative centers), and the four conditions (1.13). Some typical cases are listed below:

- C_1, C_2 and C_3 do not exist $\Leftrightarrow$ the system of rigid bodies is at rest ($\hat{l} = 0$).
- An absolute center C_j does not exist $\Leftrightarrow$ the jth body is fixed.
- The condition $C_i, C_j, C_{ij} \checkmark$ is not satisfied for some i and $j \Rightarrow$ the bodies i and j cannot move.
- The condition $C_{12}, C_{13}, C_{23} \checkmark$ is not satisfied $\Rightarrow$ the three bodies do not admit relative displacements, i.e., they act as a single rigid body.
- The centers $C_1, C_2, C_3, C_{12}, C_{13}$, and C_{23} are at different specific points in $\mathcal{E}^2$ and the conditions (1.13) are satisfied $\Rightarrow$ the system is *kinematically indeterminate* (in this case, $\hat{l} = 1$) and a linearized rigid displacement is possible for each body (Fig. 1.12).

Situations in which one or more centers are placed at the same point or all points on straight lines can be rotation centers are more complex.

For four or more rigid bodies ($K \geq 4$), the previous theorems hold. For example, for $K = 4$, the first theorem has to be applied to every pair of bodies, i.e., six times, whereas the second theorem has to be applied to every triplet, i.e., four times. In general, the first theorem has to be applied $\binom{K}{2}$ times, while $\binom{K}{3}$ times pertain to the second one.

Further details appear in the exercises.

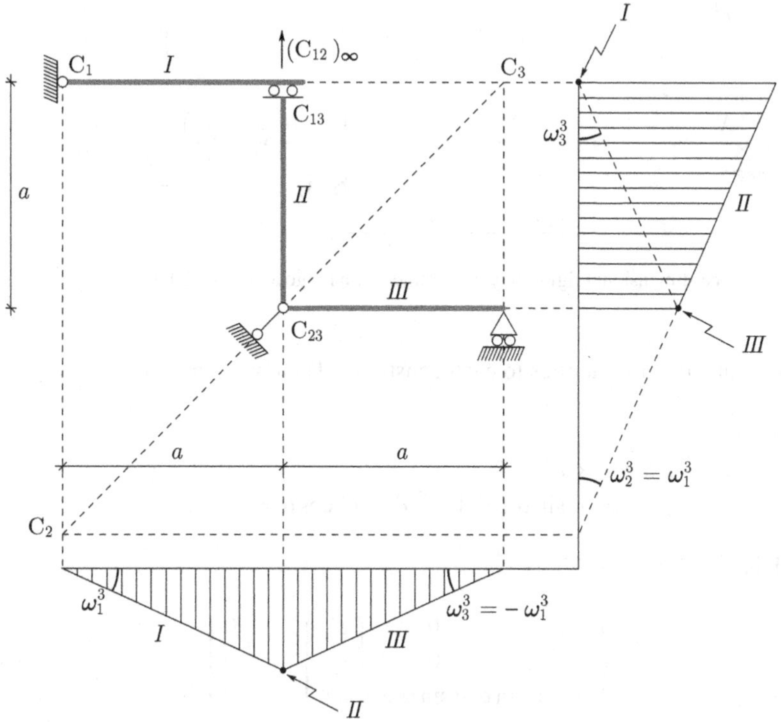

Fig. 1.12 The external constraints determine C_1 and C_3. The internal constraints fix C_{12} and C_{23}, while C_2 and C_{13} follow from the theorems on flat-link chains. Conditions (1.13) are satisfied, and the system is kinematically indeterminate ($\hat{l} = 1$). The figure shows the construction of the diagrams of rigid displacements of each body

1.11 Exercises on the Kinematics of 1-Dimensional Rigid Bodies

Exercise 4. *Analyze the kinematics of the body represented in Figure 1.13.*

Remarks and solution. The body is in $\mathcal{E}^2$, embedded in $\mathcal{E}^3$, and occupies the segment $\overline{AB}$ with length a. Its position is in reference to a frame with origin in A and axis x^1 containing the segment $\overline{AB}$. The orthogonal axis is indicated by x^2, so that the frame is, in summary, $\{A, x^1, x^2\}$. In A, there is an ideal hinge—it is equivalent to two independent simple translational constraints—while in B, there is a simple pendulum inclined by α with respect to the axis x^1. Hence we have $n_v = 3$:

- constraint 1 at $A \equiv (0,0)$, along $m = (1,0)$,
- constraint 2 at $A \equiv (0,0)$, along $m = (0,1)$,
- constraint 3 at $B \equiv (a,0)$, along $m = (\cos\alpha, \sin\alpha)$.

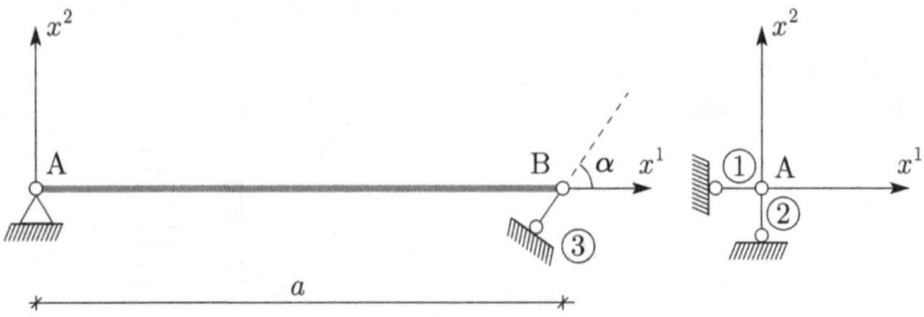

Fig. 1.13 One-dimensional rigid body constrained by a hinge and a pendulum

The equation (1.6) applies to each constraint. Then, we can write ($\bar{\delta}_P = 0$)

$$\begin{cases} c^1 = 0, \\ c^2 = 0, \\ \sin\alpha\left(c^2 + \omega^3 a\right) + c^1 \cos\alpha = 0, \end{cases}$$

which is

$$\begin{pmatrix} 1 & 0 & 0 \\ 0 & 1 & 0 \\ \cos\alpha & \sin\alpha & a\sin\alpha \end{pmatrix} \begin{Bmatrix} c^1 \\ c^2 \\ \omega^3 \end{Bmatrix} = \begin{Bmatrix} 0 \\ 0 \\ 0 \end{Bmatrix}. \tag{1.14}$$

For the determinant of the coefficient matrix ($\det \boldsymbol{A} = a\sin\alpha$), we have two prominent cases. When $\alpha \neq 0$, the three constraints are independent, and $c^1 = c^2 = \omega^3 = 0$ is the solution of (1.14): the body is kinematically isodeterminate, and no rigid displacement is possible.

Conversely, if $\alpha = 0$, we have

$$\begin{pmatrix} 1\,0\,0 \\ 0\,1\,0 \\ 1\,0\,0 \end{pmatrix} \begin{Bmatrix} c^1 \\ c^2 \\ \omega^3 \end{Bmatrix} = \begin{Bmatrix} 0 \\ 0 \\ 0 \end{Bmatrix},$$

which admits infinitely many solutions $c^1 = c^2 = 0$ and arbitrary ω^3: the body is kinematically indeterminate ($\hat{l} = 1$). The linearized rigid displacement is a rotation about A (rotation center). A diagram of the vertical rigid displacements (the horizontal ones vanish) appears in Figure 1.14. A more synthetic view can be obtained by considering the rotation center. It must be at A, due to the hinge. At the same time, it should belong to the straight line passing through B, inclined like the pendulum; hence if $\alpha \neq 0$, a unique location for the rotation center does not exist, and the body is kinematically isodeterminate, while when $\alpha = 0$, the rotation center coincides with A (the body is then labile).

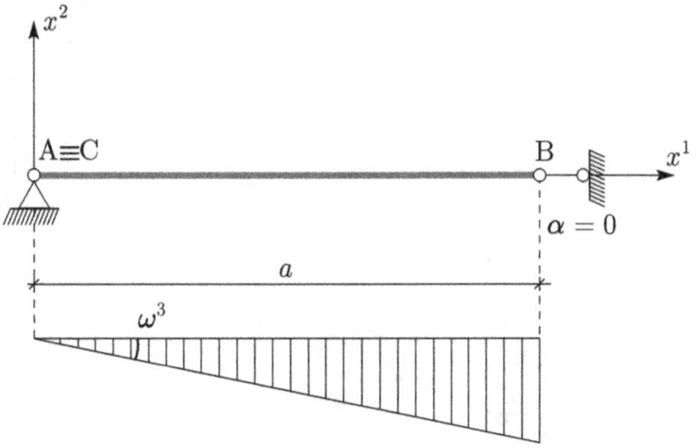

Fig. 1.14 The special case $\alpha = 0$; C is the rotation center

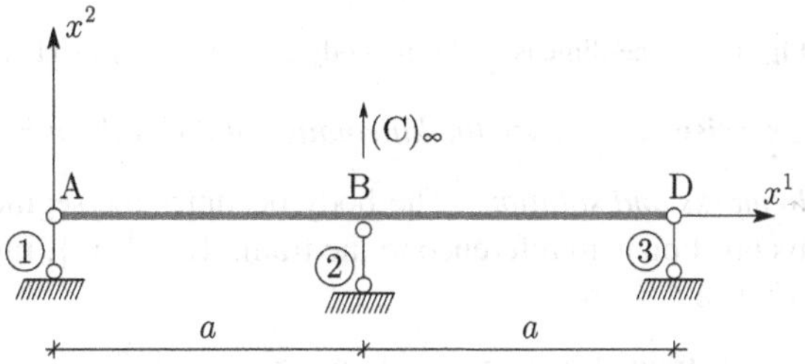

Fig. 1.15 Rigid body constrained by three vertical pendulums

Exercise 5. *Analyze the kinematics of the body in Figure 1.15.*

Remarks and solution. The body belongs to a plane in which it occupies the segment $\overline{\text{ABD}}$ with length $2a$, and no out-of-plane displacement is considered. The frame considered is the same as in the previous exercise, namely $\{\text{A}, x^1, x^2\}$. In A, B, and D, there are three ideal simple constraints: pendulums with vertical axis. Equation (1.6) reduces explicitly to the system

$$\begin{cases} c^2 = 0, \\ c^2 + \omega^3 a = 0, \\ c^2 + \omega^3 2a = 0, \end{cases}$$

which is

$$\begin{pmatrix} 0 & 1 & 0 \\ 0 & 1 & a \\ 0 & 1 & 2a \end{pmatrix} \begin{Bmatrix} c^1 \\ c^2 \\ \omega^3 \end{Bmatrix} = \begin{Bmatrix} 0 \\ 0 \\ 0 \end{Bmatrix}$$

and admits infinitely many solutions ($c^2 = 0$, $\omega^3 = 0$, and c^1 arbitrary). The body is kinematically indeterminate ($\hat{l} = 1$). In other words, although the constraints would be sufficient in number not to allow the body to move, they are arranged in a way that leaves horizontal translation free. The rotation center belongs to the three vertical straight lines passing through A, B, and D.

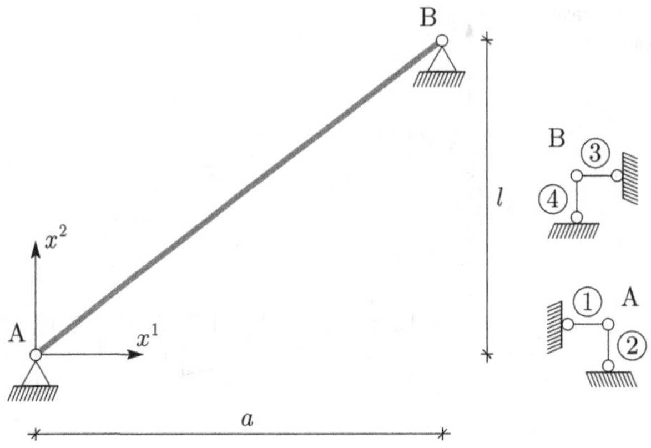

Fig. 1.16 One-dimensional rigid body constrained by two hinges

Exercise 6. *Analyze the kinematics of the body in Figure 1.16.*

Remarks and solution. The body occupies the segment $\overline{\mathrm{AB}}$ with length $\sqrt{a^2 + l^2}$. Its position is in reference to the frame $\{\mathrm{A}, x^1, x^2\}$. In A and B there are two hinges, whence $n_v = 4$:

- constraint 1 at $\mathrm{A} \equiv (0,0)$, along $m = (1,0)$,
- constraint 2 at $\mathrm{A} \equiv (0,0)$, along $m = (0,1)$,
- constraint 3 at $\mathrm{B} \equiv (a,l)$, along $m = (1,0)$,
- constraint 4 at $\mathrm{B} \equiv (a,l)$, along $m = (0,1)$.

For each constraint, equation (1.6) can be written explicitly as

$$\begin{cases} c^1 = 0, \\ c^2 = 0, \\ c^1 - \omega^3 l = 0, \\ c^2 + \omega^3 a = 0, \end{cases}$$

which is

$$\begin{pmatrix} 1 & 0 & 0 \\ 0 & 1 & 0 \\ 1 & 0 & -l \\ 0 & 1 & a \end{pmatrix} \begin{Bmatrix} c^1 \\ c^2 \\ \omega^3 \end{Bmatrix} = \begin{Bmatrix} 0 \\ 0 \\ 0 \\ 0 \end{Bmatrix}$$

and admits only the trivial solution because rank($\boldsymbol{A}$)$= 3$. There are more than three well-placed simple constraints: the body is kinematically hyperdeterminate, and no linearized rigid displacement is possible. The rotation center does not exist, because it should be placed simultaneously at A and B. If we eliminate one constraint, the body remains with three simple translational constraints, and $\boldsymbol{A}$ becomes a 3×3 matrix. Since $\det \boldsymbol{A} \neq 0$, we still have the trivial solution, and the body is kinematically isodeterminate.

Fig. 1.17 Labile rigid body

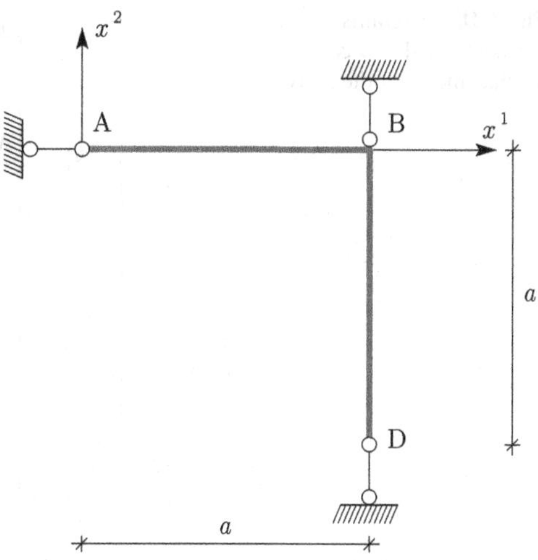

Exercise 7. *Analyze the kinematics of the body in Figure 1.17.*

Remarks and solution. The kinematics develops in the plane where the body has three simple translational constraints applied at the points A, B, and D. With respect to the reference frame indicated in the figure, we have

- constraint 1 at A $\equiv (0,0)$, along $m = (1,0)$,
- constraint 2 at B $\equiv (a,0)$, along $m = (0,1)$,
- constraint 3 at D $\equiv (a,-a)$, along $m = (0,1)$.

Equation (1.6) reduces to the system

$$\begin{cases} c^1 = 0, \\ c^2 + \omega^3 a = 0, \\ c^2 + \omega^3 a = 0, \end{cases}$$

or

$$\begin{pmatrix} 1 & 0 & 0 \\ 0 & 1 & a \\ 0 & 1 & a \end{pmatrix} \begin{Bmatrix} c^1 \\ c^2 \\ \omega^3 \end{Bmatrix} = \begin{Bmatrix} 0 \\ 0 \\ 0 \end{Bmatrix}. \tag{1.15}$$

There are just two linearly independent equations. The body has one degree of freedom ($l = 1$), and the system (1.15) has infinitely many solutions, for example $\boldsymbol{p} = (0, -a\omega^3, \omega^3)^\mathrm{T}$, with ω^3 arbitrary. The body admits a linearized rigid displacement, a rotation about C, a point with coordinates

$$x_C^1 = -\frac{c^2}{\omega^3} = a, \qquad x_C^2 = \frac{c^1}{\omega^3} = 0,$$

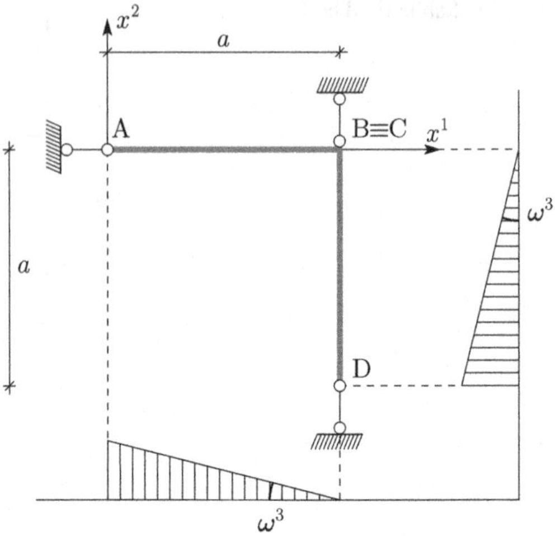

Fig. 1.18 Diagrams of horizontal and vertical displacements of the body

coinciding with B. More synthetically, the center C must be placed on the horizontal straight line crossing A and on the vertical straight line through D; hence C coincides with B. Figure 1.18 shows the displacement diagrams.

Exercise 8. *Analyze the kinematics of the body in Figure 1.19.*

Remarks and solution. The body has a rotational constraint applied at A and a translational one at B. The position of the body is in reference to the frame $\{A, x^1, x^2\}$ and we have

- constraint 1 at A $\equiv (0,0)$, about the x^3-axis,
- constraint 2 at B $\equiv (a,0)$, along $m = (0,1)$.

The kinematic conditions imposed by the constraints are then

$$\begin{cases} \omega^3 = 0, \\ c^2 + \omega^3 a = 0, \end{cases}$$

that we rewrite as

$$\begin{pmatrix} 0 & 0 & 1 \\ 0 & 1 & a \end{pmatrix} \begin{Bmatrix} c^1 \\ c^2 \\ \omega^3 \end{Bmatrix} = \begin{Bmatrix} 0 \\ 0 \end{Bmatrix} .$$

We have rank($\boldsymbol{A}$)=2, because there are two linearly independent equations. The body has one degree of lability ($\hat{l} = 1$), and the system has infinitely many solutions, $(c^1, 0, 0)^{\mathrm{T}}$, with c^1 arbitrary. The body admits horizontal translation. The rotation center C is the improper point of the vertical direction. In fact, the simple pendulum at B prescribes the center C on the vertical straight line through B, while the constraint in A imposes the center C at an arbitrary point at infinity; the two conditions are satisfied when C is the improper point of the vertical direction.

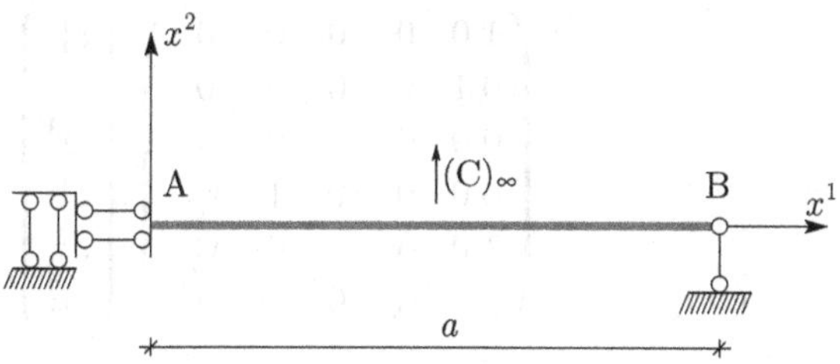

Fig. 1.19 Labile rigid body

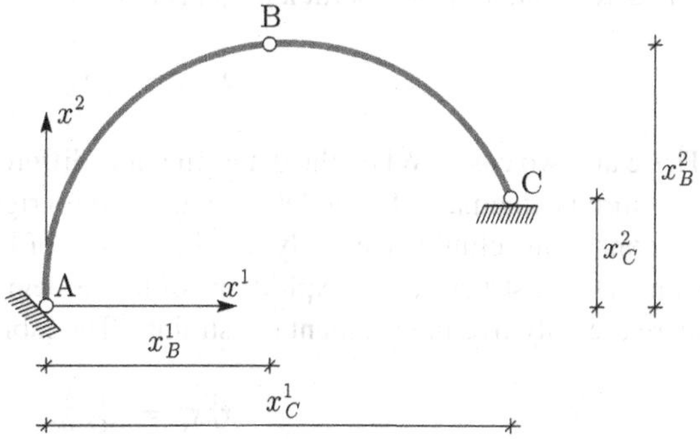

Fig. 1.20 Arch with three hinges

Exercise 9. *Analyze the kinematics of the structure in Figure 1.20 in which $x_B^1 > 0$ and $x_C^1 > x_B^1$.*

Remarks and solution. The structure is the so-called three-hinge arch. It includes six simple translational constraints, four external and two internal. The position of the bodies is in reference to the frame $\{A, x^1, x^2\}$, where for the constraints in A, C, and B (internal), we have respectively

- constraint 1 at A $\equiv (0,0)$, along $m = (1,0)$,
- constraint 2 at A $\equiv (0,0)$, along $m = (0,1)$,
- constraint 3 at C $\equiv (x_C^1, x_C^2)$, along $m = (1,0)$,
- constraint 4 at C $\equiv (x_C^1, x_C^2)$, along $m = (0,1)$,
- constraint 5 at B $\equiv (x_B^1, x_B^2)$, along $m = (1,0)$,
- constraint 6 at B $\equiv (x_B^1, x_B^2)$, along $m = (0,1)$.

The kinematic conditions imposed by the six constraints are then

$$\begin{cases} c_1^1 = 0, \\ c_1^2 = 0, \\ c_2^1 - x_C^2 \omega_2^3 = 0, \\ c_2^2 + x_C^1 \omega_2^3 = 0, \\ c_1^1 - c_2^1 - x_B^2(\omega_1^3 - \omega_2^3) = 0, \\ c_1^2 - c_2^2 + x_B^1(\omega_1^3 - \omega_2^3) = 0, \end{cases}$$

i.e.,

$$\begin{pmatrix} 1 & 0 & 0 & 0 & 0 & 0 \\ 0 & 1 & 0 & 0 & 0 & 0 \\ 0 & 0 & 0 & 1 & 0 & -x_C^2 \\ 0 & 0 & 0 & 0 & 1 & x_C^1 \\ 1 & 0 & -x_B^2 & -1 & 0 & x_B^2 \\ 0 & 1 & x_B^1 & 0 & -1 & -x_B^1 \end{pmatrix} \begin{Bmatrix} c_1^1 \\ c_1^2 \\ \omega_1^3 \\ c_2^1 \\ c_2^2 \\ \omega_2^3 \end{Bmatrix} = \begin{Bmatrix} 0 \\ 0 \\ 0 \\ 0 \\ 0 \\ 0 \end{Bmatrix}.$$

The determinant of the coefficient matrix is

$$x_B^1 x_C^2 - x_C^1 x_B^2.$$

There are two cases. When the determinant is different from zero (the first case), the structure is kinematically isodeterminate, and no rigid displacement is possible (the six constraints eliminate exactly the six degrees of freedom). When the determinant vanishes, the structure can exploit one of its six degrees of freedom ($\hat{l} = 1$), because there are only five independent constraints. The lability condition is then

$$x_B^1 x_C^2 = x_C^1 x_B^2,$$

the case in which the three hinges fall into a line. An example is shown in Figure 1.21, where $x_B^1 = x_B^2 = a$ and $x_C^1 = x_C^2 = 2a$. In this case, the previous linear system admits infinitely many solutions, i.e., $(0, 0, -\omega_2^3, 2a\omega_2^3, -2a\omega_2^3, \omega_2^3)^{\mathrm{T}}$, with ω_2^3 the rotation of the body *II*. The absolute rotation centers of the two bodies have coordinates (the relative rotation center is at B)

$$x_{C_1}^1 = 0, \qquad x_{C_1}^2 = 0, \qquad x_{C_2}^1 = 2a, \qquad x_{C_2}^2 = 2a.$$

If we consider an arbitrary small rotation of the body *II*, we have the displacement represented in Figure 1.22. The problem can be also analyzed by applying the flat-link chain method, in particular the first theorem. The possible absolute rotation center C_1 of the body *I* is located at A. Its counterpart C_2 for the body *II* is at C. Finally, the relative rotation center C_{12} is at B. Since in the present case, the three points are distinct, the structure is kinematically indeterminate if A, B, and C are collinear; otherwise, it is isodeterminate.

Exercise 10. *Analyze the kinematics of the structure in Figure 1.23.*

Remarks and solution. The structure is composed by two bodies, indicated by *I* and *II* and connected by the pendulum $\overline{\mathrm{BC}}$. There are two external and three internal simple constraints (the glyph in D and the pendulum $\overline{\mathrm{BC}}$); $n_v = 5$, and the structure has at least one degree of lability ($\hat{l} \geq 1$).

To have the possibility of a nonzero displacement, the rotation centers C_1, C_2, and C_{12} must be all on a straight line (first theorem of flat-link chains).

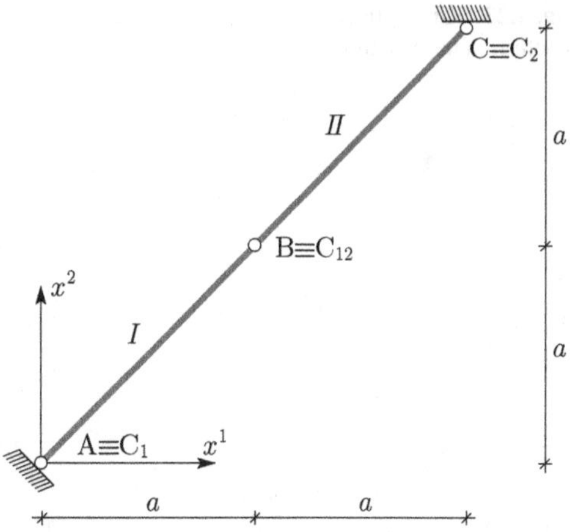

Fig. 1.21 A kinematically indeterminate structure with three hinges

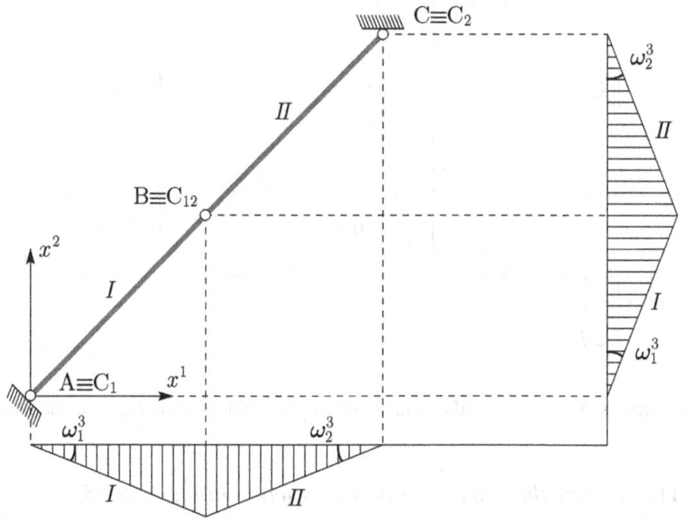

Fig. 1.22 Diagrams of possible linearized displacements

The rotation center C_1 is at B, the intersection of the lines r and s indicating the axes of the carriages. The relative rotation center C_{12} is the improper point of the vertical direction for the glyph at the point D and the pendulum $\overline{BC}$. To satisfy the alignment condition, the rotation center C_2 must be located on the line r. Hence, the position of C_2 is indeterminate on the line r. We have then double lability ($\hat{l} = 2$). Two independent linearized displacements can be obtained by fixing C_2 at two arbitrary distinct points of the line r. Figure 1.24 shows two possible choices for C_2. In the first case, C_2 coincides with C; in the second case, C_2 is the improper point of the vertical direction (a circumstance in which we have $C_2 \equiv C_{12} \neq C_1$: the body *I* is fixed and *II* may translate horizontally). The two independent rigid mechanisms are depicted in Figure 1.24.

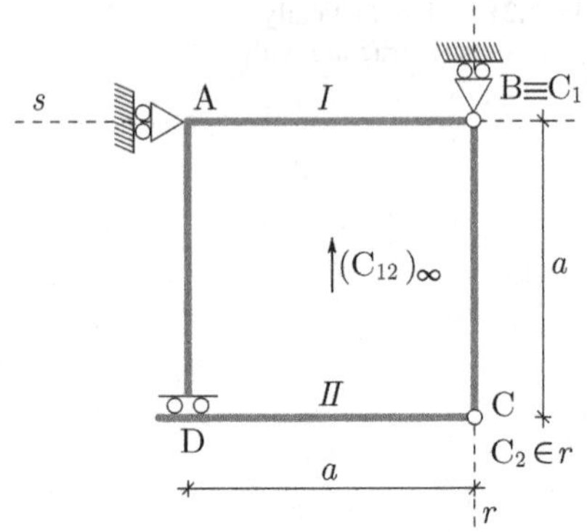

Fig. 1.23 Kinematically indeterminate structure

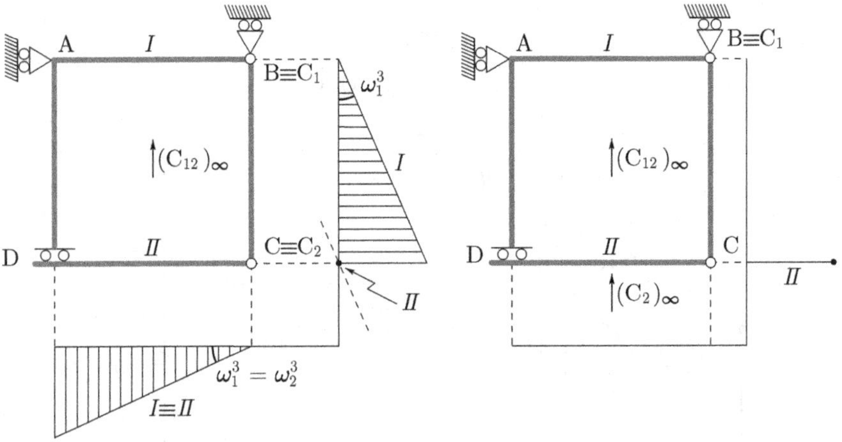

Fig. 1.24 Diagrams of the linearized displacements of two independent rigid mechanisms

Exercise 11. *Analyze the kinematics of the structure in Figure 1.25.*

Remarks and solution. The structure is composed of two bodies, *I* and *II*. There are four external and two internal simple constraints: two carriages, a hinge (the external constraints), and a glyph in B. We have $n_v = 6$, which is the number of pertinent degrees of freedom. The center C_1 of possible rotation of the body *I* is the improper point of the vertical direction, because such a point is located on the lines *r* and *s* indicating the axes of the carriages (Fig. 1.26). The rotation center C_2 coincides with D. Finally, C_{12} is the improper point of the vertical direction for the glyph in B. Hence, the three points are collinear, and

$$C_1 \equiv C_{12} \neq C_2.$$

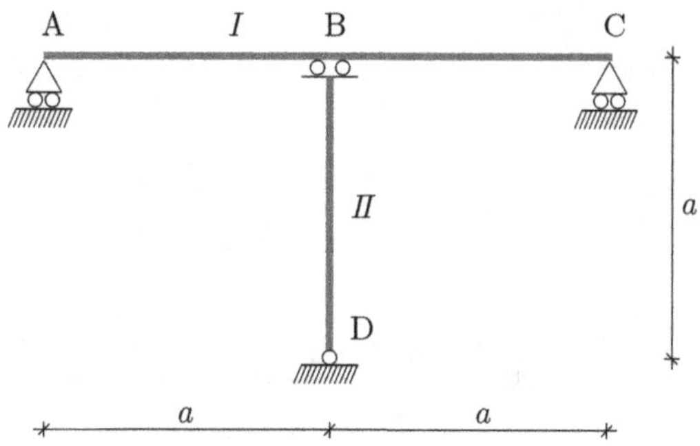

Fig. 1.25 Kinematically indeterminate structure

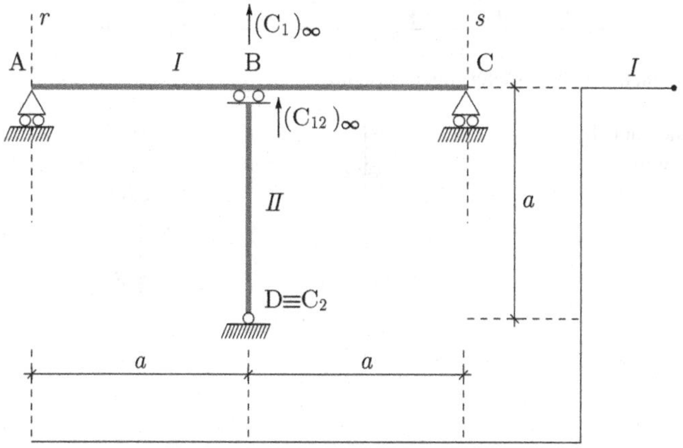

Fig. 1.26 Diagrams of possible linearized displacements

In other words, there are six constraints, but only five are linearly independent; in particular, the body *II* is fixed, and *I* translates horizontally. Figure 1.26 shows the pertinent kinematics.

Exercise 12. *Analyze the kinematics of the structure in Figure 1.27.*

Remarks and solution. We have three bodies: *I*, *II*, and *III* (nine degrees of freedom). There are six external simple constraints (two glyphs and a hinge) and four internal simple constraints (two hinges): $n_v = 10$. The centers C_1, C_2, C_3, C_{12}, and C_{23} of possible rotation are fixed by the constraints. Moreover, $C_{13} \equiv C_1$: in fact, C_{13} is on the line r (for having C_1, C_3, C_{13} aligned) and on the line s (for the alignment of C_{12}, C_{13}, and C_{23}). Then we have

$$C_1, C_2, C_{12} \quad \checkmark,$$

$$C_1, C_3, C_{13} \quad \checkmark,$$

$$C_2, C_3, C_{23} \quad \not\checkmark,$$

$$C_{12}, C_{13}, C_{23} \quad \checkmark,$$

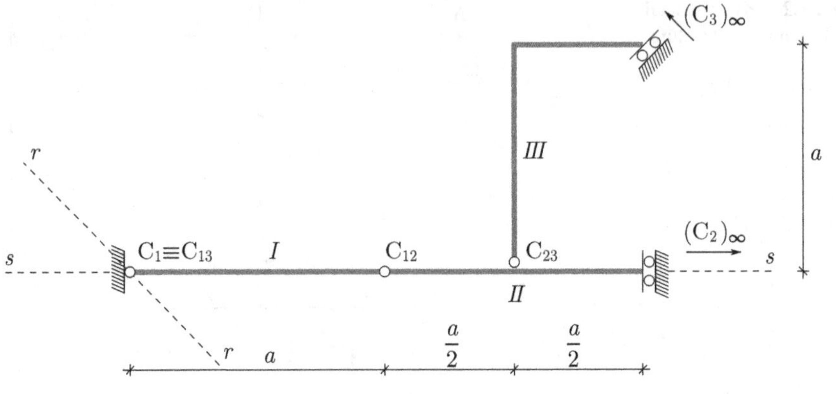

Fig. 1.27 Kinematically hyperdeterminate structure

Fig. 1.28 Kinematically impossible system

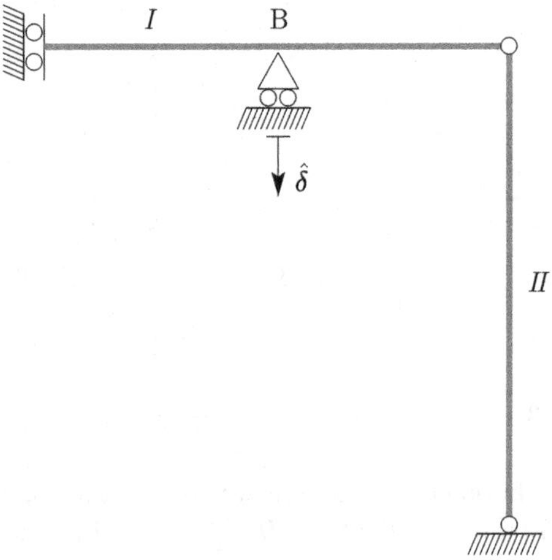

where the symbol ✓̸ indicates that the third alignment is not verified. Hence, the structure does not admit any linearized displacement and is kinematically hyperdeterminate.

Exercise 13. *Analyze the kinematics of the two-body system with a structural failure in Figure 1.28.*

Remarks and solution. The structure involves $n_v = 7$ simple constraints. At first, we do not consider the carriage in B, so we think of a system with six simple constraints and six degrees of freedom, in which the centers C_1, C_2, and C_{12} of possible rotation do not fall into line. Hence, the system is kinematically isodeterminate, and no rigid displacement is possible. If we consider also the constraint at B, we realize that the vertical displacement $\hat{\delta}$ (the structural failure) is incompatible. Consequently, the system is kinematically impossible—we cannot impose $\hat{\delta}$ at B if *I* and *II* are rigid bodies.

Exercise 14. *Analyze the kinematics of the structure in Figure 1.29.*

Remarks and solution. A frame $\{E, x^1, x^2\}$ is considered as in the figure. The structure has 15 degrees of freedom. They are collected in the vector

$$p = (c_1^1, c_1^2, \omega_1^3, c_2^1, c_2^2, \omega_2^3, c_3^1, c_3^2, \omega_3^3, c_4^1, c_4^2, \omega_4^3, c_5^1, c_5^2, \omega_5^3)^{\mathrm{T}},$$

where we have considered the numbering of the bodies in the figure. Also, there are 14 simple ideal constraints (at A and E four constraints, B involves three constraints, etc.). If we label the constraints (for example, the four external and the ten internal in sequence), the absolute and relative displacements that they prevent can be collected in the vector

$$(u_{1E}^1, u_{1E}^2, v_{3B}, u_{5C}^2, u_{12E}^1, u_{12E}^2, v_{13B}, \omega_{13}^3, v_{24D}, \omega_{24}^3, u_{34A}^1, u_{34A}^2, u_{35A}^1, u_{35A}^2)^{\mathrm{T}},$$

where by v we have denoted displacements inclined at 45°. The kinematic conditions imposed by the 14 constraints are then

$$\begin{cases} c_1^1 = 0, \\ c_3^1 + c_3^2 = 0, \\ c_1^1 - c_2^1 = 0, \\ c_1^1 + c_1^2 - c_3^1 - c_3^2 = 0, \\ c_2^1 - c_2^2 - c_4^1 + c_4^2 = 0, \\ c_3^2 - c_4^2 = 0, \\ c_3^2 - c_5^2 = 0, \end{cases} \qquad \begin{cases} c_1^2 = 0, \\ c_5^2 + 2a\omega_5^3 = 0, \\ c_1^2 - c_2^2 = 0, \\ \omega_1^3 - \omega_3^3 = 0, \\ \omega_2^3 - \omega_4^3 = 0, \\ c_3^1 + 2a\omega_3^3 - c_4^1 - 2a\omega_4^3 = 0, \\ c_3^1 + 2a\omega_3^3 - c_5^1 - 2a\omega_5^3 = 0. \end{cases}$$

The matrix A of the coefficients is of dimension 14×15. Its rank is 13: the structure has a double lability ($\hat{l} = 2$). Only 13 constraints are independent, and two arbitrary linearized displacements are possible. With $\omega_3^3 = 0$ and $c_5^2 = -\hat{\delta}$, a possible nontrivial solution is

$$(0, 0, 0, 0, 0, \hat{\delta}/a, \hat{\delta}, -\hat{\delta}, 0, -\hat{\delta}, -\hat{\delta}, \hat{\delta}/a, 0, -\hat{\delta}, \hat{\delta}/2a)^{\mathrm{T}}.$$

Figure 1.30 shows the displacements of the bodies with the position of the centers of rotation. The body *I* is fixed. With $\omega_3^3 = \Omega$ and $c_5^2 = 0$, another solution is

$$(0, 0, \Omega, 0, 0, \Omega, 0, 0, \Omega, 0, 0, \Omega, 2a\Omega, 0, 0)^{\mathrm{T}}.$$

Figure 1.31 shows the displacements with the position of the centers of rotation. In this second case, the bodies *I* to *IV* rotate around E as a single rigid body, while the body *V* translates horizontally.

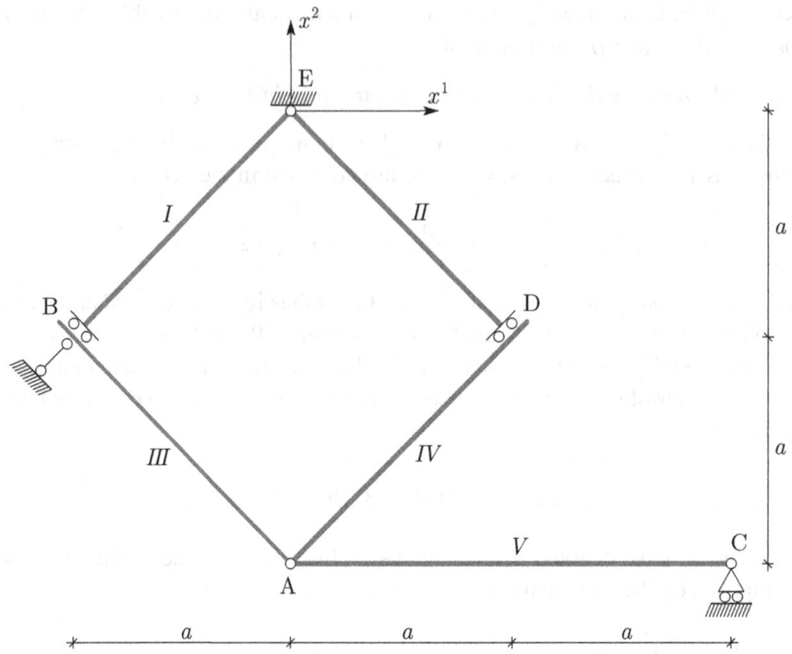

Fig. 1.29 System of five rigid bodies

1.12 Changes in Volume and the Orientation-Preserving Property

Homogeneous deformations (rigid-body displacements among them) are a very specific—although important—case. In general, F may change from place to place and this is the setting that we primarily discuss here.

Consider a point x in $\mathcal{B}$ and take there three linearly independent vectors e_1, e_2, e_3 constituting a parallelepiped in $\mathcal{B}$ with volume

$$\mathsf{V} = \langle e_1, e_2 \times e_3 \rangle_{\mathbb{R}^3} = e_1 \cdot g(e_2 \times e_3),$$

where g is the metric selected in the reference configuration.

Given a deformation and the pertinent F, at the point $y := \tilde{y}(x)$ we have another three vectors generated by the action of F over the assigned triplet e_1, e_2, e_3, namely $\tilde{e}_1 = Fe_1, \tilde{e}_2 = Fe_2, \tilde{e}_3 = Fe_3$. If we want to exclude from our description, the collapse of finite-size portions of matter into a point, we have to impose

$$\det F \neq 0. \qquad (1.16)$$

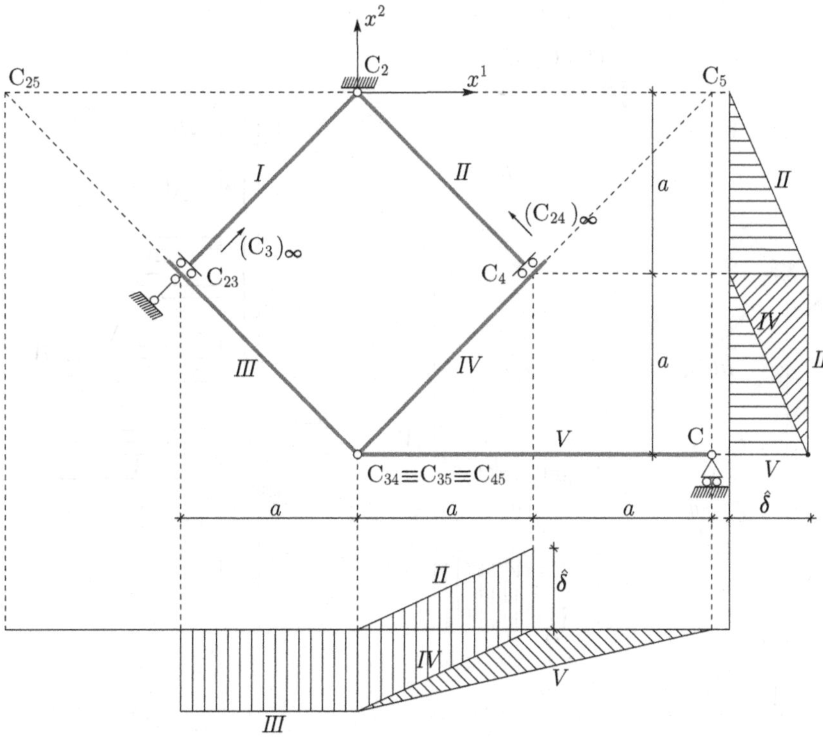

Fig. 1.30 Diagrams of the first set of possible linearized displacements

In fact, if we take a vector e in $\mathcal{B}$ with nonzero modulus to map onto a vector $\tilde{e}$ also with nonzero modulus, the relation

$$\tilde{e} = Fe,$$

considered as an algebraic system linking the components of e in a given basis to those of $\tilde{e}$ in another basis, implies that the condition (1.16) is *necessary*.

The three vectors $\tilde{e}_1, \tilde{e}_2, \tilde{e}_3$ determine another parallelepiped with volume V_a given by

$$\mathsf{V}_a = \langle \tilde{e}_1, \tilde{e}_2 \times \tilde{e}_3 \rangle_{\tilde{\mathbb{R}}^3} = \tilde{e}_1 \cdot \tilde{g}(\tilde{e}_2 \times \tilde{e}_3),$$

with $\tilde{g}$ the metric in the ambient space where there is the deformed configuration. We have

$$\mathsf{V}_a = \langle Fe_1, Fe_2 \times Fe_3 \rangle_{\tilde{\mathbb{R}}^3} = (\det F)\mathsf{V}. \tag{1.17}$$

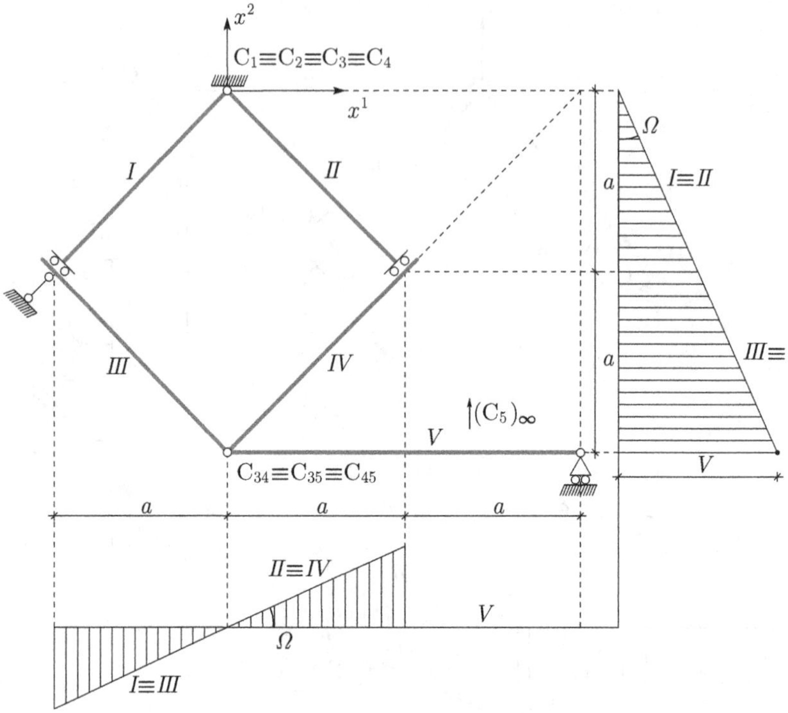

Fig. 1.31 Diagrams of the second set of possible linearized displacements

To prove the last identity, we find it useful to select the orthonormal triplet e_1, e_2, e_3, that is, the three vectors have components $(1, 0, 0)$, $(0, 1, 0)$, $(0, 0, 1)$ respectively. With these assumptions, we get

$$\mathsf{V}_a = \mathsf{e}_{ijk} F^k_1 F^j_2 F^i_3 \mathsf{V} = (\det F)\mathsf{V},$$

because the determinant of F can be computed by the mixed product of the columns of the matrix associated with F in some basis, that is, $\det F = F_1 \cdot F_2 \times F_3$, with F_i the ith column of F. The volume has its own algebraic sign: if we change the orientation of e_2 or e_3, the volume sign goes from plus to minus, and conversely for $\tilde{e}_2$ or $\tilde{e}_3$. The requirement that the deformation be orientation-preserving is then tantamount to imposing that the algebraic sign of V be preserved in the transition to V_a. To satisfy the condition, it is necessary that

$$\det F > 0,$$

which is a nonlinear constraint, the basic source of analytical difficulties in tackling some boundary value problems involving finite strains. The relation between actual and reference volumes, namely

$$\mathsf{V}_a = (\det F)\mathsf{V},$$

can be expressed in differential form:

$$d\mathsf{V}_a = (\det F)d\mathsf{V}.$$

By selecting the reference place as a setting in which we make the comparison between the two volumes, from (1.17) we get

$$\frac{\mathsf{V}_a - \mathsf{V}}{\mathsf{V}} = \det F - 1.$$

Hence $\delta_{\mathsf{V}} := \det F - 1$ represents the *local relative volume change*.

1.13 Changes in Oriented Areas: Nanson's Formula

The relation (1.17) also allows us to establish how oriented areas change as a consequence of the deformation. To this end, we write first the vector product $\tilde{e}_2 \times \tilde{e}_3$ in (1.17) as $\tilde{n}^\# A_a$, with $\tilde{n}^\#$ a unit vector that is orthogonal to the parallelogram with area A_a determined by $\tilde{e}_2$ and $\tilde{e}_3$ and considered a surface oriented by the normal $\tilde{n}^\#$. The same point of view is used for the product $e_2 \times e_3$ in the reference place, and we write $n^\# A$ for it, where A now indicates the area of the parallelogram determined by e_2 and e_3 and oriented by the unit normal $n^\#$. Using (1.17), we then have

$$\mathsf{V}_a = \left\langle Fe_1, \tilde{n}^\# A_a \right\rangle_{\tilde{\mathbb{R}}^3} = \left\langle e_1, F^{\mathrm{T}}\tilde{n}^\# A_a \right\rangle_{\mathbb{R}^3} = (\det F)\left\langle e_1, n^\# A \right\rangle_{\mathbb{R}^3} = (\det F)\mathsf{V}.$$

Hence, thanks to the arbitrariness of e_1, we get

$$F^{\mathrm{T}}\tilde{n}^\# A_a = (\det F)n^\# A,$$

from which we obtain

$$\tilde{n}^\# A_a = (\det F)F^{-\mathrm{T}}n^\# A, \tag{1.18}$$

because $\det F > 0$. However, we can also write

$$\mathsf{V}_a = \left\langle Fe_1, \tilde{n}^\# A_a \right\rangle_{\tilde{\mathbb{R}}^3} = Fe_1 \cdot \tilde{g}\tilde{n}^\# A_a = Fe_1 \cdot \tilde{n} A_a,$$

where $\tilde{n}$ is the covector associated with $\tilde{n}^\#$, namely $\tilde{n} := \tilde{g}\tilde{n}^\#$. Then, by taking into account that $\left\langle e_1, n^\# A \right\rangle = e_1 \cdot g n^\# A = e_1 \cdot nA$, with $n := gn^\#$, we can write

$$Fe_1 \cdot n^\# A_a = (\det F)e_1 \cdot nA,$$

which implies, thanks to the arbitrariness of e_1, the relation

$$\tilde{n}A_a = (\det F)F^{-\#}nA. \tag{1.19}$$

The difference between the formulas (1.18) and (1.19) rests essentially in the way we consider the normal to a surface. The relation (1.18) applies when the normal is considered a *vector*, that obtained by the vector product of two linearly independent vectors spanning the tangent plane to the surface (recall that we are considering it in three-dimensional Euclidean space) at a certain regularity point. In contrast, in (1.19) we consider the normal to be a covector, i.e., the spatial derivative of the function describing the surface, evaluated at the point under scrutiny. In components, the relations (1.18) and (1.19) can be written respectively as

$$\tilde{n}^{\#i}A_a = (\det F)(F^{-\mathrm{T}})^i_B n^{\#B}A,$$

$$\tilde{n}_i A_a = (\det F)(F^{-\#})^B_i n_B A.$$

Previous formulas also apply when we consider both A_a and A as infinitesimal areas, denoting them in this case by da_a and da, respectively. Then we can write

$$\tilde{n}^{\#}da_a = (\det F)F^{-\mathrm{T}}n^{\#}da \tag{1.20}$$

or

$$\tilde{n}da_a = (\det F)F^{-\#}nda, \tag{1.21}$$

depending on how we consider the normal. The latter point of view—the normal to a surface at a point considered as a covector—is the one that we shall use later in this book, unless otherwise specified.

Both expressions (1.20) and (1.21) are versions of what is commonly called *Nanson's formula.*

1.14 Finite Strains

Consider a vector w tangent to a smooth curve on $\mathcal{B}$ at a point x. When we impose a deformation, w changes into $\tilde{w} = Fw$. In the transition from w to $\tilde{w}$, the vector w is (at least in principle) stretched and rotated. The following theorem allows us to distinguish between the two kinematic behaviors.

Theorem 15 (Polar decomposition theorem). *Every linear operator* $G \in \mathrm{Hom}(\mathbb{R}^m, \mathbb{R}^m)$ *with* $\det G \neq 0$ *admits the following decompositions, called left and right, respectively:*

$$G = RU = VR,$$

with R an orthogonal linear operator,[11] *namely $R \in O(m)$, and U and V symmetric linear operators, i.e., $U, V \in \mathrm{Sym}(\mathbb{R}^m, \mathbb{R}^m)$. Both left and right decompositions are unique. However, if $\det G > 0$, we have $R \in SO(m)$.*

We do not give a proof here, for one can be found in standard textbooks in linear algebra. We simply apply the previous theorem to F and interpret the physical meaning of left and right decompositions

$$F = RU = VR,$$

with $R \in SO(3)$ because $\det F > 0$, as a consequence of the orientation-preserving assumption. Since $F \in \mathrm{Hom}^+(\mathbb{R}^3, \tilde{\mathbb{R}}^3)$, the superscript $+$ recalls the condition $\det F > 0$, we have to think of U and V as symmetric linear operators acting over different, although isomorphic, spaces: $U \in \mathrm{Sym}^+(\mathbb{R}^3, \mathbb{R}^3)$, $V \in \mathrm{Sym}^+(\tilde{\mathbb{R}}^3, \tilde{\mathbb{R}}^3)$.

In the case RU, we first strain the body. Then we change its location rigidly. By writing VR, we mean that we are first applying a rigid displacement. That U and V are associated with the strain can be seen by evaluating how a vector w tangent to a generic curve in $\mathcal{B}$ at x changes its length when it is transformed into $\tilde{w} = Fw$ as a consequence of the deformation $\tilde{y}$.

Write $w = ln$ and $\tilde{w} = \tilde{l}\tilde{n}$, with $l, \tilde{l} \in \mathbb{R}$, n and $\tilde{n}$ unit vectors with respect to the metrics g and $\tilde{g}$, respectively. The squared length of w is then given by

$$l^2 = \langle w, w \rangle = w \cdot gw = l^2 g \cdot n \otimes n,$$

while the squared length of $\tilde{w}$ reads

$$\tilde{l}^2 = \langle \tilde{w}, \tilde{w} \rangle = \tilde{w} \cdot \tilde{g}\tilde{w} = \tilde{l}^2 \tilde{g} \cdot \tilde{n} \otimes \tilde{n}.$$

We cannot compare them directly, because they belong to different spaces and are written with respect to different frames. Hence, we must select a common geometric setting to make the comparison. If we choose the reference place as the ambient where we compare the two lengths above, we write

$$\tilde{l}^2 = \tilde{w} \cdot \tilde{g}\tilde{w} = Fw \cdot \tilde{g}Fw = w \cdot F^*\tilde{g}Fw = l^2 n \cdot F^*\tilde{g}Fn = l^2 (F^*\tilde{g}F) \cdot n \otimes n,$$

expressing in this way $\tilde{l}$ as a function of l. We define the second-rank tensor C as

$$C := F^*\tilde{g}F$$

and call it the **right Cauchy–Green tensor**. In components, we have

$$C_{AB} = F^i_A \tilde{g}_{ij} F^j_B.$$

[11] R is said to be an orthogonal linear operator from $\mathbb{R}^m$ to itself, and we write $R \in O(m)$, when $R \in \mathrm{Hom}(\mathbb{R}^m, \mathbb{R}^m)$ and $\det R = +1$ or -1. The set of orthogonal linear operators $O(m)$ is a group. Its subset $SO(m)$, characterized by $\det R = +1$, is a group as well. When $R \in SO(m)$, it describes a rotation in $\mathbb{R}^3$. Otherwise, R represents a reflection.

Since $\det F > 0$ for the orientation-preserving condition and also $\det \tilde{g} > 0$ by the definition of a metric, C admits a positive determinant. Moreover, we have

$$C^{\mathrm{T}} = C^* = (F^* \tilde{g} F)^* = F^* \tilde{g}^* F = F^* \tilde{g} F = C,$$

i.e., C is symmetric. It is the pullback of the spatial metric $\tilde{g}$ in the reference place, and it is a new metric in $\mathcal{B}$ in this sense. The variation of length in transforming w onto $\tilde{w}$ is then given by

$$\frac{\tilde{l}^2 - l^2}{l^2} = C \cdot n \otimes n - g \cdot n \otimes n = (C - g) \cdot n \otimes n.$$

We usually call the second-rank tensor E given by

$$E := \frac{1}{2}(C - g)$$

the **strain tensor**. It has the meaning of a difference of metrics and is

$$E = E_{AB} e^A \otimes e^B,$$

with

$$E_{AB} = \frac{1}{2}(C_{AB} - g_{AB}).$$

The meaning of *relative* difference of metrics is attributed to the 1-contravariant, 1-covariant tensor

$$\tilde{E} := g^{-1} E,$$

which is

$$\tilde{E} = \tilde{E}^A_B e_A \otimes e^B,$$

with

$$\tilde{E}^A_B = \frac{1}{2}(g^{-1})^{AC}(C_{CB} - g_{CB}) = \frac{1}{2}(\tilde{C}^A_B - \delta^A_B).$$

Hence, indicating by $\tilde{C}$ the 1-contravariant, 1-covariant version of the right Cauchy–Green tensor, we write

$$\tilde{E} = \frac{1}{2}(\tilde{C} - \tilde{I}),$$

where $\bar{I}$ is the 1-contravariant, 1-covariant identity tensor $\bar{I} = \delta^A_B e_A \otimes e^B$. The strain tensor $\tilde{E}$ is symmetric. In fact, we have

$$\tilde{C} = g^{-1}C = g^{-1}F^*\tilde{g}F = F^{\mathrm{T}}F,$$

so that $\tilde{C}^{\mathrm{T}} = (F^{\mathrm{T}}F)^{\mathrm{T}} = F^{\mathrm{T}}F = \tilde{C}$. Moreover, the identity $\bar{I}$ is symmetric, too. In terms of the displacement field

$$u := \tilde{u}(x) = y - i(x) = \tilde{y}(x) - i(x),$$

already defined, with $i : \mathcal{E}^3 \to \tilde{\mathcal{E}}^3$ the identification between the two copies of the Euclidean space where we have considered reference and actual places, respectively, the right Cauchy–Green tensor $\tilde{C}$ becomes

$$\tilde{C} = F^{\mathrm{T}}F = (I + Du)^{\mathrm{T}}(I + Du),$$

where $I = \delta^i_A \tilde{e}_i \otimes e^A$, so that

$$\tilde{E} = \frac{1}{2}((I + Du)^{\mathrm{T}}(I + Du) - \bar{I}) = \frac{1}{2}(D\bar{u} + D\bar{u}^{\mathrm{T}}) + \frac{1}{2}Du^{\mathrm{T}}Du, \tag{1.22}$$

where $D\bar{u}$ and $D\bar{u}^{\mathrm{T}}$ have, respectively, components $(D\bar{u})^A_B = \delta^A_i (Du)^i_B$ and $(D\bar{u}^{\mathrm{T}})^A_B = (Du^{\mathrm{T}})^A_i \delta^i_B$. In deriving the relation (1.22), we have exploited the identity

$$I^{\mathrm{T}}I = (g^{AB}\delta^j_B \tilde{g}_{ji}\delta^i_C)e_A \otimes e^C = \delta^B_C e_B \otimes e^C = \bar{I}.$$

The transpose of I, namely $I^{\mathrm{T}} = \delta^A_i \tilde{e}^i \otimes e_A$, changes Du into

$$D\bar{u} = (D\bar{u}^A)_C e_A \otimes e^C = (g^{AB}\delta^i_B \tilde{g}_{ij}(Du^j)_C)e_A \otimes e^C.$$

By definition, in fact, the displacement u is a vector in the physical space $\tilde{\mathcal{E}}^3$, an element of its translation space that becomes $\tilde{\mathbb{R}}^3$ once we fix an origin of the coordinate frame, while E is a tensor in the reference space. The premultiplication of $Du = (Du^i)_A \tilde{e}_i \otimes e^A$ by $I^{\mathrm{T}} = (g^{AB}\delta^j_B \tilde{g}_{ji})e_A \otimes \tilde{e}^i$ implies the projection of u from $\tilde{\mathbb{R}}^3$ to $\mathbb{R}^3$. Also, the postmultiplication of $Du^{\mathrm{T}} = (g^{AB}D_B u^j \tilde{g}_{ji})e_A \otimes \tilde{e}^i$ by the identity $I = \delta^i_A \tilde{e}_i \otimes e^A$ determines also the projection of u into $\mathbb{R}^3$. In other words, the second-rank identity tensor $I = \delta^i_A \tilde{e}_i \otimes e^A$ is a *shifter* from $\mathbb{R}^3$ to $\tilde{\mathbb{R}}^3$.

The right-hand side of the relation (1.22) is commonly called the **Almansi tensor**.

There are many possible measures of strain: functions of C vanishing when C is the identity can in principle be adopted as strain measures.

Write δ_l for the ratio $\frac{\tilde{l}-l}{l}$. We have

$$\delta_l = \sqrt{2E \cdot n \otimes n + 1} - 1,$$

or in components,

$$\delta_l = \sqrt{2E_{AB}n^A n^B + 1} - 1.$$

In particular, in the direction of the vector e_A of the basis at x we obtain

$$\delta_{l_A} = \sqrt{2E_{AA} + 1} - 1.$$

Consider now two linearly independent vectors v and w tangent at x to different smooth curves in $\mathcal{B}$. When we impose a deformation, we find corresponding vectors at y in $\mathcal{B}_a$, namely $\tilde{v} = Fv$ and $\tilde{w} = Fw$. The change in angle between the two vectors is defined by

$$\gamma_\alpha := \alpha - \alpha_a,$$

in which α is the angle between v and w, and α_a is the angle between $\tilde{v}$ and $\tilde{w}$. Hence, we compute

$$\gamma_\alpha = \arccos \frac{\langle v, w\rangle}{\sqrt{\langle v, v\rangle}\sqrt{\langle w, w\rangle}} - \arccos \frac{\langle Fv, Fw\rangle}{\sqrt{\langle Fv, Fv\rangle}\sqrt{\langle Fw, Fw\rangle}}.$$

In particular, if the two vectors at x are orthogonal, the change of angle is given by

$$\gamma_\alpha = \frac{\pi}{2} - \alpha_a.$$

Assign an orthogonal basis in a neighborhood of $x \in \mathcal{B}$. Let e_A and e_B two vectors of such a basis. Write α_{AB} for the angle between the vectors Fe_A and Fe_B, obtained after deformation, and γ_{AB} for the variation in angle $\gamma_{AB} = \frac{\pi}{2} - \alpha_{AB}$. We compute (prove it as an exercise)

$$\gamma_{AB} = \arcsin \frac{2E_{AB}}{\sqrt{(1 + 2E_{AA})(1 + 2E_{BB})}}.$$

Let A be the area of the parallelogram defined by two linearly independent vectors v and w at the point x, and A_a the area of the parallelogram defined by Fv and Fw at y. Write δ_{Avw} for the ratio $\frac{A_a - A}{A}$. If we consider an orthogonal basis at x, the change of area defined by the two vectors e_A and e_B, namely $\delta_{A_{AB}}$, is

$$\delta_{A_{AB}} = \sqrt{(1 + 2E_{AA})(1 + 2E_{BB}) - 4(E_{AB})^2} - 1.$$

The local relative volume change δ_V in terms of the Almansi tensor reads

$$\delta_V = \sqrt{\det(2\tilde{E} + I)} - 1.$$

1.15 Small Strains

The condition

$$|Du| \ll 1 \tag{1.23}$$

defines the **small-strain regime**.

When the small-strain regime holds, we can *confuse* reference and actual shapes without distinguishing them as subsets of different (although isomorphic) spaces. Hence, in the small-strain regime, we shall write just x for the space variables, and we shall not distinguish between upper- and lowercase indices, using just the latter. No distinction is also made, in this setting, between the reference metric g and the actual one $\tilde{g}$, and we shall write $g = g_{ij}e^i \otimes e^j$ by adopting the convention on indices just mentioned.

Under the condition (1.23), we can neglect the term $\frac{1}{2}Du^{\mathrm{T}}Du$ in the expression of $\tilde{E}$, since it is of higher order with respect to $|Du|$, so that we can approximate $\tilde{E}$ by

$$\tilde{E} \approx \frac{1}{2}(Du + Du^{\mathrm{T}}).$$

Here we have written Du instead of $D\bar{u}$, because in the small-strain regime, we confuse reference and current configurations, so that we avoid the use of the shifter $I = \delta^i_A \tilde{e}_i \otimes e^A$.

We indicate by $\tilde{\varepsilon}$ the symmetric part of Du, namely

$$\tilde{\varepsilon} := \frac{1}{2}(Du + Du^{\mathrm{T}}) = \mathrm{Sym}(Du),$$

while by ε, we denote its fully covariant version

$$\varepsilon = \mathrm{Sym}(gDu),$$

so that $\varepsilon = \varepsilon_{ij}e^i \otimes e^j$, with the generic component ε_{ij} given by

$$\varepsilon_{ij} = \frac{1}{2}(u_{i,j} + u_{j,i}),$$

where $u_i := g_{ij}u^j$ is the ith component of the covector field $u^\flat = gu$, and the comma indicates the derivative with respect to the coordinate corresponding to the index to the right of the comma; in other words, $u_{i,j}$ is the ijth component of gDu, assuming g constant in space. When $\{O, x^1, x^2, x^3\}$ is orthogonal, we omit the superscript $\flat$ that indicates the action of g, and we have simply $u_i = \delta_{ij}u^j$. The matrix associated with ε with respect to an orthonormal frame $\{O, x^1, x^2, x^3\}$ is given by

$$[\varepsilon_{ij}] = \begin{pmatrix} \frac{\partial u_1}{\partial x^1} & \frac{1}{2}\left(\frac{\partial u_1}{\partial x^2} + \frac{\partial u_2}{\partial x^1}\right) & \frac{1}{2}\left(\frac{\partial u_1}{\partial x^3} + \frac{\partial u_3}{\partial x^1}\right) \\ \frac{1}{2}\left(\frac{\partial u_1}{\partial x^2} + \frac{\partial u_2}{\partial x^1}\right) & \frac{\partial u_2}{\partial x^2} & \frac{1}{2}\left(\frac{\partial u_2}{\partial x^3} + \frac{\partial u_3}{\partial x^2}\right) \\ \frac{1}{2}\left(\frac{\partial u_1}{\partial x^3} + \frac{\partial u_3}{\partial x^1}\right) & \frac{1}{2}\left(\frac{\partial u_2}{\partial x^3} + \frac{\partial u_3}{\partial x^2}\right) & \frac{\partial u_3}{\partial x^3} \end{pmatrix}. \tag{1.24}$$

The tensor ε is what is commonly called a **small-strain tensor**.

Consider a point x and a unit vector s. The relative variation in length along s in the small-strain regime is given by

$$\delta_l = \varepsilon \cdot (s \otimes s) = \varepsilon_{ij}s^j s^i.$$

In particular, with reference to the coordinate system $\{O, x^1, x^2, x^3\}$ introduced above, the term $\varepsilon_{11} = \frac{\partial u_1}{\partial x^1}$ represents the elongation δ_1 along the x^1-axis. An analogous meaning can be attributed to the other terms on the principal diagonal, i.e., $\frac{\partial u_2}{\partial x^2}$ and $\frac{\partial u_3}{\partial x^3}$. Hence, if we imagine an infinitesimal cube centered at the point x with edges corresponding to the three coordinate axes, the relative variation in volume in the small-strain regime, indicated by δ_V, is given by the trace of ε, namely

$$\delta_V = \operatorname{tr}\varepsilon = \varepsilon_{11} + \varepsilon_{22} + \varepsilon_{33} = \frac{\partial u_1}{\partial x^1} + \frac{\partial u_2}{\partial x^2} + \frac{\partial u_3}{\partial x^3} = \operatorname{div} u. \tag{1.25}$$

The result can be obtained formally from what we have already found in the finite-strain regime. Under the regularity assumptions for the deformation given at the beginning of these notes, the determinant of F admits the expansion

$$\det F = \det(I + Du) = 1 + \operatorname{tr}\tilde{\varepsilon} + o(|Du|).$$

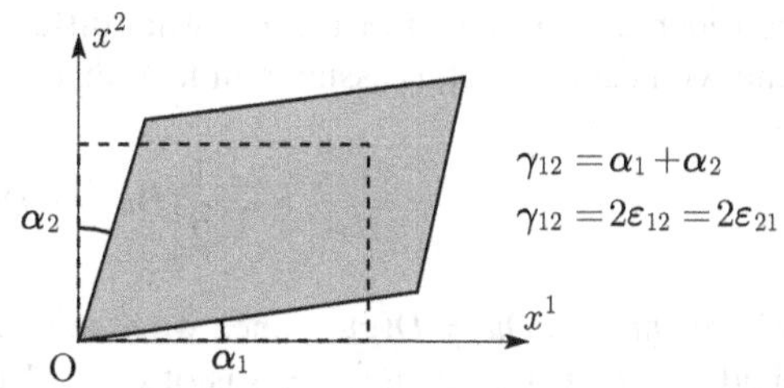

Fig. 1.32 Angle variation in the small-strain regime

By neglecting leading-order terms, as allowed by the condition $|Du| \ll 1$, and lowering the first index of $\tilde{\varepsilon}$ by multiplying $\tilde{\varepsilon}$ from the left by the metric, we obtain once again (1.25), since $\det F - 1$ represents the volume variation in the finite-strain setting.

Off-diagonal terms in (1.24) have the meaning of angle variation. Precisely, $\frac{1}{2}\left(\frac{\partial u_1}{\partial x^2} + \frac{\partial u_2}{\partial x^1}\right)$ describes half the angle variation between the x^1 and x^2 axes; see Figure 1.32.

The sums

$$\frac{\partial u_1}{\partial x^1} + \frac{\partial u_2}{\partial x^2},$$

$$\frac{\partial u_1}{\partial x^1} + \frac{\partial u_3}{\partial x^3},$$

$$\frac{\partial u_2}{\partial x^2} + \frac{\partial u_3}{\partial x^3},$$

represent the area variations over the coordinate planes x^1x^2, x^1x^3, x^2x^3, respectively, in the case of the small-strain regime.

Exercise 16. *Linearize the finite-strain area variation.*

We have determined ε so far from the knowledge of the displacement field $x \longmapsto \tilde{u}(x)$. In experiments, however, we are often able to evaluate the elongations along three independent directions and the angle variations among them at numerous points. Let us suppose—it is an idealization—that we have a field of these measures, namely a map $x \longmapsto \varepsilon(x) \in \mathrm{Sym}(\mathbb{R}^3, \mathbb{R}^3)$ assigning to every x a symmetric tensor ε. The problem now is to evaluate whether there is a displacement field $x \longmapsto \tilde{u}(x)$ such that the associated strain field matches the assigned one. In this case, we say that the strain ε is *compatible*. Formally, such a requirement is tantamount to asking for conditions of integrability of

$$\varepsilon_{ij} = \frac{1}{2}(u_{i,j} + u_{j,i})$$

considered as a system of linear partial differential equations in the unknowns u_i, and with data ε_{ij}. Let us assume, in fact, that once we know $\varepsilon(\cdot)$, it is true that

$$\varepsilon = \frac{1}{2}(Du^\flat + (Du^\flat)^{\mathrm{T}}),$$

where $Du^\flat = gDu = D(gu)$, since we assume g to be constant. When the assigned field $x \longmapsto \varepsilon(x) \in \mathrm{Sym}(\mathbb{R}^3, \mathbb{R}^3)$ is of class $\mathrm{C}^2(\mathcal{B})$, we may evaluate its curl twice. We compute

$$\operatorname{curl} \varepsilon = \frac{1}{2}\operatorname{curl}(Du^\flat + (Du^\flat)^{\mathrm{T}}) = \frac{1}{2}(\operatorname{curl} Du^\flat + \operatorname{curl}(Du^\flat)^{\mathrm{T}})$$
$$= \frac{1}{2}(0 + D\operatorname{curl} u^\flat) = \frac{1}{2}D\operatorname{curl} u^\flat,$$

so that

$$\operatorname{curl}\operatorname{curl} \varepsilon = \frac{1}{2}\operatorname{curl} D \operatorname{curl} u^\flat = 0,$$

because $\operatorname{curl} D(\cdot)$ vanishes. The condition

$$\operatorname{curl}\operatorname{curl} \varepsilon = 0,$$

in components

$$\varepsilon_{ij,hk} + \varepsilon_{hk,ij} = \varepsilon_{ik,hj} + \varepsilon_{hj,ik},$$

is a *compatibility condition* for the strain field $\varepsilon(\cdot)$ and is necessary under the assumed regularity assumptions. The compatibility is also sufficient when $\mathcal{B}$ is simply connected (rigorous proofs can be found in treatises on the integration of differential forms).

The analysis of compatibility conditions is formally simpler in the finite-strain regime when a differentiable field $x \longmapsto F(x)$ is assigned. In fact, since $F = I + Du$, we just need to have

$$\operatorname{curl} F = 0,$$

to ensure the existence of a displacement field determining F point by point.

1.16 Finite Elongation of Curves and Variations of Angles

Consider a continuous and continuously differentiable map $\varphi : [0, l] \longrightarrow \mathcal{B}$, $l > 0$, defining a curve on the reference place. Its length $l(\varphi)$ is given by

$$l(\varphi) = \int_0^l |\varphi'(s)|\, ds,$$

where the prime denotes differentiation with respect to s, so that $\varphi'(s)$ is the tangent to the curve at the point $\varphi(s)$. When we impose a deformation $\tilde{y} : \mathcal{B} \longrightarrow \tilde{\mathcal{E}}^3$, we have over the deformed shape $\mathcal{B}_a = \tilde{y}(\mathcal{B})$ another curve defined by the map $\bar{\varphi} := \tilde{y} \circ \varphi : [0, l] \longrightarrow \tilde{y}(\mathcal{B})$. The vector $\bar{\varphi}'(s) = \frac{d}{ds}\tilde{y}(\varphi(s))$ is then the tangent to the new curve on $\mathcal{B}_a$ at the point $\tilde{y}(\varphi(s))$ and is, by the chain rule, $\bar{\varphi}'(s) = F\varphi'(s)$. The length of the deformed curve is

$$l(\bar{\varphi}) = \int_0^l |\bar{\varphi}'(s)|\, ds.$$

Polar decomposition allows us to write

$$\begin{aligned} |\bar{\varphi}'(s)| &= \langle \bar{\varphi}'(s), \bar{\varphi}'(s)\rangle^{\frac{1}{2}} = \langle F\varphi'(s), F\varphi'(s)\rangle^{\frac{1}{2}} = \langle RU\varphi'(s), RU\varphi'(s)\rangle^{\frac{1}{2}} \\ &= \langle U\varphi'(s), U\varphi'(s)\rangle^{\frac{1}{2}} = |U\varphi'(s)|, \end{aligned}$$

so that

$$l(\bar{\varphi}) = \int_0^l |U\varphi'(s)|\, ds,$$

and the variation in length δl, namely

$$\delta l = \frac{l(\bar{\varphi}) - l(\varphi)}{l(\varphi)},$$

is given by

$$\delta l = \frac{\int_0^l |U\varphi'(s)|\, ds}{\int_0^l |\varphi'(s)|\, ds} - 1,$$

and it depends only on U. When U is the identity, the deformation from φ to $\bar{\varphi}$ is just a rigid-body change of place.

Consider now two such curves, say φ_1 and φ_2, crossing a point: $\varphi_1(\bar{s}) = \varphi_2(\bar{s})$, $\bar{s} \in [0, l]$. The angle α between the two tangent vectors $\varphi_1'(\bar{s})$ and $\varphi_2'(\bar{s})$ satisfies the relation

$$\cos\alpha = \frac{\langle \varphi_1'(\bar{s}), \varphi_2'(\bar{s}) \rangle}{|\varphi_1'(\bar{s})|\,|\varphi_2'(\bar{s})|}.$$

By imposing a deformation $\tilde{y}$ on $\mathcal{B}$, we determine on the current shape $\mathcal{B}_a$ two new curves $\bar{\varphi}_1 = \tilde{y} \circ \varphi_1$ and $\bar{\varphi}_2 = \tilde{y} \circ \varphi_2$ meeting at the image of the original crossing point, namely

$$\bar{\varphi}_1(\bar{s}) = \bar{\varphi}_2(\bar{s}).$$

The two tangent vectors $\bar{\varphi}_1'(\bar{s})$ and $\bar{\varphi}_2'(\bar{s})$ then form an angle $\bar{\alpha}$ such that

$$\cos\bar{\alpha} = \frac{\langle \bar{\varphi}_1'(\bar{s}), \bar{\varphi}_2'(\bar{s}) \rangle}{|\bar{\varphi}_1'(\bar{s})|\,|\bar{\varphi}_2'(\bar{s})|} = \frac{\langle U\varphi_1'(\bar{s}), U\varphi_2'(\bar{s}) \rangle}{|U\varphi_1'(\bar{s})|\,|U\varphi_2'(\bar{s})|}. \tag{1.26}$$

Hence, the variation in the angle $\gamma_\alpha := \alpha - \bar{\alpha}$ depends only on U.

1.17 Deviatoric Strain

The small strain ε can be naturally decomposed additively into *spheric* and *deviatoric* components, indicated respectively by ε^s and ε^d:

$$\varepsilon = \varepsilon^s + \varepsilon^d,$$

where

$$\varepsilon^s := \frac{1}{3}(\operatorname{tr}\varepsilon)I,$$

with I the second-rank identity tensor, and

$$\varepsilon^d := \varepsilon - \frac{1}{3}(\operatorname{tr}\varepsilon)I.$$

An immediate consequence of the definition of ε^d is

$$\operatorname{tr}\varepsilon^d = 0,$$

which specifies that ε^d describes just the isochoric strain—the shape changes, but the volume remains invariant.

When the strain is large, a multiplicative (rather than additive) decomposition of F emerges. Define first

$$\bar{F} := (\det F)^{-\frac{1}{3}} F.$$

Here $\bar{F}$ is the spatial derivative of an isochoric deformation. In fact, we get[12]

$$\det \bar{F} = \frac{\det F}{\det F} = 1.$$

Hence, we can write the decomposition

$$F = \tilde{F}\bar{F} \quad \text{or} \quad F = \bar{F}\tilde{F}$$

with

$$\tilde{F} := (\det F)^{\frac{1}{3}} I,$$

where I is once again the second-rank identity tensor. Using $\bar{F}$, we can also define a version of the Cauchy–Green tensor $\bar{C}$ not accounting for volume changes. It reads

$$\bar{C} := \bar{F}^{\mathrm{T}}\bar{F} = (\det F)^{-\frac{2}{3}} C = (\det F)^{-\frac{2}{3}} F^{\mathrm{T}} F.$$

In components, we get

$$\bar{C}^A_B = (\det F)^{-\frac{2}{3}} (F^{\mathrm{T}})^A_i F^i_B.$$

By direct computation, we obtain

$$\det \bar{C} = 1,$$

which confirms that $\bar{C}$ does not account for volume changes.

1.18 Motions

Motions are time-parameterized families of deformations

$$\tilde{y} : \mathcal{B} \times (t_0, t_1) \longrightarrow \tilde{\mathcal{E}}^3,$$

[12] For every $n \times n$ real matrix M and real number β, the identity $\det \beta M = \beta^n \det M$ holds.

which have the already discussed properties with respect to space and are at least two times piecewise differentiable with respect to time $t \in (t_0, t_1)$. The vector

$$\dot{y}(x,t) := \frac{d}{dt}\tilde{y}(x,t)$$

is the **velocity** at the instant t of a material point $y = \tilde{y}(x,t)$ that was at x in its reference place. Hence, we have the field

$$(x,t) \longmapsto \dot{y}(x,t) \in \tilde{\mathbb{R}}^3,$$

which is the so-called *Lagrangian description* of the velocity. More precisely, $\dot{y}$ is a tangent vector to $\mathcal{B}_a$ at $y = \tilde{y}(x,t)$; formally, $\dot{y} \in T_y\mathcal{B}_a$. We can also imagine the velocity as a vector attached at y at the instant t, irrespective of the reference place x. In this way, we have the field

$$(y,t) \longmapsto v(y,t) \in \tilde{\mathbb{R}}^3,$$

which is the so-called *Eulerian description* of the velocity. Once again, we have $v \in T_y\mathcal{B}_a$, and above all,

$$\dot{y} = v. \tag{1.27}$$

The Eulerian and Lagrangian velocities coincide. The same statement does not hold for the **acceleration**, for which we have, in Lagrangian description,

$$\ddot{y}(x,t) := \frac{d^2}{dt^2}\tilde{y}(x,t),$$

while in the Eulerian representation, we write a for the acceleration at y and t, defined by

$$a(y,t) := \frac{d}{dt}\tilde{v}(x,t) = \frac{\partial v}{\partial t} + (D_y v)\dot{y} = \frac{\partial v}{\partial t} + (D_y v)v,$$

where $\frac{\partial v}{\partial t}$ is computed by holding y fixed. Then we obtain

$$\ddot{y} \neq a.$$

Since $y = \tilde{y}(x,t)$, we can compute the derivative of v with respect to x, obtaining by the chain rule,

$$Dv(\tilde{y}(x,t),t) = D_y v Dy = D_y v F,$$

where D_y is the derivative with respect to y, while D is the derivative with respect to x, as adopted so far. However, due to the identity (1.27), we have $Dv = D\dot{y} = \dot{F}$ and $\dot{F} = D_y v F$, which implies

$$D_y v = \dot{F}F^{-1}$$

and

$$\mathrm{tr}(\dot{F}F^{-1}) = \mathrm{tr}(D_y v) = \mathrm{div}\, v.$$

1.19 Exercises and Supplementary Remarks

Exercise 17. *Calculate the length variation of segments by taking $\mathcal{B}_a$ as a geometric environment where we compare lengths, volumes, areas. Write such a variation in terms of the so-called* ***left Cauchy–Green tensor****, denoted by $\mathbb{B}$ and defined by $\mathbb{B} := FF^T$ with $\mathbb{B} = \mathbb{B}^i_j \tilde{e}_i \otimes \tilde{e}^j = (F^i_A (F^T)^A_j)\tilde{e}_i \otimes \tilde{e}^j$.*

Suggestion. Write l as a function of $\tilde{l}$ by exploiting the relation $w = F^{-1}\tilde{w}$, with $w \in T_x\mathcal{B}$ and $\tilde{w} \in T_y\mathcal{B}_a$.

Exercise 18. *Show that the second-rank tensor $\mathbb{A} := \mathbb{B}\tilde{g}^{-1}$, with components $\mathbb{A}^{ik} = \mathbb{B}^i_j \tilde{g}^{jk}$, is the pushforward in the current shape $\mathcal{B}_a$ of the inverse referential metric g^{-1}, and that the second-rank tensor $\hat{\mathbb{A}} := \tilde{g}\mathbb{B}^{-1}$, with components $\hat{\mathbb{A}}_{ki} = \tilde{g}_{kj}(\mathbb{B}^{-1})^j_i$, is the pushforward of the material metric g in $\mathcal{B}_a$.*

Suggestion. Use the relation $F^{\mathrm{T}} = g^{-1}F^*\tilde{g}$ and its counterpart $F^{-\mathrm{T}} = \tilde{g}^{-1}F^{-*}g$.

Exercise 19. *Express the components of ε in a spherical coordinate frame $\{\mathrm{O}, r, \theta, \varphi\}$, with r the radius from the origin O and in a cylindrical frame $\{\mathrm{O}, \bar{r}, \theta, z\}$.*

Suggestion. Recall that for spherical coordinates, we have $x^1 = r\cos\vartheta \sin\varphi$, $x^2 = r\sin\vartheta \sin\varphi$, $x^3 = r\cos\varphi$, while the cylindrical system is defined by $x^1 = \bar{r}\cos\vartheta$, $x^2 = \bar{r}\sin\vartheta$, $x^3 = z$.

Exercise 20. *Prove formally that the off-diagonal terms in (1.24) describe angle variation. Precisely, $\frac{1}{2}\left(\frac{\partial u_1}{\partial x^2} + \frac{\partial u_2}{\partial x^1}\right)$ represents half the angle variation between the x^1 and x^2 axes (Fig. 1.32).*

Suggestion. Write (1.26) in terms of u and linearize the resulting expression.

In the subsequent exercises in this section, we identify $\mathbb{R}^3$ with its dual $\mathbb{R}^{3*}$ and denote components just by covariant indices—in this way, we refer essentially to a flat metric. The choice is simply for notational convenience. There will be, in fact, some powers of the coordinates, so that we find it convenient to write, e.g., x_i^2 instead of $(x^i)^2$.

Exercise 21. *Consider a prism obtained by translating a compact open subset Ω of a plane in $\mathcal{E}^3$ along an interval $(0,l)$ of an axis orthogonal to the plane containing Ω and crossing it at the barycenter. The boundary of the prism closure is $\partial\Omega \times [0,l] \cup \Omega \times \{0\} \cup \Omega \times \{l\}$. Assume that the prism is rather slender, that is, $\operatorname{diam}\Omega \ll l$. Take an orthogonal frame of reference $\{O, x_1, x_2, x_3\}$ with origin at the barycenter of the set $\Omega \times \{0\}$, the x_3-axis coinciding with the prism axis itself, and x_1 and x_2 corresponding to Ω. Consider a displacement field having in the frame selected the following components:*

$$u_1 = -\nu a x_1 x_2,$$
$$u_2 = -a\left(\frac{x_3^2}{2} - \nu\frac{x_1^2 - x_2^2}{2}\right),$$
$$u_3 = a x_2 x_3,$$

where ν and a are positive constants. Find the components of ε and interpret their physical meaning.

Exercise 22. *For the body described in the previous exercise, consider a displacement field having in the frame selected the following components:*

$$u_1 = -\vartheta x_2 x_3,$$
$$u_2 = \vartheta x_1 x_3,$$
$$u_3 = w(x_1, x_2),$$

where ϑ is a positive constant and $w(x_1,x_2)$ a differentiable function. Find ε and interpret the physical meaning of the deformation produced.

Exercise 23. *For the body considered in the two previous exercises, consider a displacement field having the following components in the frame selected:*

$$u_1 = c_1(-\nu x_1 x_2 x_3),$$
$$u_2 = c_1\left(-\frac{l^3}{3} + \frac{x_3 l^2}{2} - \frac{x_3^3}{6} - \frac{\nu x_3\left(x_2^2 - x_1^2\right)}{2}\right),$$
$$u_3 = c_1\left(\varphi(x_1, x_2) - \frac{l^2 - x_3^2}{2}x_2 - \frac{2+\nu}{6}x_2^3 + \frac{\nu x_1^2 x_2}{2}\right),$$

where c_1 and ν are positive constants, and $\varphi(x_1,x_2)$ is a differentiable function. Find ε and interpret the physical meaning of the deformation produced.

Exercise 24. *Along the prism considered above assign a field $x \longmapsto \varepsilon(x)$ with*

$$\left[\varepsilon_{ij}\right](x_1, x_2, x_3) = \begin{pmatrix} 0 & 0 & \varepsilon_{13} \\ 0 & 0 & \varepsilon_{23} \\ \varepsilon_{13} & \varepsilon_{23} & 0 \end{pmatrix},$$

where $\varepsilon_{13} = \frac{1}{2}\left(\frac{\partial w}{\partial x_1} - \vartheta x_2 + a x_3^3 x_1\right)$, $\varepsilon_{23} = \frac{1}{2}\left(\frac{\partial w}{\partial x_2} + \vartheta x_1 + \bar{a} x_3^3 x_2\right)$, $w(\cdot)$ *is a smooth function of* x_1 *and* x_2, $\vartheta > 0$, a *and* $\bar{a}$ *are constants. Find values of* a *and* $\bar{a}$ *ensuring that the strain field* $x \longmapsto \varepsilon(x)$ *is compatible.*

Suggestion. Calculate the second derivative of ε and determine a and $\bar{a}$ satisfying $\varepsilon_{ij,hk} + \varepsilon_{hk,ij} = \varepsilon_{ik,hj} + \varepsilon_{hj,ik}$.

Exercise 25. *In three-dimensional space, consider a point* x_0, *a vector* $c \in \mathbb{R}^3$ *with unit length, and a displacement field*

$$u := \tilde{u}(x) = (\nu - 1)\,(c \cdot (x - x_0))\, c,$$

where $\nu > 0$ *(consider a flat metric). Find the finite-strain change in volume around the point* x_0 *and the small-strain change.*

Suggestion. Since we have considered a flat metric, we have identified c with it dual counterpart $c^\flat$. Consider then that $\det F = \det(Du + I)$. The rest is a matter of direct calculation.

Exercise 26. *In three-dimensional space and fixing an orthogonal frame of reference, consider a deformation given by*

$$\begin{aligned} y_1(x) &= kx_1 - \bar{k}x_2 - 1, \\ y_2(x) &= kx_1 + \bar{k}x_2, \\ y_3(x) &= 2x_3, \end{aligned}$$

with $k, \bar{k} > 0$, $x = (x_1, x_2, x_3)$. *Find* F, C, R, U, *where the last two second-rank tensors are the factors of the left polar decomposition of* F.

Suggestion. An algorithm to calculate R and U goes as follows:

1. Construct C and evaluate its eigenvalues λ_i and the associated eigenvectors (called also proper vectors) $\tilde{\tau}_i$.
2. Collect the square roots of eigenvalues in a diagonal matrix $\Lambda := \operatorname{diag}\left(\sqrt{\lambda_i}\right)$. Denote by T the matrix having as columns the eigenvectors $\tilde{\tau}_i$, namely $T = (\tilde{\tau}_i)$. By the definition of eigenvalues, we have $\Lambda = T\tilde{C}T^{\mathrm{T}}$. Also, since $\tilde{C} = U^{\mathrm{T}}U = U^2$, we have $U = T\Lambda^{\frac{1}{2}}T^{\mathrm{T}}$.
3. R follows from the polar decomposition: $R = FU^{-1}$.

The eigenvalues of C are called **principal stretches**, while the eigenvectors are called **principal stretch directions**. In three dimensions, the principal stretches are solutions of the algebraic equation

$$\lambda^3 - I_1\lambda^2 + I_2\lambda - I_3 = 0,$$

where I_1, I_2, and I_3 are the **principal invariants** of C, and $\mathbb{B} = FF^{\mathrm{T}}$. They are given by

$$
\begin{aligned}
I_1 &= \operatorname{tr} C = \operatorname{tr} \mathbb{B}, \\
I_2 &= \frac{1}{2}\left((\operatorname{tr} C)^2 - \operatorname{tr} C^2\right) = \frac{1}{2}\left((\operatorname{tr} \mathbb{B})^2 - \operatorname{tr} \mathbb{B}^2\right), \\
I_3 &= \det C = \det \mathbb{B}.
\end{aligned}
$$

Exercise 27. *Consider a deformation involving only a simple shear:*

$$
\begin{aligned}
y_1(x) &= x_1, \\
y_2(x) &= x_2 + kx_1, \\
y_3(x) &= x_3,
\end{aligned}
$$

with $k > 0$. Find the principal stretches.

Exercise 28. *Consider the cylinder represented in Figure 1.33 with the reference frame $\{\mathrm{O}, x_1, x_2, x_3\}$. Its radius is r, the height $r/\sqrt{2}$. Consider a displacement field with components*

$$
\begin{aligned}
u_1 &= kx_3^2 x_2, \\
u_2 &= -kx_3^2 x_1, \\
u_3 &= kx_3^3,
\end{aligned}
$$

where k is a positive constant. Check that the resulting deformation is orientation-preserving and evaluate k so that the elongation of the segment $\overline{\mathrm{OA}}$ (point A has coordinates $(r/\sqrt{2}, r/\sqrt{2}, r/\sqrt{2})$) is equal to $0.02 l_0$ in the small-strain regime, with $l_0 = r\sqrt{3}/\sqrt{2}$ the initial length of $\overline{\mathrm{OA}}$.

Remarks and solution. From $y = u + x$, we get

$$
\begin{aligned}
y_1 &= kx_3^2 x_2 + x_1, \\
y_2 &= -kx_3^2 x_1 + x_2, \\
y_3 &= kx_3^3 + x_3.
\end{aligned}
$$

As a consequence, F has components listed in the following matrix:

$$
\begin{pmatrix}
1 & kx_3^2 & 2kx_3x_2 \\
-kx_3^2 & 1 & -2kx_3x_1 \\
0 & 0 & 1 + 3kx_3^2
\end{pmatrix},
$$

and

$$
\det F = 1 + 3kx_3^2 + k^2x_3^4 + 3k^3x_3^6
$$

Fig. 1.33 Cylinder

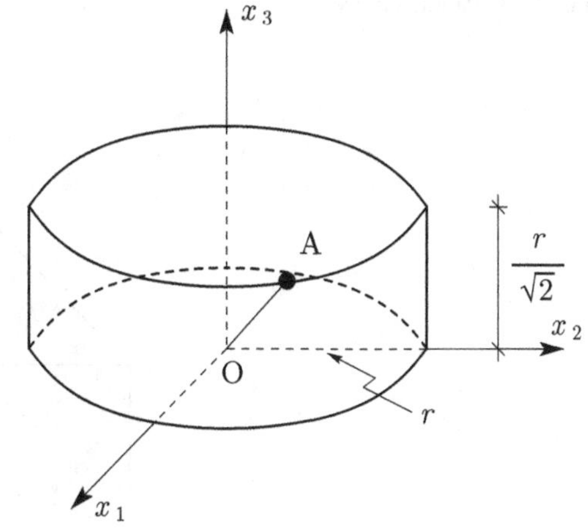

is positive at all points of the cylinder, and the deformation is orientation-preserving. Find ε. With n unit vector orienting the segment $\overline{\text{OA}}$ from O to A, the variation in length δ_l at a generic point of $\overline{\text{OA}}$ along n defined by $n_1 = n_2 = n_3 = 1/\sqrt{3}$ is

$$\delta_l = \varepsilon_{ij} n_i n_j = \frac{k(2x_2x_3 - 2x_1x_3 + 3x_3^2)}{3}.$$

By parameterizing the segment $\overline{\text{OA}}$ by $x_1 = s, x_2 = s, x_3 = s, s \in \left[0, r/\sqrt{2}\right]$, we have

$$\delta l_{OA} = \sqrt{3} \int_0^{r/\sqrt{2}} \frac{k(2x_2x_3 - 2x_1x_3 + 3x_3^2)}{3}\, ds = \frac{k\sqrt{3}}{6\sqrt{2}} r^3.$$

Finally, the data prescribe

$$\delta l_{OA} = 0.02 l_0,$$

so that we must have

$$k = \frac{12}{100 r^2}.$$

Fig. 1.34 Parallelepiped

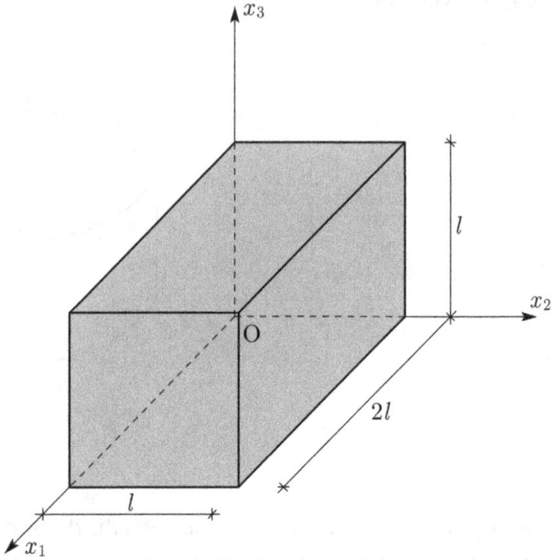

Exercise 29. *Consider the body in Figure 1.34, with the reference frame* $\{O, x_1, x_2, x_3\}$. *Take a displacement field having the following components:*

$$u_1 = \frac{1}{10^3}\left(\frac{3x_1^3}{l^2} + \frac{2x_1^2}{l}\right),$$

$$u_2 = \frac{1}{10^3}\left(\frac{2x_2^2}{l}\right),$$

$$u_3 = \frac{1}{10^3}(3x_3).$$

Calculate the volume variation δV *in the small-strain regime.*

Remarks and solution. We see that

$$\delta_V = \operatorname{tr}\varepsilon = \frac{\partial u_1}{\partial x_1} + \frac{\partial u_2}{\partial x_2} + \frac{\partial u_3}{\partial x_3} = \frac{1}{10^3}\left(\frac{9x_1^2}{l^2} + \frac{4x_1}{l} + \frac{4x_2}{l} + 3\right).$$

By integrating over the parallelepiped, we get

$$\delta V = \frac{1}{10^3}\int_0^{2l}\int_0^{l}\int_0^{l}\left(\frac{9x_1^2}{l^2} + \frac{4x_1}{l} + \frac{4x_2}{l} + 3\right)dx_1 dx_2 dx_3 = \frac{21}{500}l^3.$$

The body increases its volume.

Exercise 30. *Consider the cube of side l, represented in Figure 1.35, and the reference frame $\{O, x_1, x_2, x_3\}$. Take a displacement field with the following components:*

$$
\begin{aligned}
u_1 &= \frac{3}{400l}(x_1^2 - 2x_2^2 + x_3^2),\\
u_2 &= \frac{3}{400l}(2x_1^2 - x_2^2 - 3x_3^2),\\
u_3 &= \frac{3}{400l}(2x_2^2 + 2x_3^2).
\end{aligned}
$$

Find the variation in area of the lateral surface of the cube.

Remarks and solution. In the small-strain regime, the specific changes of area parallel to the cube faces are

$$
\delta_{A_{23}} = \left(\frac{\partial u_2}{\partial x_2} + \frac{\partial u_3}{\partial x_3}\right), \quad \delta_{A_{13}} = \left(\frac{\partial u_1}{\partial x_1} + \frac{\partial u_3}{\partial x_3}\right), \quad \delta_{A_{12}} = \left(\frac{\partial u_1}{\partial x_1} + \frac{\partial u_2}{\partial x_2}\right).
$$

Considering the cube faces 1 to 6, we have

$$
\begin{aligned}
\delta_{A_1} &= \frac{3}{400l}(-2x_2 + 4x_3),\\
\delta_{A_2} &= \frac{3}{400l}(-2x_2 + 4x_3),\\
\delta_{A_3} &= \frac{3}{400l}(2x_1 + 4x_3),\\
\delta_{A_4} &= \frac{3}{400l}(2x_1 + 4x_3),\\
\delta_{A_5} &= \frac{3}{400l}(2x_1 - 2x_2),\\
\delta_{A_6} &= \frac{3}{400l}(2x_1 - 2x_2),
\end{aligned}
$$

so that

$$
\delta A_{lat} = \frac{3}{200l}\left(\int_0^l\int_0^l (4x_3 - 2x_2)\,dx_2dx_3 + \int_0^l\int_0^l (2x_1 + 4x_3)\,dx_1dx_3\right.
$$
$$
\left. + \int_0^l\int_0^l (2x_1 - 2x_2)\,dx_1dx_2\right) = \frac{3}{50}l^2,
$$

a positive value: the body increases its lateral surface area in the deformation.

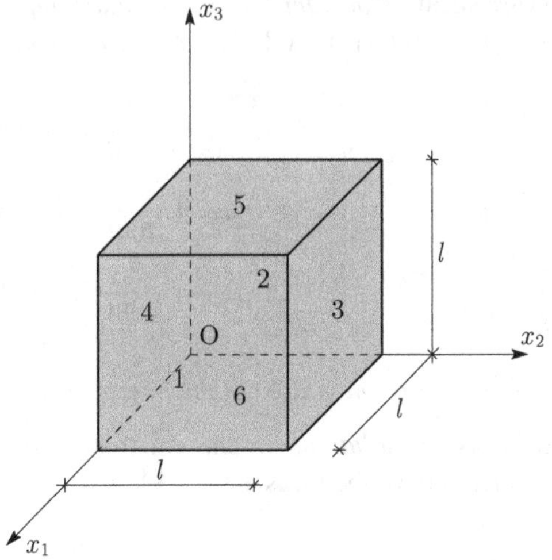

Fig. 1.35 Cube

Exercise 31. *Consider the cube represented in Figure 1.36, and the reference frame* $\{O, x_1, x_2, x_3\}$. *Take a displacement field having in the frame selected the following components:*

$$u_1 = \frac{1}{10^3 l}(3x_1^2 + x_2^2 - x_1 x_3),$$
$$u_2 = \frac{1}{10^3 l}(-x_2^2 + x_1 x_2 + x_3^2),$$
$$u_3 = \frac{1}{10^3 l}(2x_3^2 + x_1 x_3 - 3x_2^2).$$

In the small-strain regime, find the point at which the local relative change in volume δ_V *is maximum.*

Remarks and solution. In the small-strain regime, the volume variation is given by

$$\delta_V = \operatorname{tr} \varepsilon = \frac{1}{10^3 l}(8x_1 - 2x_2 + 3x_3).$$

Its maximum occurs at $P_m \equiv (l, 0, l)$, where $\delta_V^{max} = \frac{11}{10^3}$.

Exercise 32. *At a point of a body and with reference to a local and orthogonal frame* $\{O, x_1, x_2, x_3\}$, *the components of the small-strain tensor are constant and given by*

$$[\varepsilon_{ij}] = 10^{-3}\begin{pmatrix} 2 & 1 & 0 \\ 1 & 1 & 3 \\ 0 & 3 & 2 \end{pmatrix}.$$

Find the ratio between the elongation in direction n *and that along* m *if* $n_1 = 1/\sqrt{2}, n_2 = 1/\sqrt{2}, n_3 = 0$ *and* $m_1 = 0, m_2 = 1/\sqrt{2}, m_3 = 1/\sqrt{2}$.

Fig. 1.36 Cube

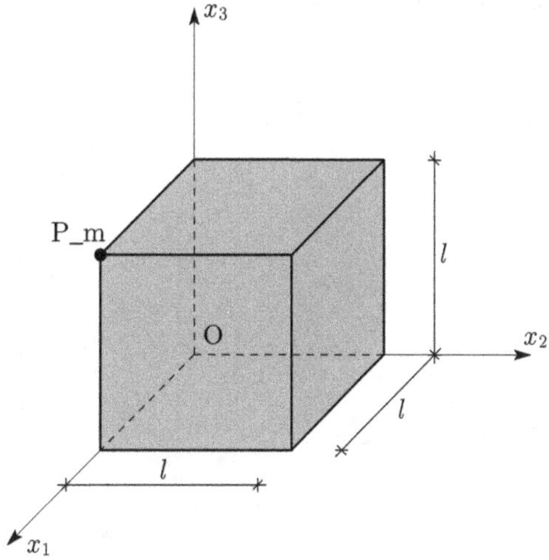

Remarks and solution. The variation in length at a generic point along n is

$$\delta_{l_n} = \varepsilon_{ij} n_i n_j = \frac{1}{10^3} \begin{pmatrix} \frac{1}{\sqrt{2}} & \frac{1}{\sqrt{2}} & 0 \end{pmatrix} \begin{pmatrix} 2 & 1 & 0 \\ 1 & 1 & 3 \\ 0 & 3 & 2 \end{pmatrix} \begin{pmatrix} \frac{1}{\sqrt{2}} \\ \frac{1}{\sqrt{2}} \\ 0 \end{pmatrix} = \frac{5}{2} \frac{1}{10^3},$$

while along m, it is

$$\delta_{l_m} = \varepsilon_{ij} m_i m_j = \frac{1}{10^3} \begin{pmatrix} 0 & \frac{1}{\sqrt{2}} & \frac{1}{\sqrt{2}} \end{pmatrix} \begin{pmatrix} 2 & 1 & 0 \\ 1 & 1 & 3 \\ 0 & 3 & 2 \end{pmatrix} \begin{pmatrix} 0 \\ \frac{1}{\sqrt{2}} \\ \frac{1}{\sqrt{2}} \end{pmatrix} = \frac{9}{2} \frac{1}{10^3}.$$

Their ratio is then $\frac{5}{9}$.

Exercise 33. *Consider the cube with side l and a reference frame $\{O, x_1, x_2, x_3\}$ as in Figure 1.37. Take a displacement field having the following components:*

$$\begin{aligned} u_1 &= k x_3^2 x_2, \\ u_2 &= -k x_3^2 x_1, \\ u_3 &= k x_3^3, \end{aligned}$$

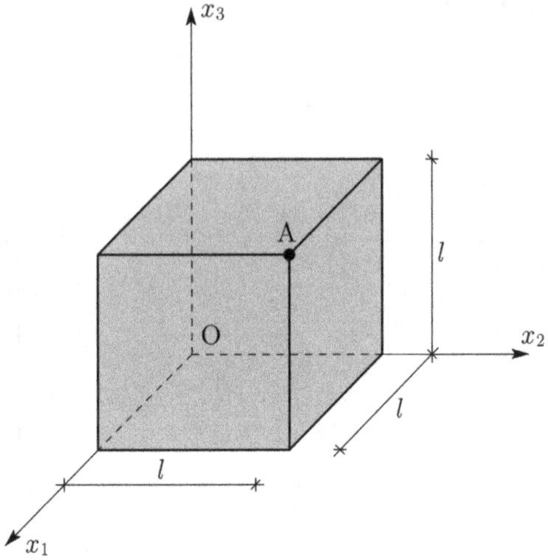

Fig. 1.37 Cube

with k a positive constant. Find the components of ε *and E. Calculate the length, angle, area, and volume variations at* A *along the coordinate axes and planes in the two cases and compare the volume variation* $\delta \mathsf{V}_{cube}$ *in the cases* $k = \dfrac{1}{10l^2}$ *and* $k = \dfrac{1}{100l^2}$.

Remarks and solution. The deformation gradient F has components (see Exercise 28)

$$\begin{pmatrix} 1 & kx_3^2 & 2kx_3x_2 \\ -kx_3^2 & 1 & -2kx_3x_1 \\ 0 & 0 & 1+3kx_3^2 \end{pmatrix},$$

and

$$\det F = 1 + 3kx_3^2 + k^2x_3^4 + 3k^3x_3^6$$

is positive at all points of the cube, so that the deformation is orientation-preserving. As regards the components of the small-strain tensor, we get

$$\left[\varepsilon_{ij}\right] = k \begin{pmatrix} 0 & 0 & x_2x_3 \\ 0 & 0 & -x_1x_3 \\ x_2x_3 & -x_1x_3 & 3x_3^2 \end{pmatrix}.$$

At A with coordinates (l, l, l), ε becomes

$$[\varepsilon_{ij}]\,(\mathrm{A}) = k\begin{pmatrix} 0 & 0 & l^2 \\ 0 & 0 & -l^2 \\ l^2 & -l^2 & 3l^2 \end{pmatrix}.$$

Hence, at A, the variations of length δ_{l_i} and angles γ_{ij}, $i \neq j$, in the axis directions, and areas $\delta_{A_{ij}}$ on the coordinate planes and the volume variation δ_V are given by

$$\begin{aligned} &\delta_{l_1} = 0, && \delta_{l_2} = 0, && \delta_{l_3} = 3kl^2, \\ &\gamma_{12} = 0, && \gamma_{13} = 2kl^2, && \gamma_{23} = -2kl^2, \\ &\delta_{A_{12}} = 0, && \delta_{A_{13}} = 3kl^2, && \delta_{A_{23}} = 3kl^2, \\ &\delta_V = 3kl^2. \end{aligned}$$

Finally, the variation of volume of the cube is

$$\delta \mathrm{V}_{cube} = \int_0^l \int_0^l \int_0^l \left(3kx_3^2\right) dx_1 dx_2 dx_3 = kl^5.$$

Let us consider finite strain. In terms of the displacement, we have

$$E_{AB} = \frac{1}{2}(u_{A,B} + u_{B,A} + u_{K,A}u_{K,B}),$$

which is, in the case treated here,

$$[E_{AB}] = \begin{pmatrix} \dfrac{k^2x_3^4}{2} & 0 & kx_2x_3 + k^2x_1x_3^3 \\ 0 & \dfrac{k^2x_3^4}{2} & -kx_1x_3 + k^2x_2x_3^3 \\ kx_2x_3 + k^2x_1x_3^3 & -kx_1x_3 + k^2x_2x_3^3 & \dfrac{kx_3^2(6+9kx_3^2)}{2} + 2k^2x_3^2(x_1^2 + x_2^2) \end{pmatrix},$$

so that at A,

$$[E_{AB}]\,(\mathrm{A}) = \begin{pmatrix} \dfrac{k^2l^4}{2} & 0 & kl^2 + k^2l^4 \\ 0 & \dfrac{k^2l^4}{2} & -kl^2 + k^2l^4 \\ kl^2 + k^2l^4 & -kl^2 + k^2l^4 & 3kl^2 + \dfrac{17}{2}k^2l^4 \end{pmatrix}.$$

The variations in length of a segment of unit length in the direction of the coordinate axes are given by

$$\delta_{l_1} = \sqrt{1 + k^2l^4} - 1, \quad \delta_{l_2} = \sqrt{1 + k^2l^4} - 1, \quad \delta_{l_3} = \sqrt{1 + 6kl^2 + 17k^2l^4} - 1.$$

The angle variations of the coordinate axes are

$$\gamma_{12} = 0,$$

$$\gamma_{13} = \arcsin\left(\frac{2(kl^2 + k^2l^4)}{\left(\sqrt{1 + k^2l^4}\right)\left(\sqrt{1 + 6kl^2 + 17k^2l^4}\right)}\right),$$

$$\gamma_{23} = \arcsin\left(\frac{2(-kl^2 + k^2l^4)}{\left(\sqrt{1 + k^2l^4}\right)\left(\sqrt{1 + 6kl^2 + 17k^2l^4}\right)}\right).$$

The area variations on the coordinate planes are given by

$$\delta_{A_{12}} = k^2l^4,$$

$$\delta_{A_{13}} = \sqrt{(1 + k^2l^4)(1 + 6kl^2 + 17k^2l^4) - 4(kl^2 + k^2l^4)^2} - 1,$$

$$\delta_{A_{23}} = \sqrt{(1 + k^2l^4)(1 + 6kl^2 + 17k^2l^4) - 4(-kl^2 + k^2l^4)^2} - 1.$$

The volume variation in a neighborhood of a point is $\delta_V = 3kx_3^2 + k^2x_3^4 + 3k^3x_3^6$ (at A it is $\delta_V = 3kl^2 + k^2l^4 + 3k^3l^6$). The total volume variation δV_{cube} is then

$$\delta V_{cube} = \int_0^l \int_0^l \int_0^l (3kx_3^2 + k^2x_3^4 + 3k^3x_3^6)\, dx_1dx_2dx_3 = kl^5 + \frac{k^2l^7}{5} + \frac{3k^3l^9}{7}.$$

Let us assume that $k = \dfrac{1}{10l^2}$ and $k = \dfrac{1}{100l^2}$. In the small-strain regime, we get

$$\delta V_{cube} = \frac{l^3}{10} = 0.1l^3, \qquad \delta V_{cube} = \frac{l^3}{100} = 0.01l^3,$$

while when the strain is finite, we have

$$\delta V_{cube} = \frac{717}{7000}l^3 \approx 0.1024l^3, \qquad \delta V_{cube} = \frac{70143}{7 \cdot 10^6}l^3 \approx 0.01002l^3.$$

1.20 Further exercises

Exercise 34. *Analyze the kinematics of the rigid body in Figure 1.38 and draw the diagrams of the displacements.*

Some elements of the solution: The body is kinematically indeterminate ($\hat{l} = 1$), $x_C^1 = a$, $x_C^2 = a$, C = center of rotation.

Exercise 35. *Analyze the kinematics of the structure of three rigid bodies in Figure 1.39 and draw the diagrams of the displacements.*

Some elements of the solution: The structure is geometrically indeterminate ($\hat{l} = 1$); the three bodies behave as a single rigid body rotating around the point C, which is the intersection of the straight lines passing through A and D with directions parallel to the axes of the pendulums.

Exercise 36. *Analyze the kinematics of the structure in Figure 1.40 and draw the diagrams of the displacements.*

Some elements of the solution: The structure is geometrically indeterminate ($\hat{l} = 1$); the body *III* does not displace; the body *II* translates horizontally; the body *I* rotates around to hinge at A.

Exercise 37. *Analyze the kinematics of the structure in Figure 1.41 and draw the diagrams of the displacements.*

Some elements of the solution: The structure is geometrically indeterminate ($\hat{l} = 1$); the two bodies translate vertically as a single rigid body.

Exercise 38. *Analyze the kinematics of the structure in Figure 1.42 and draw the diagrams of the displacements.*

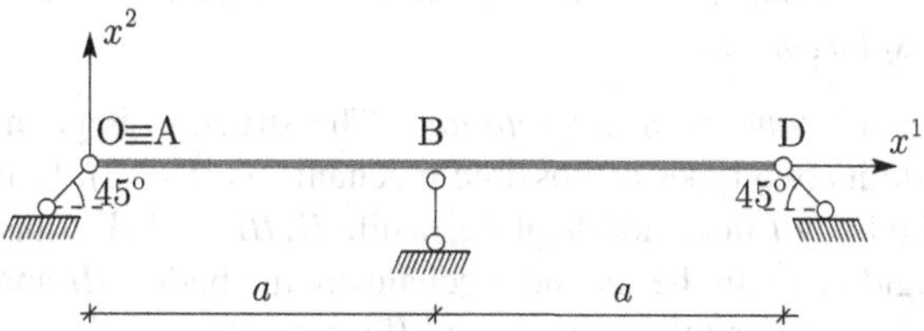

Fig. 1.38 Kinematically indeterminate one-dimensional rigid body in the plane

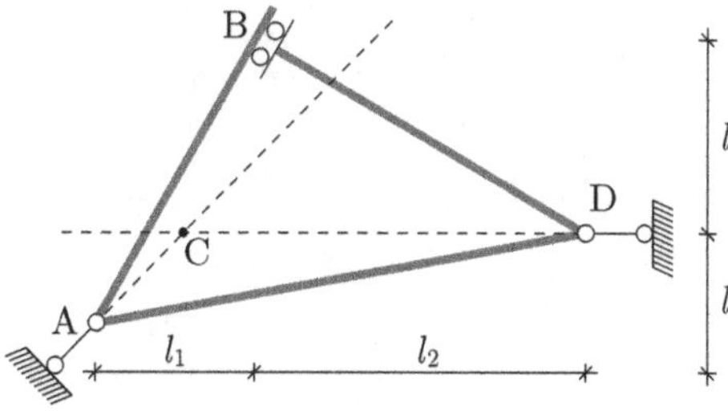

Fig. 1.39 Kinematically indeterminate planar system of three one-dimensional rigid bodies

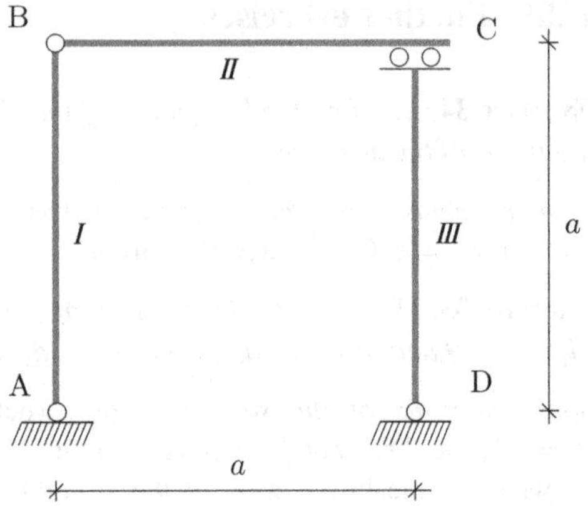

Fig. 1.40 Kinematically indeterminate planar system

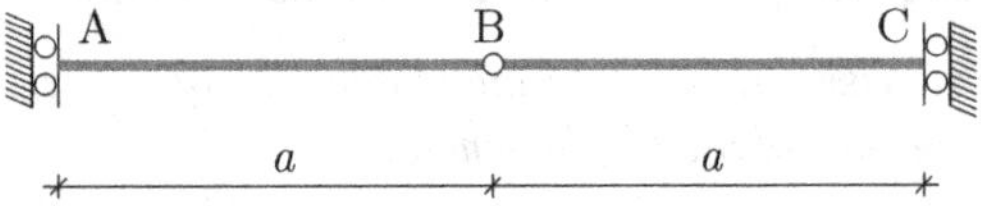

Fig. 1.41 Kinematically indeterminate planar structure

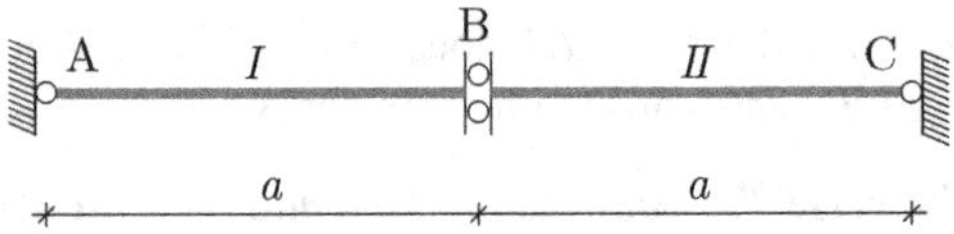

Fig. 1.42 Kinematically indeterminate planar structure

Some elements of the solution: The structure is geometrically indeterminate ($\hat{l} = 1$); the body *I* rotates around the hinge at A; the body *II* rotates around the hinge at C with the same angle magnitude as that for body *I* but of opposite sign.

Exercise 39. *Analyze the kinematics of the planar structure consisting of four one-dimensional rigid bodies as shown in Figure 1.43 and draw the diagrams of the displacements.*

Some elements of the solution: The structure is geometrically indeterminate and admits two types of possible mechanisms ($\hat{l} = 2$). In the first possible kinematics, the body *I* does not displace, while *II*, *III*, and *IV* translate horizontally as a single rigid body. In the second mechanism, the bodies *III* and *IV* do not displace, while *I* rotates around the point A, and *II* rotates through the same angle around the point B.

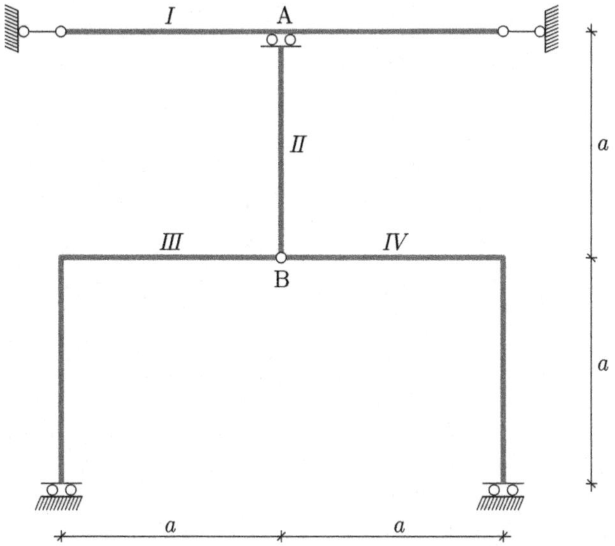

Fig. 1.43 Kinematically indeterminate planar structure consisting of four one-dimensional rigid bodies

Chapter 2
Observers

2.1 A Definition

An **observer** is a representation of (that is, the assignment of reference frames to) *all* geometric environments necessary to describe the morphology of a body and its motion.

This definition extends the standard one, which foresees the assignment of reference frames just in the ambient physical space. It is worth recalling, however, that in traditional treatments, there is no distinction between the space including the reference place and the one in which we evaluate the body shapes considered deformed. We accept, in contrast, such a distinction here for reasons, which will become evident in this chapter, related to changes in observers, and for other technical reasons appearing in the evaluation of minimizers in nonlinear elasticity and related to energy variations about them—a topic going, however, rather far from the introductory character of this book.

The word "all" in the definition suggests a way of considering observers in refined descriptions of the morphology of bodies aiming to represent the finer architecture of the matter in the aspects that appear (or that we believe are) sometimes crucial in the phenomena under consideration. In these cases, another space enters the scene: the one in which we select appropriate additional descriptors of the material morphology at various spatial scales. We do not go into details here. A *special* case of what we have just mentioned is, however, developed with a number of details in Chapter 8: it is the construction of the so-called direct models of beams or rods. We remark that it is a special case because the general setting includes in a unified view the available models of materials with prominent microstructural effects on the gross behavior such as ferroelectrics, quasicrystals, microcracked bodies, and polymeric materials. However, for the moment and to maintain the introductory character of this book, we focus attention on only the ambient physical

P.M. Mariano, L. Galano, *Fundamentals of the Mechanics of Solids*,
DOI 10.1007/978-1-4939-3133-0_2

space, the one including the reference place, and the time interval for the description of the body morphology adopted so far is based solely on the recognition of the regions in the three-dimensional point space that bodies may occupy.

2.2 Classes of Changes in Observers

When we alter in some specific way frames (i.e., coordinate systems) in the spaces where we describe the body morphology and its evolution in time, we are defining **classes of changes in observers**. We have many options. However, since the space where we place current shapes of bodies is the Euclidean one, it is natural to think of changes in observers determined by time-parameterized families of isometries, in particular those maintaining the orientation of frames. Translations and rotations can be, in fact, defined in the three-dimensional point space adopted so far, since they can be assigned together in the n-dimensional case with $n > 1$, because the one-dimensional space admits only translations.

In speaking of changes in observers, it is commonly required that all observers be able to evaluate the *same* reference place. Such a requirement forces us to distinguish between the space containing the reference place and the one in which we include the body shapes considered deformed. In fact, if reference and deformed shapes were to be described in the same space, since a change in observer alters frames in the *whole* space, two observers, i.e., two different frames, would in general evaluate different reference places. Such a circumstance would be in contrast to the standard requirement of an invariant reference place, at least in the absence of material irreversible mutations, a case not treated here.

According to traditional instances, we define **classical changes in observers** to be those changes

1. that leave invariant the reference place (then the reference space),
2. that are synchronous,
3. that alter by means of translations and rotations the (physical) space in which we record motions.

Let $\mathcal{O}$ and $\mathcal{O}'$ be two observers, i.e., two classes of frames. At time t, the observer $\mathcal{O}$ records an actual place y for a certain material element, which is in a place y' for the observer $\mathcal{O}'$. If $\mathcal{O}$ and $\mathcal{O}'$ are connected in time by classical changes as defined above, then y and y' satisfy the relation

$$y' = w(t) + Q(t)(y - y_0), \tag{2.1}$$

involving the values of smooth functions $t \longmapsto w(t) \in \tilde{\mathcal{E}}^3$ and $t \longmapsto Q(t) \in SO(3)$, and an arbitrary fixed point y_0. By differentiation, we get

$$\dot{y}' = \dot{w} + \dot{Q}(y - y_0) + Q\dot{y}. \tag{2.2}$$

The vector $\dot{y}'$ can be pulled back from $\mathcal{O}'$ to $\mathcal{O}$ by means of the inverse of Q, namely Q^{T}, and the projection into $\mathcal{O}$ is indicated here by $\dot{y}^*$ and given by

$$\dot{y}^* = Q^{\mathrm{T}}\dot{y}'. \tag{2.3}$$

To accept such a relation, consider two points y_1 and y_2, as evaluated by $\mathcal{O}$, and their counterparts y_1' and y_2' in $\mathcal{O}'$ given by (2.1). Their differences $y_1' - y_2'$ and $y_1 - y_2$ are elements of the translation space $\mathcal{V}$ over $\tilde{\mathcal{E}}^3$. $\mathcal{V}$ is an affine vector space, which reduces to $\mathcal{R}^3$ once the origin of a coordinate system is fixed. The vectors $y_1' - y_2'$ and $y_1 - y_2$ are connected by the relation

$$y_1' - y_2' = Q(y_1 - y_2),$$

which follows from (2.1) by direct calculation. The result justifies immediately the relation (2.3).

For $\dot{y}^*$, we get

$$\dot{y}^* = c(t) + q(t) \times (y - y_0) + \dot{y}, \tag{2.4}$$

where $c(t)$ and $q(t)$ belong to $\tilde{\mathbb{R}}^3$. In fact, by premultiplication by Q^{T}, from (2.3) we get first

$$\dot{y}^* = Q^{\mathrm{T}}\dot{y}' = Q^{\mathrm{T}}\dot{w} + Q^{\mathrm{T}}\dot{Q}(y - y_0) + \dot{y}.$$

Since by definition $Q^{\mathrm{T}}Q = \hat{I}$, with $\hat{I}$ the second-rank identity tensor $\hat{I} = \delta^i_j \tilde{e}_i \cdot \tilde{e}^j$, by time differentiation we obtain

$$\dot{Q}^{\mathrm{T}}Q + Q^{\mathrm{T}}\dot{Q} = 0,$$

or better,

$$Q^{\mathrm{T}}\dot{Q} = -(Q^{\mathrm{T}}\dot{Q})^{\mathrm{T}},$$

a relation declaring that the second-rank tensor $Q^{\mathrm{T}}\dot{Q}$, with components $Q^i_k\dot{Q}^k_j$, is skew-symmetric, so that for every vector $a \in \tilde{\mathbb{R}}^3$, there exists a vector $q \in \tilde{\mathbb{R}}^3$ such that $(Q^{\mathrm{T}}\dot{Q})a = q \times a$. In the common geometric jargon, q is called the *axial vector* of $Q^{\mathrm{T}}\dot{Q}$.

Since $\dot{y}(x,t)$ is equal to $v(y,t)$—the equivalence between Lagrangian and Eulerian representations of the velocity, $\dot{y}$ and v respectively, was discussed in the previous chapter—we can substitute $\dot{y}^*$ with v^* and $\dot{y}$ with v in (2.4), obtaining its equivalent relation in the Eulerian representation:

$$v^* = c(t) + q(t) \times (y - y_0) + v. \tag{2.5}$$

The vectors c and q represent the translational (c) and rotational (q) velocities at which $\mathcal{O}$ realizes that $\mathcal{O}'$ is flowing away.

In summary, we affirm that the classical changes in observers just defined are changes of frames (then of coordinates) in the ambient space $\tilde{\mathcal{E}}^3$ determined *by the action of the Eulerian group*.[1]

2.3 Objectivity

Consider a linear map G from $\tilde{\mathbb{R}}^3$ to $\tilde{\mathbb{R}}^3$, namely $G \in \mathrm{Hom}(\tilde{\mathbb{R}}^3, \tilde{\mathbb{R}}^3)$. Take a vector $v \in \tilde{\mathbb{R}}^3$ and its image Gv, a vector with components $G^i_j v^j$. Under classical changes in observers, the vector v becomes $v' = Qv$ according to (2.3). Define G' as the element of $\mathrm{Hom}(\tilde{\mathbb{R}}^3, \tilde{\mathbb{R}}^3)$ such that

$$Q(Gv) = G'(Qv).$$

By premultiplication by $Q^{\mathrm{T}} = Q^{-1}$, we get

$$G' = QGQ^{\mathrm{T}}, \tag{2.6}$$

thanks to the arbitrariness of v. In components, the previous expression can be written as

$$G'^h_k = Q^h_i G^i_j (Q^{\mathrm{T}})^j_k,$$

where the indices i, j refer to the coordinates *before* the change in observer, while h, k denote coordinates *thereafter*. By acting from the left, $Q \in SO(3)$ projects the vectorial component of G, the one indicated by the contravariant index of G, in the frame of the second observer. The action of Q^{T} from the right in (2.6) projects the covariant component of G into the frame defining the second observer. The procedure can be extended to tensors of arbitrary rank (p, q) over the ambient space. Let $\bar{G}$ be such a tensor with components $\bar{G}^{i_1 \ldots i_p}_{j_1 \ldots j_q}$. Under classical changes in observers, $\bar{G}$ becomes $\bar{G}'$ and has components

$$\bar{G}'^{h_1 \ldots h_p}_{k_1 \ldots k_q} = Q^{h_1}_{i_1} \ldots Q^{h_p}_{i_p} \bar{G}^{i_1 \ldots i_p}_{j_1 \ldots j_q} (Q^{\mathrm{T}})^{j_1}_{k_1} \ldots (Q^{\mathrm{T}})^{j_q}_{k_q}. \tag{2.7}$$

All tensors altered by isometric changes in observers in accord with (2.7) are called **objective**.

---

[1]Such a group is denoted by $\tilde{\mathbb{R}}^3 \rtimes SO(3)$, since it is the semidirect product of $\tilde{\mathbb{R}}^3$ and the special orthogonal group $SO(3)$.

- The relations (2.2) and (2.5) tell us that the velocity in both Lagrangian and Euleria representations, is *not* objective.
- By deriving (2.2) in time, we also get that the acceleration $\ddot{y}(x,t)$ in Lagrangian representation is *not objective*, because with $\ddot{y}'$ the acceleration evaluated by the second observer, we have

$$\ddot{y}' = \ddot{w} + \ddot{Q}(y - y_0) + 2\dot{Q}\dot{y} + Q\ddot{y}.$$

- An analogous result holds for the acceleration in Eulerian representation. The deformation gradient F is *not objective*. Here F has components F^i_A. The contravariant index i refers to coordinates in the ambient (physical) space $\tilde{\mathcal{E}}^3$. The covariant index A is associated with coordinates in the reference space, which remains unchanged by classical changes in observers in $\tilde{\mathcal{E}}^3$. Then with F' the deformation gradient evaluated by the second observer, we get

$$F' = QF.$$

2.4 Remarks and Generalizations

We shall assume objectivity for some elements of the framework for the mechanics of continua presented here—in fact the standard one—with consequences far from being merely technical.

We record (apprehend) physical events through the intermediary of a framing. In the setting discussed here, we can distinguish between place and time, a distinction that we do not have naturally at our disposal in a general relativistic setting. Also, we exclude here influence of the observation on the phenomena, which we include, in contrast, in quantum mechanics. In the present classical setting, a phenomenon is independent of the assignment of frames (a meter, a clock, let us say) although different framings (different observers) may in general propose different descriptions of it. The question then arises how to relate statements made by different observers.

When we require objectivity of some entity (scalar, vector, tensor of any rank greater than one), we are prescribing a type of relation between evaluations of such an entity by two different observers related by a sequence of rotations and translations.

However, objectivity, defined in the standard sense presented here, is not the sole option for the class of changes in observers discussed so far. It is just one of the possibilities, indeed a natural one, once we have selected Euclidean space as the ambient one. A list of other possible classes of changes in observers follows.

- We could imagine considering not only changes of frames based on rotations and translations in the ambient space but also in the reference space. Such a class of changes in observers is appropriate when we want to describe the mechanics of solids undergoing structural macroscopic irreversible mutations such as the creation and growth of voids and/or cracks, and the growth of inclusions or their movement with respect to the rest of the body. In such cases, in fact, the actual shape is no longer in one-to-one correspondence with the reference place; rather, it can be connected with another reference place differing from the original one by only the "defect" pattern. In this case, we have naturally a family of reference places, and we can use a vector field $x \longmapsto w(x) \in \mathbb{R}^3$ over $\mathcal{B}$ to describe the incoming change of reference place, due to the occurrence of defects, so that we may assume that $x \longmapsto w(x)$ is not always associated with a diffeomorphism, at least in a neighborhood of x.[2] Hence in principle, changes in observers may involve different evaluations of the vector field w. Changes in observers determined by the action of the Euclidean group *even* on the reference space—a class of changes enlarged with respect to what we have treated so far—would impose a rule of the type

$$w^* = w + \bar{c} + \bar{q} \times (x - x_0),$$

where $\bar{c}$ and $\bar{q}$ are different from the translational and rotational velocities c and q, respectively, in the ambient space. Also, the same concept of objectivity could be enlarged to include changes of frames in the reference space.

– It is not strictly necessary that changes in observers be isometric, like the classical ones. By leaving invariant the reference space, we could alter frames in the ambient space by means of time-parameterized families of diffeomorphisms. Then, instead of (2.4), we would have

$$\dot{y}^* = \dot{y} + \tilde{v}, \tag{2.8}$$

with $\tilde{v}$ a superposed velocity field over the current place. When $\tilde{v} = c(t) + q(t) \times (y - y_0)$, the equation (2.8) reduces to (2.4). A generalization of such a class of changes in observers involves the action of the group of diffeomorphisms over the reference space.

We do not explore in detail the possible consequences of selecting one of these changes in observers and requiring the invariance of some appropriate functional.

However, if we were to explore these possibilities, we would understand in what sense the choice of a class of changes in observers can be considered a *structural ingredient* of a mechanical model, not a rule independent of the model itself.

[2] A map $f : \mathbb{R}^3 \longrightarrow \mathbb{R}^3$ is a diffeomorphism when it is one-to-one and both f and f^{-1} are differentiable.

Chapter 3
Forces, Torques, Balances

3.1 A Preamble on the Notion of Force

In beginning the third chapter of his 1977 book *A First Course in Continuum Mechanics*, Clifford Ambrose Truesdell III (1919–2000) wrote the following:

> Forces and torques, like bodies, motion and masses are primitive elements of mechanics. They are mathematical quantities introduced a priori, represented by symbols, and subjected to mathematical axioms that delimit their properties and render them clear and useful for the description of mechanical phenomena in nature.

That point of view has been successfully pursued by the Truesdellian school. In the production of that school we find results that allow us to consider mass and balances of forces and torques (call them also couples) as results emerging from invariance properties imposed on the power they exert, as external entities with respect to the body, or some energetic principle. Such invariance requirements are considered primitive concepts. These results fall within the basic program of analysis of the foundations in continuum mechanics, based on the search for first principles allowing reduction of axioms. The program is motivated not only by aesthetic considerations, but also and primarily by the essential desire to have at one's disposal tools that can be safely managed when we want to enlarge the traditional scheme of continuum mechanics to tackle the description of macroscopic effects of complex microstructural phenomena. In any case, regardless of the setting in which we want to address our analyses, we face the problem of describing the fact that bodies interact with the environment and that their constituents display analogous behavior toward each other to maintain the integrity of the body itself or to break it up.

Here, our attention is focused on the mechanics of tangible bodies. We leave a part the description of corpuscular phenomena at atomic scale, which can be adequately described by concepts and methods pertaining to quantum theories,

P.M. Mariano, L. Galano, *Fundamentals of the Mechanics of Solids*,
DOI 10.1007/978-1-4939-3133-0_3

or we consider just indirectly the effects of such phenomena on the macroscopic deformative behavior. The mechanisms we are discussing are classical in this sense.

Within the realm of Newtonian mechanics, consider a point with constant mass m in three-dimensional space. What we call *force* on the point is, in fact, a *model* of a circumstance that alters the motion of that mass point, evaluated by some observer. Moreover, to alter that motion, we have to exert *power*.

The first descriptor of the action over the mass point considered is kinematic: the velocity. Then we have two options: the first is to assume that force is a primitive concept, indicating it by the common arrow, or to consider power to be primitive, taking it as a linear functional of the velocity. The two options have been accepted and pursued successfully. We follow the second one, finding it more flexible when we want to extend the setting treated here to more intricate situations involving, e.g., material microstructures—a matter not tackled here but falling within a general model-building framework including the special case discussed in Chapter 8.

When we write $\mathcal{P}(v)$ for the power over the mass point considered (v is the velocity), the assumed linearity with respect to v implies

$$\mathcal{P}(v) = f^{\ddagger} \cdot v$$

with $f^{\ddagger}$ a covector, an element of the vector space dual to the one including v. However, since $v \in \mathbb{R}^3$, $f^{\ddagger}$ belongs to $\mathbb{R}^{3*}$, which is isomorphic to $\mathbb{R}^3$. Such a choice defines the force $f^{\ddagger}$ through the power and justifies automatically its usual representation in vector form.

Since $\mathcal{P}(v)$ is a scalar, we find it natural to assume that it is invariant under changes in observers altering v according to

$$v^* = v + c + q \times (y - y_0),$$

with y the current place of the mass point, c and q the translational and rotational velocities, respectively, defining the relative rigid motion between the observers considered (see the previous chapter). Invariance of $\mathcal{P}(v)$ means that we impose the identity

$$\mathcal{P}(v) = \mathcal{P}(v^*)$$

for every choice of c and q. That requirement implies

$$\mathcal{P}(c + q \times (y - y_0)) = 0,$$

that is,

$$f^{\ddagger} \cdot (c + q \times (y - y_0)) = f^{\ddagger} \cdot c + ((y - y_0) \times f^{\ddagger}) \cdot q = 0.$$

The arbitrariness of c and q *imposes* the identities

$$f^{\ddagger} = 0$$

and

$$(y - y_0) \times f^{\ddagger} = 0.$$

They are *Newton's laws* requiring that the total force $f^{\ddagger}$ and the total torque $(y - y_0) \times f^{\ddagger}$ acting over the mass point be balanced.

Let us define now the quadratic form

$$\frac{1}{2}\mathsf{m}\,|v|^2 = \frac{1}{2}\mathsf{m}\,\langle v, v\rangle = \frac{1}{2}\mathsf{m}\tilde{g}v \cdot v,$$

with g the metric in space, and assume that v is a differentiable function of time. It is the *kinetic energy* of the mass point considered.

Assume that $f^{\ddagger}$ is the sum of *inertial*, f^{in}, and *noninertial*, f, components, with f^{in} a force such that its power is equal to the negative of the kinetic energy time rate, namely

$$\frac{d}{dt}\left(\frac{1}{2}\mathsf{m}\tilde{g}v \cdot v\right) + f^{in} \cdot v = 0$$

for every choice of v.

Since neither m nor g depends on time, we obtain

$$\left(\mathsf{m}\tilde{g}\frac{dv}{dt} + f^{in}\right) \cdot v = 0,$$

and the arbitrariness of v implies

$$f^{in} = -\mathsf{m}\tilde{g}\frac{dv}{dt},$$

so that Newton's law $f^{\ddagger} = f^{in} + f = 0$ acquires its common differential form written first by Euler, namely

$$f = \mathsf{m}\tilde{g}\frac{dv}{dt},$$

that is,

$$f = \mathsf{m}\frac{dv}{dt}$$

when the metric in space is flat, i.e., when $\tilde{g}$ is the second-rank identity tensor with both covariant components. However, we are analyzing bodies extended in space, not a single mass point or a finite number of copies of it. The problem then is to extend the previous reasoning to the analysis of continuous bodies extended in space.

3.2 Bulk and Contact Actions

To start, we find it expedient to render precise in what sense we consider portions of a body extended in space. We say that a **part** of a body is any portion of it occupying in the reference place $\mathcal{B}$ a subset $\mathfrak{b}$ with nonvanishing volume measure, an open set coinciding with the interior of its closure and endowed with a surface-like boundary oriented by the normal at all points but a finite number of corners and edges. We can consider parts of the body in the actual shape $\mathcal{B}_a$. The previous definition still holds, and we shall indicate by $\mathfrak{b}_a$ parts taken in $\mathcal{B}_a$.

When we imagine extracting a part $\mathfrak{b}_a$ of[1] $\mathcal{B}_a$ and to look at it only, its interactions with the surrounding medium are commonly subdivided into **bulk** and **contact** classes.

The first class includes the interactions between the interior of $\mathfrak{b}_a$ and the set of the other separate parts of $\mathcal{B}_a$ and the rest of the environment in which the body is placed (in fact a universe of bodies). Contact actions are those exerted between $\mathfrak{b}_a$ and the other separate parts of $\mathcal{B}_a$ and the rest of the environment that share with $\mathcal{B}_a$ at least part of the boundary $\partial\mathfrak{b}_a$.

The question is now the representation of such actions. A remark addressing a reasonable choice is that we have represented the morphology of the body *only* by means of the region in the Euclidean point space that it occupies, without any further description or specification of the material structure. In other words, in our modeling view, $\mathfrak{b}_a$ is occupied by an uncountable number of material points. The rate of change in the shape is described by the velocity field, which is in fact a vector field. We could then imagine reproducing for all points in $\mathfrak{b}_a$ what we have done for a single material point isolated in space. In other words, for every point, we consider a velocity (a vector) and a force (consequently a covector) exerting power over it.

To formalize such a view, we insist that the **power** be a functional depending on parts and velocity fields that is *linear* with respect to the velocity and *additive* with respect to disjoint parts. The difference between this and what we have discussed for the single material point in the previous section rests on the presence of parts, which is necessary if the body is extended in space. In principle, we can define different powers. Here, we prefer on the one hand to be minimalist—meaning that we want to include the smallest number of ingredients—while on the other hand, we have to account for both bulk and contact actions as described qualitatively above.

We define then the **external power** $\mathcal{P}^{ext}_{\mathfrak{b}_a}(v)$ exerted over $\mathfrak{b}_a$ in the velocity field v, i.e., the power exerted by all agencies external to $\mathfrak{b}_a$, by

$$\mathcal{P}^{ext}_{\mathfrak{b}_a}(v) := \int_{\mathfrak{b}_a} b_a^{\ddagger} \cdot v \, d\mu + \int_{\partial\mathfrak{b}_a} \mathfrak{t} \cdot v \, d\mathcal{H}^2,$$

[1]When we speak of "part $\mathfrak{b}$ of $\mathcal{B}$" or "part $\mathfrak{b}_a$ of $\mathcal{B}_a$," we mean "part of the body occupying the subset $\mathfrak{b}$ in $\mathcal{B}$," with the same interpretation for $\mathfrak{b}_a$.

where $d\mu$ is the volume measure and $d\mathcal{H}^2$ is the area measure, generically indicated this way throughout the book, leaving to the domain of integration the specification of the space in which they are considered. Here $b_a^\ddagger$ is a covector field over $\mathfrak{b}_a$, so that the product $b_a^\ddagger \cdot v$ is naturally defined without specifying a scalar product structure. Once one such a product has been defined, essentially by the assignment of a metric in space, $b_a^\ddagger$ can be identified with a vector at every point when it is premultiplied by the metric. Analogous remarks hold also for $\mathfrak{t}$, defined uniquely at almost all points of $\partial\mathfrak{b}_a$.

Since $\mathcal{P}_{\mathfrak{b}_a}^{ext}(v)$ is a scalar, we find it natural to accept the following axiom:

Axiom 1 (Power invariance). The external power $\mathcal{P}_{\mathfrak{b}_a}^{ext}(v)$ is invariant under classical (isometry-based) changes in observers.

Formally, the axiom can be expressed as follows:

$$\mathcal{P}_{\mathfrak{b}_a}^{ext}(v) = \mathcal{P}_{\mathfrak{b}_a}^{ext}(v^*),$$

for every part $\mathfrak{b}_a$ and all velocities c and q appearing in (2.5). Since $v^* = v + v_R$, with $v_R = c + q \times (y - y_0)$, the axiom implies

$$\mathcal{P}_{\mathfrak{b}_a}^{ext}(v_R) = 0,$$

which is

$$\int_{\mathfrak{b}_a} b_a^\ddagger \cdot (c + q \times (y - y_0))d\mu + \int_{\partial\mathfrak{b}_a} \mathfrak{t} \cdot (c + q \times (y - y_0))d\mathcal{H}^2 = 0.$$

Since c and q are constant in space, the previous identity becomes

$$c \cdot \left(\int_{\mathfrak{b}_a} b_a^\ddagger d\mu + \int_{\partial\mathfrak{b}_a} \mathfrak{t}\, d\mathcal{H}^2 \right) + q \cdot \left(\int_{\mathfrak{b}_a} (y - y_0) \times b_a^\ddagger d\mu + \int_{\partial\mathfrak{b}_a} (y - y_0) \times \mathfrak{t}\, d\mathcal{H}^2 \right) = 0.$$

The arbitrariness of c and q implies

$$\int_{\mathfrak{b}_a} b_a^\ddagger d\mu + \int_{\partial\mathfrak{b}_a} \mathfrak{t}\, d\mathcal{H}^2 = 0, \quad (3.1)$$

$$\int_{\mathfrak{b}_a} (y - y_0) \times b_a^\ddagger d\mu + \int_{\partial\mathfrak{b}_a} (y - y_0) \times \mathfrak{t}\, d\mathcal{H}^2 = 0 \quad (3.2)$$

independently. These two equations are the **integral balances** of **forces** and **couples** in Eulerian representation, because the fields involved are defined over the current shape $\mathcal{B}_a$ of the body.

1. The integral balances of forces and couples can both be derived from a *unique* source: the invariance of the external power of actions over a generic part $\mathfrak{b}_a$ of $\mathcal{B}_a$ under classical (isometry-based) changes in observers. It is not necessary to postulate them a priori as *two* independent axioms.
2. If $\partial\mathfrak{b}_a$ is partially in common with the external boundary $\partial\mathcal{B}_a$, over $\partial\mathfrak{b}_a \cap \partial\mathcal{B}_a$ the density t has to be understood as that of the external forces applied to the body boundary.
3. Contact actions described by t do not include couples. In other words, we assume that no actions along the boundary $\partial\mathfrak{b}_a$ develop power in curl v. The assumption is Augustin-Louis Cauchy's (1979-1857). Imagine that we divide the body in $\mathcal{B}_a$ virtually into two distinct pieces by an orientable (smooth) surface Σ. The adjective *orientable* indicates that the normal is uniquely defined over the cut point by point. Denote by A the intersection of Σ with $\mathcal{B}_a$, i.e., $A = \Sigma \cap \mathcal{B}_a$. The action of the atomic bonds linking the two portions of the body across A and ensuring coherence of the body can be summarized as a force F and a couple M. Cauchy's axiom prescribes that when we consider a sequence $\{A_k\}$ of subsets of A, shrinking to a point $y \in A$ as k goes to infinity, the averages $\frac{\mathsf{F}}{|A_k|}$ and $\frac{\mathsf{M}}{|A_k|}$, with $|A_k|$ the area of A_k, tend to $\mathsf{t}(y)$ and 0, namely

$$\lim_{k\to\infty} \frac{\mathsf{F}}{|A_k|} = \mathsf{t}, \qquad \lim_{k\to\infty} \frac{\mathsf{M}}{|A_k|} = 0. \tag{3.3}$$

It is a common usage to call t **traction**. Cauchy's assumption has in a sense its origin in the geometric description of the body morphology. We have, in fact, identified the body just with the region $\mathcal{B}$ that it occupies in three-dimensional Euclidean point space. In this way, for us a body is just a set containing infinitely many material points connected to one another. When we cut the body in $\mathcal{B}_a$ through Σ, we are ideally breaking at each point of $\Sigma \cap \mathcal{B}_a$ the links bridging across Σ pairs of material points; Figure 3.1 contains a schematic picture. Such links can carry, in principle, forces and couples. Material elements are, in fact, just points, and their instantaneous changes of place are described by the velocity, a vector field. When we break (ideally, we repeat) the material bond at a point,

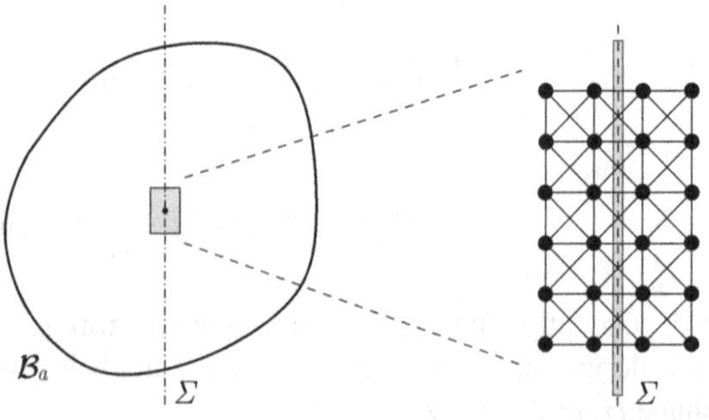

Fig. 3.1 Schematic interpretation of Cauchy's cut

the possible couple there is reduced to a force, because the two forces in the couple are subdivided by symmetry, so one force goes to one side of the cut, its companion to the other side. This picture justifies the choice of the second identity in (3.3). The actions described by this scheme are of first-neighbor type.

4. When we cross $y \in \mathcal{B}_a$ with another cut, say $\tilde{\Sigma}$, oriented at y by a normal different from the one pertaining to Σ, in principle the limit process in (3.3) will generate a traction different from that associated with Σ, and we can assume that

$$\mathfrak{t} = \tilde{\mathfrak{t}}(y, n),$$

with n the normal to the cut at y. This assumption is the called **Cauchy's postulate**. It is assumed to hold at every instant along every admissible motion; we have left implicit the dependence on time. George Karl Wilhelm Hamel (1877–1954) and Walter Noll later *proved* that the same $\mathfrak{t}$ pertains to all cuts having at y a common normal n. Also, Roger Fosdick and Epifanio G. Virga have shown in 1989 the independence of $\mathfrak{t}$ on the curvature of the surface Σ.

The bulk actions $b_a^{\ddagger}$ depend on y in addition to the time t. Assume that

$$\sup_{y \in \mathcal{B}_a} \left| b_a^{\ddagger} \right| < +\infty.$$

The validity of the integral balance of forces implies then that there exists $k > 0$ such that

$$\left| \int_{\partial \mathfrak{b}_a} \mathfrak{t} \, d\mathcal{H}^2 \right| \leq k \operatorname{vol}(\mathfrak{b}_a).$$

Theorem 2 (Action–reaction principle). *Assume that $\tilde{\mathfrak{t}}(\cdot, n)$ is continuous on $\overline{\mathcal{B}}_a$ (the overbar indicating closure), and $b_a^{\ddagger}$ is bounded as above. For every unit vector n, we have*

$$\tilde{\mathfrak{t}}(y, n) = -\tilde{\mathfrak{t}}(y, -n).$$

Proof. Consider a square $\mathcal{D}$ with side length $\alpha < 1$(the length is, in fact, 1α, with 1 endowed with dimension, so that α is dimensionless), oriented by the normal n and with the center at y. Take the parallelepiped given by

$$\mathfrak{P}_\alpha := \left\{ \hat{y} = \bar{y} \pm \xi n \mid \bar{y} \in \mathcal{D}, -\frac{\alpha^2}{2} \leq \xi \leq \frac{\alpha^2}{2} \right\}.$$

Call the squares given by

$$\mathcal{D}_{\pm\alpha} := \left\{ \hat{y} = \bar{y} \pm \xi n \mid \bar{y} \in \mathcal{D}, \xi = \pm\frac{\alpha^2}{2} \right\}$$

$\mathcal{D}_{+\alpha}$ and $\mathcal{D}_{-\alpha}$.

The boundary $\partial\mathfrak{P}_\alpha$ is the union $\mathcal{D}_{+\alpha} \cup \mathcal{D}_{-\alpha} \cup \left(\partial\mathcal{D} \times \left[-\frac{\alpha^2}{2}, \frac{\alpha^2}{2}\right]\right)$. If $y \notin \partial\mathcal{B}_a$, then for appropriate choices of α we can take $\mathfrak{P}_\alpha$ all inside $\mathcal{B}_a$. The volume of $\mathfrak{P}_\alpha$ is α^4; the areas of both $\mathcal{D}_{+\alpha}$ and $\mathcal{D}_{-\alpha}$ are α^2; while the area of $\left(\partial\mathcal{D} \times \left[-\frac{\alpha^2}{2}, \frac{\alpha^2}{2}\right]\right)$, the lateral boundary with respect to n, is $4\alpha^3$. Since the interior of $\mathfrak{P}_\alpha$ is a part of $\mathcal{B}_\alpha$, the integral balance of forces (3.1) holds for it. The assumed boundedness of $b_a^{\ddagger}$ implies also the inequality

$$\left|\int_{\partial\mathfrak{P}_\alpha} \mathfrak{t}\, d\mathcal{H}^2\right| \leq k\alpha^4,$$

which is

$$\frac{1}{\alpha^2}\left|\int_{\partial\mathfrak{P}_\alpha} \mathfrak{t}\, d\mathcal{H}^2\right| \leq k\alpha^2,$$

and tends to

$$\frac{1}{\alpha^2}\int_{\partial\mathfrak{P}_\alpha} \mathfrak{t}\, d\mathcal{H}^2 = 0$$

as $\alpha \to 0$. The last identity becomes

$$\frac{1}{\alpha^2}\int_{\mathcal{D}_{+\alpha}} \mathfrak{t}\, d\mathcal{H}^2 + \frac{1}{\alpha^2}\int_{\mathcal{D}_{-\alpha}} \mathfrak{t}\, d\mathcal{H}^2 + \frac{1}{\alpha^2}\int_{\partial\mathcal{D}\times\left[-\frac{\alpha^2}{2},+\frac{\alpha^2}{2}\right]} \mathfrak{t}\, d\mathcal{H}^2 = 0.$$

Thanks to the assumed continuity for $\mathfrak{t}$ with respect to y, by letting α go to zero, shrinking $\mathcal{D}$ to y, we see that the first integral in the previous sum tends to $\tilde{\mathfrak{t}}(y, n)$, the second to $\tilde{\mathfrak{t}}(y, -n)$, while the third goes to zero because the integral is of order α^3. Hence in the limit obtained by shrinking $\mathfrak{P}_\alpha$ to y, the relation in the statement of the theorem follows. □

A shorter proof can be developed, though we achieve conciseness at the cost of greater abstraction.

Alternative Proof. Consider a plane π oriented by the normal n that intersects $\mathcal{B}_a$ and select an arbitrary compact set $\mathcal{D}$ with regular boundary in the intersection of the plane with $\mathcal{B}_a$. Write $\mathcal{C}_\alpha$ for the set

$$\mathcal{C}_\alpha := \left\{y = \bar{y} + \xi n \mid \bar{y} \in \mathcal{D}, -\frac{\alpha}{2} \leq \xi \leq \frac{\alpha}{2}\right\}.$$

For α sufficiently small, we can take $\mathcal{C}_\alpha$ such that its interior is a part of $\mathcal{B}_a$. Then the integral balance of forces (3.1) holds for it. By letting α go to 0, from the inequality

$$\left|\int_{\partial\mathcal{C}_\alpha} \tilde{\mathfrak{t}}(y,n)d\mathcal{H}^2\right| \leq k\,\mathrm{vol}(\mathcal{C}_\alpha),$$

determined by the boundedness of $b_a^\ddagger$, we get

$$\lim_{\alpha\to 0}\int_{\partial\mathcal{C}_\alpha} \tilde{\mathfrak{t}}(y,n)d\mathcal{H}^2 = 0. \tag{3.4}$$

In the limit $\alpha \to 0$, the two copies of $\mathcal{D}$ at height $\frac{\alpha}{2}$ and $\frac{-\alpha}{2}$ with respect to the normal n of $\mathcal{D}$ both collapse onto $\mathcal{D}$, and the equation (3.4) reduces to

$$\int_{\mathcal{D}} \left(\tilde{\mathfrak{t}}(y,n) - \tilde{\mathfrak{t}}(y,-n)\right) d\mathcal{H}^2 = 0.$$

The arbitrariness of $\mathcal{D}$ implies that the integrand vanishes.

□

The previous theorem is a version of the **action–reaction principle**. It has a key role in proving Cauchy's theorem stating the linearity of $\tilde{\mathfrak{t}}(y,n)$ with respect to n.

Theorem 3 (Augustin-Louis Cauchy). *If $\tilde{\mathfrak{t}}(\cdot,n)$ is continuous, then $\tilde{\mathfrak{t}}(y,\cdot)$ is homogeneous and additive, i.e., a second-rank tensor field $y \longmapsto \sigma(y)$ exists and is such that*

$$\tilde{\mathfrak{t}}(y,n) = \sigma(y)n(y).$$

The version in components of the previous expression depends on whether we consider the normal n as a vector or a covector. In the latter case, it is the normalized derivative of the function defining the surface oriented at y by n, namely $n = \frac{(Df)_i}{|Df|}\tilde{e}^i$. When we transform Df into the gradient ∇f, we get $n = \frac{(\nabla f)^i}{|\nabla f|}\tilde{e}_i$. Hence in the first case,

$$\mathfrak{t}_i = \sigma_i^j n_j,$$

and $\sigma(y)$ is a linear operator mapping covectors at $y \in \mathcal{B}_a$ into other covectors at the same point, and we write $\sigma(y) \in \mathrm{Hom}(T_y^*\mathcal{B}_a, T_y^*\mathcal{B}_a)$ for short. In the second case,

$$\mathfrak{t}_i = \sigma_{ij}n^j,$$

and $\sigma(y)$ is now a linear operator mapping vectors at $y \in \mathcal{B}_a$ into covectors at the same point, and we write $\sigma(y) \in \mathrm{Hom}(T_y\mathcal{B}_a, T_y^*\mathcal{B}_a)$.

Proof. The proof here is the standard one by Cauchy. Consider a point y in $\mathcal{B}_a$, an orthogonal frame (e_1, e_2, e_3) with origin in y, and a unit *vector* n (with covector counterpart $n^\flat$) at y not coinciding with any base vector e_i. Take a tetrahedron $\mathcal{T}$ with oblique side $\tilde{\tau}$ of area $a(\tilde{\tau})$, oriented by n; the other sides $\tilde{\tau}_i$, with area $a(\tilde{\tau}_i)$, are oriented by the normals $n_i = -\operatorname{sgn}(n^\flat \cdot e_i)e_i$; in this sense, these faces are mutually orthogonal. The assumed boundedness of the bulk actions implies, as above, the validity of the inequality

$$\frac{1}{a(\tilde{\tau})}\left|\int_{\partial\mathcal{T}} \mathfrak{t}\, d\mathcal{H}^2\right| \leq \frac{k\,\mathrm{vol}(\mathcal{T})}{a(\tilde{\tau})} \tag{3.5}$$

with k the constant already mentioned above. If we let the diameter[2] of $\mathcal{T}$ go to zero, then the right-hand side of the inequality (3.5) goes to zero, too, and we obtain

$$\lim_{\mathrm{diam}\mathcal{T}\to 0} \frac{1}{a(\tilde{\tau})}\int_{\partial\mathcal{T}} \mathfrak{t}\, d\mathcal{H}^2 = 0. \tag{3.6}$$

Such an integral is the sum of the integrals over the faces of the tetrahedron. In particular, due to the continuity of $\tilde{\mathfrak{t}}(\cdot, n)$, we get

$$\lim_{\mathrm{diam}\mathcal{T}\to 0} \frac{1}{a(\tilde{\tau})}\int_{\tilde{\tau}} \mathfrak{t}\, d\mathcal{H}^2 = \tilde{\mathfrak{t}}(y, n)$$

and, thanks to Theorem 2,

$$\lim_{\mathrm{diam}\mathcal{T}\to 0} \frac{1}{a(\tilde{\tau})}\int_{\tilde{\tau}_i} \mathfrak{t}\, d\mathcal{H}^2 = -(n^\flat \cdot n_i)\tilde{\mathfrak{t}}(y, n).$$

Since n_i is the normal to the ith coordinate side of the tetrahedron, we compute

$$\begin{aligned}
&-(n^\flat \cdot n_i)\,\tilde{\mathfrak{t}}(y, n_i)\\
&= \operatorname{sgn}(n^\flat \cdot e_i)(n^\flat \cdot e_i)\,\tilde{\mathfrak{t}}(y, -\operatorname{sgn}(n^\flat \cdot e_i)e_i)\\
&= -\operatorname{sgn}(n^\flat \cdot e_i)(n^\flat \cdot e_i)\,\tilde{\mathfrak{t}}(y, \operatorname{sgn}(n^\flat \cdot e_i)e_i),
\end{aligned}$$

but once again, Theorem 2 implies

$$\tilde{\mathfrak{t}}(y, \operatorname{sgn}(n^\flat \cdot e_i)e_i) = \operatorname{sgn}(n^\flat \cdot e_i)\,\tilde{\mathfrak{t}}(y, e_i),$$

and we get

$$-(n^\flat \cdot n_i)\,\tilde{\mathfrak{t}}(y, n_i) = -(n^\flat \cdot e_i)\,\tilde{\mathfrak{t}}(y, e_i)$$

---

[2]The diameter of a set $\mathcal{A}$ in a metric space is the maximum distance between two points taken over all pairs of points in $\mathcal{A}$.

for $(\text{sgn}(n \cdot e_i))^2 = 1$. As a consequence, as diam$\mathcal{T} \to 0$, equation (3.6) can be written as

$$\tilde{\mathfrak{t}}(y,n) - \sum_{i=1}^{3} (n^\flat \cdot e_i)\,\tilde{\mathfrak{t}}(y,e_i) = 0,$$

which is

$$\tilde{\mathfrak{t}}(y,n) = \sum_{i=1}^{3} (\tilde{\mathfrak{t}}(y,e_i) \otimes e_i) n^\flat .$$

Then $\mathfrak{t}$ is a linear function of n, the proportional factor being the second-rank tensor

$$\sigma(y) = \sum_{i=1}^{3} \tilde{\mathfrak{t}}(y,e_i) \otimes e_i . \qquad \square$$

The tensor σ defined by the previous expression has components σ_i^j. The first is determined by the covector component $\tilde{\mathfrak{t}}_i(y,e_j)$, the traction over the side with normal e_j. The second component is given by e_j itself.

By considering a part $\mathfrak{b}_a$ and taking the normal to $\partial\mathfrak{b}_a$ as a covector, the integral balances then read

$$\int_{\mathfrak{b}_a} b_a^\ddagger d\mu + \int_{\partial\mathfrak{b}_a} \sigma n \, d\mathcal{H}^2 = 0,$$

$$\int_{\mathfrak{b}_a} (y - y_0) \times b_a^\ddagger d\mu + \int_{\partial\mathfrak{b}_a} (y - y_0) \times \sigma n \, d\mathcal{H}^2 = 0.$$

3.3 Inertial Effects and Mass Balance

We adopt the following assumptions:

1. $b_a^\ddagger$ admits additive decomposition into *inertial*, b_a^{in}, and *noninertial*, b_a, components, namely

$$b_a^\ddagger = b_a + b_a^{in} .$$

2. The inertial component satisfies the identity

$$\frac{d}{dt}(\text{kinetic energy of } \mathfrak{b}_a) + \int_{\mathfrak{b}_a} b_a^{in} \cdot v \, d\mu = 0$$

for every $\mathfrak{b}_a$ and velocity v at all instants where the velocity is time differentiable. The *kinetic energy* pertaining to $\mathfrak{b}_a$ with reference to the velocity v is given by

$$\{\text{kinetic energy } \mathfrak{b}_a\} = \frac{1}{2} \int_{\mathfrak{b}_a} \varrho_a \langle v, v \rangle \, d\mu,$$

where $\rho_a := \tilde{\rho}_a(y, t)$ is the mass density, with $\tilde{\rho}_a$ a continuous and differentiable function of space y and time t such that the mass $M(\mathcal{B}_a)$ of the body in the actual configuration $\mathcal{B}_a$ at the instant t is given by

$$M(\mathcal{B}_a) := \int_{\mathcal{B}_a} \tilde{\rho}_a(y, t) d\mu.$$

3. The **mass** of the body in its reference place at the instant t is given by

$$M(\mathcal{B}) := \int_{\mathcal{B}} \tilde{\rho}(x) d\mu.$$

If the body does not lose its mass, $M(\mathcal{B})$ and $M(\mathcal{B}_a)$ coincide. Moreover, we have $d\mu(y) = (\det F) d\mu(x)$ by a change of coordinates, namely

$$\int_{\mathcal{B}_a} \varrho_a d\mu = \int_{\mathcal{B}} \varrho_a \det F d\mu.$$

If not only is the total mass preserved but also mass is not redistributed within the body, meaning that no generic part loses mass to the rest of the body or acquires mass from it, we have

$$M(\mathfrak{b}_a) = M(\mathfrak{b})$$

when $\mathfrak{b}_a = \tilde{y}(\mathfrak{b})$, which is (with $\rho := \tilde{\rho}(x)$

$$\int_{\mathfrak{b}_a} \varrho_a d\mu = \int_{\mathfrak{b}} \varrho_a \det F d\mu = \int_{\mathfrak{b}} \varrho \, d\mu$$

and implies

$$\varrho = \varrho_a \det F \tag{3.7}$$

thanks to the arbitrariness of $\mathfrak{b}$. Equation (3.7) is the **local mass balance** in referential (or Lagrangian) representation. The **reference mass density** ϱ

depends only on space variables, it is, in fact, $\varrho = \tilde{\varrho}(x)$, while the **actual mass density** ϱ_a depends also on time, and we have $\varrho_a = \tilde{\varrho}_a(\tilde{y}(x,t),t)$. Under the assumption that ϱ_a is differentiable with respect to its entries, the time derivative of the local referential mass balance (3.7) can be written as

$$0 = \dot{\varrho}_a \det F + \varrho_a \overline{\det F}^{\cdot}.$$

Lemma 4 (Euler identity). *The following relation holds:*

$$\frac{\overline{\det F}^{\cdot}}{\det F} = \operatorname{div} v,$$

where div *is the divergence evaluated with respect to y, namely* $\operatorname{div} v := \operatorname{tr} D_y v$.

Proof. The proof is based on a direct calculation. We have, in fact,

$$\begin{aligned}\overline{\det F}^{\cdot} &= \frac{\partial \det F}{\partial F} \cdot \dot{F} = (\det F) F^{-*} \cdot D_y v F \\ &= (\det F) F^{-*} F^{*} \cdot D_y v \\ &= (\det F) \hat{I} \cdot D_y v = (\det F) \operatorname{div} v,\end{aligned}$$

with[3] $\hat{I}$ the second-rank, 1−contravariant, 1−covariant tensor $\hat{I} = \delta^i_j \tilde{e}_i \otimes \tilde{e}^j$. □

The Euler identity implies

$$\dot{\varrho}_a + \varrho_a \operatorname{div} v = 0, \tag{3.8}$$

which is the **local mass balance** in Eulerian representation. However, $\dot{\varrho}_a = \frac{d}{dt}\tilde{\rho}_a(\tilde{y}(x,t),t)$, that is,

$$\dot{\varrho}_a = D_y \varrho_a \cdot \frac{dy}{dt} + \left.\frac{\partial \varrho_a}{\partial t}\right|_{y \text{ fixed}} = \left.\frac{\partial \varrho_a}{\partial t}\right|_{y \text{ fixed}} + (D_y \varrho_a) \cdot v.$$

Hence another way to express the local balance of mass in Eulerian representation is given by

$$\frac{\partial \varrho_a}{\partial t} + \operatorname{div}(\varrho_a v) = 0,$$

where we have left understood that the partial derivative $\frac{\partial \varrho_a}{\partial t}$ is evaluated with y held fixed.

---

[3] We have $\det F = \mathrm{e}_{ijk} F^k_1 F^j_2 F^i_3$ so that we compute $\frac{\partial \det F}{\partial F^k_1} = \mathrm{e}_{kij} F^j_2 F^i_3 = \mathrm{e}_{kij} F^j_2 F^i_3 F^r_1 (F^{-1})^1_r = \mathrm{e}_{kij} F^j_2 F^i_3 F^k_1 \delta^r_k (F^{-1})^1_r = (\det F)(F^{-1})^1_k$, which is $\frac{\partial \det F}{\partial F} = (\det F) F^{-*}$.

The pointwise balance of mass helps us in evaluating the consequence of the assumed link between the rate of the kinetic energy and the power of the inertial action over $\mathcal{B}_a$ for any choice of the velocity field. In fact, by changing coordinates from the actual to the reference place, we have

$$\frac{d}{dt}\int_{\mathcal{B}_a} \varrho_a \langle v, v\rangle \, d\mu = \frac{d}{dt}\int_{\mathcal{B}_a} \frac{1}{2}\varrho_a \tilde{g} v \cdot v \, d\mu = \frac{d}{dt}\int_{\mathcal{B}} \frac{1}{2}\varrho_a \tilde{g} v \cdot v \det F \, d\mu,$$

but now the integration domain $\mathcal{B}$ does not depend on time, in contrast to $\mathcal{B}_a$, and we can commute time derivation with integration, obtaining, with the notation $a^\flat := \tilde{g}\frac{dv}{dt}$ for the covector associated with the acceleration $a := \frac{dv}{dt}$,

$$\frac{d}{dt}\int_{\mathcal{B}} \frac{1}{2}\varrho_a \tilde{g} v \cdot v \det F \, d\mu$$
$$= \int_{\mathcal{B}} \left(\frac{1}{2}\dot{\varrho}_a \tilde{g} v \cdot v \det F + \varrho_a a^\flat \cdot v \det F + \frac{1}{2}\varrho_a \tilde{g} v \cdot v \, \overline{\det F}^{\cdot}\right) d\mu$$
$$= \int_{\mathcal{B}} \left(\frac{1}{2}\dot{\varrho}_a \tilde{g} v \cdot v \det F + \varrho_a a^\flat \cdot v \det F + \frac{1}{2}\varrho_a \tilde{g} v \cdot v \det F \operatorname{div} v\right) d\mu$$
$$= \int_{\mathcal{B}} \left(\frac{1}{2}\dot{\varrho}_a \tilde{g} v \cdot v \det F + \varrho_a a^\flat \cdot v \det F - \frac{1}{2}\varrho_a \tilde{g} v \cdot v \frac{\dot{\varrho}_a}{\varrho_a} \det F\right) d\mu$$
$$= \int_{\mathcal{B}} \varrho_a a^\flat \cdot v \det F \, d\mu = \int_{\mathcal{B}_a} \varrho_a a^\flat \cdot v \, d\mu,$$

where we have used the identities $d\mu(y) = (\det F)\, d\mu(x)$ and $\operatorname{div} v = \frac{\dot{\varrho}_a}{\varrho_a}$. As a consequence, the assumed validity of the identity

$$\frac{d}{dt}\int_{\mathcal{B}_a} \frac{1}{2}\varrho_a \langle v, v\rangle \, d\mu + \int_{\mathcal{B}_a} b_a^{in} \cdot v \, d\mu = 0$$

for any choice of the velocity field implies

$$b_a^{in} = -\varrho_a a^\flat.$$

3.4 Pointwise Balances in Eulerian Representation

Assume that the map $y \longmapsto b_a^\ddagger(y)$ is continuous, while $y \longmapsto \sigma(y)$ is continuously differentiable. Gauss's theorem for tensor fields implies

$$\int_{\partial \mathfrak{b}_a} \sigma n \, d\mathcal{H}^2 = \int_{\mathfrak{b}_a} \operatorname{div} \sigma \, d\mu,$$

so the integral balance of forces can be written as

$$\int_{\mathfrak{b}_a} b_a^{\ddagger} d\mu + \int_{\partial \mathfrak{b}_a} \sigma n \, d\mathcal{H}^2 = \int_{\mathfrak{b}_a} \left(b_a^{\ddagger} + \operatorname{div} \sigma \right) d\mu = 0,$$

and the arbitrariness of $\mathfrak{b}_a$ implies the **local balance of forces**

$$b_a^{\ddagger} + \operatorname{div} \sigma = 0 \quad \text{in } \mathcal{B}_a,$$

or, by considering the additive decomposition of $b_a^{\ddagger}$ and the expression of inertial actions deduced above,

$$b_a + \operatorname{div} \sigma = \varrho_a a^{\flat}. \tag{3.9}$$

The integral balance of couples can be written as

$$\int_{\mathfrak{b}_a} (y - y_0) \times b_a^{\ddagger} d\mu + \int_{\partial \mathfrak{b}_a} (y - y_0) \times \sigma n \, d\mathcal{H}^2 = 0.$$

With the previous assumptions, Gauss's theorem implies

$$\int_{\partial \mathfrak{b}_a} (y - y_0) \times \sigma n \, d\mathcal{H}^2 = \int_{\mathfrak{b}_a} \operatorname{div} \left((y - y_0) \times \sigma \right) d\mu. \tag{3.10}$$

If we write p for the difference $(y - y_0)$, the second-rank tensor $(y - y_0) \times \sigma$ reads $\mathsf{e} p \sigma$, with e Ricci's symbol. By considering, for example, the version of σ with both contravariant components (which is an assumption that is not essential for the subsequent results, but is useful for the readability of the formula) we obtain $(\mathsf{e} p \sigma)_i^j = \mathsf{e}_{ikl} p^l \sigma^{kj}$.

The divergence of $\mathsf{e} p \sigma$ implies derivatives with respect to y. Then we get in components

$$\begin{aligned} (\mathsf{e} p \sigma)_{i,j}^{j} &= \mathsf{e}_{ikl} p^l \, \sigma^{kj}{}_{,j} + \mathsf{e}_{ikl} \, p^l{}_{,j} \sigma^{jk} \\ &= \mathsf{e}_{ikl} p^l \, \sigma^{kj}{}_{,j} + \mathsf{e}_{ikl} \delta_j^l \sigma^{jk} = \mathsf{e}_{ikl} p^l \, \sigma^{kj}{}_{,j} - \mathsf{e}_{ikl} \sigma^{lk}, \end{aligned}$$

or more concisely,

$$\operatorname{div}(\mathsf{e} p \sigma) = \mathsf{e} p \operatorname{div} \sigma - \mathsf{e} \sigma^{\mathrm{T}} = p \times \operatorname{div} \sigma - \mathsf{e} \sigma^{\mathrm{T}}.$$

Substituting into (3.10) and introducing the result in the integral balance of couples, we get

$$\int_{\mathfrak{b}_a} \left((y - y_0) \times (b_a^{\ddagger} + \operatorname{div} \sigma) - \mathsf{e} \sigma^{\mathrm{T}} \right) d\mu = 0.$$

The validity of (3.9) and the arbitrariness of $\mathfrak{b}_a$ imply also

$$\mathsf{e}\sigma = 0,$$

and, by premultiplication by Ricci's symbol e,

$$\mathsf{e}\mathsf{e}\sigma = 0. \tag{3.11}$$

An algebraic property is

$$\mathsf{e}_{ijk}\mathsf{e}_{klm} = \delta_{il}\delta_{jm} - \delta_{im}\delta_{jl},$$

so that

$$\mathsf{e}\mathsf{e}\sigma = \sigma - \sigma^{\mathrm{T}},$$

and (3.11) can be written as

$$\sigma = \sigma^{\mathrm{T}}, \tag{3.12}$$

which is the **local balance of couples.**

The symmetry of the stress tensor—the consequence of the integral balance of couples—and the fact that its components in every given basis are real numbers imply that σ can be expressed in diagonal form: three linearly independent vectors e_I, e_{II}, e_{III} exist with their covector counterparts e^I, e^{II}, e^{III}, and three real numbers σ_I, σ_{II}, σ_{III} such that

$$\sigma = \sigma_I e_I \otimes e^I + \sigma_{II} e_{II} \otimes e^{II} + \sigma_{III} e_{III} \otimes e^{III}.$$

The numbers σ_I, σ_{II}, and σ_{III} are called the **principal components of the stress**, or the **principal stresses** for short. Their existence means that there is a basis at y, the one given by the eigenvectors of σ, in which the matrix of the components of σ is diagonal, namely

$$\left[\sigma_i^j\right] = \begin{pmatrix} \sigma_I & 0 & 0 \\ 0 & \sigma_{II} & 0 \\ 0 & 0 & \sigma_{III} \end{pmatrix}.$$

In other words, in the frame defined by the eigenvectors, we see just normal stresses in the coordinate planes, shear stresses appearing explicitly only when we rotate the frame. In any case, from the knowledge of the principal stresses we may deduce the maximum of the shear stresses emerging in the representations of σ when the frame rotates with respect to the one determined by the eigenvectors. Such a maximum shear stress is given by

$$\frac{1}{2}\max\left(\left|\sigma_I - \sigma_{II}\right|, \left|\sigma_{II} - \sigma_{III}\right|, \left|\sigma_{III} - \sigma_I\right|\right),$$

as emerges clearly in the graphic representation of the properties of σ involving the so-called Mohr circles, a topic not touched here but common to all second-rank symmetric tensors.

3.5 Inner Power

Let us consider the external power in Eulerian representation on $\mathfrak{b}_a$ and apply Gauss's theorem to the integral over the surface $\partial\mathfrak{b}_a$. For that integral, we then get

$$\int_{\partial\mathfrak{b}_a} \sigma n \cdot v \, d\mathcal{H}^2 = \int_{\partial\mathfrak{b}_a} \sigma^* v \cdot n \, d\mathcal{H}^2 = \int_{\mathfrak{b}_a} \operatorname{div}(\sigma^* v) d\mu .$$

In components, we have $\operatorname{div}(\sigma^* v) = (\sigma_i^j v^i)_{,j}$, so that by evaluating the derivative with respect to y^j, we get

$$\operatorname{div}(\sigma^* v) = v \cdot \operatorname{div} \sigma + \sigma \cdot D_y v .$$

By inserting this last relation into the bulk integral above, we then obtain

$$\begin{aligned} \mathcal{P}^{ext}_{\mathfrak{b}_a}(v) &= \int_{\mathfrak{b}_a} b^{\ddagger} \cdot v \, d\mu + \int_{\partial\mathfrak{b}_a} \sigma n \cdot v \, d\mathcal{H}^2 \\ &= \int_{\mathfrak{b}_a} b^{\ddagger} \cdot v \, d\mu + \int_{\mathfrak{b}_a} \left(v \cdot \operatorname{div} \sigma + \sigma \cdot D_y v \right) d\mu \\ &= \int_{\mathfrak{b}_a} \left(b^{\ddagger} + \operatorname{div} \sigma \right) \cdot v \, d\mu + \int_{\mathfrak{b}_a} \sigma \cdot D_y v \, d\mu . \end{aligned}$$

The validity of the pointwise balance (3.9) implies the identity

$$\mathcal{P}^{ext}_{\mathfrak{b}_a}(v) = \int_{\mathfrak{b}_a} \sigma \cdot D_y v \, d\mu = \int_{\mathfrak{b}_a} \sigma \cdot \left(\operatorname{Sym} D_y v + \operatorname{Skw} D_y v \right) d\mu .$$

Since σ is symmetric—a consequence of the balance of couples and the validity of the balance of forces—the product $\sigma \cdot \operatorname{Skw} D_y v$ vanishes (as always occurs in the scalar product between symmetric and skew-symmetric real second-rank tensors), and we obtain

$$\mathcal{P}^{ext}_{\mathfrak{b}_a}(v) = \int_{\mathfrak{b}_a} \sigma \cdot \operatorname{Sym} D_y v \, d\mu .$$

The integral on the right-hand side is commonly called the **inner power in Eulerian representation** and is denoted by $\mathcal{P}^{inn}_{\mathfrak{b}_a}(v)$.

The identity

$$\mathcal{P}^{ext}_{\mathfrak{b}_a}(v) = \mathcal{P}^{inn}_{\mathfrak{b}_a}(v)$$

can be chosen as a first principle. We can require, in fact, a priori that the identity

$$\mathcal{P}^{ext}_{\mathcal{B}_a}(v) = \mathcal{P}^{inn}_{\mathcal{B}_a}(v) \tag{3.13}$$

hold for every choice of the velocity fields assumed to be differentiable in space and compactly supported over $\mathcal{B}_a$.

The arbitrariness in the choice of the velocity implies that it is not strictly related to the specific motion under investigation—the one along which the stress involved in the expression of the power occurs—and in this sense, the velocity field in the identity (3.13) can be called **virtual**. The previous identity, accepted as a first principle, could be then called the **principle of virtual power** for the presence of the virtual velocity. Its immediate consequence would then be the validity of the pointwise balance equations of forces and couples.

There is a difference between obtaining balance equations from the principle of the virtual power or deriving them from the invariance of the external power alone, as done so far.

1. The assumption of the principle of virtual power as a first principle requires that we prescribe an expression of the inner power, an additional assumption to those we have made so far. For it, we have also to know the existence of the stress σ, which is proved here on the basis of the knowledge of the integral balances, reducing in this way the role of the principle of virtual power as primary origin of the pointwise balances.
2. The acceptance a priori of (3.13) for every choice of the velocity field is the prescription of the (so-called) *weak form* of the balance equations. In other words, when we write (3.13) as a primary assumption, we should have already in mind the structure of the balance equations, in contrast to what we do when we postulate at the beginning the sole expression of the external power, which is based only on the presumed classification of the actions over a generic part of the body.

3.6 Balance Equations in Lagrangian Representation

The relations describing how volumes and oriented areas change along a deformation allow us to write balance equations in terms of fields defined over the reference shape $\mathcal{B}$. In fact, we have

$$\int_{\mathfrak{b}_a} b_a^{\ddagger} d\mu = \int_{\mathfrak{b}} b_a^{\ddagger} \det F d\mu$$

and

$$\int_{\partial \mathfrak{b}_a} \sigma n \, d\mathcal{H}^2 = \int_{\partial \mathfrak{b}} (\det F)\sigma F^{-*} \bar{n} \, d\mathcal{H}^2,$$

where $\mathfrak{b}_a = \tilde{y}(\mathfrak{b})$, with $\mathfrak{b}$ a part of $\mathcal{B}$. As a consequence, if we define $b^\ddagger$ and P by

$$b^\ddagger := b_a^\ddagger \det F$$

and

$$P := (\det F)\sigma F^{-*},$$

the integral balances of forces and couples become respectively

$$\int_{\mathfrak{b}} b^\ddagger d\mu + \int_{\partial \mathfrak{b}} P\bar{n} \, d\mathcal{H}^2 = 0 \tag{3.14}$$

and

$$\int_{\mathfrak{b}} (y - y_0) \times b^\ddagger d\mu + \int_{\partial \mathfrak{b}} (y - y_0) \times P\bar{n} \, d\mathcal{H}^2 = 0. \tag{3.15}$$

In this case, we affirm that the integral balance equations are expressed in **Lagrangian (or referential) description (or representation)**. From (3.14), using Gauss's theorem and exploiting the arbitrariness of $\mathfrak{b}$, we get

$$b^\ddagger + \mathrm{Div} P = 0, \tag{3.16}$$

where the uppercase letter denoting the divergence operator reminds us that the derivatives are evaluated with respect to coordinates x^A in $\mathcal{B}$. The quantity P is called the **first Piola–Kirchhoff stress** after Gabrio Piola (1794–1850) and Gustav Robert Georg Kirchhoff (1824–1887).

The same technique—a combination of Gauss's theorem and the arbitrariness of $\mathfrak{b}$—can also be applied to the referential balance of couples (3.15). However, without repeating all calculations already developed, we just observe that the identity

$$(\det F)\sigma = PF^*,$$

deriving from the definition of P, implies that

$$PF^* \in \mathrm{Sym}(\mathbb{R}^3, \mathbb{R}^3),$$

since σ is symmetric. In components, we write $P_i^A F_A^j = F_A^i P_j^A$ for $P = P_i^A \tilde{e}^i \otimes e_A$. The same reasoning holds if we consider the normal as a vector (so in this case, σ

has both covariant components: $\sigma = \sigma_{ij}\tilde{e}^i \otimes \tilde{e}^j$) and define a version of P with both covariant components by

$$P := (\det F)\sigma F^{-\mathrm{T}},$$

yielding

$$PF^{\mathrm{T}} \in \mathrm{Sym}(\mathbb{R}^3, \mathbb{R}^3)$$

once again from the symmetry of σ. In this case, $P = P_{iA}\tilde{e}^i \otimes e^A$.

The external power in the referential description, denoted by $\mathcal{P}^{ext}_{\mathfrak{b}}(\dot{y})$, in terms of the first Piola–Kirchhoff stress, is given by

$$\mathcal{P}^{ext}_{\mathfrak{b}}(\dot{y}) := \int_{\mathfrak{b}} b^{\ddagger} \cdot \dot{y}\, d\mu + \int_{\partial\mathfrak{b}} P\bar{n} \cdot \dot{y}\, d\mathcal{H}^2.$$

By applying Gauss's theorem to the second integral, we obtain

$$\int_{\partial\mathfrak{b}} P\bar{n} \cdot \dot{y}\, d\mathcal{H}^2 = \int_{\partial\mathfrak{b}} (P^*\dot{y}) \cdot \bar{n}\, d\mathcal{H}^2 = \int_{\mathfrak{b}} \mathrm{Div}(P^*\dot{y}) d\mu = \int_{\mathfrak{b}} (P \cdot \dot{F} + \dot{y} \cdot \mathrm{Div}P) d\mu.$$

Using the pointwise balance (3.16), we eventually get

$$\mathcal{P}^{ext}_{\mathfrak{b}}(\dot{y}) = \int_{\mathfrak{b}} P \cdot \dot{F} d\mu.$$

The right-hand side is what is called an **inner (or internal) power in referential description (or representation)** and is denoted by $\mathcal{P}^{inn}_{\mathfrak{b}}(\dot{y})$.

In a rigid motion of the whole body, the points of the body (i.e., the material elements) do not change their relative positions, so that the inner power (the **stress power**) must vanish. As a consequence, we recover once again the symmetry of PF^*, namely the pointwise balance of couples. To obtain the result, consider a rigid velocity field

$$v_R := c(t) + q(t) \times (y - y_0),$$

or equivalently,

$$v_R = c(t) + W(t)(y - y_0),$$

with y_0 an arbitrary point in the physical space $\tilde{\mathcal{E}}^3$; $c(t)$ and $q(t)$ translational and rotational velocity vectors, respectively; $W(t)$ a skew-symmetric tensor having $q(t)$ as characteristic vector, so that $Wp = q \times p$ for every $p \in \mathbb{R}^3$. For v_R, we have

$$D_y v_R = W.$$

Since

$$P \cdot \dot{F} = P \cdot D_y v F = PF^* \cdot D_y v,$$

imposing the absence of inner power along every rigid motion of the whole body corresponds to the condition

$$PF^* \cdot D_y v_R = PF^* \cdot W = 0$$

for every W. The arbitrariness of W implies the symmetry of PF^*, since the scalar product between symmetric and a skew-symmetric real tensors always vanishes. The same result can be obtained by considering the velocity in referential description, namely

$$\dot{y}_R = c(t) + W(t)(y - y_0),$$

for which, in fact, we have

$$F = D\dot{y}_R = WF.$$

The consequence is that

$$0 = P \cdot WF = PF^* \cdot W$$

for every choice of W. Hence, the symmetry of PF^* follows once again. Previous calculations show also that in the Eulerian description,

$$\sigma \cdot W = 0$$

for every skew-symmetric second-rank tensor W. The arbitrariness of W implies once again the symmetry of σ.

In referential description, the inertia terms can be rederived by requiring that their power equal the negative of the time variation of the kinetic energy for every choice of velocity field compactly supported over $\mathcal{B}$. The formal reasoning retraces steps already taken in the Eulerian representation.

1. Since $b^\ddagger = (\det F) b_a^\ddagger$ by definition, the assumed additive decomposition of $b_a^\ddagger$ into inertial, b_a^{in}, and noninertial, b_a, components translates into a similar decomposition of $b^\ddagger$, namely

$$b^\ddagger = b^{in} + b,$$

where $b^{in} := (\det F) b_a^{in}$ and $b := (\det F) b_a$.

2. As in the Eulerian case, b^{in} is identified by imposing the identity

$$\frac{d}{dt}\{\text{kinetic energy of } \mathcal{B}\} + \int_{\mathcal{B}} b^{in} \cdot \dot{y}\, d\mu = 0 \qquad (3.17)$$

for every velocity field $(x,t) \longmapsto \dot{y}(x,t)$ compactly supported over $\mathcal{B}$ at all instants t. The kinetic energy of $\mathcal{B}$ is here given by

$$\int_{\mathcal{B}} \frac{1}{2}\varrho \langle \dot{y}, \dot{y} \rangle \, d\mu,$$

which can be written also as

$$\int_{\mathcal{B}} \frac{1}{2}\varrho \dot{y} \cdot \tilde{g}\dot{y}\, d\mu.$$

Since $\mathcal{B}$ is fixed in time, by substituting this last integral into (3.17) and evaluating the time derivative, we eventually get

$$\int_{\mathcal{B}} (\varrho \tilde{g}\ddot{y} - b^{in}) \cdot \dot{y}\, d\mu = 0,$$

which implies, thanks to the arbitrariness of $\dot{y}$, compactly supported over $\mathcal{B}$, the identification

$$b^{in} = -\varrho \tilde{g}\ddot{y},$$

so that the pointwise balance of forces in Lagrangian (referential) representation can be written as

$$b + \text{Div}P = \varrho \tilde{g}\ddot{y},$$

or alternatively,

$$b + \text{Div}P = \varrho \ddot{y}^{\flat},$$

with $\ddot{y}^{\flat} := \tilde{g}\ddot{y}$ the covector naturally associated with $\ddot{y}$ by the metric $\tilde{g}$ in the ambient space. Even in the Lagrangian representation, to derive balance equations we could accept the identity

$$\mathcal{P}^{ext}_{\mathcal{B}}(\dot{y}) = \mathcal{P}^{inn}_{\mathcal{B}}(\dot{y})$$

as a first principle holding for every choice of differentiable velocity field compactly supported over $\mathcal{B}$, or equivalently, we could write

$$\mathcal{P}^{ext}_{\mathfrak{b}}(\dot{y}) = \mathcal{P}^{inn}_{\mathfrak{b}}(\dot{y}),$$

assuming its validity for every part $\mathfrak{b}$ and square-integrable velocity field, with bulk and contact actions at least with the same regularity in space. However, as in the case of the Eulerian representation, if we followed this point of view, we should once again postulate not only the expression of the external power but also the form of the internal power and the existence of the stress, which is, in contrast, a derived quantity. Moreover, we should also impose that the internal power vanishes when it is evaluated along any rigid-body motion. These assumptions correspond even in this case to an a priori assumption of the weak form of the balance equations, in other words, to have already in mind the structure of such equations.

3.7 Virtual Work in the Small-Strain Setting

In terms of the displacement field $u := \tilde{u}(x,t) := \tilde{y}(x,t) - i(x)$, and in the small-strain regime, namely when $|Du| \ll 1$, also $P \approx \sigma$, and we do not distinguish between reference and current configurations, the **external work** over a generic part $\mathfrak{b}$ of $\mathcal{B}$ is denoted by $L^{ext}(u;\mathfrak{b})$ and defined by

$$L^{ext}(u;\mathfrak{b}) := \int_{\mathfrak{b}} b^{\ddagger} \cdot u \, d\mu + \int_{\partial\mathfrak{b}} \sigma n \cdot u \, d\mathcal{H}^2,$$

where σ is now considered a function of x rather than y. The reason for this choice relies on the absence of an explicit distinction between $\mathcal{B}$ and $\mathcal{B}_a$, and therefore between x and y, and we write impartially Div or div "confusing" the derivatives with respect to x with those with respect to y. Moreover, we have taken into account that in the small-strain setting, $b^{\ddagger} \approx b_a^{\ddagger}$.

By applying Gauss's theorem to the second integral in the definition of L^{ext}, we obtain

$$\begin{aligned} L^{ext}(u;\mathcal{B}) &:= \int_{\mathcal{B}} b^{\ddagger} \cdot u \, d\mu + \int_{\mathcal{B}} (u \cdot \operatorname{div}\sigma + \sigma \cdot Du) d\mu \\ &= \int_{\mathcal{B}} \left((b^{\ddagger} + \operatorname{div}\sigma) \cdot u + \sigma \cdot Du \right) d\mu . \end{aligned}$$

If $b^{\ddagger}$ and σ satisfy the balance of forces, the last integrand reduces to

$$\sigma \cdot Du = \sigma \cdot (\operatorname{Sym} Du + \operatorname{Skw} Du) .$$

The validity of the balance of couples implies

$$\sigma \cdot Du = \sigma \cdot \operatorname{Sym} Du = \sigma \cdot \varepsilon ,$$

so that

$$L^{ext}(u; \mathcal{B}) = \int_{\mathcal{B}} \sigma \cdot \varepsilon \, d\mu.$$

The last integral in called the **inner (or internal) work** and is denoted by L^{inn}. The identity

$$L^{ext} = L^{inn} \tag{3.18}$$

can be interpreted in three ways that are different versions of what is called the **principle of virtual work**.

To have a concise expression for the three versions of the principle, as a matter of notation, we state that a system of actions $(b^{\ddagger}, \sigma)$ is *balanced* when it satisfies the balance equations $b^{\ddagger} + \text{Div}\sigma = 0$ and σ is symmetric, and a displacement–strain pair (u, ε) is *compatible* when $\varepsilon = \text{Sym}\, Du$, irrespective of the contravariant or covariant character of the components of the tensors involved.

1. For **every** pair of balanced actions and compatible displacement–strain fields, even independent of one another, in the small-strain setting, the identity

$$L^{ext} = L^{inn}$$

holds.
2. Given a pair of fields $(b^{\ddagger}, \sigma)$, if the identity

$$L^{ext} = L^{inn}$$

holds for **all** compatible displacement–strain fields, then the pair $(b^{\ddagger}, \sigma)$ is balanced.
3. Given a displacement–strain pair (u, ε) of fields, if the identity

$$L^{ext} = L^{inn}$$

holds for **every** field of balanced actions, the pair (u, ε) is compatible.

3.8 Remarks on the External Power Invariance Procedure

The procedure based on power-invariance, used so far, is indeterminate with respect to powerless terms. In particular, the identification of inertia terms furnishes a result holding to within powerless addenda (Coriolis-type terms). This circumstance suggests a way of defining inertial frames. In particular, we may **assume** that *the result of the power-based identification of the inertial terms is exact in the reference frame of at least one observer*—exactness means here that powerless terms vanish. We apply the term **inertial** to all the observers satisfying the previous statement and for which a body with uniform velocity remains in that state until external actions appear.

Traditionally, inertial observers are identified as those in the three-dimensional ambient space oriented toward the stars that were considered (rather erroneously) fixed. Relativity theory, in particular its general version, has furnished a way to overcome the difficulty of identifying privileged frames that we can call inertial.

Another aspect of the procedure is that the representation of the actions and the balance equations emerge independently of the specification of the state variables. Moreover, integral balances of forces and couples can be derived, because the Euclidean group acts naturally over the ambient space selected. Such balances can be also defined because that space can be naturally equipped with a linear structure. In fact, when we fix a point in $\mathcal{E}^3$ and consider the vectors determined by the difference of the other points and the fixed one, we construct $\mathbb{R}^3$, a linear space. The integrands of the integral balances of forces and torques are elements of $\mathbb{R}^{3*}$, the dual space, which can be identified with $\mathbb{R}^3$. Their integrals make sense, because the integrands belong to a linear space, so that the sum, even of infinitely many terms, as the integral implies, is well defined, and the result is in the same space as that of the addenda.

3.9 Discontinuity Surfaces

3.9.1 *Classification*

The assumption that the stress fields $x \longmapsto P(x)$ and $y \longmapsto \sigma(y)$ are continuous and continuously differentiable can be relaxed by accepting that they may experience bounded jumps across a finite number of surfaces or lines or points in the domain where they are defined ($\mathcal{B}$ or $\mathcal{B}_a$). In this case, we state that they are piecewise C^1 and write PC^1 for short.

The assumption is dictated by physical circumstances. A body composed of two different pieces of material glued across a surface is an example. If we think of corrosion of metallic bodies, one such discontinuity surface is the one distinguishing the corroded portion from the rest. These examples, two among many, suggest a first classification of discontinuity surfaces: *those exhibiting relative motion with respect to the rest of the body* (second example) and *those not exhibiting such motion* (first example).

Another distinction is between **coherent** and **incoherent** discontinuity surfaces (interfaces for short). In the first class we find interfaces across which the rest of the body can be just folded. When, in contrast, the portions of the body on one side and the other can slide one with respect to the other, the interface is called incoherent. Furthermore, we distinguish **structured** interfaces from the **unstructured** ones. The former class includes interfaces that can sustain surface stresses, thin layers modeled by surfaces endowed with their own energy. The absence of particular interface stress defines the latter class.

Below we shall consider only unstructured interfaces moving relatively to the rest of the body. As a further restriction, accepted just for the sake of simplicity, we shall consider interfaces that can be represented by means of C^2 functions. However, the extension of the result presented to faceted interfaces does not present special difficulties.

3.9.2 Geometry and Analysis

Let us consider in $\mathcal{B}$ a surface $\mathcal{C}$ defined by

$$\mathcal{C} := \{x \in \mathcal{B} \,|\, f(x) = 0\},$$

with $x \longmapsto f(x) \in \mathbb{R}$ a $C^2(\mathcal{B}, \mathbb{R})$ function. At $x \in \mathcal{C}$, the surface is oriented by the normal

$$\bar{m}^{\#} := \frac{\nabla f(x)}{|\nabla f(x)|}$$

considered as a vector or by

$$\bar{m} := \frac{Df(x)}{|Df(x)|},$$

taken as a covector.

By projecting ∇ or D over the surface $\mathcal{C}$, we obtain respectively the **surface gradient** and the **surface derivative** of a given differentiable field, denoted respectively by ∇_s and D_s. We have, in fact,

$$\nabla_s a := \nabla a(\bar{I} - \bar{m} \otimes \bar{m}^{\#}), \qquad D_s a := Da(\bar{I}^{\#} - \bar{m}^{\#} \otimes \bar{m}),$$

where $\bar{I}$ is here the 1-covariant, 1-contravariant identity tensor $\bar{I} = \delta^A_B e_A \otimes e^B$, and $\bar{I}^{\#}$ its dual counterpart 1-contravariant, 1-covariant.

The second-rank tensor

$$\mathsf{L} := -\nabla_s \bar{m}^{\#}$$

with contravariant components is called the **curvature tensor** of $\mathcal{C}$ at x. The negative of its trace, namely

$$\mathcal{K} := -\mathrm{tr}\,\mathsf{L},$$

is the **Gaussian curvature**.

Consider a field $x \longmapsto a(x)$ taking values in some linear space $\mathcal{L}$, namely $a(x) \in \mathcal{L}$, $x \in \mathcal{B}$, continuous everywhere but on a smooth surface $\mathcal{C}$ oriented by the normal $\bar{m}^{\#}$ at every point x in $\mathcal{C}$. For $\delta \in \mathbb{R}^{+}$, we denote by $a^{\pm}(x)$, x in $\mathcal{C}$, the limits

$$a^{\pm}(x) := \lim_{\delta \to 0+} a(x \pm \delta \bar{m}^{\#}),$$

if they exist. We define the **jump** of $x \longmapsto a(x)$ at $x \in \mathcal{C}$ to be the difference

$$[a](x) := a^{+}(x) - a^{-}(x),$$

and call

$$\langle a \rangle (x) := \frac{1}{2}\left(a^{+}(x) + a^{-}(x)\right)$$

the **average**.

If in $\mathcal{L}$, a product $(a_1(x), a_2(x)) \longmapsto a_1(x)a_2(x)$ is defined and is distributive with respect to the sum, for the limit values of a_1 and a_2 at $\mathcal{C}$ we can write

$$[a_1 a_2] = [a_1] \langle a_2 \rangle + \langle a_1 \rangle [a_2] .$$

The jump, as just defined, plays a role in Gauss's theorem. In fact, consider, for example, the stress field $y \longmapsto \sigma(y)$ on $\mathcal{B}_a$ and assume that it is of class C^1 over $\mathcal{B}_a$ except on a surface $\mathcal{C}_a$, where it has bounded jumps. In this case, Gauss's theorem prescribes that for every part $\mathfrak{b}_{aC}$ crossed completely by $\mathcal{C}_a$ and divided into two distinguished parts with nonzero volume, we have

$$\int_{\partial \mathfrak{b}_{aC}} \sigma n \, d\mathcal{H}^2 = \int_{\mathfrak{b}_{aC}} \operatorname{div} \sigma \, d\mu + \int_{\mathfrak{b}_{aC} \cap \mathcal{C}_a} [\sigma] \, m \, d\mathcal{H}^2, \tag{3.19}$$

where m is the normal orienting $\mathcal{C}_a$ point by point. Assume that $\mathcal{C}_a$ evolves within $\mathcal{B}_a$ with velocity field $\bar{v}$ relative to the velocity field v pertaining to $\mathcal{B}_a$ in some motion. A pertinent physical example is that of a solidification front. Other examples are the evolution of paraelectric–ferroelectric interfaces in ferroelectrics and ordered–isotropic interfaces in liquid crystals or polymeric classes such as nematic elastomers. Write $\bar{v}_m$ for the component of $\bar{v}$ along the normal m and consider a field $(y, t) \longmapsto \phi_a := \tilde{\phi}_a(y, t)$ taking values in a finite-dimensional vector space. Assume also that ϕ_a might have bounded discontinuities across $\mathcal{C}_a$ and is of class C^1 elsewhere at each t. In this case, a transport theorem states that

$$\begin{aligned} &\frac{d}{dt} \int_{\mathfrak{b}_{aC}} \phi_a d\mu \\ &= \int_{\mathfrak{b}_{aC}} \left(\frac{\partial \phi_a}{\partial t} + \operatorname{div}(\phi_a v) \right) d\mu + \int_{\mathfrak{b}_{aC} \cap \mathcal{C}_a} \left([\phi_a v] \cdot m - [\phi_a] \, \bar{v}_m \right) d\mathcal{H}^2, \end{aligned} \tag{3.20}$$

provided we give meaning to the products $\phi_a v$ and $[\phi_a v]\cdot m$, depending on the tensor nature of ϕ_a, a meaning that is obvious when ϕ_a is a scalar. Detailed proofs of the relations (3.19) and (3.20) can be found in several classical treatises of mathematical analysis.

We are here interested in two significant special cases. The first is the case in which ϕ_a coincides with the density of mass ρ_a. By applying the previous transport theorem to the integral balance

$$\frac{d}{dt}\int_{\mathfrak{b}_{aC}} \rho_a d\mu = 0 \tag{3.21}$$

and using the pointwise balance of mass (3.8), we obtain

$$\int_{\mathfrak{b}_{aC}\cap\mathcal{C}_a} \left([\rho_a v]\cdot m - [\rho_a]\,\bar{v}_m\right) d\mathcal{H}^2 = 0.$$

The arbitrariness of $\mathfrak{b}_{aC}$ implies the **pointwise balance of mass along a moving discontinuity surface** $\mathcal{C}_a$:

$$[\rho_a v]\cdot m - [\rho_a]\,\bar{v}_m = 0.$$

Notice that when $\bar{v}_m = 0$, i.e., $\mathcal{C}_a$ has no normal motion, and the previous balance reduces to

$$[\rho_a v]\cdot m = 0,$$

which implies the continuity of the normal component of the momentum across the fixed $\mathcal{C}_a$. Such a property is lost when $\mathcal{C}_a$ moves relatively to the rest of the body, with normal motion, as we are considering here. To evaluate the details—and this is the second case mentioned above—we have to choose ϕ_a coincident with the momentum $\rho_a v^\flat$. Hence by the transport theorem, we obtain

$$\frac{d}{dt}\int_{\mathfrak{b}_{aC}} \rho_a v^\flat d\mu = \int_{\mathfrak{b}_{aC}} \rho_a a\, d\mu - \int_{\mathfrak{b}_{aC}\cap\mathcal{C}_a} j[v^\flat] d\mathcal{H}^2,$$

where j is shorthand notation defined by

$$j := \langle\rho_a\rangle\,\bar{v}_m - \left\langle\rho_a v^\flat\right\rangle\cdot m^\#.$$

Hence, the balance of forces in Eulerian representation,

$$\frac{d}{dt}\int_{\mathfrak{b}_{aC}} \rho_a v^\flat d\mu = \int_{\mathfrak{b}_{aC}} b_a d\mu + \int_{\partial\mathfrak{b}_{aC}} \sigma n\, d\mathcal{H}^2,$$

becomes

$$\int_{\mathfrak{b}_{aC}} (b_a + \operatorname{div}\sigma - \rho_a a)\, d\mu + \int_{\mathfrak{b}_{aC}\cap C_a} ([\sigma]\, m + j[v^\flat])\, d\mathcal{H}^2 = 0,$$

thanks to the transport and divergence theorems.

The arbitrariness of $\mathfrak{b}_{aC}$ implies the pointwise balance of forces (3.9) at points not belonging to C_a, where, in addition, we obtain

$$[\sigma]\, m + j[v^\flat] = 0, \tag{3.22}$$

which is the **balance of forces along a moving discontinuity surface**. Notice that when $v = 0$, or alternatively, $j = 0$, such a balance reduces to the equation

$$[\sigma]\, m = 0,$$

which declares the continuity of the normal component of the stress.

The surface C_a has a counterpart in the reference configuration $\mathcal{B}$, because the map $x \longmapsto y$ is one-to-one. Let us write C for the preimage in $\mathcal{B}$ of C_a under the deformation $x \longmapsto y$. The surface balance of forces (3.22) has a referential (Lagrangian) counterpart in terms of the first Piola–Kirchhoff stress. To get it, we have first to consider a version of the transport theorem on $\mathcal{B}$, involving fields of type $(x, t) \longmapsto \phi := \tilde{\phi}(y, t)$, taking values on finite-dimensional linear spaces. If $\tilde{\phi}$ is a C^1 field over $\mathcal{B}$ with bounded discontinuity over C, we get

$$\frac{d}{dt}\int_{\mathfrak{b}_C} \phi\, d\mu = \int_{\mathfrak{b}_C} \dot{\phi}\, d\mu - \int_{\mathfrak{b}_C\cap C} [\phi]\bar{v}^r_m d\mathcal{H}^2, \tag{3.23}$$

where $\mathfrak{b}_C$ is the counterpart of $\mathfrak{b}_{aC}$ and is completely crossed by C, which divided it into two disjoint pieces with nonzero volumes. Here $\bar{v}^r_m$ is the referential normal velocity of C. In $\mathcal{B}$, the density of mass is assumed constant. Using Gauss's theorem and taking into account the transport theorem (3.23), the referential version of the balance of forces,

$$\frac{d}{dt}\int_{\mathfrak{b}_C} \rho \dot{y}^\flat d\mu = \int_{\mathfrak{b}_C} b\, d\mu + \int_{\partial\mathfrak{b}_C} P\bar{n}\, d\mathcal{H}^2,$$

becomes

$$\int_{\mathfrak{b}_C} (b + \operatorname{Div}P - \rho\ddot{y}^\flat)\, d\mu + \int_{\mathfrak{b}_C\cap C} ([P]\bar{m} + \rho[\dot{y}^\flat]\bar{v}^r_m)\, d\mathcal{H}^2 = 0.$$

The arbitrariness of $\mathfrak{b}_C$ then implies the pointwise balance of forces (3.16) at points far from C and the **surface balance**

$$[P]\bar{m} = -\rho[\dot{y}^\flat]\bar{v}^r_m$$

along C.

3.10 Exercises

In the following exercises, we identify $\mathbb{R}^3$ with its dual and write all components with the indices in covariant position. The choice is just a matter of convenience, motivated by the same reasons expressed in Chapter 1.

Exercise 5. *At a point* P *of the actual shape of a body, the components of the stress tensor with respect to an orthogonal frame are given by*

$$[\sigma_{ij}] = \begin{pmatrix} 1 & 0 & 3 \\ 0 & 2 & 0 \\ 3 & 0 & 1 \end{pmatrix}$$

(physical units are N/mm^2*). Calculate the principal stresses and the principal directions, the maximum and the minimum normal stresses, the maximum modulus of the shear stresses.*

Remarks and solution. The principal stresses σ_I, σ_{II}, and σ_{III} are the eigenvalues of the stress tensor, as defined in Section 3.4, and are the roots of the so-called *characteristic equation*, i.e., the third-rank polynomial equation

$$\sigma^3 - I_1\sigma^2 + I_2\sigma - I_3 = 0,$$

in which I_1, I_2, and I_3 are the invariants of σ. Here I_1 is the trace σ_{ii} of the stress tensor; I_2 is the sum of the determinants of the principal minors of σ; I_3 is the determinant of σ:

$$\begin{aligned} I_1 &= \sigma_{11} + \sigma_{22} + \sigma_{33}, \\ I_2 &= \sigma_{11}\sigma_{22} + \sigma_{11}\sigma_{33} + \sigma_{22}\sigma_{33} - \sigma_{12}^2 - \sigma_{13}^2 - \sigma_{23}^2, \\ I_3 &= \det(\sigma_{ij}). \end{aligned}$$

In the special case of the exercise, we have $I_1 = 4$, $I_2 = -4$, and $I_3 = -16$, and the characteristic equation is

$$\sigma^3 - 4\sigma^2 - 4\sigma + 16 = 0,$$

which can be rewritten as

$$(\sigma^2 - 2\sigma - 8)(\sigma - 2) = 0.$$

It admits three real solutions: $\sigma = -2$, $\sigma = 2$, and $\sigma = 4$, so that

$$\sigma_I = -2, \qquad \sigma_{II} = 2, \qquad \sigma_{III} = 4$$

are the principal stresses in N/mm^2. Using them, the invariants of σ can be rewritten as

$$I_1 = \sigma_I + \sigma_{II} + \sigma_{III},$$
$$I_2 = \sigma_I\sigma_{II} + \sigma_I\sigma_{III} + \sigma_{II}\sigma_{III},$$
$$I_3 = \sigma_I\sigma_{II}\sigma_{III}.$$

The eigenvectors associated with σ_I, σ_{II}, and σ_{III} respectively are called *principal directions* and can be determined by substituting for σ in the algebraic system

$$\begin{pmatrix} \sigma_{11}-\sigma & \sigma_{12} & \sigma_{13} \\ \sigma_{21} & \sigma_{22}-\sigma & \sigma_{23} \\ \sigma_{31} & \sigma_{32} & \sigma_{33}-\sigma \end{pmatrix} \begin{pmatrix} n_1 \\ n_2 \\ n_3 \end{pmatrix} = \begin{pmatrix} 0 \\ 0 \\ 0 \end{pmatrix}$$

the relevant principal stress. For example, taking $\sigma = \sigma_I = -2$, we have

$$\begin{pmatrix} 3 & 0 & 3 \\ 0 & 4 & 0 \\ 3 & 0 & 3 \end{pmatrix} \begin{pmatrix} n_1 \\ n_2 \\ n_3 \end{pmatrix} = \begin{pmatrix} 0 \\ 0 \\ 0 \end{pmatrix}$$

in the specific case of the exercise. Such an algebraic system admits as a solution the vector $(n_1, 0, -n_1)$. Similarly, we obtain for $\sigma = \sigma_{II} = 2$ the solution $(0, n_2, 0)$, and for $\sigma = \sigma_{III} = 4$, the solution $(n_1, 0, n_1)$. Normalizing to have unitary lengths, we obtain the three eigenvectors n_I, n_{II}, and n_{III}, namely

$$n_I = \left(\frac{1}{\sqrt{2}}, 0, -\frac{1}{\sqrt{2}}\right), \qquad n_{II} = (0, 1, 0), \qquad n_{III} = \left(\frac{1}{\sqrt{2}}, 0, \frac{1}{\sqrt{2}}\right).$$

We also realize that $n_{III} = n_I \times n_{II}$. We can call the stress state in the exercise *triaxial*, because all the principal stresses are nonzero. The minimum normal stress is $\sigma_I = -2$ (compressive stress), and the maximum is $\sigma_{III} = 4$ (tensile stress). Hence, in this case there are planes on which the normal stress is zero. Denoting by n the generic vector normal to one of these planes, we have $\sigma_{ij}n_in_j = 0$. The maximum modulus of the shear stresses around P is (in N/mm^2)

$$\frac{1}{2}\left|\sigma_{III} - \sigma_I\right| = 3.$$

Exercise 6. *At a point* P *of the actual shape of a body, the components of the stress tensor with respect to an orthogonal frame are given by*

$$[\sigma_{ij}] = \begin{pmatrix} 2 & -1 & 1 \\ -1 & 0 & -1 \\ 1 & -1 & 0 \end{pmatrix}$$

(physical units are N/mm^2). *Find the principal stresses, the normal stress component*

$$\sigma_n := \sigma \cdot (\hat{n} \otimes \hat{n}) \tag{3.24}$$

acting on the plane with normal $\hat{n} = \left(\frac{1}{\sqrt{3}}, \frac{1}{\sqrt{3}}, \frac{1}{\sqrt{3}}\right)$ *and the modulus of the shear stress*

$$\tau_n := \sigma\hat{n} - \sigma_n\hat{n} \tag{3.25}$$

on the same plane.

Remarks and solution. The invariants of the stress tensor are

$$I_1 = 2, \qquad I_2 = -3, \qquad I_3 = 0,$$

and the characteristic equation is

$$\sigma^3 - 2\sigma^2 - 3\sigma = 0.$$

Its roots are $\sigma_I = -1$, $\sigma_{II} = 0$, and $\sigma_{III} = 3$. The principal stresses are distinct, so that there are exactly three principal planes. The stress state is called *biaxial*, because one principal stress is zero. We have from the definition (3.24),

$$\sigma_n = \sigma_{ij}\hat{n}_i\hat{n}_j = 0,$$

and from the definition (3.25),

$$\tau_n = \sigma\hat{n} = \frac{2}{\sqrt{3}}e_1 - \frac{2}{\sqrt{3}}e_2.$$

Its modulus is equal to $\frac{2\sqrt{2}}{\sqrt{3}}$ N/mm^2.

Exercise 7. *At a point* P *of the actual shape of a body, the components of the stress tensor with respect to an orthogonal frame are given by*

$$[\sigma_{ij}] = \begin{pmatrix} 2\alpha & -1 & 0 \\ -1 & 2\alpha & 2 \\ 0 & 2 & 3 \end{pmatrix},$$

with α *a parameter. Decompose the stress into its spherical and deviatoric parts. If* $\sigma_m(\alpha)$ *and* $J_2(\alpha)$ *denote the average stress and the quadratic invariant of the deviatoric stress, calculate the value of* α *such that* $\sigma_m(\alpha) = J_2(\alpha)$, *if any.*

Remarks and solution. The stress tensor admits the decomposition

$$\sigma_{ij} = \delta_{ij}\sigma_m + s_{ij} \tag{3.26}$$

with $\delta_{ij}\sigma_m$ the *isotropic* or *spherical* component,

$$\sigma_m = \frac{1}{3}\sigma_{ii} = \frac{1}{3}I_1,$$

and s_{ij} the *deviatoric* component. The decomposition in unique. The principal directions of s_{ij} are the same as those of the total stress, while the eigenvalues of the tensor s are given by

$$s_I = \sigma_I - \sigma_m, \qquad s_{II} = \sigma_{II} - \sigma_m, \qquad s_{III} = \sigma_{III} - \sigma_m.$$

Then we obtain

$$\begin{aligned}
J_1 &= s_{11} + s_{22} + s_{33} = 0, \\
J_2 &= -\frac{1}{2}\left(s_{11}^2 + s_{22}^2 + s_{33}^2 + 2s_{12}^2 + 2s_{13}^2 + 2s_{23}^2\right), \\
J_3 &= \det(s_{ij}),
\end{aligned}$$

or, in terms of the principal stresses,

$$\begin{aligned}
J_1 &= s_I + s_{II} + s_{III} = 0, \\
J_2 &= -\frac{1}{2}\left(s_I^2 + s_{II}^2 + s_{III}^2\right), \\
J_3 &= \frac{1}{3}\left(s_I^3 + s_{II}^3 + s_{III}^3\right).
\end{aligned}$$

In the case under consideration, the isotropic stress is

$$\sigma_m = \frac{4}{3}\alpha + 1,$$

and the deviatoric component is given by

$$[s_{ij}] = \begin{pmatrix} \frac{2}{3}\alpha - 1 & -1 & 0 \\ -1 & \frac{2}{3}\alpha - 1 & 2 \\ 0 & 2 & -\frac{4}{3}\alpha + 2 \end{pmatrix}.$$

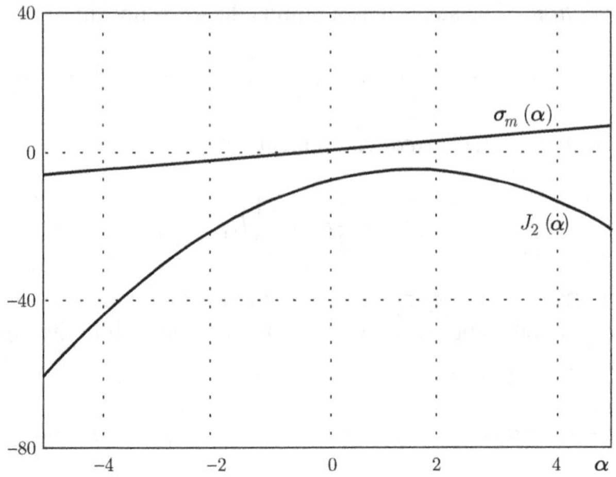

Fig. 3.2 Isotropic stress and quadratic invariant of s_{ij} versus the parameter α

The quadratic invariant $J_2 = -\frac{1}{2} s_{ij} s_{ij}$ is then

$$\frac{4}{3}\left(-\alpha^2 + 3\alpha - 6\right).$$

By representing the functions $\sigma_m(\alpha)$ and $J_2(\alpha)$ (Fig. 3.2), we see that the equation $\sigma_m(\alpha) = J_2(\alpha)$ has no real solutions, since

$$\frac{4}{3}\alpha + 1 = \frac{4}{3}\left(-\alpha^2 + 3\alpha - 6\right) \quad \Rightarrow \quad \alpha^2 - 2\alpha + \frac{27}{4} = 0,$$

which has discriminant equal to $-23 < 0$.

Exercise 8. *At a point* P *of the actual shape of a body, the components of the stress tensor with respect to an orthogonal frame are the same as in Exercise 5. Calculate the octahedral tangential stress, namely the modulus* τ_{oct} *of the tangential stress on the plane with normal* $n_d = \frac{1}{\sqrt{3}}(n_I + n_{II} + n_{III})$.

Remarks and solution. We have already calculated the principal stresses and the principal directions n_I, n_{II}, n_{III} of the stress tensor. Writing σ_{nd} for $\sigma \cdot (n_d \otimes n_d)$, we have by definition

$$\tau_{oct} = |\tau_d| = |\sigma n_d - \sigma_{nd} n_d|.$$

Consequently, we obtain

$$\tau_d = \frac{-2n_I + 2n_{II} + 4n_{III}}{\sqrt{3}} - \frac{4}{3}\frac{(n_I + n_{II} + n_{III})}{\sqrt{3}} = \frac{-10n_I + 2n_{II} + 8n_{III}}{3\sqrt{3}}$$

and

$$\tau_{oct} = |\tau_d| = \frac{2\sqrt{14}}{3} \approx 2.494\ \mathrm{N/mm^2}.$$

3.11 Further Exercises

Exercise 9. *Consider a body in a three-dimensional space. Fixing an orthogonal frame, a generic point y has coordinates* (y_1, y_2, y_3). *In this frame, the stress has components*

$$[\sigma_{ij}] = k_0 \begin{pmatrix} 20y_1 + y_2 & y_3 & k_1 y_3 \\ y_3 & 30y_1 + 200 & y_1 \\ k_1 y_3 & y_1 & 30y_2 + k_2 y_3 \end{pmatrix}.$$

In the absence of inertial effects, find the components of the bulk forces b *equilibrating the stress field and the values of* k_1 *and* k_2 *so that* $b = 0$ *for every* k_0.

Elements of the solution. $k_1 = -20, k_2 = 0$.

Exercise 10. *At a point* P *of the actual shape of a body, the components of the stress tensor with respect to an orthogonal frame are given by*

$$[\sigma_{ij}] = \begin{pmatrix} -5/6 & -1/3 & 13/6 \\ -1/3 & 5/3 & -1/3 \\ 13/6 & -1/3 & -5/6 \end{pmatrix}$$

(physical units are $\mathrm{N/mm^2}$*). Calculate the principal stresses of the deviatoric part of the stress.*

Solution. $s_I = -3, s_{II} = 1, s_{III} = 2$.

Exercise 11. *At a point* P *of the actual shape of a body, the components of the stress tensor with respect to an orthogonal frame are given by*

$$[\sigma_{ij}] = \begin{pmatrix} 0 & 1 & 1 \\ 1 & 1/2 & 1/2 \\ 1 & 1/2 & 1/2 \end{pmatrix}$$

(physical units are N/mm^2*). Find a plane* π_n *such that the moduli of normal and tangential stresses are* $\sigma_n = 3/2$, $\tau_n = 4/5$ *or* $\sigma_n = 3/2$, $\tau_n = 9/10$.

An element of the solution: In the first case, the plane does not exist. In the second case, the plane is available.

Chapter 4
Constitutive Structures: Basic Aspects

4.1 Motivations and Principles

Consider an orthogonal coordinate frame $\{O, x^1, x^2, x^3\}$ in the reference space $\mathcal{E}^3$ and write explicitly the components of the balance equations

$$\varrho \ddot{u}^\flat = b + \text{Div}P,$$

where we have considered that $\ddot{y}(x,t) = \ddot{u}(x,t)$, with $u := \tilde{u}(x,t)$ the displacement at the point x and the instant t. We obtain

$$\varrho \ddot{u}_1^\flat = b_1 + \frac{P_1^1}{\partial x^1} + \frac{P_1^2}{\partial x^2} + \frac{P_1^3}{\partial x^3},$$

$$\varrho \ddot{u}_2^\flat = b_2 + \frac{P_2^1}{\partial x^1} + \frac{P_2^2}{\partial x^2} + \frac{P_2^3}{\partial x^3},$$

$$\varrho \ddot{u}_3^\flat = b_3 + \frac{P_3^1}{\partial x^1} + \frac{P_3^2}{\partial x^2} + \frac{P_3^3}{\partial x^3},$$

i.e., a system of partial differential equations with three unknown components of the displacement, namely u_i, $i = 1, 2, 3$, and nine unknown stress components P_i^A, $i, A = 1, 2, 3$. The number of independent stress components is further reduced by the balance of couples, which implies the symmetry of PF^*. The number of stress components exceeds that of the balance equations. Such is also the case in the spatial representation of the balance equations.

To overcome the problem, it suffices to link in some way the stress components with the displacement. The relations establishing these links are what we call **constitutive structures**. They are **state functions** expressing the inner actions at a point in the matter—the traction in the scheme that we are analyzing here—with the variables describing the **state** of the material element at the place considered.

P.M. Mariano, L. Galano, *Fundamentals of the Mechanics of Solids*,
DOI 10.1007/978-1-4939-3133-0_4

The notion of *state* is a primitive concept in the continuum theories that we know. The specification of the list of state variables indicates the way in which we think that a material element—the organization of the matter at a geometric point—feels the rest of the body.

- The simplest way to describe the interactions of a material element at x with the rest of the body is to assume that the stress at x depends on the state of the matter in an *immediate* neighborhood of x, meaning that the state of material elements at finite distance from x in some shape of the body may be disregarded in calculating the stress at x. We generally call this point of view the **principle of local action**. It excludes long-range interactions in space.
- Another question is the behavior in time. For this, our experience of the macroscopic world suggests that we assume that the stress at the place x and the instant t is determined by the history of the state up to the time t. We commonly call such an assumption the **principle of determinism**.
- Also, it seems natural to assume that when a given observer perceives a body as made by a certain material class, any other observer *must* record the *same* type of material.

The last statement in the list above has an intrinsic vagueness in the word "same" and has to be specified with respect to changes in observers. To this end, consider, for example, two observers $\mathcal{O}$ and $\mathcal{O}'$ connected by a classical change. We have already assumed that the Cauchy stress σ is objective, meaning that σ', namely the tensor evaluated by $\mathcal{O}'$, is given by $Q\sigma Q^{\mathrm{T}}$, with Q the orthogonal tensor with positive determinant characterizing the isometry from $\mathcal{O}$ to $\mathcal{O}'$. Let us assume that σ depends only on the motion $\tilde{y}(\cdot,\cdot)$. The assumption that $\mathcal{O}$ and $\mathcal{O}'$ evaluate the same material class is here interpreted by stating that σ' depends then on $\tilde{y}'(\cdot,\cdot)$, the motion perceived by $\mathcal{O}'$. This point of view is what we commonly call the **principle of frame indifference**. An analogous interpretation can be given when instead of isometric changes in observers we consider the enlarged class determined by diffeomorphisms. In this case, the principle of frame indifference is called the **principle of covariance of the constitutive equations**.

In what follows, we shall accept the principles of determinism, locality, and frame indifference. In this way, we exclude cases in which the stress at a point could depend on some weighted average of the states of points in a neighborhood of the considered point that is finitely extended in space.

The axioms that we accept here are intuitive. Their role in the foundations of continuum mechanics was pointed out by Walter Noll in 1950s.

4.2 Examples of Material Classes

The essential step in defining a material class of bodies is to specify the nature of the **state** ς of a generic material element or the whole body. Independently of its specific expression, we say in general that ς is an element of a topological space, where for each pair of distinct states we can find nonintersecting neighborhood of them. In short, we say that the state space is Hausdorff. Moreover, when we consider the state of an entire body, the state space is infinite-dimensional, since ς is a collection of fields. For a single material element, the state space is infinite-dimensional when ς is a history (the past history, according to the axiom of determinism), i.e., ς is nonlocal in time, while it is finite-dimensional when ς depends only on the current time and ς is the value of a field mapping the point and instant considered into some finite-dimensional state space.

In what follows, within the setting of the axioms of determinism, locality, and frame indifference, we shall refer just to the state of a point and shall consider it *an element of a finite-dimensional space*. A list of special cases follows.

- The choice $\varsigma = F$ defines what we commonly call **simple materials**. In this case, since F is defined over the reference place, the natural stress measure we have to refer to is the first Piola–Kirchhoff stress P. In particular, we write

  $$P = \tilde{P}(x, F),$$

  reducing the expression to

  $$P = \tilde{P}(F)$$

  when the material does not change its type of response from place to place. In this case, the material is said to be *homogeneous*. Moreover, always within the range of simple materials, we could have

  $$P = \tilde{P}(x, t, F)$$

  if we are describing aging effects.
- Another choice could be $\varsigma = (F, \dot{F})$, and the first Piola–Kirchhoff stress would be

  $$P = \tilde{P}(x, F, \dot{F}).$$

  That choice leads to the simplest approach to the description of the viscous behavior. We have $P = \tilde{P}(F, \dot{F})$ for homogeneous viscous materials and $P = \tilde{P}(x, t, F, \dot{F})$ in the case we need to describe aging effects. In general, however, viscosity is more completely described by assuming the dependence of the stress on the entire history of F or portions of it that are finitely extended in time.
- In the previous examples, we have implicitly assumed an isothermal environment. When such a condition is not satisfied, the temperature enters the list of state variables. We must ask what the word *temperature* in fact refers to. We have a clear definition of temperature for a perfect gas in terms of the molecular kinetic energy. In the case of solids, that definition can be applied to the phonons, particles determined by the quantization of the traveling waves,

in crystals, for phonons can be considered a gas flowing through the atomic lattice. However, such a concept—it is neither exploited nor detailed here—is not completely pertinent to all possible changes in solids. A standard way to introduce temperature in continuum mechanics is to refer to it as a phenomenological quantity. A thermometric scale or a heating measure is a way of ordering states. In this way, an empirical temperature ϑ is commonly introduced, with the idea that it is specified by the laws it satisfies, specifically by the second law of thermodynamics and its consequences. Once we have accepted this view, for elastic and/or viscoelastic processes along which heat is absorbed or emitted, we write

$$P = \tilde{P}(x, F, \vartheta)$$

or

$$P = \tilde{P}(x, F, \dot{F}, \vartheta)$$

respectively for what we call thermoelastic or thermoviscoelastic materials. In writing such expressions, we are conscious that the temperature is a concept with a clear meaning at equilibrium, a concept making sense also in a number of nonequilibrium processes, with related limitations in the interpretation of its character.

The previous list does not exhaust all possible material classes but gives an idea of what we are talking about.

4.3 A Priori Constitutive Restrictions and the Mechanical Dissipation Inequality

Once we have decided on the list of the state variables, a problem is the assignment of the explicit constitutive structure. Experiments address possible choices. Experiments are, however, necessarily limited and developed in special conditions. We consider basic principles in determining a priori constitutive restrictions to the constitutive structures.

The first is the assumption that different observers must perceive the same material class. Formally, we express this assumption by imposing objectivity or covariance, as already mentioned.

Another principle limiting possible constitutive relations is the second law of thermodynamics. Any constitutive relation must be compatible with the second law of thermodynamics. If we accept such a statement, we have to render explicit the expression of the second law we are referring to. In the isothermal setting considered here, we commonly call such a law the **mechanical dissipation inequality** or the

isothermal version of the Clausius–Duhem inequality. For every part $\mathfrak{b}$ of $\mathcal{B}$, it reads

$$\frac{d}{dt}(\text{free energy of } \mathfrak{b}) - \mathcal{P}_{\mathfrak{b}}^{ext}(\dot{y}) \leq 0.$$

We assume that the inequality holds for every part $\mathfrak{b}$ of $\mathcal{B}$ and choice of the velocity fields. For the free energy, we accept the expression

$$\{\text{free energy of } \mathfrak{b}\} = \int_{\mathfrak{b}} \psi \, d\mu,$$

calling ψ the **free energy density**,[1] a function of the state variables that is assumed to be differentiable, so that for every part $\mathfrak{b}$ of $\mathcal{B}$, we write

$$\frac{d}{dt}\int_{\mathfrak{b}} \psi \, d\mu - \mathcal{P}_{\mathfrak{b}}^{ext}(\dot{y}) \leq 0,$$

and we assume that it is valid for every choice of $\mathfrak{b}$ and $\dot{y}$. The assumption accepted in the previous chapter that $\mathcal{P}_{\mathfrak{b}}^{ext}(\dot{y})$ is invariant under classical changes in observers and the validity of the regularity assumptions justifying the pointwise balances imply the identity $\mathcal{P}_{\mathfrak{b}}^{ext}(\dot{y}) = \mathcal{P}_{\mathfrak{b}}^{inn}(\dot{y})$. Then we can write

$$\frac{d}{dt}\int_{\mathfrak{b}} \psi \, d\mu - \int_{\mathfrak{b}} P \cdot \dot{F} d\mu \leq 0.$$

Since $\mathfrak{b}$ does not vary in time, since it belongs to $\mathcal{B}$, the time derivative and integral commute, and we obtain

$$\int_{\mathfrak{b}} \left(\dot{\psi} - P \cdot \dot{F}\right) d\mu \leq 0.$$

---

[1] The existence of the free energy can be determined in an abstract setting, without reference to a specific material class, by resorting to the notions of *state*, *process*, and *action*. We do not need to specify otherwise the state, except to declare that it is an element of a Hausdorff topological space. Paths in the state space represent state transformations. They are determined by processes, operators acting over the state space and representing ways in which the body "perceives" the external environment. Finally, actions are functionals assigning to every state transformation a number. Actions depend on the initial state of the path and the process. We commonly require that actions be continuous with respect to the states and additive with respect to the prolongations of paths by means of "subsequent" state transformations (the external power that we have defined previously is an action, for example). In this setting, the free energy is a concept associated with an action or a class of actions. It is a state function such that the difference of its values over an arbitrary pair of states is bounded by the infimum of the action considered, provided that the two states can be connected by a path. We do not go into details. We just mention these aspects for the sake of completeness and to open a window onto a wider landscape. For further details, the reader can refer to Miroslav Šilhavý's treatise referenced at the end of this book as a suggestion for further reading. Here, we just assume the possibility of defining the free energy as an integral over the volume of a differentiable density.

The arbitrariness of $\mathfrak{b}$ implies the **local form of the mechanical dissipation inequality**:

$$\dot{\psi} - P \cdot \dot{F} \leq 0. \tag{4.1}$$

To exploit it, we have to determine not only the list of state variables entering P but also those on which the free energy density depends. Below, we discuss again the examples from the previous section.

4.3.1 Simple Bodies

Let us assume

$$\psi = \tilde{\psi}(x, F)$$

and

$$P = \tilde{P}(x, F).$$

At fixed x we presume that both the free energy and the stress are defined on the whole set of all possible F's. An analogous assumption holds when we insert other state variables. Then the inequality (4.1) reduces to

$$\left(\frac{\partial \psi}{\partial F} - P\right) \cdot \dot{F} \leq 0.$$

The arbitrariness of $\dot{F}$, justified by that of $\dot{y}$ and the absence of constraints over $\dot{F}$, implies the a priori restriction

$$P = \frac{\partial \psi}{\partial F} \tag{4.2}$$

on the first Piola–Kirchhoff stress, which is transferred to the Cauchy stress by the standard relation

$$\sigma = (\det F)^{-1} PF^* = (\det F)^{-1} \frac{\partial \psi}{\partial F} F^*,$$

where, we recall,

$$\det F = \frac{\sqrt{\det \tilde{g}}}{\sqrt{\det g}} \det \left(\frac{\partial y^i}{\partial x^A}\right).$$

The Cauchy stress above is of type $\sigma = \sigma_i^j \tilde{e}^i \cdot \tilde{e}_j$.

With the assumptions made here, the free energy density appears as a potential for the whole stress. It is not always so, as is evident in the presence of viscous effects, where the previous property holds just for a part of the stress.

4.3.2 *Viscous Bodies: Nonconservative Stress Component*

We say that a body manifests **viscous** behavior when the actual value of the stress depends on portions in time of the strain history. Among possible dependencies, the simplest choice is to include in the list of state variables the strain rate. Let us tentatively assume

$$\psi = \tilde{\psi}(x, F, \dot{F})$$

and

$$P = \tilde{P}(x, F, \dot{F}),$$

as we have already mentioned. Substitution into (4.1) implies the inequality

$$\left(\frac{\partial \psi}{\partial F} - P\right) \cdot \dot{F} + \frac{\partial \psi}{\partial \dot{F}} \cdot \ddot{F} \leq 0, \tag{4.3}$$

provided that the map $(x, t) \longmapsto F := D\tilde{y}(x, t)$ is twice differentiable in time. Unless we do not impose constraints on $\dot{F}$ and $\ddot{F}$, we can choose arbitrarily both $\dot{F}$ and $\ddot{F}$, independently of each other. Since the inequality (4.3) is assumed to hold for every choice of the rates involved, the arbitrariness of $\ddot{F}$ implies

$$\frac{\partial \psi}{\partial \dot{F}} = 0.$$

Hence the free energy density cannot depend on $\dot{F}$ in this setting. The inequality (4.3) then does not imply (4.2), because the dependence of P on $\dot{F}$ would be incompatible with ψ, which is independent of the same variable. A way to solve the controversy is to assume that the first Piola–Kirchhoff stress admits the decomposition

$$P = \tilde{P}(x, F, \dot{F}) = \tilde{P}^c(x, F) + \tilde{P}^d(x, F, \dot{F}),$$

so that by defining $P^c := \tilde{P}^c(x, F)$ and $P^d := \tilde{P}^d(x, F, \dot{F})$, the inequality (4.3) reduces to

$$\left(\frac{\partial \psi}{\partial F} - P^c\right) \cdot \dot{F} - P^d \cdot \dot{F} \leq 0,$$

and the arbitrariness of $\dot{F}$ implies

$$P^c = \frac{\partial \psi}{\partial F},$$

which is analogous to (4.2), and

$$P^d \cdot \dot{F} \geq 0. \tag{4.4}$$

The last inequality verifies the dissipative nature of P^d, implicitly that of the viscous processes. Its validity for every choice of $\dot{F}$ implies

$$P^d = a(x, F, \dot{F})\dot{F} \tag{4.5}$$

with $a(\cdot)$ a positive definite scalar function. If so, we obtain

$$P^d \cdot \dot{F} = a(x, F, \dot{F})|\dot{F}|^2,$$

which is greater than or equal to zero and vanishes when $\dot{F} = 0$. Accordingly, the overall expression for the first Piola–Kirchhoff stress is

$$P = \frac{\partial \tilde{\psi}(x, F)}{\partial F} + a(x, F, \dot{F})\dot{F},$$

and the related Cauchy tensor is

$$\sigma = (\det F)^{-1} \frac{\partial \tilde{\psi}(x, F)}{\partial F} F^* + (\det F)^{-1} a(x, F, \dot{F})\dot{F}F^*.$$

A structure for P^d of the type

$$P^d = \mathbb{A}(x, F, \dot{F})\dot{F},$$

with $\mathbb{A}$ a second-rank positive definite tensor, would also be compatible with (4.4). However, although the second law permits such a choice, it could have further undesirable consequences, the analysis of which falls outside the scope of this book, in which we pay attention essentially to elastic behavior. We consider, in fact, just conservative processes. For them, the mechanical dissipation inequality reduces to an identity; the free energy coincides with the **internal energy**, which is also called **elastic energy**.

4.3.3 *Elastic Materials in an Isothermal Setting: Further Constitutive Restrictions*

Consider a body in an isothermal bath displaying conservative behavior under external action. Assume that the body is simple. By rendering formally explicit what is expressed in the last lines of the previous section, we write

$$\frac{d}{dt}\int_{\mathfrak{b}} \tilde{e}(x,F)d\mu - \mathcal{P}_{\mathfrak{b}}^{ext}(\dot{y}) = 0, \tag{4.6}$$

where, as anticipated above, the **internal energy density** $e = \tilde{e}(x,F)$ appears in place of the free energy. The two densities are related by the following standard expression:

$$\psi = e + \vartheta\eta,$$

where ϑ is the temperature, η the entropy. When the temperature and entropy are constant, the free energy reduces to the internal energy, modulo a constant: the mechanical dissipation inequality becomes (4.6).

The identity (4.6) excludes the possibility of a dissipative component of the stress, as can be proved by following the same steps of the previous section. The structural constitutive choice for P is then the same as the internal energy density—what we commonly call the **elastic energy density** in this setting—so that the first Piola–Kirchhoff stress derives from a potential, and (4.3) reduces to

$$P = \tilde{P}(x,F) = \frac{\partial e}{\partial F}(x,F). \tag{4.7}$$

A problem is then the explicit assignment of $\tilde{e}(x,F)$. Help in reducing the arbitrariness of the choice is given by the requirement of objectivity of the elastic energy density, which pertains to the inner structure of the material. Classical changes in observers alter by isometries the *entire* ambient space in which we describe the current (macroscopic) configurations of the body. Such changes do not alter the material structure. It is then natural to require objectivity of the elastic energy, with an essential consequence.

Definition 1. Let $\mathcal{L}$ be a linear space. A function

$$f : \mathcal{L} \longrightarrow \mathbb{R}$$

is said to be **convex** if for every $\xi \in [0,1]$ and $A, B \in \mathcal{L}$, the inequality

$$f(\xi A + (1-\xi)B) \leq \xi f(A) + (1-\xi) f(B)$$

holds.

Lemma 2. *Let $f : \mathcal{L} \longrightarrow \mathbb{R}$ be differentiable. Then f is convex over $\mathcal{L}$ if and only if*

$$f(A) - f(B) \geq \frac{\partial f}{\partial B}(B) \cdot (A - B), \tag{4.8}$$

for every A and B in $\mathcal{L}$.

The proof proceeds in two steps: (1) We assume that f is convex, and we deduce the inequality (4.8). (2) Then we take a generic differentiable function f satisfying the inequality (4.8) and show that as a consequence, it is convex.

Proof. If f is convex, from the definition we get

$$f(B + \xi(A - B)) \leq f(B) + \xi\,(f(A) - f(B)),$$

which is

$$\frac{f(B + \xi(A - B)) - f(B)}{\xi} \leq f(A) - f(B). \tag{4.9}$$

We then find the limit

$$\frac{f(B + \xi(A - B)) - f(B)}{\xi} \longrightarrow \left.\frac{d}{d\xi} f(B + \xi(A - B))\right|_{\xi=0} = \frac{\partial f}{\partial B}(B) \cdot (A - B)$$

as $\xi \longrightarrow 0$. By inserting the result in the inequality (4.9), we get (4.8). To prove the converse, let us use Z to indicate the sum $\xi A + (1 - \xi)B$. If f is differentiable and satisfies the inequality (4.8), we get

$$f(A) - f(Z) \geq \frac{\partial f}{\partial Z}(Z) \cdot (A - Z) = (1 - \xi)\frac{\partial f}{\partial Z}(Z) \cdot (A - B)$$

and

$$f(B) - f(Z) \geq \frac{\partial f}{\partial Z}(Z) \cdot (B - Z) = -\xi\frac{\partial f}{\partial Z}(Z) \cdot (A - B).$$

By multiplying the first inequality above by ξ and the second by $(1 - \xi)$, and summing the results, we get

$$\xi f(A) + (1 - \xi) f(B) \geq f(Z) = f(\xi A + (1 - \xi)B),$$

i.e., f is convex. □

Theorem 3. *In the finite-strain regime, the objectivity of the elastic energy density of simple bodies, $\tilde{e}(x, F)$, and its possible convexity with respect to F are physically incompatible.*

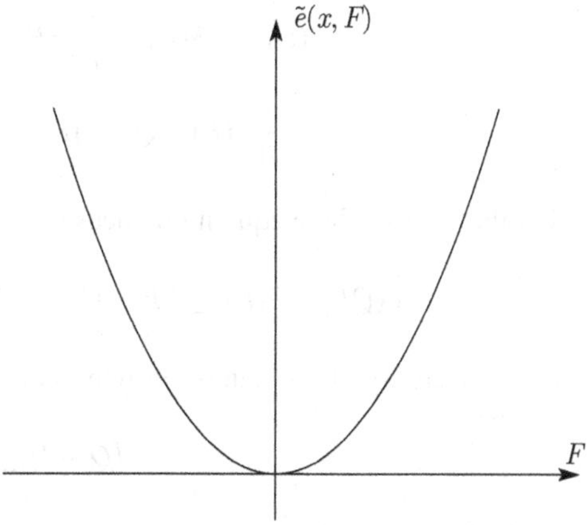

Fig. 4.1 If $\tilde{e}(x, F)$ were quadratic with respect to F, it would have a unique minimum

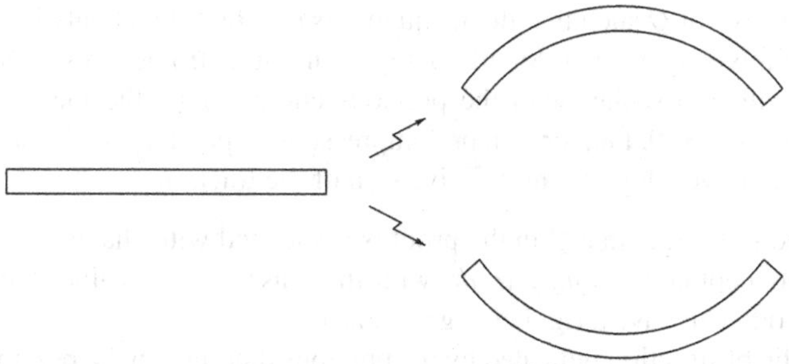

Fig. 4.2 Section of an elastic plate reaching two possible equilibrium configurations under the same boundary conditions

The theorem excludes the choice of $\tilde{e}(x, F)$ as a quadratic function of F. In other words, thanks to the a priori restriction (4.7), in the finite-strain regime, the first Piola–Kirchhoff stress P cannot depend linearly on the deformation gradient F. Informal evidence of the result emerges, however, when we accept that an equilibrium configuration of an elastic body satisfies the minimum of the elastic energy. Hence, if $\tilde{e}(x, F)$ were quadratic (i.e., strictly convex) with respect to F, there would be a unique minimum (see Figure 4.1). Consider now an elastic plate undergoing bending as a consequence of some boundary conditions expressed in terms of displacements. A deformed configuration and its reverse counterpart appear in Figure 4.2. Both shapes are at equilibrium under the same boundary conditions, so that we do not have uniqueness. A rigorous proof of the theorem is, however, necessary.

Proof. Write $\tilde{e}(F)$ instead of $\tilde{e}(x, F)$ for the sake of brevity, because the explicit dependence on x does not play a role in the proof. If $\tilde{e}(\cdot)$ were a convex function of F, since it is differentiable by assumption, the previous lemma would allow us to write

$$\tilde{e}(QF) - \tilde{e}(F) \geq \frac{\partial e}{\partial F}(F) \cdot (QF - F)$$
$$= \frac{\partial e}{\partial F}(F) \cdot (Q - I)F = \frac{\partial e}{\partial F}F^* \cdot (Q - I).$$

In other words, the inequality reduces to

$$\tilde{e}(QF) - \tilde{e}(F) \geq PF^* \cdot (Q - I) = (\det F)\sigma \cdot (Q - I).$$

The objectivity of $\tilde{e}(\cdot)$ can be written as $\tilde{e}(QF) = \tilde{e}(F)$, so that

$$\sigma \cdot (Q - I) \leq 0,$$

since $\det F > 0$. The inequality must hold for every choice of Q. However, we could find some choices of Q such that the inequality is satisfied if and only if σ admits at least one negative eigenvalue. Should σ represent an isotropic stress state, namely $\sigma = -pI$, with p a scalar, with the previous choice of Q, the inequality would exclude the state $p \leq 0$, i.e., tension or compression, depending on the choice of the convenction that we adopt on the positive sign of the traction. □

We should not forget that Q in the proof is associated with changes in observers, and it has conceptually *nothing to do* with the class of admissible deformations, although a body can have a rigid change of place.

Beyond the obstruction indicated by the previous theorem on the possible choices of the elastic energy density, objectivity implies further restrictions. To satisfy the objectivity condition

$$\tilde{e}(QF) = \tilde{e}(F),$$

if and only if it depends on the symmetric tensor U appearing in the left polar decomposition or, equivalently, on the square power of U, i.e., $\tilde{C} = F^{\mathrm{T}}F$, which is not affected by rotations in the ambient space, being defined on the reference place. By the left polar decomposition, we have

$$\tilde{C} = F^{\mathrm{T}}F = (RU)^{\mathrm{T}}(RU) = U^{\mathrm{T}}R^{\mathrm{T}}RU = U^{\mathrm{T}}U.$$

Then we write

$$\tilde{e}(x, F) = \check{e}(x, \tilde{C}).$$

However, since $\tilde{C} = g^{-1}C$, with $C = F^*\tilde{g}F$, and in the elastic setting that we treat here, g does not change, we can also write

$$\tilde{e}(x, F) = \hat{e}(x, C),$$

or more explicitly,

$$\tilde{e}(x, F) = \hat{e}(x, C(F, \tilde{g})).$$

Write also

$$\tilde{e}(x, F) = \rho\bar{e}(x, C(F, \tilde{g}))$$

to express the density of mass, and consider once again the spatial metric $\tilde{g}$ independent of time. We assume here also that the mass is pointwise conserved.

Theorem 4. *For an elastic material undergoing large strains, the following relations hold:*

$$P = 2\rho\tilde{g}F\frac{\partial\bar{e}}{\partial C}, \qquad \sigma = 2\rho_a\tilde{g}\frac{\partial\bar{e}}{\partial\tilde{g}}.$$

Proof. First we notice that

$$\begin{aligned} P \cdot \dot{F} &= FF^{-1}\tilde{g}^{-1}P \cdot \dot{F} = F^{-1}\tilde{g}^{-1}P \cdot F^*\tilde{g}\dot{F} \\ &= F^{-1}\tilde{g}^{-1}P \cdot \left(\mathrm{Sym}(F^*\tilde{g}\dot{F}) + \mathrm{Skw}(F^*\tilde{g}\dot{F})\right). \end{aligned}$$

Moreover, since $\tilde{g}$ is independent of time, we compute

$$\dot{C} = F^*\tilde{g}\dot{F} + \dot{F}^*\tilde{g}F = F^*\tilde{g}\dot{F} + (F^*\tilde{g}\dot{F})^{\mathrm{T}}.$$

Hence, by defining the second-rank tensor S by

$$S := F^{-1}\tilde{g}^{-1}P,$$

a tensor commonly called the **second Piola–Kirchhoff stress tensor**, we can write[2]

$$P \cdot \dot{F} = \frac{1}{2}S \cdot \dot{C},$$

because S is symmetric (prove the property as an exercise), so that $S \cdot \mathrm{Skw}(F^*\tilde{g}\dot{F}) = 0$. Also, we consider S a function of x and C, as we are forced to do once the objectivity restricts the dependence of e on F to that on C, suggesting the same choice for P, which is fully connected with the energy in the way we have already shown.

Consequently, from equation (4.6), thanks to the arbitrariness of $\mathfrak{b}$, we get

$$\dot{e} = P \cdot \dot{F} = \frac{1}{2}S \cdot \dot{C},$$

[2]Notice that the choice $P = \tilde{P}(x, C)$ corresponds to $S = \tilde{S}(x, C)$.

i.e.,

$$\left(\rho\frac{\partial \bar{e}}{\partial C}(x, C) - \frac{1}{2}\tilde{S}(x, C)\right)\cdot \dot{C} = 0,$$

for every choice of $\dot{C}$, which is tantamount to stating that

$$\tilde{S}(x, C) = 2\rho\frac{\partial \bar{e}}{\partial C}(x, C).$$

From the definition, however,

$$S = F^{-1}\tilde{g}^{-1}P = 2\rho\frac{\partial \bar{e}}{\partial C},$$

and the first relation of the theorem is proved. For the second relation, we first notice that

$$\frac{\partial \bar{e}\,(x, C(F, \tilde{g}))}{\partial \tilde{g}_{ij}} = \frac{\partial \bar{e}}{\partial C_{AB}}\frac{\partial C_{BA}}{\partial \tilde{g}_{ij}} = \frac{\partial \bar{e}}{\partial C_{AB}}F^i_{\ B}F^j_{\ A},$$

that is

$$\frac{\partial \bar{e}}{\partial \tilde{g}} = F\frac{\partial \bar{e}}{\partial C}F^*. \tag{4.10}$$

Then, using the first relation in the theorem, we compute

$$\sigma = \frac{1}{\det F}PF^* = \frac{2\rho}{\det F}\tilde{g}F\frac{\partial \bar{e}}{\partial C}F^*,$$

and using the pointwise balance of mass $\rho_a \det F = \rho$ and the identity (4.10), we get finally the second relation in the theorem, commonly called the **Doyle–Eriksen formula**. □

We make the following comments on the previous results.

- When we require objectivity of the elastic energy, we are requiring that we pay internal energy (inside the material) only when we change the shape of the body by crowding and/or shearing the material elements. The energy density e depends, in fact, just on C or, by extension, on any strain measure in the isothermal setting that we are considering.
- The stress σ can be viewed as a consequence of the change of energy due to the variation of the metric in space: the geometric meaning of the Doyle–Eriksen formula. The result is not in contrast with the standard view that the stress is a consequence of the elongation or the shortening of material bonds. Consider, in fact, a segment in $\mathcal{B}_a$. By fixing $\tilde{g}$, we can elongate the segment and evaluate the

superposed strain by computing the new length of the segment with respect to $\tilde{g}$. Stress emerges in the process. Conversely, we can evaluate the initial length of the segment; we change $\tilde{g}$ and compute the length of the same segment with respect to the varied metric. If we have changed $\tilde{g}$ in a way producing the length of the segment previously elongated, we have simulated the same process, so we must have stress once again.

4.3.4 *The Small-Strain Regime*

In the small-strain regime, since $\sigma \approx P$, the mechanical dissipation inequality can be written on $\mathcal{B}$ as

$$\frac{d}{dt}\int_{\mathcal{B}} \psi \, d\mu - \int_{\mathcal{B}} \sigma \cdot \dot{\varepsilon} \, d\mu \leq 0, \tag{4.11}$$

and we assume that it holds for every choice of $\dot{\varepsilon}$. For inhomogeneous simple bodies in an isothermal setting, we assume $\psi = \tilde{\psi}(x, \varepsilon)$ and $\sigma = \tilde{\sigma}(x, \varepsilon)$. We write $\psi = \tilde{\psi}(\varepsilon)$ and $\sigma = \tilde{\sigma}(\varepsilon)$ in the homogeneous case.

The arbitrariness of $\dot{\varepsilon}$ implies that of the part of $\mathcal{B}$, because we can select arbitrary fields $x \longmapsto \dot{\varepsilon}(x)$ compactly supported over $\mathcal{B}$, the local form

$$\left(\frac{\partial \psi}{\partial \varepsilon} - \sigma\right) \cdot \dot{\varepsilon} \leq 0,$$

i.e.,

$$\sigma = \frac{\partial \psi}{\partial \varepsilon}, \tag{4.12}$$

or

$$\sigma = \frac{\partial e}{\partial \varepsilon},$$

once we substitute the free energy with the elastic one, a substitution permitted in the elastic isothermal setting. Such a substitution is inappropriate when viscous effects occur, because the entropy varies as a consequence of the viscous dissipation. For viscous inhomogeneous materials in the same isothermal conditions, we write $\psi = \tilde{\psi}(x, \varepsilon, \dot{\varepsilon})$ and $\sigma = \tilde{\sigma}(x, \varepsilon, \dot{\varepsilon})$. The additional assumption

$$\sigma = \sigma^e + \sigma^d,$$

with $\sigma^e = \tilde{\sigma}^e(x, \varepsilon)$ and $\sigma^d = \tilde{\sigma}^d(x, \varepsilon, \dot{\varepsilon})$, is also necessary, because the mechanical dissipation inequality requires that ψ be independent of $\dot{\varepsilon}$, analogously to what we have discussed in the finite-strain regime. In this case, the relation (4.12) holds only for the energy-dependent part σ^e of the stress, namely

$$\sigma^e = \frac{\partial \psi}{\partial \varepsilon},$$

while the dissipative part satisfies the inequality

$$\sigma^d \cdot \dot{\varepsilon} \geq 0, \tag{4.13}$$

which implies, for example,

$$\sigma^d = a(x, \varepsilon, \dot{\varepsilon})\dot{\varepsilon}, \tag{4.14}$$

with $a(\cdot)$ a positive definite scalar function. Once again, the relation (4.14) is not the sole solution to the dissipation inequality (4.13). However, we prefer to refer just to it, avoiding here a discussion about the ancillary consequences of solutions of the type

$$\sigma^d = \mathbb{A}(x, \varepsilon, \dot{\varepsilon})\dot{\varepsilon},$$

with $\mathbb{A}$ a positive definite fourth-rank tensor, which is symmetric in the first two and the last two components, namely $\mathbb{A}_{ijhk} = \mathbb{A}_{jihk}$ and $\mathbb{A}_{ijhk} = \mathbb{A}_{ijkh}$.

4.3.5 Linear Elastic Constitutive Relations

Let us assume that the condition $|Du| \ll 1$ applies, ensuring the small-strain regime. The argument pictured in Figure 4.2 to visualize the physical incompatibility evidenced in Theorem 3 no longer holds, since in the small-strain regime, we do not distinguish between reference and actual shapes: they can be (approximately) considered superposed, modulo the identification of the reference space with the ambient one. The elastic energy is a function of the small-strain tensor ε and can be chosen to be quadratic:

$$\tilde{e}(x, \varepsilon) := \frac{1}{2}(\mathbb{C}\varepsilon) \cdot \varepsilon + \sigma_0 \cdot \varepsilon,$$

where $\mathbb{C}$ is a fourth-rank tensor depending possibly on x. When $\mathbb{C}$ is independent of x, we say that the material is **homogeneous**. The relation (4.12) implies the linear constitutive equation

$$\sigma = \mathbb{C}\varepsilon + \sigma_0,$$

where σ_0 is the so-called **prestress**, which is a possible initial stress state, like the one appearing in extruded metallic artifacts.

We do not consider here the prestress σ_0, so that the elastic energy density $\tilde{e}(x, \varepsilon)$ has the form

$$\tilde{e}(x, \varepsilon) = \frac{1}{2}(\mathbb{C}\varepsilon) \cdot \varepsilon, \tag{4.15}$$

and the constitutive restriction (4.12) implies simply

$$\sigma = \mathbb{C}\varepsilon.$$

Without distinguishing between covariant and contravariant components,[3] we can write

$$\sigma_{ij} = \mathbb{C}_{ijhk}\varepsilon_{hk}.$$

The symmetry of σ requires the symmetry of $\mathbb{C}$ with respect to its first two indices, i.e.,

$$\mathbb{C}_{ijhk} = \mathbb{C}_{jihk}.$$

Moreover, since

$$\mathbb{C}_{ijhk} = \frac{\partial^2 e}{\partial \varepsilon_{ij} \partial \varepsilon_{hk}},$$

by Schwartz's theorem, $\mathbb{C}$ is endowed with major symmetries, namely

$$\mathbb{C}_{ijhk} = \mathbb{C}_{hkij}.$$

Symmetry with respect to the last two indices then follows, namely

$$\mathbb{C}_{ijhk} = \mathbb{C}_{ijkh}.$$

The physical assumption that the energy density is non negative implies the need of considering $\mathbb{C}$ a positive definite tensor:

$$(\mathbb{C}A) \cdot A \geq 0 \tag{4.16}$$

for every second-rank tensor A. The equality sign holds only when A vanishes. In this case, we say that $\mathbb{C}$ is **pointwise stable**. A special choice of A is $A = a \otimes c$, with a and c in $\mathbb{R}^3$, and the condition (4.16) becomes

[3] We exploit the identification of $\mathbb{R}^3$ with its dual.

$$\mathbb{C}_{ijhk}a^k c^h a^j c^i \geq 0.$$

In particular, for a and c different from the null vector, we have

$$\mathbb{C}_{ijhk}a^k c^h a^j c^i > 0, \tag{4.17}$$

a condition called **strong positive ellipticity**. *Pointwise stability implies strong positive ellipticity, but the converse does not hold.*

4.4 Material Isomorphisms and Symmetries

Implicit in the discussion so far about the constitutive relations is the assignment of a reference shape for the body. A question is whether and in which sense constitutive relations are sensitive to the choice of the reference place. Preliminary notions appear necessary to answer this, at least in the specific case of simple bodies considered above and in what follows.

4.4.1 *Objectivity of the Cauchy Stress*

In deriving above the Doyle–Ericksen formula, we have obtained the relation

$$\sigma = 2\rho_a \tilde{g} F \frac{\partial \bar{e}(x, C(F, \tilde{g}))}{\partial C} F^*. \tag{4.18}$$

We have considered σ as the value at (y, t) of a symmetric-tensor-valued function. This point of view still remains, but now, the relation (4.18) allows us to consider σ as a value at a certain instant t of a function depending on x, namely, fixing $\tilde{g}$,

$$\sigma = \tilde{\sigma}(F(x, t), x). \tag{4.19}$$

Write now $\bar{\sigma}$ for $\tilde{g}^{-1}\sigma$. The second-rank contravariant tensor $\bar{\sigma}$ is then

$$\bar{\sigma} = \bar{\sigma}^{ij}\tilde{e}_i \otimes \tilde{e}_j,$$

so that by (4.18),

$$\bar{\sigma} = 2\rho_a F \frac{\partial \bar{e}}{\partial C} F^*. \tag{4.20}$$

The value $\bar{\sigma}'$ measured after a classical change in observer in the ambient space is then

$$\bar{\sigma}' = 2\rho_a QF \frac{\partial \bar{e}}{\partial C} F^* Q^* = Q\sigma Q^*,$$

since F becomes QF.

Let us now write $\hat{\sigma}$ for the second-rank tensor $\bar{\sigma}\tilde{g}$. The tensor $\hat{\sigma}$ is obtained from $\bar{\sigma}$ just by lowering by $\tilde{g}$ the second index of $\bar{\sigma}$, namely

$$\hat{\sigma} = \hat{\sigma}^i_j \tilde{e}_i \otimes \tilde{e}^j.$$

For the value $\bar{\sigma}'$ evaluated after a classical change in observer, from the identity (4.20) we obtain

$$\hat{\sigma}' = Q\hat{\sigma}Q^{\mathrm{T}}.$$

All developments in this section are based on the assumption that $Q \in SO(3)$ is of the form

$$Q = Q^i_j \tilde{e}_i \otimes \tilde{e}^j.$$

The choice is natural because F is 1-contravariant and 1-covariant, so that the product QF appearing after a classical change in observer can be written as

$$QF = Q^i_j F^j_A \tilde{e}_i \otimes \tilde{e}^A.$$

If we consider in addition even $Q \in SO(3)$ of the form

$$Q = Q^j_i \tilde{e}^i \otimes \tilde{e}_j,$$

we can write

$$\sigma' = Q\sigma Q^{\mathrm{T}} \tag{4.21}$$

for the stress tensor σ evaluated after a classical (isometry-based) change in observer, a value indicated by σ'.

This relation follows directly from (4.18). To get it, we have first to recall that under classical changes in observers in the ambient space, we have

$$\tilde{g}_{ij} \longmapsto Q^h_i \tilde{g}_{hk} Q^k_j,$$

since $\tilde{g}$ is a fully covariant second-rank (symmetric) tensor. By insertion in (4.18), we get

$$\sigma_i'^j = 2\rho_a Q_i^h \tilde{g}_{hk} Q_s^k Q_r^s F_A^r \frac{\partial \bar{e}}{\partial C_{AB}} F_B^{*l} Q_l^j$$
$$= 2\rho_a Q_i^h \tilde{g}_{hk} \delta_r^k F_A^r \frac{\partial \bar{e}}{\partial C_{AB}} F_B^{*l} Q_l^j$$
$$= 2\rho_a Q_i^h \tilde{g}_{hk} F_A^k \frac{\partial \bar{e}}{\partial C_{AB}} F_B^{*l} Q_l^j = Q_i^h \sigma_h^k Q_k^j,$$

which is exactly the relation (4.21) in components. It declares that under the validity of (4.18), the stress tensor σ is objective. We generally assume that the relation (4.21) holds beyond the validity of the relation (4.18). In other words, *we assume that σ is objective independently of the material class it is referred to.* We can call the assumption **the principle of frame indifference of the stress**.

4.4.2 *Material Isomorphisms*

Let us restrict once again our attention to the class of simple materials. The constitutive structure indicated by (4.19) is referred to a reference place $\mathcal{B}$, as we have already mentioned. To put in evidence the dependence on the reference configuration, we could write explicitly $\mathcal{B}$ as a subscript of σ, namely

$$\sigma = \tilde{\sigma}_{\mathcal{B}}(F(x,t), x).$$

The previous relation states that the reference place $\mathcal{B}$ is occupied by a body consisting of a simple material, and nothing more. Consider two distinct points x_1 and x_2 in $\mathcal{B}$. They are occupied by two different material elements. A question is how we can give formal expression to the statement "the two elements at x_1 and x_2 are made of the *same* simple material" by exploiting only the concepts discussed so far. A viewpoint on this problem was suggested by Walter Noll in 1958. We say that the two elements are made of the same type of material elements when (1) $\tilde{\rho}(x_1) = \tilde{\rho}(x_2)$ and (2) under the same strain, they undergo in time the same stress. In other words, the two material elements at x_1 and x_2 have the same mass and are indistinguishable by *local* measures of stress under the same strain. In short, we say that the two elements are materially isomorphic.

The same question can be asked about two points x_1 and x_2 pertaining to two distinct reference configurations $\mathcal{B}_1$ and $\mathcal{B}_2$ of different simple bodies. To allow the comparison, we assume the possibility of establishing between at least a pair of neighborhoods $\mathcal{I}(x_1)$ and $\mathcal{I}(x_2)$ of x_1 and x_2 a one-to-one differentiable mapping $f : \mathcal{I}(x_1) \longrightarrow \mathcal{I}(x_2)$, with differentiable inverse, a diffeomorphism. Let us write H for $Df^{-1}(x)$.

Definition 5. We say that there is a **material isomorphism** between x_1 and x_2 when at every instant,

1. $\tilde{\rho}(x_1) = \tilde{\rho}(x_2)$,
2. $\tilde{\sigma}(F) = \tilde{\sigma}(FH)$, where we leave understood the dependence on x.

Condition (1) in the previous definition implies that $\det H$ is *unitary*, thanks to the balance of mass. We summarize formally the previous statement by writing $H \in \mathrm{Unim}^+(\mathbb{R}^3, \mathbb{R}^3)$ and reduce further the notation to Unim to indicate the unimodular group of linear operators from $\mathbb{R}^3$ into $\mathbb{R}^3$. In particular, we need Unim^+, i.e., the subgroup of second-rank tensors with determinant equal to 1. The restriction is necessary because if we assume $\det F > 0$, we need also $\det FH > 0$. When H is the identity, the second condition in the previous definition is trivially satisfied. However, there could be other linear operators H satisfying that condition. Write H and K for two of them. We should have then

$$\tilde{\sigma}(F) = \tilde{\sigma}(FH)$$

and

$$\tilde{\sigma}(F) = \tilde{\sigma}(FK),$$

which implies

$$\tilde{\sigma}(F) = \tilde{\sigma}(FHK).$$

As a special case, the map f can be selected so as to map at $x \in \mathcal{B}$ the neighborhood $\mathcal{I}(x)$ onto itself. With this choice, if H satisfies condition (2), then H^{-1} does also. In this case, the set of H satisfying condition (2) contains the identity and is closed under the composition $(H, K) \longmapsto HK \in \mathrm{Unim}$ and the inverse operation. Such a set is then a group, and we call it the **symmetry group** *of the material occupying the reference place $\mathcal{B}$ at a point x*. Let us denote it by $\mathcal{G}_{\mathcal{B}}(x)$. It is a subgroup of Unim^+. The special orthogonal group $SO(3)$ is a maximal subgroup of Unim^+. When $\mathcal{G}_{\mathcal{B}}(x)$ is a proper subgroup of Unim^+ and contains $SO(3)$, then it coincides with $SO(3)$ itself. In Noll's view, the nature of the symmetry group discriminate between simple solids and simple fluids.

Definition 6. Along orientation-preserving motions we call a material element at x *solid* when at every instant, its symmetry group $\mathcal{G}_{\mathcal{B}}(x)$ is included in $SO(3)$ or coincides with it, while we call it *fluid* when the symmetry group is the whole of Unim^{+}.[4]

Imagine that a body is solid at x. Choose $H = \bar{Q}^{\mathrm{T}} \in SO(3)$. Condition (2) in the definition of material isomorphism can then be written as

$$\tilde{\sigma}(F) = \tilde{\sigma}(F\bar{Q}^{\mathrm{T}}),$$

with $\bar{Q}$ of the type $\bar{Q}^{A}_{B} e_A \otimes e^B$, i.e., a linear operator mapping vectors over the reference space onto covectors of the same space. When such a $\bar{Q}$ acts on the right as in the previous formula involving the stress, it maps covectors of the reference space onto covectors of the same space. We use the transpose of $\bar{Q}$ when it is applied on the right to allow the action of the same component (although now in contravariant position) operating when it acts on the left.

4.5 Symmetry Group and Changes of the Reference Place

Consider another reference place $\mathcal{B}'$, obtained from $\mathcal{B}$ by means of a one-to-one differentiable mapping $h : \mathcal{B} \longrightarrow \mathcal{B}'$ with derivative having nonzero determinant. Write G for $Dh(x)$ (recall that we take $\det G \neq 0$, so that G^{-1} exists). Write also x' for $h(x)$. Assume that it is possible to reach the same actual configuration $\mathcal{B}_a$ from $\mathcal{B}'$ and $\mathcal{B}$ using different deformations, and write F' and F for the relevant deformation gradients evaluated at x' and x, respectively. Figure 4.3 explains the situation.

Since the deformed configurations are the same, if we are handling a body made of a simple material, we should have

$$\tilde{\sigma}(F') = \tilde{\sigma}(FG^{-1}).$$

This remark allows us to find a relationship between the two symmetry groups $\mathcal{G}_{\mathcal{B}'}(x')$ and $\mathcal{G}_{\mathcal{B}}(x)$. We have first

[4]The definition is different from that proposed in 1959 (and refined in 1972) by Noll. In fact, he calls a material element at x solid when its symmetry group is included in the full orthogonal group and not only its special subgroup of rotations. In other words, in the definition of solids, Noll includes reflections. Then he calls a material element having the full unimodular group as symmetry group fluid. In proposing such a definition, however, Noll does not impose the orientation-preserving nonlinear constraint $\det F > 0$. In this way, he can select as changes in observers those involving the full orthogonal group. In other words, with y and y' the actual placements of the same material element evaluated by two observers, $\mathcal{O}$ and $\mathcal{O}'$ respectively, which differ from each other by a time-parameterized family of isometries, we have always $y' = y + w(t) + Q(t)(y - y_0)$, with $w(t) \in \mathbb{R}^3$, but now $Q(t) \in O(3)$ instead of $Q(t)$ belonging just to $SO(3)$. In this setting, even the notion of objectivity changes, because it involves the full orthogonal group, i.e., reflections in addition to rotations.

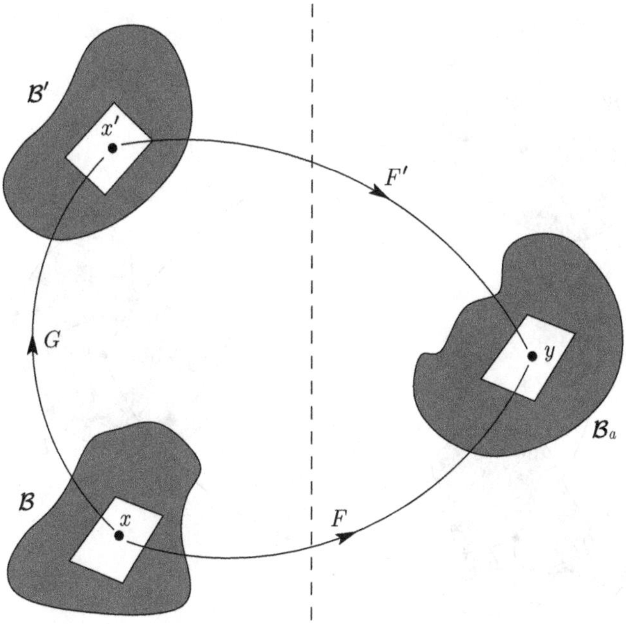

Fig. 4.3 Reaching the same actual configuration from different reference places

$$\tilde{\sigma}(FG^{-1}) = \tilde{\sigma}(FHG^{-1}) \tag{4.22}$$

with $H \in \mathcal{G}_{\mathcal{B}}(x)$. Figure 4.4 clarifies the identity.

We then obtain

$$\tilde{\sigma}(F') = \tilde{\sigma}(FHG^{-1}) = \tilde{\sigma}(FG^{-1}GHG^{-1}) = \tilde{\sigma}(F'GHG^{-1}) \tag{4.23}$$

and

$$\tilde{\sigma}(F') = \tilde{\sigma}(F'H') \tag{4.24}$$

with $H' \in \mathcal{G}_{\mathcal{B}'}(x')$.

Notice that GHG^{-1} is unimodular, because H is. As a consequence, the two symmetry groups $\mathcal{G}_{\mathcal{B}}(x)$ and $\mathcal{G}_{\mathcal{B}'}(x')$ are related, and we have

$$\mathcal{G}_{\mathcal{B}'}(x') = G\mathcal{G}_{\mathcal{B}}(x)G^{-1}.$$

In particular, when the reference place is just rotated so that $G = \bar{Q} \in SO(3)$, the previous relation reduces to

$$\mathcal{G}_{\mathcal{B}'}(x') = \bar{Q}\mathcal{G}_{\mathcal{B}}(x)\bar{Q}^{\mathrm{T}}. \tag{4.25}$$

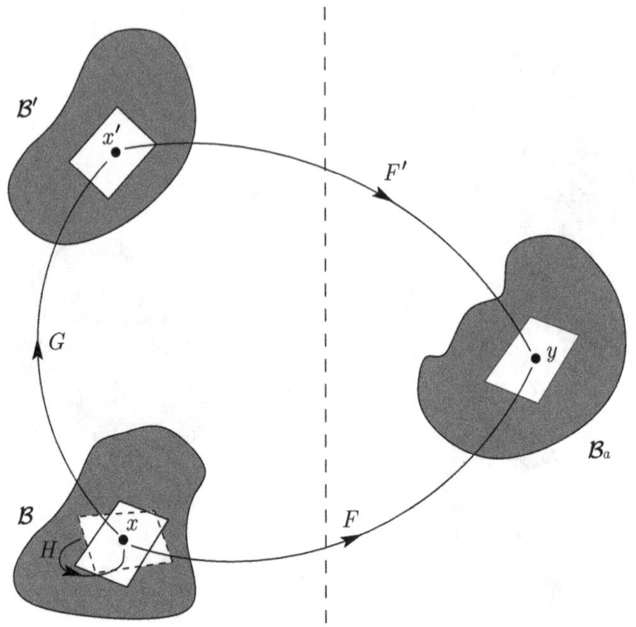

Fig. 4.4 Sketch justifying the identity (4.22)

The orthogonal tensor $\bar{Q}$ appearing in (4.25) acts on the reference place. Let us now consider a classical change in observer: an isometric transformation of the ambient space onto itself determined by the action of the Euclidean group over the physical ambient space. The assumed objectivity of the stress tensor can be written—as we already know—as

$$\tilde{\sigma}(QF) = Q\tilde{\sigma}(F)Q^{\mathrm{T}},$$

with now $Q = Q^i_j \tilde{e}_i \otimes \tilde{e}^j$, and $\tilde{\sigma}(F) = \sigma^i_j \tilde{e}_i \otimes \tilde{e}^j$. We could ask now for a condition ensuring that Q belongs to the symmetry group $\mathcal{G}_{\mathcal{B}}(x)$, provided that we have the identification of the reference space with the ambient one. If Q belonged to $\mathcal{G}_{\mathcal{B}}(x)$, we should have

$$\tilde{\sigma}(QF) = \tilde{\sigma}(QFQ^{\mathrm{T}}).$$

Hence, the required condition is

$$\tilde{\sigma}(QFQ^{\mathrm{T}}) = Q\tilde{\sigma}(F)Q^{\mathrm{T}}, \tag{4.26}$$

where we recall that Q^{T} appearing on the right-hand side of F acts on the reference space, while Q in premultiplication positions maps vectors in the ambient space into vectors of the same space. The condition (4.26) is in general not necessarily satisfied by all the elements of $SO(3)$, those that can be involved in classical changes in observers. When $\tilde{\sigma}(\cdot)$ satisfies (4.26) for all the elements of $SO(3)$, we call it isotropic.[5] In fact, tensor-valued functions are defined isotropic with respect to the full orthogonal group $O(3)$. Here, since the a priori condition $\det F > 0$ excludes any role for the subgroup of reflections (the one composed of orthogonal second-rank tensors with determinant equal to -1), we involve just $SO(3)$ in the definition of isotropic functions. Isotropic functions are advantageous because there are explicit representation theorems for them.

4.6 Isotropic Simple Materials

4.6.1 Simple Bodies

Definition 7. We call a simple material **isotropic** at a point x in a reference shape $\mathcal{B}$ when its symmetry group $\mathcal{G}_{\mathcal{B}}(x)$ contains the full special orthogonal group $SO(3)$. When a simple material is isotropic at all points x in $\mathcal{B}$, we call it isotropic without further specifications.

- Since $SO(3)$ is maximal in Unim^+, if the symmetry group of a simple *isotropic* material is a proper subgroup of Unim^+, the material is necessarily a solid.
- Every simple fluid is isotropic.

Definition 8. When $F = I$, we say that a material element placed at x in $\mathcal{B}$ is undistorted after the deformation determining such an F.

Proposition 9. *In a simple isotropic material, the stress pertaining to an undistorted state of a material element is spherical,*[6] *namely* $\sigma = -pI$, *with* $p \in \mathbb{R}$ *the so-called* ***pressure,***[7] *and* I *the second-rank identity tensor.*

Proof. In the special case given in the proposition, the condition (4.26) reduces to $\tilde{\sigma}(I) = Q\tilde{\sigma}(I)Q^{\mathrm{T}}$, since $QIQ^{\mathrm{T}} = QQ^{\mathrm{T}} = I$. The identity states that σ must commute with every $Q \in SO(3)$. Then σ has to be of the form αI, with α a scalar that we indicate by $-p$ for future convenience. □

Proposition 10. *For a compressible nonviscous simple fluid we have* $\sigma = -pI$, *where the pressure* p *is a function of the mass density.*

---

[5]The definition is justified here by the exclusion of reflections due to the orientation-preserving constraint. In the absence of it, we usually call the tensor-valued functions satisfying (4.26) for all elements of $O(3)$ *isotropic*.

[6]Recall that a second-rank tensor is called **spherical** when it is of the form αI, with α a scalar and I the unit second-rank tensor.

[7]The pressure appears also in different contexts. An example is nonviscous compressible fluids.

Proof. By the definition of a fluid, the stress σ must be such that $\tilde{\sigma}(F) = \tilde{\sigma}(FH)$ for every $H \in \text{Unim}^+$. A special choice of H is $H = (\det F)^{\frac{1}{3}} F^{-1}$, which is unimodular, since we have $\det H = \left((\det F)^{\frac{1}{3}}\right)^3 \det F^{-1} = \frac{\det F}{\det F} = 1$. The choice is justified, however, after identification of the reference space with the ambient one. A consequence is then

$$\tilde{\sigma}(F) = \hat{\sigma}\left((\det F)^{\frac{1}{3}} I\right).$$

A simple fluid is also isotropic. Then the argument followed in the proof of the previous proposition implies the spherical structure of the stress, but now the pressure p is a function of $\det F$, i.e., of the actual density of mass ρ, thanks to the mass balance $\rho = \rho_a \det F$. □

The pressure appears also as a reactive stress determined by the constraint of incompressibility in both solids and fluids. The next section clarifies the point. In any case, when the Cauchy stress is spherical, the balance of momentum becomes

$$\rho_a a^\flat = b_a - D_y p,$$

or more explicitly,

$$\rho \frac{\partial v^\flat}{\partial t} + \rho (D_y v^\flat) v = b_a - D_y p,$$

which is the system of **Euler equations**.

4.6.2 *A Digression on Incompressible Viscous Fluids*

Constraints between a body and its surrounding environment appeared already in Chapter 1, where we also defined rigid bodies as those admitting just motions of the type $\tilde{y}(x,t) = w(t) + Q(t)(x - x_0)$, with $w(t) \in \mathbb{R}^3$ and $Q(t) \in SO(3)$. This last requirement can be viewed as an **internal constraint**, for it deals with the inner structure of the material, i.e., the possibility of the material elements changing their places relative to one another.

More generally, we can think of an internal constraint of the type

$$\gamma(F) = 0,$$

with γ a differentiable function of its argument. Physical requirements restrict its generality.

– **Assumption 1**: The function γ is objective.

This assumption implies that the constraint has to be considered a function of $\tilde{C}$, or equivalently, C, i.e.,

$$\gamma(F(x,t)) = \hat{\gamma}(C(x,t)).$$

By time differentiation, we write

$$\frac{\partial \hat{\gamma}}{\partial C} \cdot \dot{C} = 0.$$

We have also (see Chapter 1)

$$\dot{C} = 2F^*(\mathrm{Sym}(\tilde{g}D_y v))F = 2F^*\mathsf{D}F$$

with $\mathsf{D} := \mathrm{Sym}(\tilde{g}D_y v)$, so that

$$\frac{\partial \hat{\gamma}}{\partial C} \cdot \dot{C} = 2\frac{\partial \hat{\gamma}}{\partial C} \cdot F^*\mathsf{D}F = 2F\frac{\partial \hat{\gamma}}{\partial C}F^* \cdot \mathsf{D} = 0. \tag{4.27}$$

The condition $\hat{\gamma}(C) = 0$ restricts the class of simple bodies that we are discussing. Associated with such a constraint we could consider it natural to imagine a **reactive stress** σ^r.

– **Assumption 2**: σ^r determines the additive decomposition

$$\sigma = \sigma^a + \sigma^r, \tag{4.28}$$

where σ^a is the **active stress** endowed with energetic and/or dissipative factors.
– **Assumption 3**: σ^r is powerless, i.e.,

$$\sigma^r \cdot \mathsf{D} = 0$$

for every D. The constraint is, in this sense, holonomic. This assumption and equation (4.27) imply that σ^r must be proportional to $F\frac{\partial \hat{\gamma}}{\partial C}F^*$. Write $-p$ for the scalar proportionality factor. We must have then

$$\sigma^r = -pF\frac{\partial \hat{\gamma}}{\partial C}F^*. \tag{4.29}$$

– **Assumption 4**: We consider a homogeneous viscoelastic simple material, i.e.,

$$\sigma^a = \sigma^e(F) + \sigma^d(F, \dot{F}). \tag{4.30}$$

– **Assumption 5**: The symmetry group is Unim^+, i.e., we are handling a fluid. Moreover, we assume

$$\sigma^e = 0, \tag{4.31}$$

i.e., there is no energetic component of the stress.

This assumption implies from the second law of thermodynamics that the free energy is constant with respect to the strain—we could have taken such a result as an assumption in place of $\sigma^e = 0$. Moreover, we obtain also (see previous sections)

$$\sigma^d \cdot \mathsf{D} \geq 0.$$

A possible solution of this inequality is

$$\sigma^d = \bar{\nu}\mathsf{D}^{\#}, \tag{4.32}$$

with $\bar{\nu} > 0$ and $\mathsf{D}^{\#} = \tilde{g}^{-1}\mathsf{D}\tilde{g}^{-1}$, i.e., $\mathsf{D}^{\#} = \mathsf{D}^{ij}\tilde{e}_i \otimes \tilde{e}_j$.

– **Assumption 6**: The material under consideration is incompressible, i.e.,

$$\hat{\gamma}(C) = \det C - 1.$$

Recall that $\det C = (\det F)^2$. The volume does not vary when $\det F = 1$, i.e., $\det C = 1$. This last assumption implies

$$\frac{\partial \hat{\gamma}}{\partial C} = \frac{\partial \det C}{\partial C} = C^{-\mathrm{T}} \det C = C^{-\mathrm{T}} = C^{-1},$$

the last identity being justified by the symmetry of C. Consequently, we compute

$$F\frac{\partial \hat{\gamma}}{\partial C}F^* = FC^{-1}F^* = \tilde{g}^{-1}.$$

– **Assumption 7**: The spatial metric $\tilde{g}$ is flat (i.e., $\tilde{g}$ coincides with the identity). Then equation (4.29) can be written as

$$\sigma^r = -pI. \tag{4.33}$$

The previous assumptions imply that when we change the shape of the body satisfying them, we just have friction among neighboring material elements. We exclude, in fact, elastic material bonds (Assumption 5) and mass density-dependent pressure (Assumption 6). Consequently, in physical terms, we are handling what we commonly call a **viscous incompressible fluid**.

Moreover, we compute

$$\operatorname{div}\sigma^d = \bar{\nu}\operatorname{div}\mathsf{D}^\# = \bar{\nu}\Delta v + \bar{\nu}\nabla_y \operatorname{div} v = \bar{\nu}\Delta v,$$

the last identity being justified, since the incompressibility assumption above implies $\operatorname{div} v = 0$ as a consequence of the Euler identity (see Chapter 3).

By taking into account the relations (4.28), (4.30), (4.31), (4.32), and (4.33), under the validity of the previous assumptions, the pointwise balance of forces becomes

$$\rho_a a = \bar{\nu}\Delta_y v - \nabla_y p + b,$$

where the Laplacian Δ_y and gradient ∇_y operators are evaluated with respect to y; the variable y appears in subscript position just to indicate this circumstance. When noninertial body forces are absent, i.e., when $b = 0$, the previous equation reduces to

$$\rho_a\left(\frac{\partial v}{\partial t} + (\nabla_y v)v\right) = \bar{\nu}\Delta_y v - \nabla_y p,$$

which is the system of the **Navier–Stokes equations**, after Claude-Louis Navier (1785–1836) and George Gabriel Stokes (1819–1903).

4.6.3 Isotropic Elastic Solids in the Large-Strain Regime

Objectivity requires that the elastic energy density be a function of the right Cauchy–Green tensor:

$$e = \tilde{e}(x, C).$$

When the material is isotropic, for any $\bar{Q} \in SO(3)$ we have

$$\tilde{\sigma}(F) = \tilde{\sigma}(F\bar{Q}).$$

Under the transformation $F \longrightarrow F\bar{Q}$, we obtain

$$C = F^*\tilde{g}F \longrightarrow \bar{Q}^*C\bar{Q} = \bar{Q}^*F^*\tilde{g}F\bar{Q}.$$

The isotropy can be interpreted by stating that the energy density does not change under rotations of the reference place. We say, in short, that the energy density e is *rotationally* invariant. Then, since C is symmetric, we can find a rotation bringing it into a diagonal form. In this way, the energy density is *in this case* a function of the eigenvalues of C, namely the solutions of the algebraic equation

$$\xi^3 - I_1(C)\xi^2 + I_2(C)\xi - I_3(C) = 0,$$

where $I_1(C)$ is the trace of C, $I_2(C) = (\det C)\operatorname{tr} C^{-1}$, $I_3(C) = \det C$. Since the coefficients of the previous equation are real and their algebraic signs alternate, we obtain three real eigenvalues ξ_1, ξ_2, ξ_3 such that

$$I_1(C) = \xi_1 + \xi_2 + \xi_3, \quad I_2(C) = \xi_1\xi_2 + \xi_2\xi_3 + \xi_1\xi_3, \quad I_3(C) = \xi_1\xi_2\xi_3.$$

Moreover, by the Cayley–Hamilton theorem, C itself satisfies the equation

$$C^3 - I_1(C)C^2 + I_2(C)C - I_3(C) = 0,$$

and the circumstance allows us to evaluate both C^{-1} and C^{-2}. In fact, if we multiply the previous equation by C^{-2} and C^{-1} respectively, we get

$$\hat{C}^{-2} = \frac{1}{I_3(C)}(\hat{C} - I_1\hat{I} + I_2\hat{C}^{-1}) \tag{4.34}$$

and

$$\hat{C}^{-1} = \frac{1}{I_3(C)}(\hat{C}^2 - I_1\hat{C} + I_2\hat{I}), \tag{4.35}$$

where the hat indicates that the tensors involved are 1-contravariant, 1-contravariant, namely

$$\hat{C}^{-1} := C^{-1}g = (\hat{C}^{-1})^A_B e_A \otimes e^B,$$

so $\hat{C}$ is defined by

$$\hat{C} := g^{-1}C.$$

The eigenvalues ξ_i of C are complicated functions of C. Then it appears convenient to assume that the elastic energy is a differentiable function of the invariants of C, namely $I_i(C)$, rather than its eigenvalues, namely

$$e = \tilde{e}(x, I_1(C), I_2(C), I_3(C)).$$

The derivative of e with respect to C is given by

$$\frac{\partial e}{\partial C} = \frac{\partial e}{\partial I_1(C)}\frac{\partial I_1(C)}{\partial C} + \frac{\partial e}{\partial I_2(C)}\frac{\partial I_2(C)}{\partial C} + \frac{\partial e}{\partial I_3(C)}\frac{\partial I_3(C)}{\partial C}, \tag{4.36}$$

since the invariants $I_i(C)$ are differentiable functions of C.

Lemma 11. *The following relations hold:*

$$\frac{\partial I_1(C)}{\partial C} = g^{-1}, \tag{4.37}$$

$$\frac{\partial I_2(C)}{\partial C} = \frac{I_2(C)}{I_2(C)}(C^{\#2} - I_1 C^{\#} + I_2 I^{\#}) + (C^{\#} - I_1 I^{\#} + I_2 C^{-1}) \tag{4.38}$$

$$\frac{\partial I_3(C)}{\partial C} = (C^{\#2} - I_1(C)C^{\#} + I_2(C)I^{\#}), \tag{4.39}$$

where $C^{\#} = C^{AB} e_A \otimes e_B = g^{-1}Cg^{-1}$ *and* $I^{\#}$ *is the 1-covariant, 1-contravariant identity*

Proof. We write C_{AA} for the trace of C when the metric in the ambient space is flat (i.e., g_{AB} coincides with δ_{AB}). When the metric is not flat, we recall that

$$\operatorname{tr} C = C_{AB} g^{BA} = C \cdot g^{-1},$$

which is identical to the standard notation when g is the identity. Consequently, by taking the derivative with respect to C, we obtain the relation (4.37). The determinant of C is given by

$$\det C = \mathsf{e}^{EDB} C_{A_1B} C_{A_2D} C_{A_3E},$$

where A_1, A_2, A_3 is a fixed even permutation of $1, 2, 3$. We compute

$$\begin{aligned}
\frac{\partial \det C}{\partial C_{A_1B}} &= \mathsf{e}^{BED} C_{A_2D} C_{A_3E} = \mathsf{e}^{BED} C_{A_2D} C_{A_3E} C_{A_1R} (C^{-1})^{RA_1} \\
&= \mathsf{e}^{BED} C_{A_2D} C_{A_3E} C_{A_1B} \delta^B_R (C^{-1})^{RA_1} = (\det C)(C^{-1})^{BA_1} \\
&= (\det C)(C^{-1})^{A_1B},
\end{aligned}$$

which implies from (4.39) from (4.35). The proof of the relation (4.38) follows also from a direct calculation. We get, in fact,

$$\begin{aligned}
\frac{\partial I_2(C)}{\partial C} &= \frac{\partial}{\partial C} \det C \operatorname{tr} C^{-1} = (\det C) C^{-1} \operatorname{tr} C^{-1} + (\det C) \frac{\partial \operatorname{tr} C^{-1}}{dC} \\
&= (\det C) C^{-1} \operatorname{tr} C^{-1} + (\det C) \operatorname{tr} \frac{\partial C^{-1}}{\partial C}.
\end{aligned} \tag{4.40}$$

We have also $CC^{-1} = I^{\#}$, with $I^{\#}$ the 1-covariant, 1-contravariant identity, i.e.,

$$C_{AB}(C^{-1})^{BD} = \delta^D_A,$$

and we compute

$$\frac{\partial}{\partial C_{AE}} C_{AB}(C^{-1})^{BD} = \delta_B^E (C^{-1})^{BD} + C_{AB}\frac{\partial (C^{-1})^{BD}}{\partial C_{AE}} = 0.$$

Then we obtain

$$\frac{\partial (C^{-1})^{BD}}{\partial C_{AE}} = -(C^{-1})^{AB}(C^{-1})^{ED}.$$

The derivative $\frac{\partial C^{-1}}{\partial C}$ is a fourth-rank tensor. Its trace is a second-rank tensor obtained by saturating the indices A and D. Then we get

$$\mathrm{tr}\frac{\partial (C^{-1})}{\partial C} = -C^{-2}.$$

The second-rank tensors C^{-2} and C^{-1} follow from (4.34) and (4.35) respectively once we multiply from the right by g^{-1}. Insertion of the results into (4.40) gives (4.38). □

Using the lemma, from the derivative (4.36), we get

$$P = 2\tilde{g}F\frac{\partial e}{\partial C} = 2\tilde{g}F(r_1 g^{-1} + r_2 C^{\#} + r_3 C^{\#2}), \tag{4.41}$$

where we have included the mass density in e, for the sake of conciseness, and r_1, r_2, and r_3 are coefficients emerging from the introduction of (4.37), (4.38), and (4.39) into (4.36). We leave to the reader the explicit derivation of the coefficients. We just want to note here the general expression of the derivative of the energy density with respect to C.

If we do not use (4.34) and (4.35) in the proof of the previous lemma and insert (4.40) into (4.36), we get

$$\frac{\partial e}{\partial C} = \frac{\partial e}{\partial I_1} g^{-1} + \left(\frac{\partial e}{\partial I_2} + \frac{\partial e}{\partial I_3} I_3\right) C^{-1} - \frac{\partial e}{\partial I_2} I_3 C^{-2}.$$

4.6.4 Isotropic Simple Materials in the Small-Strain Regime

Consider in the small-strain regime the elastic energy as a quadratic form in ε. Isotropy implies the dependence of the energy on the invariants of ε, excluding $I_3(\varepsilon) = \det \varepsilon$, because the determinant of a 3×3 matrix is cubic. For the second invariant of ε, we can consider the expression

$$I_2(\varepsilon) = \frac{1}{2}\left((\operatorname{tr}\varepsilon)^2 - \operatorname{tr}\varepsilon^2\right).$$

This formula can be proven by writing ε in terms of its eigenvalues and computing directly $I_2(\varepsilon)$. The result does not depend on the vector basis chosen because $I_2(\varepsilon)$ is rotationally invariant. Consider[8]

$$e(\varepsilon) = \mu|\varepsilon|^2 + \frac{\lambda}{2}(\operatorname{tr}\varepsilon)^2.$$

The resulting stress–strain relationship is of the form

$$\sigma_{ij} = \frac{\partial e}{\partial \varepsilon^{ij}} = \lambda(\operatorname{tr}\varepsilon)g_{ij} + 2\mu\varepsilon_{ij},$$

and it emerges from the linearization of (4.41), and the use of g. Notice that we do not distinguish between g and $\tilde{g}$ as well, since we confuse $\mathcal{B}$ with $\mathcal{B}_a$ on identification of $\mathcal{E}^3$ and $\tilde{\mathcal{E}}^3$. The previous relation can also be written as

$$\sigma_{ij} = \mathbb{C}_{ijhk}\varepsilon_{hk}, \tag{4.42}$$

provided that

$$\mathbb{C}_{ijhk} = \lambda g_{ij}g_{hk} + \mu(g_{ik}g_{jh} + g_{ih}g_{jk}). \tag{4.43}$$

When the metric is flat, we get

$$\sigma_{ij} = \lambda(\operatorname{tr}\varepsilon)\delta_{ij} + 2\mu\varepsilon_{ij}$$

and

$$\mathbb{C}_{ijhk} = \lambda\delta_{ij}\delta_{hk} + \mu(\delta_{ik}\delta_{jh} + \delta_{ih}\delta_{jk}).$$

If the material is homogeneous, then λ and μ do not depend on x, an assumption that we accept from now on, unless otherwise specified. In this case, we call λ and μ the **Lamé constants** after Gabriel Lamé (1795–1870). They define in the small-strain regime the linear elastic constitutive relation (4.42) with $\mathbb{C}$ given by (4.43). In this case, by evaluating the trace of σ, we get

$$\operatorname{tr}\sigma = (2\mu + 3\lambda)\operatorname{tr}\varepsilon,$$

i.e.,

$$\operatorname{tr}\varepsilon = \frac{1}{2\mu + 3\lambda}\operatorname{tr}\sigma,$$

[8]Here μ has nothing to do with $d\mu$, which denotes the volume measure.

which implies

$$\varepsilon = \frac{1}{2\mu}\left(\sigma - \frac{\lambda}{2\mu + 3\lambda}\text{tr}\,\sigma I\right). \tag{4.44}$$

Consider a test in which $\sigma_{11} \neq 0$ and $\sigma_{ij} = 0$, $\{i,j\} \neq \{1,1\}$. We get

$$\varepsilon_{11} = \frac{1}{2\mu}\left(1 - \frac{\lambda}{2\mu + 3\lambda}\right)\sigma_{11},$$

i.e.,

$$\sigma_{11} = E\varepsilon_{11},$$

where E is given by

$$E = \frac{\mu(2\mu + 3\lambda)}{\mu + \lambda}$$

and is called **Young's modulus** after Thomas Young (1773–1829). Moreover, for the strain components ε_{22} and ε_{33}, we compute

$$\varepsilon_{22} = \varepsilon_{33} = -\frac{\lambda}{2\mu(2\mu + 3\lambda)}\sigma_{11}$$

and define

$$\nu := -\frac{\varepsilon_{22}}{\varepsilon_{11}} = -\frac{\varepsilon_{33}}{\varepsilon_{11}} = \frac{\lambda}{2(\mu + \lambda)},$$

calling it **Poisson's ratio** after Simon-Denis Poisson (1781–1840). In a linear elastic isotropic material, the elongation along a given direction implies then a transversal contraction. Algebra leads us to the identity

$$\frac{\nu}{1 + \nu} = \frac{\lambda}{2\mu + 3\lambda},$$

so that we can write the stress–strain linear elastic isotropic relation as

$$\varepsilon = \frac{1}{2\mu}\left(\sigma - \frac{\nu}{1 + \nu}\text{tr}\,\sigma I\right),$$

an expression that we shall find useful in the subsequent development.

Theorem 12. *Take $\mathbb{C}$ isotropic in a Cartesian frame. Consider $\mathbb{C}$ as a linear map from the space of symmetric second-rank tensors onto itself. Provided that $\lambda \neq 0$, $\mathbb{C}$*

has two eigenvalues, namely $2\mu + 3\lambda$ and 2μ. The relevant eigenspaces are those of spherical and traceless second-rank symmetric tensors.

Proof. We consider first two independent mechanisms: uniform pressure and pure shear. In the first case, the stress σ is of the form $\sigma = -pI$, with p a scalar. Consequently, from (4.44) we get

$$\varepsilon = \frac{1}{2\mu}\left(\frac{3\lambda}{2\mu + 3\lambda} - 1\right) pI.$$

Then ε is spherical: it is of the form $\mathfrak{e}I$. We have also

$$\operatorname{tr}\varepsilon = 3\mathfrak{e} = -\frac{3}{2\mu + 3\lambda}p,$$

i.e.,

$$-p = 3\mathrm{k}\mathfrak{e}, \tag{4.45}$$

with

$$\mathrm{k} = \lambda + \frac{2}{3}\mu,$$

called the **bulk modulus**. Then 3k is an eigenvalue of $\mathbb{C}$, as stated by (4.45).

Consider now

$$\varepsilon = \beta\,\mathrm{Sym}(m \otimes n),$$

with β a scalar and m and n two linearly independent vectors. In this case, we get

$$\sigma = 2\mu\beta\,\mathrm{Sym}(m \otimes n),$$

i.e., 2μ is an eigenvalue of $\mathbb{C}$, which is here isotropic in general. Let α be a generic eigenvalue of $\mathbb{C}$. By definition, we have

$$\mathbb{C}\varepsilon = \alpha\varepsilon.$$

Since the tensor $\mathbb{C}$ considered here is isotropic, we can write the previous relation as

$$(2\mu - \alpha)\varepsilon + \lambda(\operatorname{tr}\varepsilon)I = 0. \tag{4.46}$$

When $\alpha = 2\mu$ (one of the two eigenvalues of $\mathbb{C}$), we obtain $\operatorname{tr}\varepsilon = 0$, since $\lambda \neq 0$. In other words, the eigenspace corresponding to the eigenvalue $\alpha = 2\mu$ is the space of traceless tensors. When $\alpha = 2\mu + 3\lambda$, equation (4.46) becomes

$$-3\lambda\varepsilon + \lambda\operatorname{tr}\varepsilon I = 0,$$

which is satisfied when $\varepsilon = \frac{1}{3}(\operatorname{tr}\varepsilon)I$, since $\lambda \neq 0$, so that the eigenspace associated with the eigenvalue $2\mu + 3\lambda$ is the space of spherical tensors. □

The inequalities

$$\mu > 0 \quad \text{and} \quad 2\mu + 3\lambda > 0$$

ensure the positive definiteness of $\mathbb{C}$. Alternatively, we can impose the following pairs of inequalities:

$$\mu > 0 \quad \text{and} \quad \mathrm{k} > 0,$$

$$\mu > 0 \quad \text{and} \quad -1 < \nu < \frac{1}{2},$$

$$E > 0 \quad \text{and} \quad -1 < \nu < \frac{1}{2},$$

by taking into account previous definitions of k, ν, E. The positive definiteness of $\mathbb{C}$ implies the positivity of the elastic energy density

$$e(\varepsilon) = \frac{1}{2}\varepsilon \cdot \mathbb{C}\varepsilon.$$

4.7 Additional Remarks

In contrast with what we have done so far, we could assume a priori the constitutive relation

$$\sigma = \mathbb{C}\varepsilon$$

with $\mathbb{C}$ endowed *only* with *minor* symmetries, imagining that this choice defines the range of linear elasticity. Then we could prove the existence of the elastic energy under the restrictive condition that $\mathbb{C}$ is endowed with *major* symmetry. According to tradition, we should then call this case **hyperelastic**. We have followed the opposite approach, showing a preference for the hyperelastic setting that we call simply elastic, for we find it thermodynamically natural and sound. The reader, of course, can have a different opinion.

Chapter 5
Topics in Linear Elasticity

5.1 Minimum of the Total Energy

Under small-strain linear elastic behavior, we call a triplet of fields $\hat{u}(\cdot)$, $\hat{\varepsilon}(\cdot)$, $\hat{\sigma}(\cdot)$, denoted by ς, such that

- $\hat{u} \in C^2(\mathcal{B}) \cap C(\bar{\mathcal{B}})$,
- $\mathrm{Sym}\, D\hat{u} \in C(\bar{\mathcal{B}})$, with $\bar{\mathcal{B}}$ the closure of $\mathcal{B}$,
- $\hat{\varepsilon}(\cdot) \in C(\mathcal{B})$,
- $\hat{\sigma}(\cdot) \in C^1(\mathcal{B}) \cap C(\bar{\mathcal{B}})$,
- $\mathrm{div}\, \hat{\sigma} \in C(\bar{\mathcal{B}})$,

an **admissible state**.

We write u, ε, σ for the values $\hat{u}(x)$, $\hat{\varepsilon}(x)$, and $\hat{\sigma}(x)$, respectively. We then have $u \in \mathbb{R}^3$, ε, and $\sigma \in \mathrm{Sym}(\mathbb{R}^3, \mathbb{R}^3)$.

We denote by $\mathcal{A}$ the **space of admissible states**. It is a function space.

We consider *mixed boundary conditions*, which means that we subdivide the boundary $\partial\mathcal{B}$ of the region $\mathcal{B}$ that the body occupies into two distinct pieces $\partial\mathcal{B}_u$ and $\partial\mathcal{B}_f$, i.e., $\partial\mathcal{B}_u \cup \partial\mathcal{B}_f = \partial\mathcal{B}$ and $\partial\mathcal{B}_u \cap \partial\mathcal{B}_f = \varnothing$, and we assign surface forces f along $\partial\mathcal{B}_f$ through a piecewise smooth map $x \longmapsto f(x) \in \mathbb{R}^3, x \in \partial\mathcal{B}_f$, and prescribe $\tilde{u}(x) = \bar{u}(x)$ for $x \in \partial\mathcal{B}_u$, with $\bar{u}(\cdot)$ an assigned continuous function. The body is also subjected to noninertial body forces b.

For the bulk elastic energy density $\mathcal{E}(\varepsilon)$, we consider the quadratic expression

$$\mathcal{E}(\varepsilon) = \frac{1}{2}\int_{\mathcal{B}} \varepsilon \cdot \mathbb{C}\varepsilon \, d\mu,$$

with $\mathbb{C}$ endowed with the symmetries already discussed.

P.M. Mariano, L. Galano, *Fundamentals of the Mechanics of Solids*,
DOI 10.1007/978-1-4939-3133-0_5

Lemma 1. *If $x \longmapsto \varepsilon(x) \in \mathrm{Sym}(\mathbb{R}^3, \mathbb{R}^3)$ and $x \longmapsto \bar{\varepsilon}(x) \in \mathrm{Sym}(\mathbb{R}^3, \mathbb{R}^3)$ are square-integrable fields, we have*

$$\mathcal{E}(\varepsilon + \bar{\varepsilon}) = \mathcal{E}(\varepsilon) + \mathcal{E}(\bar{\varepsilon}) + \int_{\mathcal{B}} \varepsilon \cdot \mathbb{C}\bar{\varepsilon} d\mu.$$

Proof. Due to the symmetry of $\mathbb{C}$, we obtain

$$\bar{\varepsilon} \cdot \mathbb{C}\varepsilon = \varepsilon \cdot \mathbb{C}\bar{\varepsilon}.$$

Then we compute

$$(\varepsilon + \bar{\varepsilon}) \cdot \mathbb{C}(\varepsilon + \bar{\varepsilon}) = \varepsilon \cdot \mathbb{C}\varepsilon + \bar{\varepsilon} \cdot \mathbb{C}\bar{\varepsilon} + 2\varepsilon \cdot \mathbb{C}\bar{\varepsilon}.$$ □

Notice that in the previous proof, $\mathbb{C}$ can depend on x, i.e., the material can be inhomogeneous. Here we assume that the map $x \longmapsto \mathbb{C}(x)$ is smooth over $\mathcal{B}$ and continuous over its closure $\bar{\mathcal{B}}$.

We say that a triplet of fields $\hat{u}(\cdot)$, $\hat{\varepsilon}(\cdot)$, and $\hat{\sigma}(\cdot)$, a **state**, is **kinematically admissible** when

$$\varepsilon = \mathrm{Sym}\, Du, \qquad \sigma = \mathbb{C}\varepsilon, \qquad u = \bar{u} \quad \text{along } \partial\mathcal{B}_u.$$

The *set* $\mathcal{A}^{(k)}$ of *kinematically admissible states* contains stress fields that do not necessarily satisfy the balance equations with the data b and f prescribed. States satisfying the balance equations are characterized by the following theorem.

Theorem 2. *Let $\mathcal{E}_{tot} : \mathcal{A}^{(k)} \longmapsto \mathbb{R}$ be the **total energy** defined by*

$$\mathcal{E}_{tot}(\varsigma) := \mathcal{E}(\varepsilon) - \int_{\mathcal{B}} b \cdot u \, d\mu - \int_{\partial\mathcal{B}_f} f \cdot u \, d\mathcal{H}^2.$$

Let also ς be a solution of the mixed boundary value problem (as specified above). Then ς realizes the minimum of the total energy, namely

$$\mathcal{E}_{tot}(\varsigma) \leq \mathcal{E}_{tot}(\tilde{\varsigma})$$

for every $\tilde{\varsigma} \in \mathcal{A}^{(k)}$, with equality holding when $\tilde{\varsigma} = \varsigma$ modulo a rigid displacement.

Proof. Let the state ς be a solution of the mixed boundary value problem. For $\tilde{\varsigma} \in \mathcal{A}^{(k)}$, define

$$\varsigma' := \tilde{\varsigma} - \varsigma.$$

This new state is such that

$$\varepsilon' = \operatorname{Sym} Du', \qquad \sigma' = \mathbb{C}\varepsilon', \qquad u' = 0 \quad \text{on } \partial\mathcal{B}_u,$$

with

$$u' = \tilde{u} - u, \qquad \varepsilon' = \tilde{\varepsilon} - \varepsilon, \qquad \sigma' = \tilde{\sigma} - \sigma.$$

By the previous lemma, we get

$$\mathcal{E}(\tilde{\varepsilon}) = \mathcal{E}(\varepsilon + \varepsilon') = \mathcal{E}(\varepsilon) + \mathcal{E}(\varepsilon') + \int_{\mathcal{B}} \varepsilon' \cdot \mathbb{C}\varepsilon \, d\mu = \mathcal{E}(\varepsilon) + \mathcal{E}(\varepsilon') + \int_{\mathcal{B}} \sigma \cdot \varepsilon' d\mu,$$

and Gauss's theorem implies

$$\int_{\mathcal{B}} \sigma \cdot \varepsilon' d\mu = \int_{\partial\mathcal{B}_f} \sigma n \cdot u' d\mathcal{H}^2 - \int_{\mathcal{B}} u' \cdot \operatorname{div} \sigma d\mu.$$

The surface integral in the previous expression is limited to $\partial\mathcal{B}_f$, because u' vanishes identically over $\partial\mathcal{B}_u$. The definition of $\mathcal{E}_{tot}$ implies then

$$\mathcal{E}_{tot}(\tilde{\varsigma}) - \mathcal{E}_{tot}(\varsigma) = \mathcal{E}(\tilde{\varepsilon}) - \mathcal{E}(\varepsilon) - \int_{\mathcal{B}} b \cdot (\tilde{u} - u) d\mu - \int_{\partial\mathcal{B}_f} f \cdot (\tilde{u} - u) d\mathcal{H}^2$$

$$= \mathcal{E}(\varepsilon') - \int_{\mathcal{B}} (\operatorname{div} \sigma + b) \cdot u' d\mu + \int_{\partial\mathcal{B}_f} (\sigma n - f) \cdot u' d\mathcal{H}^2.$$

Since ς is a solution of the boundary value problem, the relevant stress field is such that

$$\operatorname{div} \sigma + b = 0 \quad \text{and} \quad \sigma n = f \quad \text{along } \partial\mathcal{B}_f.$$

Consequently, we obtain

$$\mathcal{E}_{tot}(\tilde{\varsigma}) - \mathcal{E}_{tot}(\varsigma) = \mathcal{E}(\varepsilon') = \frac{1}{2} \int_{\mathcal{B}} \varepsilon' \cdot \mathbb{C}\varepsilon' d\mu.$$

Then, since $\mathbb{C}$ is positive definite,

$$\mathcal{E}_{tot}(\tilde{\varsigma}) - \mathcal{E}_{tot}(\varsigma) = 0$$

if and only if

$$\varepsilon' = \tilde{\varepsilon} - \varepsilon = 0,$$

i.e., when $\tilde{u}$ equals u modulo a rigid displacement. In fact, take

$$\tilde{u} = u + w + q \times (x - x_0),$$

with w and q constant vectors in space. We get

$$D\tilde{u} = Du + \mathsf{e}q$$

with $\mathsf{e}q$ a skew-symmetric second-rank tensor with ijth component given by $\mathsf{e}_{ijk}q^k$, with e_{ijk} the Ricci symbol. The symmetric part of the previous relation is then $\tilde{\varepsilon} = \varepsilon$. □

5.2 Minimum of the Complementary Energy

Write $\mathcal{A}^{(\varsigma)}$ for the set of **statically admissible states**, i.e., those with stress fields satisfying the balance equations and the constitutive relation, namely

$$\operatorname{div} \sigma + b = 0, \qquad \sigma \in \mathrm{Sym}, \qquad \sigma n = f \quad \text{along } \partial\mathcal{B}_f, \sigma = \mathbb{C}\varepsilon$$

with b, f, and $\partial\mathcal{B}_f$ pertaining to the mixed boundary value problem defined in the previous section. Let $\mathcal{E}_c(\sigma)$ be defined by

$$\mathcal{E}_c(\sigma) := \frac{1}{2}\int_{\mathcal{B}} \sigma \cdot \mathbb{K}\sigma \, d\mu$$

with $\mathbb{K} := \mathbb{C}^{-1}$. Since $\mathbb{C}$ is positive definite, $\mathbb{K}$ is also positive definite. The quantity $\mathcal{E}_c$ is generally called the **complementary energy**.

Define also $\mathcal{E}_c^+ : \mathcal{A}^{\varsigma} \longmapsto \mathbb{R}$ by

$$\mathcal{E}_c^+(\varsigma) := \mathcal{E}_c(\sigma) - \int_{\partial\mathcal{B}_u} \sigma n \cdot \bar{u} \, d\mathcal{H}^2.$$

Theorem 3. *Let ς be a solution of the mixed boundary value problem, which implies the inclusion $\varsigma \in \mathcal{A}^{(k)} \cap \mathcal{A}^{(\varsigma)}$. Then ς minimizes $\mathcal{E}_c^+$, i.e.,*

$$\mathcal{E}_c^+(\varsigma) \leq \mathcal{E}_c^+(\tilde{\varsigma})$$

for every $\tilde{\varsigma}$ in $\mathcal{A}^{(\varsigma)}$. The equality sign holds only when $\tilde{\varsigma} = \varsigma$.

Proof. For every $\tilde{\varsigma}$ in $\mathcal{A}^{(\varsigma)}$, the admissible state $\varsigma' = \tilde{\varsigma} - \varsigma$ has a stress field $\sigma' = \tilde{\sigma} - \sigma$ such that

$$\operatorname{div} \sigma' = 0 \quad \text{and} \quad \sigma' n = 0 \quad \text{along } \partial\mathcal{B}_f, \tag{5.1}$$

since both $\tilde{\sigma}$ and σ satisfy the balance equations with the same data. By the lemma in the previous section, we write

$$\mathcal{E}_c(\tilde{\sigma}) = \mathcal{E}_c(\sigma + \sigma') = \mathcal{E}_c(\sigma) + \mathcal{E}_c(\sigma') + \int_{\mathcal{B}} \sigma' \cdot \varepsilon \, d\mu. \tag{5.2}$$

Since ς is also kinematically compatible, since it is a solution of the mixed boundary value problem considered here, ε coincides with the symmetric part of Du, and we have, by Gauss's theorem,

$$\int_{\mathcal{B}} \sigma' \cdot \varepsilon \, d\mu = \int_{\partial\mathcal{B}} \sigma' n \cdot u \, d\mathcal{H}^2 - \int_{\mathcal{B}} u \cdot \operatorname{div} \sigma' d\mu,$$

which reduces to

$$\int_{\mathcal{B}} \sigma' \cdot \varepsilon \, d\mu = \int_{\partial\mathcal{B}_u} \sigma' n \cdot u \, d\mathcal{H}^2 = \int_{\partial\mathcal{B}_u} (\tilde{\sigma} - \sigma) n \cdot u \, d\mathcal{H}^2,$$

due to (5.1). Then by substitution in (5.2), we find

$$\mathcal{E}_c^+(\tilde{\varsigma}) - \mathcal{E}_c^+(\varsigma) = \mathcal{E}_c(\tilde{\sigma}) - \mathcal{E}_c(\sigma) - \int_{\partial\mathcal{B}_u} \tilde{\sigma} n \cdot u \, d\mathcal{H}^2$$

$$+ \int_{\partial\mathcal{B}_u} \sigma n \cdot u \, d\mathcal{H}^2 = \mathcal{E}_c(\sigma') = \frac{1}{2} \int_{\mathcal{B}} \sigma' \cdot \mathbb{K} \sigma' d\mu \geq 0,$$

since $\mathbb{K}$ is positive definite. The equality sign holds when $\sigma' = 0$, that is, when $\tilde{\sigma} = \sigma$. □

5.3 The Hellinger–Prange–Reissner stationarity principle

We need to satisfy four conditions for solving the mixed problem in linearized elastostatics: (1) the strain-displacement compatibility, (2) the validity of the stress–strain linear constitutive equations, (3) the identity $u = \bar{u}$ along the portion $\partial\mathcal{B}_u$ of the boundary $\partial\mathcal{B}$ where the displacement is prescribed, (4) the balance equations in the bulk and along the boundary $\partial\mathcal{B}_f$ where forces are applied. Recall the assumption $\partial\mathcal{B}_f \cap \partial\mathcal{B}_u = \varnothing$ with $\partial\mathcal{B}_f \cup \partial\mathcal{B}_u = \partial\mathcal{B}$.

In both theorems discussed in the two previous sections, some conditions are assumed to hold. A state realizing the minimum of the functionals considered there satisfies all four conditions. Options are possible. We can start, for example, by considering the subset of $\mathcal{A}$ including admissible states for which just the strain-displacement compatibility is granted a priori. Let us denote by $\mathcal{A}^{sd}$ such a set. Notice that for u and $\tilde{u}$ admissible displacement fields, with α a scalar, we have

$\mathrm{Sym}\, D(u + \alpha\tilde{u}) = \mathrm{Sym}\, Du + \alpha \mathrm{Sym}\, D\tilde{u}$. Hence, if ς and $\tilde{\varsigma}$ belong to $\mathcal{A}^{\varsigma d}$, then $\varsigma + \alpha\tilde{\varsigma} \in \mathcal{A}^{\varsigma d}$, the sum of two states being defined by the state composed by the sum of the relevant fields, namely $\varsigma + \alpha\tilde{\varsigma} = (u + \alpha\tilde{u}, \varepsilon + \alpha\tilde{\varepsilon}, \sigma + \alpha\tilde{\sigma})$.

Theorem 4 (Ernst David Hellinger, 1914, G. Prange, 1916, Eric Reissner, 1950). *Let* $\mathcal{E}_{HPR}(\varsigma) : \mathcal{A}^{\varsigma d} \longmapsto \mathbb{R}$ *be defined by*

$$\mathcal{E}_{HPR}(\varsigma) := \mathcal{E}_c(\sigma) - \int_{\mathcal{B}} \sigma \cdot \varepsilon \, d\mu + \int_{\mathcal{B}} b \cdot u \, d\mu + \int_{\partial\mathcal{B}_u} \sigma n \cdot (u - \bar{u}) d\mathcal{H}^2 + \int_{\partial\mathcal{B}_f} f \cdot u \, d\mathcal{H}^2.$$

A state renders $\mathcal{E}_{HPR}$ stationary if and only if it is a solution of the mixed boundary value problem in linearized elastostatics.

Before going into the details of the formal proof, we recall that the stationarity of $\mathcal{E}_{HPR}$ is ensured at all states where the first variation of $\mathcal{E}_{HPR}$, denoted by $\delta\mathcal{E}_{HPR}$, vanishes. We say that the condition is ensured at $\varsigma \in \mathcal{A}^{\varsigma d}$ when the derivative

$$\delta_{\tilde{\varsigma}} \mathcal{E}_{HPR}(\varsigma) := \left. \frac{d}{d\alpha} \mathcal{E}_{HPR}(\varsigma + \alpha\tilde{\varsigma}) \right|_{\alpha=0} \tag{5.3}$$

exists and vanishes for *every* $\tilde{\varsigma} \in \mathcal{A}^{\varsigma d}$.

The definition has a geometric interpretation. We can think of $\delta_{\tilde{\varsigma}} \mathcal{E}_{HPR}(\varsigma)$ as the derivative of $\mathcal{E}_{HPR}$ along the "direction" $\tilde{\varsigma}$, evaluated at ς. In this way, when we state that the first variation of $\mathcal{E}_{HPR}$ vanishes at a state ς, we are saying that the first derivatives of that functional along all the possible "directions," evaluated at ς, vanish.

Proof. Take ς and $\tilde{\varsigma}$ in $\mathcal{A}^{sd}$. By the lemma in Section 5.1, we get

$$\mathcal{E}_c(\sigma + \alpha\tilde{\sigma}) = \mathcal{E}_c(\sigma) + \alpha^2 \mathcal{E}_c(\tilde{\sigma}) + \alpha \int_{\mathcal{B}} \tilde{\sigma} \cdot \mathbb{K}\sigma \, d\mu.$$

Hence, by computing the derivative in (5.3), we obtain

$$\begin{aligned} \delta_{\tilde{\varsigma}} \mathcal{E}_{HPR}(\varsigma) &= \int_{\mathcal{B}} ((\mathbb{K}\sigma - \varepsilon) \cdot \tilde{\sigma} - \sigma \cdot \tilde{\varepsilon} + b \cdot \tilde{u}) d\mu \\ &+ \int_{\partial\mathcal{B}_u} (\tilde{\sigma} n \cdot (u - \bar{u}) + \sigma n \cdot \tilde{u}) d\mathcal{H}^2 + \int_{\partial\mathcal{B}_f} (f \cdot \tilde{u}) d\mathcal{H}^2. \end{aligned} \tag{5.4}$$

Since σ is symmetric and the compatibility between $\tilde{\varepsilon}$ and $\tilde{u}$ is ensured a priori as $\tilde{\varsigma} \in \mathcal{A}^{\varsigma d}$, by Gauss's theorem we obtain

$$\int_{\mathcal{B}} \sigma \cdot \tilde{\varepsilon} \, d\mu = \int_{\mathcal{B}} \sigma \cdot D\tilde{u} \, d\mu = \int_{\partial\mathcal{B}} \sigma n \cdot \tilde{u} \, d\mathcal{H}^2 - \int_{\mathcal{B}} \tilde{u} \cdot \mathrm{div}\, \sigma \, d\mu. \tag{5.5}$$

By inserting the result into (5.4), $\delta_{\tilde{\varsigma}}\mathcal{E}_{HPR}(\varsigma)$ reduces to

$$\begin{aligned}\delta_{\tilde{\varsigma}}\mathcal{E}_{HPR}(\varsigma) &= \int_{\mathcal{B}}((\mathbb{K}\sigma-\varepsilon)\cdot\tilde{\sigma}+(b+\operatorname{div}\sigma)\cdot\tilde{u})d\mu \\ &+\int_{\partial\mathcal{B}_u}(\tilde{\sigma}n\cdot(u-\bar{u})d\mathcal{H}^2+\int_{\partial\mathcal{B}_f}(f-\sigma n)\cdot\tilde{u}\,d\mathcal{H}^2.\end{aligned} \tag{5.6}$$

If ς is a solution of the mixed boundary value problem, we get

$$\delta_{\tilde{\varsigma}}\mathcal{E}_{HPR}(\varsigma)=0 \tag{5.7}$$

for every choice of $\tilde{\varsigma}$, which is

$$\delta\mathcal{E}_{HPR}(\varsigma)=0. \tag{5.8}$$

Conversely, if ς satisfies the equations (5.8), i.e., (5.7) for every choice of $\tilde{\varsigma}=(\tilde{u},\tilde{\varepsilon},\tilde{\sigma})$, by the fundamental theorem of calculus, from the expressions (5.5) and (5.6) it follows that ς is a solution of the mixed boundary value problem in linearized elastostatics. □

Although the stress fields appearing in the definition of admissible states are symmetric, they do not correspond necessarily to the tractions satisfying the integral balances with assigned data b and f.

5.4 The Hu–Washizu Variational Principle

We can weaken further the approach by assuming that none of the conditions of the mixed boundary value problem in linearized elastostatics is satisfied a priori. Even in this case, we can show that a solution of the mixed boundary value problem renders stationary a certain functional, let us denote it by $\mathcal{E}_{HW}$, defined over the whole space of admissible states.

Theorem 5 (M. Fraeijs de Veubekele Baudouin, 1951, Haichang Hu, 1955, K. Washizu, 1955, 1968). *Let $\mathcal{E}_{HW}(\varsigma):\mathcal{A}\longmapsto\mathbb{R}$ be defined by*

$$\begin{aligned}\mathcal{E}_{HW}(\varsigma) &:= \mathcal{E}(\varepsilon)-\int_{\mathcal{B}}\sigma\cdot\varepsilon\,d\mu-\int_{\mathcal{B}}(b+\operatorname{div}\sigma)\cdot u\,d\mu \\ &+\int_{\partial\mathcal{B}_u}\sigma n\cdot\bar{u}\,d\mathcal{H}^2+\int_{\partial\mathcal{B}_f}(\sigma n-f)\cdot u\,d\mathcal{H}^2.\end{aligned}$$

Then $\mathcal{E}_{HW}$ becomes stationary at ς if and only if ς itself is a solution of the mixed boundary value problem in linearized elastostatics.

Proof. With $\varsigma = (u, \varepsilon, \sigma)$ and $\tilde{\varsigma} = (\tilde{u}, \tilde{\varepsilon}, \tilde{\sigma})$ in $\mathcal{A}$, $\varsigma + \alpha\tilde{\varsigma} \in \mathcal{A}$ for every $\alpha \in \mathbb{R}$. By the lemma in Section 5.1, we also get

$$\mathcal{E}(\varepsilon + \alpha\tilde{\varepsilon}) = \mathcal{E}(\varepsilon) + \alpha^2\mathcal{E}(\tilde{\varepsilon}) + \alpha \int_{\mathcal{B}} \tilde{\varepsilon} \cdot \mathbb{C}\varepsilon \, d\mu.$$

Then we can compute

$$\begin{aligned}\delta_{\tilde{\varsigma}}\mathcal{E}_{HW}(\varsigma) &:= \left.\frac{d}{d\alpha}\mathcal{E}_{HW}(\varsigma + \alpha\tilde{\varsigma})\right|_{\alpha=0} \\ &= \int_{\mathcal{B}} ((\mathbb{C}\varepsilon - \sigma) \cdot \tilde{\varepsilon} - (b + \operatorname{div}\sigma) \cdot \tilde{u} - \tilde{\sigma} \cdot \varepsilon - u \cdot \operatorname{div}\tilde{\sigma}) d\mu \\ &+ \int_{\partial\mathcal{B}_u} \tilde{\sigma}n \cdot \bar{u} \, d\mathcal{H}^2 + \int_{\partial\mathcal{B}_f} (\tilde{\sigma}n \cdot u + (\sigma n - f) \cdot \tilde{u}) d\mathcal{H}^2.\end{aligned}$$

Gauss's theorem and the symmetry of $\tilde{\sigma}$ allow us to write

$$\int_{\mathcal{B}} u \cdot \operatorname{div}\tilde{\sigma} \, d\mu = \int_{\partial\mathcal{B}} \tilde{\sigma}n \cdot u \, d\mathcal{H}^2 - \int_{\mathcal{B}} \tilde{\sigma} \cdot \operatorname{Sym} Du \, d\mu,$$

so that

$$\begin{aligned}\delta_{\tilde{\varsigma}}\mathcal{E}_{HW}(\varsigma) &= \int_{\mathcal{B}} ((\mathbb{C}\varepsilon - \sigma) \cdot \tilde{\varepsilon} - (b + \operatorname{div}\sigma) \cdot \tilde{u}) d\mu \\ &+ \int_{\mathcal{B}} (\operatorname{Sym} Du - \varepsilon) \cdot \tilde{\sigma} \, d\mu + \int_{\partial\mathcal{B}_u} \tilde{\sigma}n \cdot (\bar{u} - u) d\mathcal{H}^2 \\ &+ \int_{\partial\mathcal{B}_f} (\sigma n - f) \cdot \tilde{u} \, d\mathcal{H}^2.\end{aligned} \tag{5.9}$$

If ς is a solution of the mixed boundary value problem, then $\delta_{\tilde{\varsigma}}\mathcal{E}_{HW}(\varsigma) = 0$ for every $\tilde{\varsigma} \in \mathcal{A}$, i.e.,

$$\delta\mathcal{E}_{HW}(\varsigma) = 0. \tag{5.10}$$

Conversely, assume that (5.10) holds. By taking $\tilde{\varsigma} \in \mathcal{A}$ of the form $\tilde{\varsigma} = (\tilde{u}, 0, 0)$, equation (5.9) implies

$$\int_{\mathcal{B}} (b + \operatorname{div}\sigma) \cdot \tilde{u} \, d\mu = 0 \tag{5.11}$$

for every $\tilde{u} \in \mathcal{A}$, i.e., $b + \operatorname{div}\sigma = 0$ thanks to the arbitrariness of $\tilde{u}$. The choice that $\tilde{u}$ vanishes near $\partial\mathcal{B}_f$ and the previous result imply also

$$\int_{\partial\mathcal{B}_f} (\sigma n - f)\cdot \tilde{u}\, d\mathcal{H}^2 = 0 \tag{5.12}$$

for every $\tilde{u} \in \mathcal{A}$, i.e., $\sigma n = f$ along $\partial\mathcal{B}_f$. Also, for $\tilde{\varsigma} \in \mathcal{A}$ of the form $\tilde{\varsigma} = (0, \tilde{\varepsilon}, 0)$, from equations (5.9) and (5.10) we obtain

$$\int_{\mathcal{B}} (\mathbb{C}\varepsilon - \sigma)\cdot \tilde{\varepsilon}\, d\mu = 0 \tag{5.13}$$

for every $\tilde{\varepsilon}$, i.e., $\sigma = \mathbb{C}\varepsilon$. For $\tilde{\varsigma} \in \mathcal{A}$ of the form $\tilde{\varsigma} = (0, 0, \tilde{\sigma})$, with $\tilde{\sigma}$ vanishing near $\partial\mathcal{B}$, from equations (5.9) and (5.10), we get

$$\int_{\mathcal{B}} (\mathrm{Sym}\, Du - \varepsilon)\cdot \tilde{\sigma}\, d\mu = 0 \tag{5.14}$$

for every $\tilde{\sigma}$, i.e., the strain–displacement relation $\varepsilon = \mathrm{Sym}\, Du$. Moreover, if $\tilde{\sigma}$ vanishes everywhere but $\partial\mathcal{B}_u$, from equations (5.9) and (5.10) we obtain

$$\int_{\partial\mathcal{B}_u} \tilde{\sigma} n \cdot (u - \bar{u}) d\mathcal{H}^2 = 0 \tag{5.15}$$

for every $\tilde{\sigma}$, i.e., $u = \bar{u}$ along $\partial\mathcal{B}_u$. The relations (5.11) to (5.15) then imply that $\varsigma = (u, \varepsilon, \sigma)$ is a solution of the mixed boundary value problem in linearized elastostatics. □

5.5 The Betti Reciprocal Theorem

Consider a body composed of a linear-elastic material with constitutive tensor $\mathbb{C}$ depending continuously on x and having all major and minor symmetries.

Consider a first pair of bulk and surface actions, namely $b_1^{\ddagger}$ and f_1, acting over the body, and denote by u_1 the displacement field solving the mixed boundary value problem with boundary data f_1 along $\partial\mathcal{B}_f$ and $u = \bar{u}$ along $\partial\mathcal{B}_f$.

Take now a second pair of actions $b_2^{\ddagger}$ and f_2, with the pertinent displacement field u_2 as above. Notice that both $b_1^{\ddagger}$ and $b_2^{\ddagger}$ may include inertial actions, so that what we discuss here holds also in elastodynamics.

Theorem 6 (Enrico Betti, 1872). *Under the previous conditions, the external work performed by the actions $b_1^{\ddagger}$ and f_1 over the displacements u_2 equals the analogous work of the pair $b_2^{\ddagger}$, f_2 along u_1.*

Proof. We have already shown in Chapter 3 that the invariance of the external power under classical changes in observers implies the identity with the internal power under appropriate regularity conditions of the fields involved. That expression has its counterpart in terms of work, as emerges, e.g., by multiplying the pointwise balances by u and integrating over $\mathcal{B}$. As a consequence, we can write

$$\int_{\mathcal{B}} b_1^{\ddagger} \cdot u_2 \, d\mu + \int_{\partial\mathcal{B}_f} f_1 \cdot u_2 \, d\mathcal{H}^2 = \int_{\mathcal{B}} \sigma_1 \cdot \varepsilon_2 \, d\mu$$

and

$$\int_{\mathcal{B}} b_2^{\ddagger} \cdot u_1 \, d\mu + \int_{\partial\mathcal{B}_f} f_2 \cdot u_1 \, d\mathcal{H}^2 = \int_{\mathcal{B}} \sigma_2 \cdot \varepsilon_1 \, d\mu .$$

Due to the major symmetry of $\mathbb{C}$, we have also

$$\sigma_1 \cdot \varepsilon_2 = \mathbb{C}\varepsilon_1 \cdot \varepsilon_2 = \varepsilon_1 \cdot \mathbb{C}\varepsilon_2 = \varepsilon_1 \cdot \sigma_2 ,$$

which proves the theorem, namely

$$\int_{\mathcal{B}} b_1^{\ddagger} \cdot u_2 \, d\mu + \int_{\partial\mathcal{B}_f} f_1 \cdot u_2 \, d\mathcal{H}^2 = \int_{\mathcal{B}} b_2^{\ddagger} \cdot u_1 \, d\mu + \int_{\partial\mathcal{B}_f} f_2 \cdot u_1 \, d\mathcal{H}^2 . \qquad \square$$

5.6 Kirchhoff's Theorem

Theorem 7 (Gustav Robert Kirchhoff, 1859). *The solution of the mixed boundary value problem for the equilibrium of a linear-elastic body is unique to within a rigid change of place.*

Proof. Let u' and u'' be two solutions of the mixed boundary value problem specified above; ε' and ε'' are the pertinent strain fields, while σ' and σ'' are the respective stress fields associated with them by the constitutive tensor $\mathbb{C}$. By assumption, we have

$$\operatorname{div} \sigma' + b = 0 \ \text{ in } \mathcal{B}, \qquad \sigma' n = f \ \text{ on } \partial\mathcal{B}_f, \qquad u' = \bar{u} \ \text{ on } \partial\mathcal{B}_u,$$
$$\sigma' = \sigma'^{\mathrm{T}}$$

and

$$\operatorname{div} \sigma'' + b = 0 \ \text{ in } \mathcal{B}, \qquad \sigma'' n = f \ \text{ on } \partial\mathcal{B}_f, \quad u'' = \bar{u} \ \text{ on } \partial\mathcal{B}_u,$$
$$\sigma'' = \sigma''^{\mathrm{T}}.$$

By subtraction and after denoting by $\bar{\bar{\sigma}}$ and $\bar{\bar{u}}$ the differences $\bar{\bar{\sigma}} = \sigma' - \sigma''$ and $\bar{\bar{u}} = u' - u''$, we obtain

$$\operatorname{div} \bar{\bar{\sigma}} = 0, \qquad \bar{\bar{\sigma}} n = 0 \ \text{ on } \partial\mathcal{B}_f, \qquad \bar{\bar{u}} = 0 \ \text{ on } \partial\mathcal{B}_u .$$

By multiplication by $\bar{\bar{u}}$ and integration, we get also

$$0 = \int_{\mathcal{B}} \bar{\bar{u}} \cdot \operatorname{div} \bar{\bar{\sigma}} \, d\mu = \int_{\mathcal{B}} \bar{\bar{\sigma}} \cdot D\bar{\bar{u}} \, d\mu = \int_{\mathcal{B}} \bar{\bar{\sigma}} \cdot \bar{\bar{\varepsilon}} \, d\mu, \tag{5.16}$$

where $\bar{\bar{\varepsilon}} := \operatorname{Sym} D\bar{\bar{u}}$ and we have used the symmetry of $\bar{\bar{\sigma}}$ and the boundary conditions $\bar{\bar{\sigma}} n = 0$ on $\partial\mathcal{B}_f$ and $\bar{\bar{u}} = 0$ on $\partial\mathcal{B}_u$, so that

$$\int_{\partial\mathcal{B}} \bar{\bar{\sigma}} n \cdot \bar{\bar{u}} \, d\mathcal{H}^2 = 0.$$

Equation (5.16) can then be written as

$$\int_{\mathcal{B}} \bar{\bar{\varepsilon}} \cdot \mathbb{C} \bar{\bar{\varepsilon}} \, d\mu = 0,$$

since the material composing the body is linear-elastic. Since $\mathbb{C}$ is positive definite, equality holds when $\bar{\bar{\varepsilon}} = 0$, i.e., $\varepsilon' = \varepsilon''$, so u' differs by u'' just by a rigid displacement. □

5.7 The Navier Equations and the Biharmonic Problem

In this section, we consider a flat space, i.e., the metric is orthogonal, namely $g_{ij} = \delta_{ij}$. We then use systematically the natural identification of $\mathbb{R}^3$ with its dual $\mathbb{R}^{3*}$. For this reason, we shall not distinguish covariant and contravariant components, and we shall write the standard symbol of gradient, ∇, instead of derivative D. The choice is dictated by the desire to express the results in this section and the following one in traditional fashion.

For the reader's convenience, we recall here some identities in tensor analysis before going into details.

- Let $x \longmapsto a(x) \in \mathbb{R}^3$ be a twice differentiable vector field. The following identities hold:

$$\operatorname{div}((\operatorname{div} a) I) = \nabla \operatorname{div} a, \tag{5.17}$$

$$\operatorname{div}(\nabla a)^{\mathrm{T}} = \nabla \operatorname{div} a, \tag{5.18}$$

$$\operatorname{div} \operatorname{curl} a = 0, \tag{5.19}$$

$$\Delta a = \nabla \operatorname{div} a - \operatorname{curl} \operatorname{curl} a. \tag{5.20}$$

- If $a(\cdot)$ is such that

$$\operatorname{div} a = 0 \quad \text{and} \quad \operatorname{curl} a = 0,$$

then a is harmonic, namely

$$\Delta a = 0.$$

– If $a(x)$ is a covector instead of a vector, then the previous identities hold, provided that we make the substitution of ∇ with D, even in the definition of the Laplacian operator and the curl.

We disregard here inertia and assume that the displacement is four times differentiable. We consider a body composed of a linear-elastic isotropic material, i.e., we recall

$$\sigma = \mathbb{C}\varepsilon = 2\mu\varepsilon + \lambda \operatorname{tr}\varepsilon I = \mu(\nabla u + \nabla u^{\mathrm{T}}) + \lambda(\operatorname{div} u)I.$$

Here λ and μ are differentiable functions of x. By inserting such an expression into the balance of forces, we get

$$\operatorname{div}(\mu(\nabla u + \nabla u^{\mathrm{T}}) + \lambda(\operatorname{div} u)I) + b = 0,$$

and using the identities (5.17) to (5.19), we compute

$$\mu\Delta u + 2(\operatorname{Sym}\nabla u)\nabla\mu + (\lambda + \mu)\nabla\operatorname{div} u + (\operatorname{div} u)\nabla\lambda + b = 0.$$

When λ and μ are constant, the previous balance reduces to

$$\mu\Delta u + (\lambda + \mu)\nabla\operatorname{div} u + b = 0, \tag{5.21}$$

which is the system of **Navier equations**. In coordinates, we have

$$\mu u_{i,kk} + (\lambda + \mu)u_{k,ki} + b_i = 0,$$

where the choice of not distinguishing between covariant and contravariant components is evident.

Using the identity (5.20), we can rewrite equation (5.21) as

$$(\lambda + 2\mu)\nabla\operatorname{div} u - \mu \operatorname{curl}\operatorname{curl} u + b = 0. \tag{5.22}$$

If we evaluate first the divergence of equation (5.22) and then its curl, we obtain respectively

$$(\lambda + 2\mu)\Delta\operatorname{div} u = -\operatorname{div} b \tag{5.23}$$

and

$$-\mu \operatorname{curl}\operatorname{curl}\operatorname{curl} u = -\operatorname{curl} b. \tag{5.24}$$

Since $\operatorname{curl} u$ is still a vector, taking into account the identity (5.19), we get

$$\operatorname{curl}\operatorname{curl}\operatorname{curl} u = \Delta\operatorname{curl} u.$$

As a consequence, equation (5.24) can be written as

$$\mu\Delta\operatorname{curl} u = -\operatorname{curl} b. \tag{5.25}$$

When $\operatorname{curl} b = 0$ and $\operatorname{div} b = 0$, with $1 + 2\mu \neq 0$ and $\mu \neq 0$, the equations (5.23) and (5.25) reduce to

$$\Delta\operatorname{div} u = 0 \quad \text{and} \quad \Delta\operatorname{curl} u = 0,$$

that is,

$$\operatorname{div}\Delta u = 0 \quad \text{and} \quad \operatorname{curl}\Delta u = 0,$$

since the Laplacein differential operator commutes with div and curl. These last relations imply then that Δu is a harmonic vector function, namely

$$\Delta\Delta u = 0,$$

i.e., u is biharmonic.

5.8 The Beltrami–Donati–Michell Equations

The one-to-one stress–strain correspondence in linear elasticity allows us to express the strain compatibility conditions $\operatorname{curl}\operatorname{curl}\varepsilon = 0$ in terms of the stress σ when the map $x \longmapsto \mathbb{K}(x) = \mathbb{C}^{-1}(x)$ is $C^2(\mathcal{B})$. The resulting expression

$$\operatorname{curl}\operatorname{curl}(\mathbb{K}\sigma) = 0 \tag{5.26}$$

assumes a form useful in special analyses developed later when the material is isotropic, i.e., when

$$\varepsilon = \frac{1}{2\mu}\left(\sigma - \frac{\nu}{1+\nu}(\operatorname{tr}\sigma)I\right). \tag{5.27}$$

To get such a special form of equation (5.26), we primarily make use of an identity in tensor analysis that is valid for every twice differentiable symmetric tensor-valued field:

$$\operatorname{curl}\operatorname{curl}\varepsilon = -\Delta\varepsilon + 2\operatorname{Sym}\nabla\operatorname{div}\varepsilon - \nabla\nabla\operatorname{tr}\varepsilon + I(\Delta\operatorname{tr}\varepsilon - \operatorname{div}\operatorname{div}\varepsilon),$$

provided that ε is twice differentiable.[1] Then the strain compatibility condition curl curl $\varepsilon = 0$ implies

$$-\Delta\varepsilon + 2\,\mathrm{Sym}\nabla\mathrm{div}\,\varepsilon - \nabla\nabla\mathrm{tr}\,\varepsilon + I(\Delta\mathrm{tr}\,\varepsilon - \mathrm{div}\,\mathrm{div}\,\varepsilon) = 0. \tag{5.28}$$

Its trace reads

$$\Delta\mathrm{tr}\,\varepsilon - \mathrm{div}\,\mathrm{div}\,\varepsilon = 0, \tag{5.29}$$

an expression that can be derived by taking into account the identities

$$\begin{aligned}
&\mathrm{tr}\Delta\varepsilon = \mathrm{tr}\frac{\partial^2\varepsilon_{ij}}{\partial x^h\partial x^h} = \frac{\partial^2\varepsilon_{ii}}{\partial x^h\partial x^h} = \Delta\mathrm{tr}\,\varepsilon,\\
&\mathrm{tr}\nabla\mathrm{div}\,\varepsilon = \mathrm{div}\,\mathrm{div}\,\varepsilon,\\
&\mathrm{tr}\nabla\nabla\mathrm{tr}\,\varepsilon = \Delta\mathrm{tr}\,\varepsilon.
\end{aligned}$$

Equation (5.29) then implies

$$\Delta\varepsilon - 2\,\mathrm{Sym}\nabla\mathrm{div}\,\varepsilon + \nabla\nabla\mathrm{tr}\,\varepsilon = 0. \tag{5.30}$$

In the case of a linear-elastic homogeneous and isotropic material, using the constitutive relation in the form (5.27) and taking into account the pointwise balance of forces in the absence of inertia, we can write

$$\begin{aligned}
&2\mu\Delta\varepsilon = \Delta\left(\sigma - \frac{\nu}{1+\nu}(\mathrm{tr}\,\sigma)I\right) = \Delta\sigma - \frac{\nu}{1+\nu}(\Delta\mathrm{tr}\,\sigma)I,\\
&-2\mu 2\,\mathrm{Sym}\nabla\mathrm{div}\,\varepsilon = -2\,\mathrm{Sym}\nabla\mathrm{div}\left(\sigma - \frac{\nu}{1+\nu}(\mathrm{tr}\,\sigma)I\right)\\
&= 2\,\mathrm{Sym}\nabla b + \frac{2\nu}{1+\nu}\nabla\nabla\mathrm{tr}\,\sigma,\\
&2\mu\nabla\nabla\mathrm{tr}\,\varepsilon = \nabla\nabla\mathrm{tr}\,\sigma - \frac{3\nu}{1+\nu}\nabla\nabla\mathrm{tr}\,\sigma.
\end{aligned}$$

As a consequence, we rewrite equation (5.30) as

$$\Delta\sigma - \frac{\nu}{1+\nu}(\Delta\mathrm{tr}\,\sigma)I + 2\,\mathrm{Sym}\nabla b + \frac{1}{1+\nu}\nabla\nabla\mathrm{tr}\,\sigma = 0. \tag{5.31}$$

---

[1] As we stated in the previous section, we exploit systematically the natural identification of $\mathbb{R}^3$ with its dual $\mathbb{R}^{3*}$, confusing in this way covariant and contravariant components, so ∇ with D, and adopt a flat metric. However, the result holds also in a more general setting (nonflat metric), provided some care is taken in writing appropriately the intermediate steps and writing ∇ or D depending on the initial choice of the contravariant, covariant, or mixed nature of the components of the strain ε.

Its trace is given by

$$\Delta \operatorname{tr} \sigma - \frac{3\nu}{1+\nu} \Delta \operatorname{tr} \sigma + 2 \operatorname{div} b + \frac{1}{1+\nu} \Delta \operatorname{tr} \sigma = 0,$$

i.e.,

$$\frac{1-\nu}{1+\nu} \Delta \operatorname{tr} \sigma = -\operatorname{div} b.$$

Then equation (5.30) becomes

$$\Delta \sigma + \frac{1}{1+\nu} \nabla\nabla \operatorname{tr} \sigma + \frac{\nu}{1+\nu} \operatorname{div} b + 2 \operatorname{Sym} \nabla b = 0, \tag{5.32}$$

which reduces to

$$\Delta \sigma + \frac{1}{1+\nu} \nabla\nabla \operatorname{tr} \sigma = 0 \tag{5.33}$$

when b is constant. Both (5.32) and (5.33) are called the **Beltrami–Donati–Michell equations** after the work of Eugenio Beltrami (1892), L. Donati (1894), and John Henry Michell (1899, 1900).

5.9 Plane Problems

5.9.1 Plane-Strain States

Take a narrow right but not necessarily circular cylinder loaded in a way that the displacement along the cylinder's axis vanishes, while its orthogonal components remain in the cross-section plane at each point. In this case, we say that the cylinder is in what we commonly call a **plane-strain state**. Here we discuss this state without accounting for inertial effects, i.e., focusing attention on the statics. We refer analysis to an orthonormal frame $\{O, x_1, x_2, x_3\}$ with the x_3-axis coinciding with the cylinder's axis.

Formally, the plane-strain state in the plane $x_1 x_2$ is defined by the conditions

$$u_1 = \tilde{u}_1(x_1, x_2), \qquad u_2 = \tilde{u}_2(x_1, x_2), \qquad u_3 = 0,$$

where u_1, u_2, u_3 are the displacement components. The associated small-strain tensor components are

$$\varepsilon_{11} = \frac{\partial u_1(x_1, x_2)}{\partial x_1}, \qquad \varepsilon_{22} = \frac{\partial u_2(x_1, x_2)}{\partial x_2}, \tag{5.34}$$

$$\varepsilon_{12} = \frac{1}{2}\left(\frac{\partial u_1(x_1,x_2)}{\partial x_2} + \frac{\partial u_2(x_1,x_2)}{\partial x_1}\right), \tag{5.35}$$

$$\varepsilon_{13} = \varepsilon_{23} = \varepsilon_{33} = 0. \tag{5.36}$$

Moreover, we realize immediately that

$$\frac{\partial \varepsilon_{11}}{\partial x_3} = \frac{\partial \varepsilon_{22}}{\partial x_3} = \frac{\partial \varepsilon_{12}}{\partial x_3} = 0.$$

The question of strain compatibility in the planar case replicates what has been discussed in the three-dimensional ambient. Given functions

$$(x_1,x_2) \longmapsto \varepsilon_{11} := \tilde{\varepsilon}_{11}(x_1,x_2),$$

$$(x_1,x_2) \longmapsto \varepsilon_{22} := \tilde{\varepsilon}_{22}(x_1,x_2),$$

$$(x_1,x_2) \longmapsto \varepsilon_{12} := \tilde{\varepsilon}_{12}(x_1,x_2),$$

are there differential functions

$$(x_1,x_2) \longmapsto u_1(x_1,x_2), \qquad (x_1,x_2) \longmapsto u_2(x_1,x_2),$$

satisfying (5.34) and (5.35) pointwise over the body?

A necessary condition can be derived when the functions $\tilde{\varepsilon}_{11}$, $\tilde{\varepsilon}_{22}$, $\tilde{\varepsilon}_{12}$ are twice differentiable. In this case, if the strain were compatible, we would have

$$\frac{\partial^2 \varepsilon_{11}}{\partial x_2^2} = \frac{\partial^3 u_1(x_1,x_2)}{\partial x_1 \partial x_2^2}, \qquad \frac{\partial^2 \varepsilon_{22}}{\partial x_1^2} = \frac{\partial^3 u_2(x_1,x_2)}{\partial x_2 \partial x_1^2},$$

$$2\frac{\partial^2 \varepsilon_{12}}{\partial x_1 x_2} = \frac{\partial^3 u_1(x_1,x_2)}{\partial x_1 \partial x_2^2} + \frac{\partial^3 u_2(x_1,x_2)}{\partial x_1^2 \partial x_2},$$

from which we deduce what we call the **plane compatibility condition**:

$$\frac{\partial^2 \varepsilon_{11}}{\partial x_2^2} + \frac{\partial^2 \varepsilon_{22}}{\partial x_1^2} = 2\frac{\partial^2 \varepsilon_{12}}{\partial x_1 x_2},$$

or, writing more synthetically,

$$\varepsilon_{11,22} + \varepsilon_{22,11} = 2\varepsilon_{12,12}, \tag{5.37}$$

where once again, the comma denotes differentiation.

Consider a homogeneous linear-elastic isotropic material. Taking into account the simplification (5.36), for α, β, $\gamma = 1, 2$ we write

$$\sigma_{\alpha\beta} = 2\mu\varepsilon_{\alpha\beta} + \lambda\varepsilon_{\gamma\gamma}\delta_{\alpha\beta}, \tag{5.38}$$

$$\sigma_{23} = 0, \qquad \sigma_{13} = 0, \qquad \sigma_{33} = \lambda\varepsilon_{\gamma\gamma}, \tag{5.39}$$

so that

$$\sigma_{\alpha\alpha} = 2(\mu + \lambda)\varepsilon_{\alpha\alpha}$$

and

$$\sigma_{33} = \frac{\lambda}{2(\mu + \lambda)}\sigma_{\alpha\alpha} = \nu\,\sigma_{\alpha\alpha}.$$

The last relation allows us to write

$$2\mu\varepsilon_{11} = (1 - \nu)\sigma_{11} - \nu\sigma_{22},$$
$$2\mu\varepsilon_{22} = (1 - \nu)\sigma_{22} - \nu\sigma_{11},$$
$$2\mu\varepsilon_{12} = \sigma_{12},$$

so that the compatibility condition (5.37) becomes

$$\sigma_{11,22} - \nu\sigma_{11,22} - \nu\sigma_{22,22} + \sigma_{22,11} - \nu\sigma_{22,11} - \nu\sigma_{11,11} = 2\sigma_{12,12}.$$

The left-hand side term equals

$$(1 - \nu)\Delta(\sigma_{11} + \sigma_{22}) - \sigma_{11,11} - \sigma_{22,22},$$

as can be shown by computing the Laplacian, so that the compatibility condition becomes

$$(1 - \nu)\Delta(\sigma_{11} + \sigma_{22}) = \sigma_{11,11} + \sigma_{22,22} + 2\sigma_{12,12}. \tag{5.40}$$

Since σ_{23} and σ_{13} vanish in the plane-strain conditions, the balance equations can be written as

$$\sigma_{11,1} + \sigma_{12,2} + b_1 = 0,$$
$$\sigma_{21,1} + \sigma_{22,2} + b_2 = 0,$$
$$\sigma_{33,3} + b_3 = 0,$$

with boundary conditions

$$\sigma_{11}n_1 + \sigma_{12}n_2 = f_1,$$
$$\sigma_{21}n_1 + \sigma_{22}n_2 = f_2,$$
$$\sigma_{33}n_3 = f_3,$$

where f_1, f_2, f_3 are the components of the applied forces f distributed over the body's boundary.

Since σ_{33} depends only on x_1 and x_2, we must have

$$b_3 = 0,$$

while since $n_3 = 0$ for the x_3-axis coincides with the axis of the cylinder, we must also have

$$f_3 = 0.$$

The two last equations are the conditions to be satisfied by the load (bulk and surface forces) to ensure the possibility of the plane-strain state.

Assume that b_1 and b_2 are differentiable functions of the space variables. By taking the derivatives of the first two balance equations above with respect to x_1 and x_2 respectively and summing up the results, we get

$$\sigma_{11,11} + \sigma_{22,22} + 2\sigma_{12,12} + b_{1,1} + b_{2,2} = 0,$$

so that the compatibility equation (5.40) becomes

$$(1 - \nu)\Delta(\sigma_{11} + \sigma_{22}) = -(b_{1,1} + b_{2,2}), \tag{5.41}$$

which is the reduction of the Beltrami–Donati–Michell equation to the plane-strain case.

5.9.2 Plane-Stress States

Consider a planar thin film loaded in its plane, say x_1x_2. In the small-strain regime, we can reasonably claim that the stress components $\sigma_{13}, \sigma_{23}, \sigma_{33}$ vanish, determining what we commonly call a **plane-stress state**.

Formally, we write

$$\sigma_{13} = \sigma_{23} = \sigma_{33} = 0, \qquad \sigma_{11,3} = \sigma_{22,3} = \sigma_{12,3} = 0.$$

The balance equations then reduce to

$$\sigma_{11,1} + \sigma_{12,2} + b_1 = 0,$$

$$\sigma_{21,1} + \sigma_{22,2} + b_2 = 0,$$

with boundary conditions

$$\sigma_{11}n_1 + \sigma_{12}n_2 = f_1, \tag{5.42}$$

$$\sigma_{21}n_1 + \sigma_{22}n_2 = f_2. \tag{5.43}$$

They imply the same conditions on the load that we have found in the plane-strain case:

$$b_3 = 0, \qquad f_3 = 0.$$

But there is something more. Under conditions of plane stress, the constitutive relations for a homogeneous, linear-elastic, isotropic body read explicitly

$$\sigma_{\alpha\beta} = 2\mu\varepsilon_{\alpha\beta} + \lambda \mathrm{tr}\,\varepsilon\delta_{\alpha\beta}, \tag{5.44}$$

$$0 = 2\mu\varepsilon_{33} + \lambda \mathrm{tr}\,\varepsilon, \qquad 0 = 2\mu\varepsilon_{13}, \qquad 0 = 2\mu\varepsilon_{23}. \tag{5.45}$$

From equation $(5.45)_1$, we get

$$\varepsilon_{33} = -\frac{\lambda}{(2\mu+\lambda)}(\varepsilon_{11} + \varepsilon_{22}). \tag{5.46}$$

Consequently, with the notation

$$\lambda^* := \frac{2\mu\lambda}{2\mu+\lambda},$$

we can write

$$\sigma_{\alpha\beta} = 2\mu\varepsilon_{\alpha\beta} + \lambda^*\varepsilon_{\gamma\gamma}\delta_{\alpha\beta}, \tag{5.47}$$

from which we compute

$$\sigma_{\alpha\alpha} = 2(\mu + \lambda^*)\varepsilon_{\alpha\alpha}. \tag{5.48}$$

The relations already obtained for planar strain states then hold, provided we substitute ν with

$$\nu^* := \frac{\lambda^*}{2(\mu+\lambda^*)}.$$

By substituting equation (5.48) into (5.47), we obtain, in fact,

$$2\mu\varepsilon_{\alpha\beta} = \sigma_{\alpha\beta} - \nu^*\sigma_{\gamma\gamma}\delta_{\alpha\beta}, \tag{5.49}$$

with $\alpha, \beta, \gamma = 1, 2$ once more.

For the plane-stress states the Beltrami–Donati–Michell equations can be written componentwise as

$$\varepsilon_{11,22} + \varepsilon_{22,11} = 2\varepsilon_{12,12}, \tag{5.50}$$

$$\varepsilon_{33,11} = 0, \qquad \varepsilon_{33,22} = 0, \qquad \varepsilon_{33,12} = 0. \tag{5.51}$$

Substituting the relation (5.49) into (5.50) and using the balance equations as in the previous section, we derive an equation differing from (5.41) by the replacement of ν with ν^*:

$$(1 - \nu^*)\Delta(\sigma_{11} + \sigma_{22}) = -(b_{1,1} + b_{2,2}) = -\operatorname{div} b.$$

From (5.51), with the use of (5.46) and (5.49), we obtain also

$$(\sigma_{11} + \sigma_{22})_{,11} = 0, \qquad (\sigma_{11} + \sigma_{22})_{,22} = 0, \qquad (\sigma_{11} + \sigma_{22})_{,12} = 0.$$

In other words, the Beltrami–Donati–Michell equations above require that in plane-stress states, the trace of the stress tensor must be a degree-one polynomial:

$$\sigma_{\alpha\alpha} = \sigma_{11} + \sigma_{22} = a_1 x_1 + a_2 x_2 + a_3,$$

with a_1, a_2, a_3 constants to be determined. Such a polynomial is a harmonic function, so that a necessary condition to realize a plane-stress state in a homogeneous, linear-elastic, isotropic material is

$$\operatorname{div} b = 0,$$

in addition to $b_3 = 0$, as already derived from the balance equations.

5.9.3 The Airy Stress Function

Let us consider a body consisting of a homogeneous, linear-elastic, and isotropic material. We assume here that the body forces vanish.

In a state of plane stress or plane strain, the stress components can be expressed in terms of the second derivatives of a scalar-valued function of x_1 and x_2 that is biharmonic when the body actions vanish. The evaluation of such a function depends on the body shape and the boundary conditions. It is called an **Airy stress function** after a 1862–1863 work by Sir George Biddell Airy (1801–1892).

Assume $b = 0$ and the absence of inertial effects.

Theorem 8. *Let $(x_1, x_2) \longmapsto \chi = \tilde{\chi}(x_1, x_2) \in \mathbb{R}$ be of class C^3 in $\mathcal{B}$ and such that*

$$\sigma_{11} = \chi_{,22}, \qquad \sigma_{22} = \chi_{,11}, \qquad \sigma_{12} = -\chi_{,12}. \tag{5.52}$$

The local balance of forces

$$\sigma_{\alpha\beta,\beta} = 0 \tag{5.53}$$

holds. Further, the version of the Beltrami–Donati–Michell equations reading

$$\Delta\sigma_{\alpha\alpha} = 0 \tag{5.54}$$

is satisfied if and only if $\tilde{\chi}$ *is biharmonic. Conversely, if* $(x_1, x_2) \longmapsto \sigma_{\alpha\beta} \in \mathrm{Sym}(\mathbb{R}^3, \mathbb{R}^3)$ *is a single-valued map of class* C^N $(N \geq 1)$ *on* $\mathcal{B}$ *and (5.53) holds, then there exists a* C^{N+2} *scalar-valued function satisfying (5.52).*

Proof. By computing the divergence of the stress field under the assumption (5.52), we realize directly that the local balance of forces (5.53) follows. Moreover, since

$$\sigma_{11} + \sigma_{22} = \chi_{,22} + \chi_{,11} = \Delta\chi,$$

the compatibility equation (5.54) supports the biharmonicity condition

$$\Delta\Delta\chi = 0$$

for χ. To prove the converse assertion, assume that equation (5.54) holds. The balance of forces means that

$$\sigma_{11,1} = -\sigma_{12,2},$$
$$\sigma_{22,2} = -\sigma_{21,1},$$

i.e., that there exist C^{N+1} functions $(x_1, x_2) \longmapsto l := \tilde{l}(x_1, x_2) \in \mathbb{R}$ and $(x_1, x_2) \longmapsto h := \tilde{h}(x_1, x_2) \in \mathbb{R}$ such that

$$\sigma_{11} = l_{,2}, \qquad \sigma_{12} = -l_{,1},$$
$$\sigma_{22} = h_{,1}, \qquad \sigma_{21} = -h_{,2}.$$

The symmetry of σ implies

$$l_{,1} = h_{,2},$$

i.e., the existence of a C^{N+2} function $(x_1, x_2) \longmapsto \chi := \tilde{\chi}(x_1, x_2) \in \mathbb{R}$ defined on $\mathcal{B}$ such that

$$l = \chi_{,2}, \qquad h = \chi_{,1},$$

a property implying the stress representation (5.52). □

The previous proof does not exclude the possibility that the **Airy stress function** $\tilde{\chi}$ could be multivalued, provided that its second derivatives, i.e., the stress components, are single-valued. In any case, when $\mathcal{B}$ is simply connected, the previous conditions require that $\tilde{\chi}$ be single-valued.

Given a biharmonic function $\hat{\chi}$, a possible Airy stress function $\tilde{\chi}$ defined by

$$\tilde{\chi}(x_1, x_2) = \hat{\chi}(x_1, x_2) + a_1 x_1 + a_2 x_2 + a_3,$$

with a_1, a_2, a_3 constants, is biharmonic and generates the same stress state determined by $\hat{\chi}$. Given a point $\bar{x} = (x_1, x_2)$, we can always choose the constants generating $\tilde{\chi}$ to be such that

$$\tilde{\chi}(\bar{x}) = 0, \qquad \frac{\partial \tilde{\chi}(\bar{x})}{\partial x_1} = \frac{\partial \tilde{\chi}(\bar{x})}{\partial x_2} = 0. \tag{5.55}$$

In terms of the Airy stress function, the boundary conditions (5.42) and (5.43) can be written

$$\chi_{,22} n_1 - \chi_{,12} n_2 = f_1, \tag{5.56}$$

$$-\chi_{,12} n_1 + \chi_{,11} n_2 = f_2. \tag{5.57}$$

Consider a simply connected body with boundary in the plane $x_1 x_2$ a simple smooth curve that we parameterize by arc length s. Assume also that the condition (5.55) holds at a point $\bar{x} = \tilde{x}(0)$ of the body's boundary described by a map $s \longmapsto \tilde{x}(s) = (\tilde{x}_1(s), \tilde{x}_2(s))$. With n the normal to that curve at a point, the tangent t is characterized by

$$\frac{d\tilde{x}_1(s)}{ds} = \mathsf{t}_1 = -n_2, \qquad \frac{d\tilde{x}_2(s)}{ds} = \mathsf{t}_2 = n_1.$$

Consequently, the conditions (5.56) and (5.57) become

$$\chi_{,12} \frac{dx_1}{ds} + \chi_{,22} \frac{dx_2}{ds} = f_1,$$

$$-\chi_{,11} \frac{dx_1}{ds} - \chi_{,12} \frac{dx_2}{ds} = f_2,$$

where $x_1 = \tilde{x}_1(s)$ and $x_2 = \tilde{x}_2(s)$, and they can be written as

$$\frac{d\chi_{,2}}{ds} = f_1, \qquad -\frac{d\chi_{,1}}{ds} = f_2.$$

If we integrate these last two equations from 0 to s and set the integration constant to zero, thanks to the condition $(5.55)_2$, we get

$$\chi_{,2}(s) = \int_0^s f_1(\zeta) d\zeta =: H_1(s),$$

$$-\chi_{,1}(s) = \int_0^s f_2(\zeta) d\zeta =: H_2(s),$$

i.e.,

$$\chi(s) = \int_0^s (H_1(\zeta)\, dx_2(\zeta) - H_2(\zeta)\, dx_1(\zeta)) =: H(s) + \hat{a},$$

with an integration constant $\hat{a}$ that we can set to zero thanks to the assumption $(5.55)_1$.

Moreover, we can also compute

$$\frac{d\chi}{dn} = \chi_{,1} n_1 + \chi_{,2} n_2 = -H_2(s)\frac{dx_2}{ds} + H_1(s)\frac{dx_1}{ds} =: G(s).$$

The biharmonic problem involving the Airy stress function χ is completed by the boundary conditions

$$\chi(s) = H(s), \qquad \frac{d\chi}{dn}(s) = G(s),$$

when the body is simply connected. For multiply connected bodies, the boundary in the plane x_1x_2 is the union of a finite number of (disjoint) closed curves. The condition (5.55) can be set only on one of them, so that in deriving boundary conditions, integration constants appear and can be evaluated by requiring that the displacements be single-valued. We do not discuss here the details of the pertinent analyses, which are by now a classical matter, as is all of the material in this chapter. We just remark that determining the Airy stress function depends on the shape of the body and the boundary conditions determined by the applied forces. Explicit solutions useful in applications can be found in standard treatises on linear elasticity.

5.9.4 Further Remarks

Imagine a plane-strain state characterized by the additional condition

$$\varepsilon_{\alpha\alpha} = 0.$$

In this case, the relation $(5.39)_3$ implies

$$\sigma_{33} = 0,$$

so that we have plane strain with in-plane stress. In this case, the sole non-a-priori vanishing constitutive relation is (5.38).

Chapter 6
The de Saint-Venant Problem

6.1 Statement of the Problem

6.1.1 Geometry

We consider a long straight cylinder with constant cross section that can even have holes—the cylinder can be a type of tube—provided that their closures have no points in common and their boundaries have nonzero minimum distance from the cross-sectional boundary.

To formalize the previous description, we assign at a point O in $\mathcal{E}^3$ three orthogonal unit vectors e_1, e_2, e_3, identifying $\mathcal{E}^3$ with $\mathbb{R}^3$ in this way.

Within the plane $\operatorname{span} e_1 \otimes \operatorname{span} e_2$, we select a region Ω with the following properties:

1. Ω is open and bounded.
2. Ω is linearly connected, which means that for every pair of points in Ω, there is a path in Ω connecting the two.
3. Ω is of the type

$$\Omega = \hat{\Omega} \backslash \hat{\Omega}_\beta,$$

P.M. Mariano, L. Galano, *Fundamentals of the Mechanics of Solids*,
DOI 10.1007/978-1-4939-3133-0_6

with $\beta = 1, 2, \ldots, m$, $\hat{\Omega}_\beta$ a closed set diffeomorphic[1] to the closed unit circle in $\mathbb{R}^2$ for every β and $\hat{\Omega}$ an open set diffeomorphic to the open unit circle in $\mathbb{R}^2$. Moreover, we assume $\hat{\Omega}_\beta \cap \hat{\Omega}_\alpha = \varnothing$ for $\alpha \neq \beta$ and

$$\min_\beta \operatorname{dist}(\partial\hat{\Omega}, \partial\hat{\Omega}_\beta) > 0,$$

where $\operatorname{dist}(\cdot, \cdot)$ is the standard Euclidean distance.

The body that we consider then occupies a region $\mathcal{B} := \Omega \times (0, l)$ with $l \gg \operatorname{diam}\Omega$, where $\operatorname{diam}\Omega$ is the *diameter* of Ω, i.e., the maximum distance between two arbitrary points in Ω; l is the length of the cylinder, which can be infinite, a case considered in the last section of this chapter. The axis of the cylinder is in $\operatorname{span} e_3$. The closure of $\mathcal{B}$ is $\bar{\mathcal{B}} = \bar{\Omega} \times [0, l]$. Its boundary is then the union $\partial\Omega \times [0, l] \cup \Omega \times \{0\} \cup \Omega \times \{l\}$; $\Omega \times \{0\}$ and $\Omega \times \{l\}$ are the *bases* of the cylinder, while $\partial\Omega \times [0, l]$ is its *lateral boundary*. We denote by $\Omega(x_3)$ the *cross section* at x_3.

1. We assume also that the cylinder has constant density of mass and that the origin of the frame determined by e_1, e_2, and e_3 coincides with the center of mass of $\Omega(0)$. Also, $\operatorname{span} e_1$ and $\operatorname{span} e_2$ are principal inertial axes of $\Omega(0)$.
2. The cylinder undergoes small strains.

6.1.2 Loads

The following assumptions apply:

1. The lateral boundary does not experience the action of external forces.
2. Bulk actions on the cylinder are neglected.
3. Forces are applied just to the bases of the cylinder.

Consequently, the balance equations read

$$\operatorname{div}\sigma = 0, \qquad \sigma \in \operatorname{Sym}(\tilde{\mathbb{R}}^3, \tilde{\mathbb{R}}^3)$$

with

$$\sigma n = 0 \quad \text{on} \quad \partial\Omega \times (0, l).$$

[1]A set $\mathcal{X}$ is diffeomorphic to another set $\mathcal{Y}$ if it is possible to establish a one-to-one differentiable mapping $f : \mathcal{X} \longrightarrow \mathcal{Y}$ from $\mathcal{X}$ onto $\mathcal{Y}$ such that its inverse $f^{-1} : \mathcal{Y} \longrightarrow \mathcal{X}$ is differentiable as well.

6.1.3 Material

The cylinder is composed of a linear-elastic homogeneous and isotropic material.

For the theorem discussed in the last section of this chapter we shall relax this assumption, eliminating the requirement of isotropy.

6.1.4 Assumption on the Structure of the Stress

We write $\{O, x_1, x_2, x_3\}$ for the coordinate system determined by e_1, e_2, e_3. In this frame of reference, the matrix of the stress components has the form

$$\begin{pmatrix} 0 & 0 & \sigma_{13} \\ 0 & 0 & \sigma_{23} \\ \sigma_{31} & \sigma_{32} & \sigma_{33} \end{pmatrix}. \tag{6.1}$$

Such an assumption, due to Adhémard Jean-Claude Barré de Saint-Venant (1797–1886), who first proposed and analyzed the problem discussed here (an assumption sometimes referred to as Clebsch–de Saint Venant's), has also a more intrinsic representation. Write $\bar{\mathcal{V}}$ for the two-dimensional vector space orthogonal to span e_3. Consider also the decomposition of the real 3-dimensional vector space into the tensor sum $(\text{span}\, e_3) \oplus \bar{\mathcal{V}}$, as depicted in Figure 6.1.

The Cauchy stress along the cylinder can be decomposed in this way:

$$\sigma = \sigma_{33} e_3 \otimes e_3 + \text{Sym}(a \otimes e_3) + \text{L}, \tag{6.2}$$

where $a \in \bar{\mathcal{V}}$ and $\text{L} \in \text{Sym}(\bar{\mathcal{V}}, \bar{\mathcal{V}})$. Consequently, de Saint-Venant's assumption corresponds to

$$\text{L} = 0. \tag{6.3}$$

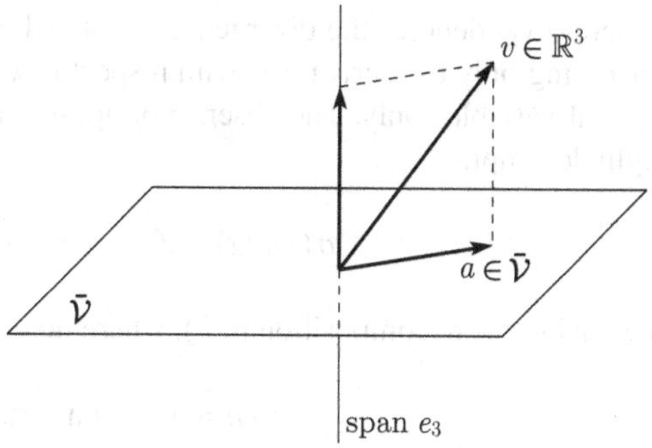

Fig. 6.1 Natural decomposition of 3-dimensional real vector spaces

Since we have fixed a global frame of reference for the cylinder, the two expressions (6.1) and (6.3) of de Saint-Venant's assumption are exactly the same: there is no additional generality in (6.3) with respect to (6.1). Hence, we follow tradition and use the expression (6.1).

6.2 First Consequences of de Saint-Venant's assumption

By writing explicitly the balance equations in the orthogonal frame chosen and taking into account the assumption (6.1), we get

$$\frac{\partial \sigma_{13}}{\partial x_3} = 0, \tag{6.4}$$

$$\frac{\partial \sigma_{23}}{\partial x_3} = 0, \tag{6.5}$$

$$\frac{\partial \sigma_{31}}{\partial x_1} + \frac{\partial \sigma_{32}}{\partial x_2} + \frac{\partial \sigma_{33}}{\partial x_3} = 0, \tag{6.6}$$

$$\sigma_{31} = \sigma_{13}, \qquad \sigma_{32} = \sigma_{23}. \tag{6.7}$$

The first two equations establish that σ_{13} and σ_{23} are functions of x_1 and x_2 alone: they do not depend on x_3. Consequently, by defining τ as the vector with components σ_{31} and σ_{32}, namely

$$\tau = \sigma_{31} e_1 + \sigma_{32} e_2,$$

the third balance equation (6.6) can be written

$$\operatorname{div}_\Omega \tau + \frac{\partial \sigma_{33}}{\partial x_3} = 0, \tag{6.8}$$

where $\operatorname{div}_\Omega$ denotes the divergence evaluated over the region Ω, i.e., a divergence involving only the derivatives with respect to x_1 and x_2, and τ is a function of these spatial variables only. The absence of applied loads on the lateral boundary of the cylinder implies

$$\sigma(x) n(x) = 0, \qquad \forall x \in \partial\Omega \times (0, l),$$

i.e., using the decomposition (6.2), where now a is τ, we obtain

$$\tau \cdot n = 0, \qquad \text{on } \partial\Omega \times (0, l), \tag{6.9}$$

at all places where n is defined, which is

$$\sigma_{31}n_1 + \sigma_{32}n_2 = 0, \quad \text{on} \quad \partial\Omega \times (0, l). \tag{6.10}$$

In fact, as stated by (6.8), de Saint-Venant's assumption foresees that the sole nonzero components of the stress are just σ_{33} and those defining the vector τ. Consequently, at each $x_3 \in (0, l)$, the sole stress components on $\partial\Omega(x_3)$, within the plane containing $\Omega(x_3)$, are those involved in the definition of τ. And τ is tangent to $\partial\Omega(x_3)$ at all points where the normal is uniquely defined. If τ were not tangential to $\partial\Omega(x_3)$ at any one of these points, it would have a component along the normal n that would be not balanced, since external forces along the lateral boundary of the cylinder are absent by assumption. Consequently, since the cylinder is at equilibrium, the boundary condition (6.9) is necessary for it.[2]

6.3 Global Balances for a Portion of the Cylinder

Define

$$N(x_3) := \int_{\Omega} \sigma_{33}\, d\mathcal{H}^2,$$

$$T_1(x_3) := \int_{\Omega} \sigma_{31} d\mathcal{H}^2,$$

$$T_2(x_3) := \int_{\Omega} \sigma_{32}\, d\mathcal{H}^2,$$

$$M_1(x_3) := \int_{\Omega} \sigma_{33}x_2\, d\mathcal{H}^2,$$

$$M_2(x_3) := -\int_{\Omega} \sigma_{33}x_1 d\mathcal{H}^2,$$

$$M_t(x_3) := \int_{\Omega} (\sigma_{32}x_1 - \sigma_{31}x_2) d\mathcal{H}^2,$$

and call them collectively **action characteristics**. Here $N(x_3)$ is the **normal traction** over the cylinder cross section at x_3; $T_1(x_3)$ and $T_2(x_3)$ are the relevant global **shear forces**; $M_1(x_3)$ and $M_2(x_3)$ are the global **bending moments** around the x_1 and x_2 axes, respectively; M_t is the **torsion moment** around the x_3-axis. We write also N_0, T_{01}, T_{02}, M_{01}, M_{02}, and M_{0t} for the counterparts of N, T_1, T_2, M_1, M_2, and M_t over the basis of the cylinder at $x_3 = 0$.

---

[2] *Regarding notation*: In the rest of this chapter, for the sake of notational conciseness, we shall use a comma in a subscript to denote differentiation with respect to the coordinates of the orthogonal frame considered for the cylinder, so that, e.g., we shall write $\sigma_{13,3}$ for (6.4), and so on.

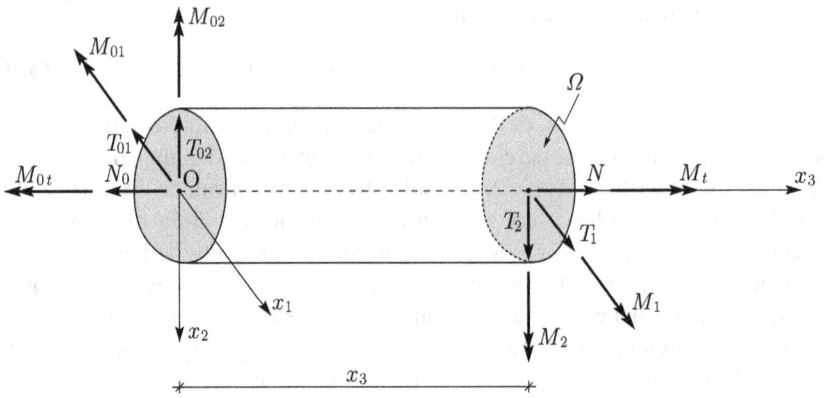

Fig. 6.2 The portion of the cylinder from 0 to x_3 with the relevant action characteristics, and the convention on positive algebraic signs

For a portion of the cylinder between 0 and x_3, the global balances—those in terms of action characteristics—reduce to

$$N_0 = N, \tag{6.11}$$

$$T_{01} = T_1, \tag{6.12}$$

$$T_{02} = T_2, \tag{6.13}$$

$$M_1 = M_{01} + x_3 T_{02}, \tag{6.14}$$

$$M_2 = M_{02} - x_3 T_{01}, \tag{6.15}$$

$$M_{0t} = M_t, \tag{6.16}$$

according to the convention on algebraic signs depicted in Figure 6.2.

6.4 An Explicit Expression for σ_{33}

Let us assume that the stress field is twice differentiable. The assumption (6.1) on its structure allows us to reduce the Beltrami–Donati–Michell equations to

$$\sigma_{33,11} = 0, \tag{6.17}$$

$$\sigma_{33,22} = 0, \tag{6.18}$$

$$\sigma_{33,12} = 0, \tag{6.19}$$

$$\sigma_{33,33} = 0, \tag{6.20}$$

$$(1 + \nu)(\sigma_{31,11} + \sigma_{31,22}) + \sigma_{33,13} = 0, \tag{6.21}$$

$$(1 + \nu)(\sigma_{32,11} + \sigma_{32,22}) + \sigma_{33,23} = 0. \tag{6.22}$$

Equations (6.17), (6.18), and (6.20) imply that σ_{33} can be at most linear in x_1, x_2, and x_3. Equation (6.19) excludes the dependence of σ_{33} on the product x_1x_2, while it allows dependence on the products x_1x_3 and x_2x_3. As a consequence, σ_{33} admits the form

$$\sigma_{33} = a_0 + a_1x_1 + a_2x_2 - x_3(c_0 + c_1x_1 + c_2x_2), \tag{6.23}$$

where a_0, a_1, a_2, c_0, c_1, c_2, are integration constants to be determined. The algebraic sign in front of x_3 is selected for the sake of convenience, as will become apparent in the subsequent developments.

6.5 Values of the Integration Constants

6.5.1 $c_0 = 0$

On inserting (6.23) into (6.6), we obtain

$$\sigma_{31,1} + \sigma_{32,2} = c_0 + c_1x_1 + c_2x_2, \tag{6.24}$$

and after differentiation, which is permissible, since we are under conditions ensuring the validity of the Beltrami–Donati–Michell equations, we obtain

$$\sigma_{31,11} + \sigma_{32,21} = c_1, \tag{6.25}$$

$$\sigma_{31,12} + \sigma_{32,22} = c_2. \tag{6.26}$$

Moreover, by inserting (6.23) into (6.21) and (6.22), we respectively find

$$\sigma_{31,11} + \sigma_{31,22} = \frac{c_1}{1+\nu}, \tag{6.27}$$

$$\sigma_{32,11} + \sigma_{32,22} = \frac{c_2}{1+\nu}. \tag{6.28}$$

By subtracting (6.25) from (6.27) and (6.26) from (6.28), we get

$$\sigma_{31,22} - \sigma_{32,21} = -\frac{\nu}{1+\nu} c_1,$$

$$\sigma_{32,11} - \sigma_{31,12} = -\frac{\nu}{1+\nu} c_2,$$

and we can rewrite them as

$$(\sigma_{32,1} - \sigma_{31,2})_{,2} = \frac{\nu}{1+\nu} c_1,$$

$$(\sigma_{32,1} - \sigma_{31,2})_{,1} = -\frac{\nu}{1+\nu} c_2,$$

by exploiting the symmetry of σ and the possibility of interchanging derivatives as ensured by Schwartz's theorem. These last two equations can be summarized in the differential form

$$d(\sigma_{32,1} - \sigma_{31,2}) = \frac{\nu}{1+\nu}(c_1 dx_2 - c_2 dx_1),$$

which implies the equation

$$\sigma_{32,1} - \sigma_{31,2} = \frac{\nu}{1+\nu}(c_1 x_2 - c_2 x_1) + \bar{c}, \tag{6.29}$$

after integration, with $\bar{c}$ a constant to be determined, and the use of the symmetry of σ.

We can write

$$\sigma_{31} = \sigma_{31}^0 + \bar{\sigma}_{31},$$

$$\sigma_{32} = \sigma_{32}^0 + \bar{\sigma}_{32},$$

where $\bar{\sigma}_{31}$ and $\bar{\sigma}_{32}$ are special solutions of the equations (6.24) and (6.29), while σ_{31}^0 and σ_{32}^0 are the general solutions to their homogeneous counterparts, namely

$$\sigma_{31,1}^0 + \sigma_{32,2}^0 = 0, \tag{6.30}$$

$$\sigma_{32,1}^0 - \sigma_{31,2}^0 = 0. \tag{6.31}$$

We shall discuss two ways of tackling the analysis of such a system of linear partial differential equations, since the two paths give us different kinds of physical information at least in the case of pure torsion. For the purpose of this section, we mention briefly just one of the two approaches, postponing the other to subsequent sections. The way we follow here is based on the assumption that σ_{31}^0 and σ_{32}^0 admit a **shear stress potential function** $\varphi(x_1, x_2)$, i.e., a differentiable scalar function such that

$$\sigma_{31}^0 = \frac{\partial \varphi}{\partial x_1}, \qquad \sigma_{32}^0 = \frac{\partial \varphi}{\partial x_2}. \tag{6.32}$$

With this choice, equation (6.31) is trivially satisfied, while (6.30) reduces to

$$\sigma_{31,1}^0 + \sigma_{32,2}^0 = \frac{\partial^2 \varphi}{\partial x_1^2} + \frac{\partial^2 \varphi}{\partial x_2^2} = 0,$$

or in short,

$$\Delta_\Omega \varphi = 0, \tag{6.33}$$

where Δ_Ω is the Laplacian operator over the set Ω. The boundary condition (6.10) can then be written as

$$\sigma_{31}^0 n_1 + \sigma_{32}^0 n_2 = -(\bar{\sigma}_{31} n_1 + \bar{\sigma}_{32} n_2),$$

or, in terms of the shear stress potential function φ,

$$\frac{d\varphi}{dn} = -\left(\bar{\sigma}_{31} n_1 + \bar{\sigma}_{32} n_2\right), \tag{6.34}$$

where

$$\frac{d\varphi}{dn} = \frac{\partial \varphi}{\partial x_1} n_1 + \frac{\partial \varphi}{\partial x_2} n_2.$$

For the class of regions Ω selected here, equation (6.33) admits the existence of a harmonic shear stress potential function φ in the subset of the space of square-integrable functions satisfying the condition (6.34) once the right-hand-side term is assigned. We do not go into the details of the proof, since they are matter of standard textbooks in functional analysis. We just mention here the result.

Integration of (6.33) over Ω and Gauss's theorem imply

$$\int_\Omega \Delta_\Omega \varphi \, d\mathcal{H}^2 = \int_{\partial\Omega} \frac{d\varphi}{dn} ds = 0,$$

where ds is the line measure over $\partial\Omega$. However, equation (6.34) holds, so that we get

$$0 = \int_{\partial\Omega} (\bar{\sigma}_{31} n_1 + \bar{\sigma}_{32} n_2) ds = \int_{\partial\Omega} \bar{\tau} \cdot n \, ds,$$

where $\bar{\tau}$ is the vector with components $\bar{\sigma}_{31}$ and $\bar{\sigma}_{32}$. Once again, by applying Gauss's theorem, we obtain

$$\int_{\partial\Omega} \bar{\tau} \cdot n \, ds = \int_\Omega \mathrm{div}_\Omega \bar{\tau} \, d\mathcal{H}^2 = \int_\Omega (\bar{\sigma}_{31,1} + \bar{\sigma}_{32,2}) d\mathcal{H}^2. \tag{6.35}$$

Equation (6.24) then implies (after substituting it into (6.35))

$$0 = \int_{\Omega} (c_0 + c_1 x_1 + c_2 x_2) d\mathcal{H}^2 = c_0 |\Omega| + c_1 \mathrm{S}_2 + c_2 \mathrm{S}_1,$$

where $|\Omega|$ is the area of Ω, while S_1 and S_2 are the **static momenta** of the cylinder cross section Ω with respect to the axes x_1 and x_2, respectively. Since these axes determine over Ω a central principal inertial frame, by our initial choice, we get

$$\mathrm{S}_1 = 0, \qquad \mathrm{S}_2 = 0, \tag{6.36}$$

which implies

$$c_0 = 0. \tag{6.37}$$

6.5.2 The Navier Polynomial

With the previous result, the expression (6.23) reduces to

$$\sigma_{33} = a_0 + a_1 x_1 + a_2 x_2 - x_3 (c_1 x_1 + c_2 x_2). \tag{6.38}$$

Due to the balance (6.11), we also have

$$\begin{aligned} N_0 &= \int_{\Omega} \sigma_{33} \, d\mathcal{H}^2 = \int_{\Omega} (a_0 + a_1 x_1 + a_2 x_2 - x_3 (c_1 x_1 + c_2 x_2)) d\mathcal{H}^2 \\ &= a_0 |\Omega| + a_1 \mathrm{S}_2 + a_2 \mathrm{S}_1 - x_3 (c_1 \mathrm{S}_2 + c_2 \mathrm{S}_1). \end{aligned}$$

Since in the central principal inertial frame that we have chosen, $\mathrm{S}_1 = 0$ and $\mathrm{S}_2 = 0$, by writing A for $|\Omega|$ to follow traditional notation, we get

$$a_0 = \frac{N_0}{A}. \tag{6.39}$$

By definition of M_1, we compute also

$$\begin{aligned} M_1(x_3) &= \int_{\Omega} \sigma_{33} x_2 \, d\mathcal{H}^2 \\ &= \int_{\Omega} \left(\frac{N_0}{A} x_2 + a_1 x_1 x_2 + a_2 x_2^2 - x_3 (c_1 x_1 x_2 + c_2 x_2^2) \right) d\mathcal{H}^2 \\ &= \frac{N_0}{A} \mathrm{S}_1 + a_1 \mathrm{I}_{12} + a_2 \mathrm{I}_1 - x_3 (c_1 \mathrm{I}_{12} + c_2 \mathrm{I}_1), \end{aligned}$$

where I_1 is the **moment of inertia** of Ω with respect to axis x_1, and I_{12} is the **mixed moment of inertia** of Ω with respect to the axes x_1 and x_2. However, since such axes constitute a central principal inertial frame for Ω, we have by definition

$$\mathrm{I}_{12} = 0$$

together with $\mathrm{S}_1 = 0$. Consequently, we have

$$M_1(x_3) = (a_2 - x_3 c_2)\mathrm{I}_1.$$

By letting x_3 go to zero, we get

$$M_{01} = a_2 \mathrm{I}_1,$$

i.e.,

$$a_2 = \frac{M_{01}}{\mathrm{I}_1}. \tag{6.40}$$

In an analogous way, for $M_2(x_3)$ we get

$$M_2(x_3) = -\int_\Omega \sigma_{33} x_1 \, d\mathcal{H}^2 = -(a_1 - x_3 c_1)\mathrm{I}_2,$$

where I_2 is the moment of inertia of Ω with respect to the axis x_2. By letting x_3 go to zero we obtain

$$M_{02} = -a_1 \mathrm{I}_2,$$

i.e.,

$$a_1 = -\frac{M_{02}}{\mathrm{I}_2}. \tag{6.41}$$

From the balance (6.14), we get also

$$x_3 T_{02} = M_1 - M_{01} = (a_2 - x_3 c_2)\mathrm{I}_1 - a_2 \mathrm{I}_1 = -x_3 c_2 \mathrm{I}_1,$$

so that

$$c_2 = -\frac{T_{02}}{\mathrm{I}_1}. \tag{6.42}$$

Finally, the balance (6.15) implies

$$x_3 T_{01} = M_{02} - M_2 = -a_1 \mathrm{I}_2 + (a_1 - x_3 c_1)\mathrm{I}_2 = -x_3 c_1 \mathrm{I}_2,$$

i.e.,

$$c_1 = -\frac{T_{01}}{\mathrm{I}_2}. \tag{6.43}$$

On inserting (6.37) and (6.39) through (6.43) into (6.23), we obtain

$$\sigma_{33} = \frac{N_0}{A} + (M_{01} + x_3 T_{02})\frac{x_2}{\mathrm{I}_1} - (M_{02} - x_3 T_{01})\frac{x_1}{\mathrm{I}_2},$$

i.e., using (6.11), (6.14), and (6.15), we have

$$\sigma_{33} = \frac{N}{A} + \frac{M_1}{\mathrm{I}_1}x_2 - \frac{M_2}{\mathrm{I}_2}x_1, \tag{6.44}$$

which is called the **Navier polynomial.**

To get σ_{33}, we used both the balance and the Beltrami–Donati–Michell equations, all the ingredients defining the problem in linear elasticity at hand. In this sense, we can assert that the formula (6.44) is exact.

6.6 The Neutral Axis

The locus of points where σ_{33} vanishes is a straight line with equation

$$\frac{N}{A} + \frac{M_1}{\mathrm{I}_1}x_2 - \frac{M_2}{\mathrm{I}_2}x_1 = 0. \tag{6.45}$$

It is called the **neutral axis**. It has a particular expression when the action over the cylinder reduces to a unique force normal to Ω, applied at a point not coinciding with the center of mass of Ω.

Consider this case and denote by $\bar{x}_1$ and $\bar{x}_2$ the coordinates of the point where $N_0 = N$ is applied. To use the Navier polynomial, we have to transfer N to the barycenter of Ω. Two transport couples arise:

$$M_1 = N\bar{x}_2, \qquad M_2 = -N\bar{x}_1.$$

In this case, equation (6.45) becomes

$$\frac{N}{A}\left(1 + \frac{\bar{x}_2 A}{\mathrm{I}_1}x_2 + \frac{\bar{x}_1 A}{\mathrm{I}_2}x_1\right) = 0,$$

i.e.,

$$1 + \frac{\bar{x}_2}{\rho_1^2} x_2 + \frac{\bar{x}_1}{\rho_2^2} x_1 = 0,$$

where

$$\rho_1^2 := \frac{\mathrm{I}_1}{A}, \qquad \rho_2^2 := \frac{\mathrm{I}_2}{A}.$$

The neutral axis is independent of the intensity of the applied traction. It depends on the point where the traction N is applied and the geometry of Ω through the factors ρ_1^2 and ρ_2^2. Moreover, we can show by polar duality that the barycenter of Ω is between the point $(\bar{x}_1, \bar{x}_2)$ and the neutral axis. However, we do not go further into details.

When N is applied at the barycenter of Ω and is the sole action on the cylinder, the neutral axis does not exist, since we do not have bending effects and σ_{33} is constant over Ω.

6.7 A Scheme for Evaluating the σ_{33} Distribution over Cross Sections

The general scheme for evaluating the stress σ_{33} over a cylinder's cross section includes the following main steps:

1. Determine the barycenter of Ω.
2. Find the principal inertial frame for Ω, centered at the barycenter.
3. Reduce the action characteristics to the center of mass of Ω.
4. Determine the neutral axis, if available.
5. Draw two straight lines that are parallel to the neutral axis and tangent to Ω so that the cross section is entirely contained in the strip they determine.
6. By using the Navier polynomial, compute the maximum and minimum values of σ_{33}: they are at the tangential points of the two straight lines, when the nautral axis crosses Ω. Once they are known, the σ_{33} distribution is given by the straight line connecting the two values.

6.8 Exercises on the Determination of σ_{33}

Exercise 1. *The section Ω depicted in Figure 6.3 consists of a square with side l and a concentrated mass equal to $\rho k l^2$, located at the barycenter of the square. The section is subjected to a bending couple M and a traction $N = 10M/l$ applied at the barycenter. Find the values of k such that the stress σ_{33} at the point* P *in Figure 6.3 vanishes. Assume $\rho = 1$.*

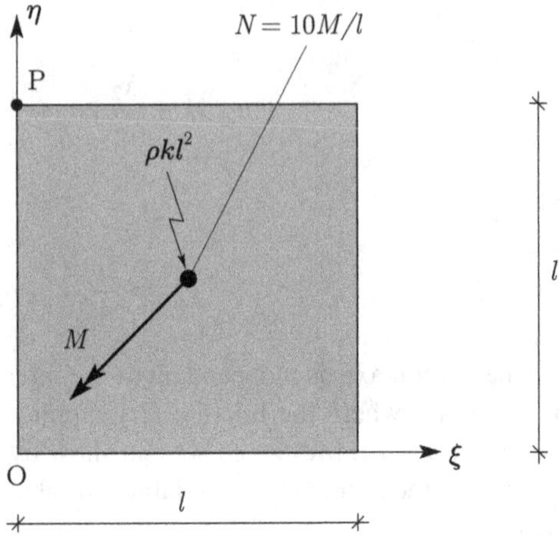

Fig. 6.3 Square cross section with a concentrated mass

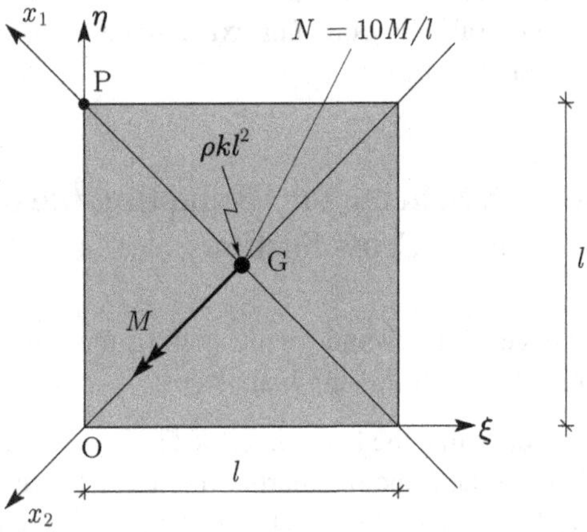

Fig. 6.4 Reference frame considered

Remarks and solution. The barycenter G coincides with the center of mass of the square. The particular configuration of the section Ω implies that the central ellipse of inertia is a circumference and that every pair of orthogonal axes intersecting at G determines a principal inertial frame. An example is given by the pair $\{x_1, x_2\}$ (Fig. 6.4).

The moments of inertia are

$$\mathrm{I}_1 = \mathrm{I}_2 = \frac{l^4}{12} = \mathrm{I},$$

and the section with concentrated mass is equivalent, in terms of the total mass, to an enlarged square of constant density of mass equal to 1 and area

$$A = l^2(1 + k).$$

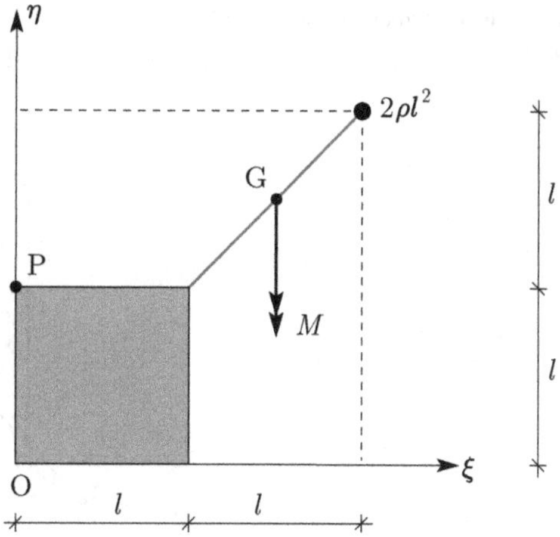

Fig. 6.5 Square cross section with a concentrated mass subjected to a bending moment

The bending moment acts "around" x_2. Then the normal stress σ_{33} at the point P is given by

$$\sigma_{33}(\mathrm{P}) = \frac{N}{A} - \frac{M}{\mathrm{I}}\frac{l\sqrt{2}}{2}.$$

By considering the datum $N = \dfrac{10M}{l}$ and imposing $\sigma_{33}(\mathrm{P}) = 0$, we obtain

$$k = \frac{10}{6\sqrt{2}} - 1 \approx 0.179.$$

Exercise 2. *The section depicted in Figure 6.5 consists of a square with side l and a concentrated mass equal to $2\rho l^2$ connected with the square by a thin plate that we can consider massless. The section is subjected to a bending couple M applied as shown in figure. Find the value of the stress σ_{33} at the point* P. *Assume $\rho = 1$, $l = 20\,\mathrm{cm}$, $M = 250\,\mathrm{Nm}$.*

Remarks and solution. Let us denote by (1) and (2) quantities pertaining to the square and the concentrated mass respectively, the latter being equivalent in terms of the total mass to a square of unit density of mass and side $l\sqrt{2}$. The whole section is equivalent to a section with total area

$$A = A^{(1)} + A^{(2)} = l^2 + 2l^2 = 3l^2,$$

endowed with the same unit density of mass.

The barycenters of the square and the concentrated mass are at $\xi^{(1)} = \dfrac{l}{2}$ and $\xi^{(2)} = 2l$ so that the coordinates of the section barycenter G in the coordinate system $\{\mathrm{O}, \xi, \eta\}$ in Figure 6.5 are

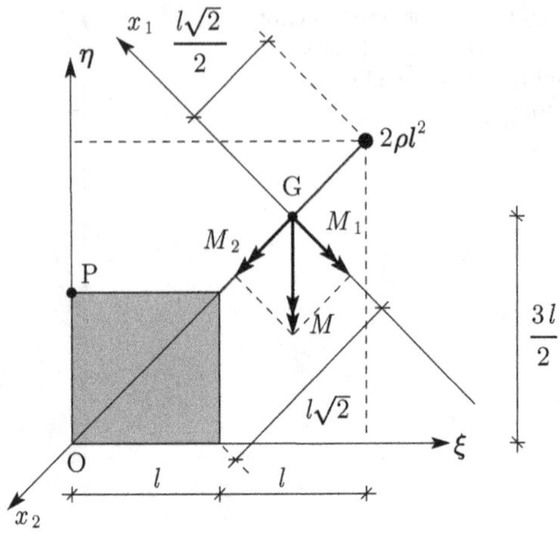

Fig. 6.6 Principal inertial axes

$$\xi_G = \eta_G = \frac{A^{(1)}\xi^{(1)} + A^{(2)}\xi^{(2)}}{A} = \frac{3l}{2}.$$

The section's symmetry determines the principal inertia axes x_1 and x_2 (Fig. 6.6). The moments of inertia with respect to the frame $\{G, x_1, x_2\}$ are

$$I_1^{(1)} = \frac{l^4}{12} + 2l^4, \quad I_1^{(2)} = l^4, \quad I_1 = \frac{37}{12}l^4 \approx 493333 \text{ cm}^4,$$
$$I_2^{(1)} = \frac{l^4}{12}, \quad I_2^{(2)} = 0, \quad I_2 = \frac{l^4}{12} \approx 13333 \text{ cm}^4.$$

In the same frame, the applied bending moment has components

$$M_1 = -M_2 = -\frac{M}{\sqrt{2}},$$

the point P has coordinates

$$x_1(\mathrm{P}) = \frac{l\sqrt{2}}{2}, \qquad x_2(\mathrm{P}) = l\sqrt{2},$$

and $\sigma_{33}(\mathrm{P})$ follows from the Navier polynomial:

$$\sigma_{33}(\mathrm{P}) = \frac{M_1}{I_1}l\sqrt{2} - \frac{M_2}{I_2}\frac{l\sqrt{2}}{2} = -\frac{234}{37}\frac{M}{l^3} \approx -19.76\,\frac{\mathrm{N}}{\mathrm{cm}^2}.$$

The negative sign denotes compression at P.

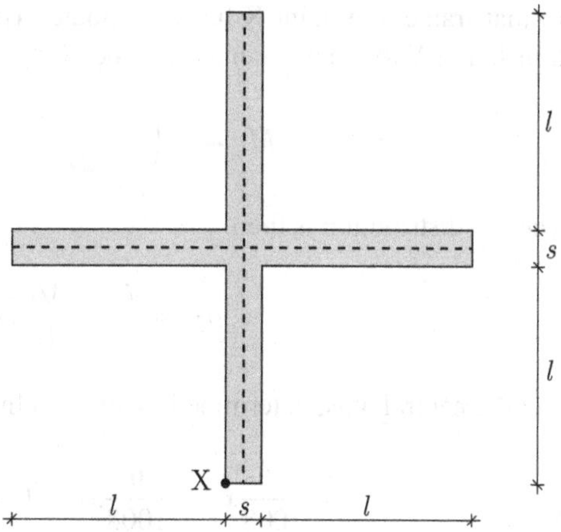

Fig. 6.7 Symmetric thin cross section subjected to eccentric traction

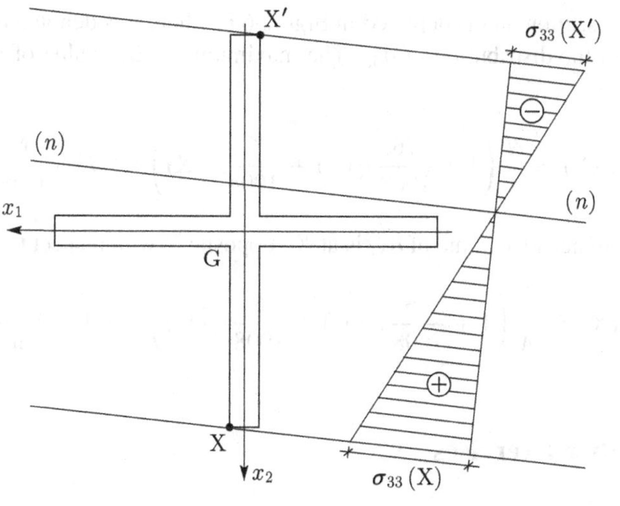

Fig. 6.8 Distribution of σ_{33} and neutral axis (n)—(n)

Exercise 3. *The section in Figure 6.7 is subjected to a traction N applied at the point* X. *Find the distribution of the stresses σ_{33} and the neutral axis. Suppose that* $l = 70\,\text{mm}$, $s = 12\,\text{mm}$, $N = 90\,\text{kN}$.

Remarks and solution. The section has two axes of symmetry that imply the barycenter G and the principal axes of inertia x_1, x_2 (Fig. 6.8). The moments of inertia in the frame $\{G, x_1, x_2\}$ and the area are

$$I_1 = I_2 = \frac{(2l+s)s^3}{12} + 2\left(\frac{sl^3}{12} + ls\left(\frac{l+s}{2}\right)^2\right), \qquad A = 4ls + s^2.$$

In that frame, the point X has coordinates $x_1(\mathrm{X}) = s/2$ and $x_2(\mathrm{X}) = l + s/2$. In translating N to the barycenter, two bending couples arise:

$$M_1 = N\left(l + \frac{s}{2}\right), \qquad M_2 = -N\frac{s}{2}.$$

The σ_{33} distribution is then

$$\sigma_{33} = \frac{N}{A} + \frac{M_1}{\mathrm{I}_1}x_2 - \frac{M_2}{\mathrm{I}_2}x_1,$$

and the neutral axis, determined by the condition $\sigma_{33} = 0$, is

$$1 + \frac{76}{1008}x_2 + \frac{6}{1008}x_1 = 0 \qquad (x_1 \text{ and } x_2 \text{ in mm}).$$

It crosses the section and is depicted in Figure 6.8, where it is denoted by (n)—(n), together with the distribution of σ_{33}. The maximum positive value of σ_{33} is at the point X:

$$\sigma_{33}(\mathrm{X}) = \frac{N}{A}\left(1 + \frac{76}{1008}x_2(\mathrm{X}) + \frac{6}{1008}x_1(\mathrm{X})\right) \approx 173.8\,\frac{\mathrm{N}}{\mathrm{mm}^2}.$$

The minimum negative value of σ_{33} is at X′, opposite X with respect to G:

$$\sigma_{33}(\mathrm{X}') = \frac{N}{A}\left(1 + \frac{76}{1008}x_2(\mathrm{X}') + \frac{6}{1008}x_1(\mathrm{X}')\right) \approx -122.4\,\frac{\mathrm{N}}{\mathrm{mm}^2}.$$

6.9 Further Exercises

Exercise 4. *Calculate the normal stress σ_{33} at the point* A *in the section in Figure 6.9. The moment M is inclined at* 45°.

Exercise 5. *For the section in Figure 6.10, find the positions of the point* C *where a traction applied is such that the extension of the σ_{33} distribution to the whole plane vanishes at* A ($s = 1$ cm).

Exercise 6. *The section represented in Figure 6.11 is subjected to a normal traction at* C. *Find the position of the neutral axis* ($a = 50$ mm, $l = 100$ mm, $s = 5$ mm).

Exercise 7. *The section represented in Figure 6.12 is subjected to a normal compression force N_1 applied at C_1 and to a normal traction force N_2 applied at C_2. Find the normal stress σ_{33} at* A. *The points C_1 and C_2 are on a straight line inclined* 45° *with respect to the horizontal axis* ($N_1 = 100$ kN, $N_2 = 70$ kN, $R = 150$ mm, $r = 70$ mm).

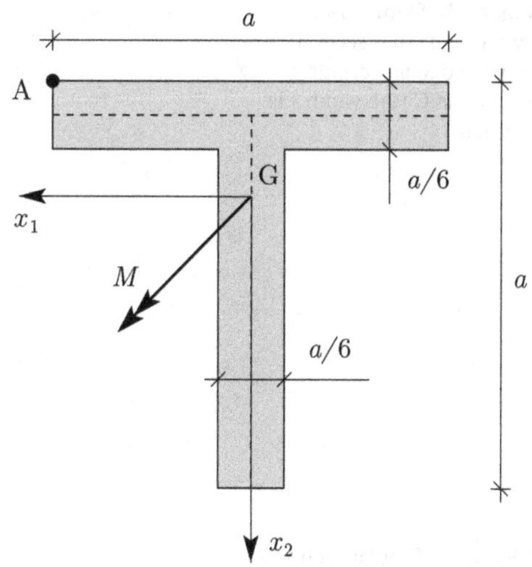

Fig. 6.9 T-shaped cross section subjected to a bending couple

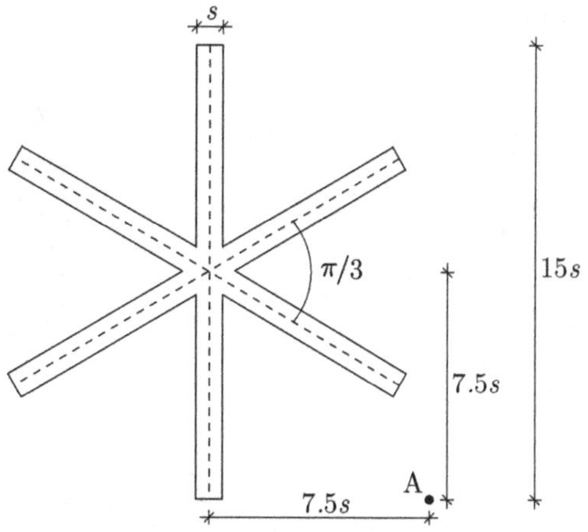

Fig. 6.10 Symmetric cylindrical cross section subjected to an eccentric traction at a point to be determined (the width s is constant for every rectangular portion)

Exercise 8. *The doubly symmetric T-shaped section in Figure 6.13 is subjected to a normal traction force N applied at* C. *Find the* σ_{33} *distribution, the equation of the neutral axis and the minimum and maximum values of* σ_{33} ($l = 100$ mm, $h = 200$ mm, $a = 5.6$ mm, $e = 8.5$ mm, $N = 150$ kN).

Exercise 9. *The* Ω*-shaped section in Figure 6.14 is subjected to a traction at* C. *Find the equation of the neutral axis* ($l = 100$ mm, $s = 10$ mm).

Exercise 10. *Two compressive forces,* N_1 *and* N_2*, act on the rectangular section in Figure 6.15. Find the points over the axis* x_2 *such that on applying* $N_2 = 3N_1$ *there, the whole section does not experience traction at any point.*

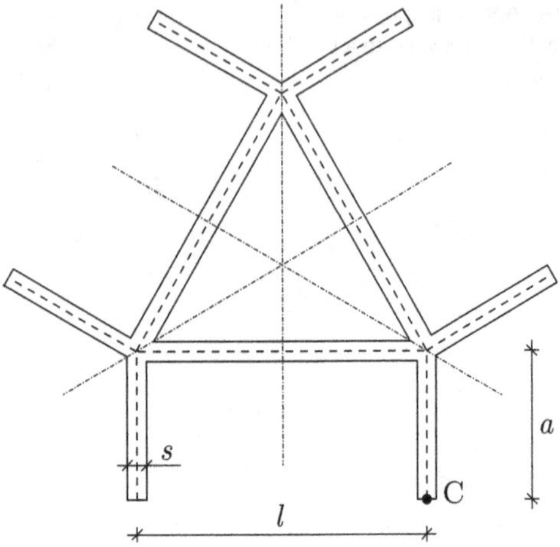

Fig. 6.11 Symmetric cylindrical cross section subjected to an eccentric traction at C (the width s is constant)

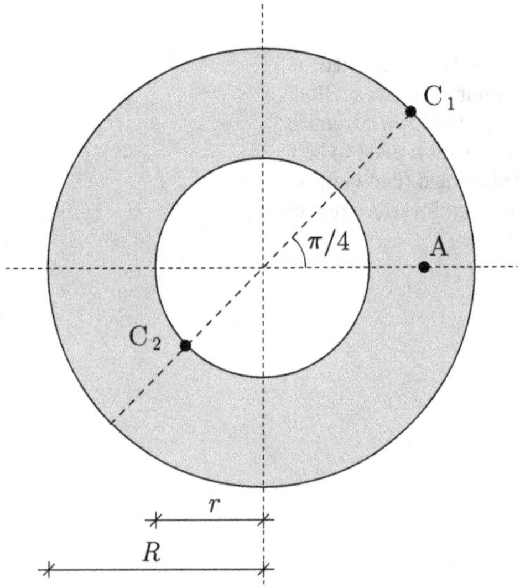

Fig. 6.12 Tubular section subjected to two normal forces

Exercise 11. *A tensile force acts on the T-shaped section in Figure 6.16. Find all points in the section where on applying there such a force,* σ_{33} *vanishes at the point* A ($l = 40$ cm, $s = 5$ cm).

6.10 The Jourawski Formula

Equations (6.24) and (6.29) furnish the shear stress components σ_{31} and σ_{32} to within the constants $\bar{c}$, to be determined. And the results are exact in the sense already mentioned. In the presence of T_1 and/or T_2 over Ω, an approximate

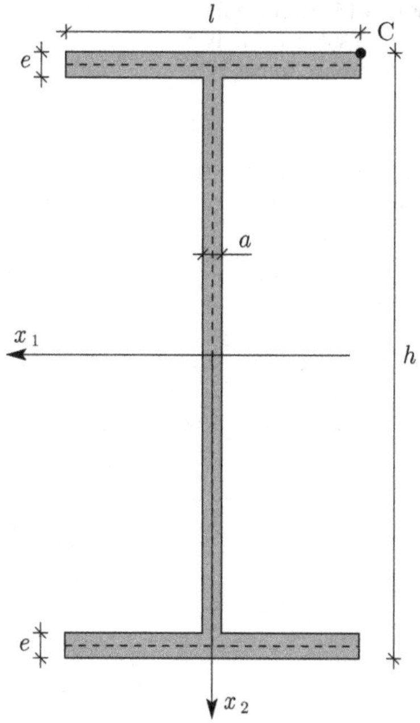

Fig. 6.13 Double T-shaped section under traction

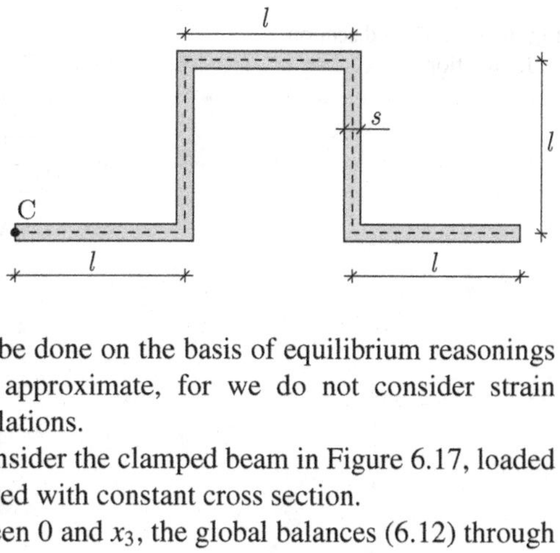

Fig. 6.14 Ω-shaped section under traction

evaluation of the shear stresses can be done on the basis of equilibrium reasonings only—in this sense, the result is approximate, for we do not consider strain compatibility and the constitutive relations.

The analysis goes as follows: Consider the clamped beam in Figure 6.17, loaded by a terminal shear force and endowed with constant cross section.
For every portion of the beam between 0 and x_3, the global balances (6.12) through (6.15) reduce to

$$T_1(x_3) = 0, \qquad T_2(x_3) = T, \qquad M_1(x_3) = Tx_3, \qquad M_2(x_3) = 0,$$

where the algebraic sign depends on the assumed positive directions for the action characteristics as denoted on the right-hand side of Figure 6.17 (we have tacitly assumed that the beam's cross section Ω is such that the principal axis x_1 is orthogonal to the plane of Figure 6.17 and is then out of the page). The analysis is straightforward, since the constraints are equal to the degrees of freedom and are well posed.

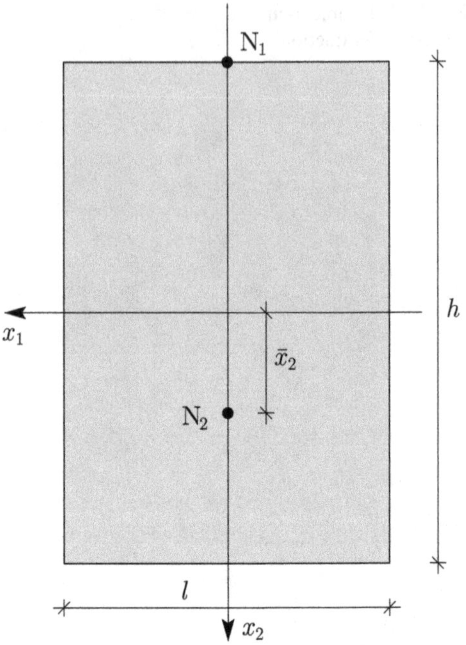

Fig. 6.15 Rectangular section subjected to two compressive forces

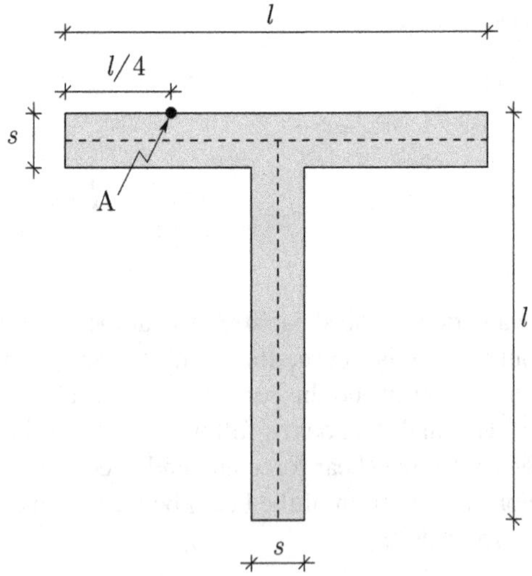

Fig. 6.16 T-shaped section under traction

Using the Navier polynomial, we obtain

$$\sigma_{33} = \frac{M_1(x_3)}{\mathrm{I}_1}x_2 = \frac{T}{\mathrm{I}_1}x_3x_2 = \frac{T_2}{\mathrm{I}_1}x_3x_2.$$

Then the balance equation (6.8) becomes

$$\operatorname{div}_\Omega \tau = -\frac{T_2}{\mathrm{I}_1}x_2. \tag{6.46}$$

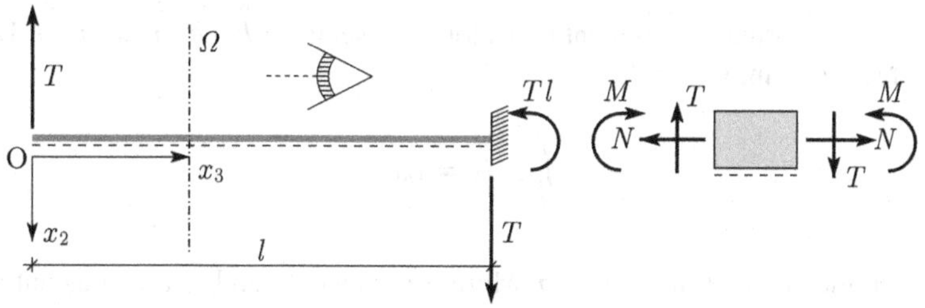

Fig. 6.17 Clamped beam loaded with a shear force. Assumed positive signs on the sides of a portion of the beam

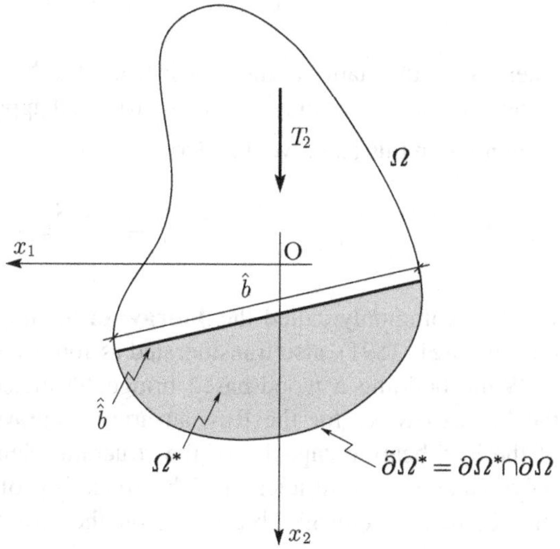

Fig. 6.18 Cross section of the beam with a special part Ω^* of it highlighted

Take a part Ω^* of Ω (a planar region in Ω with nonzero area) such that the boundary of Ω^* is partially in $\partial\Omega$, and the rest is a straight segment $\hat{\hat{b}}$, namely $\partial\Omega^* = \tilde{\partial}\Omega^* \cup \hat{\hat{b}}$ with $\tilde{\partial}\Omega^* := \partial\Omega^* \cap \partial\Omega$. An example is shown in Figure 6.18. Write $\hat{b}$ for the length of $\hat{\hat{b}}$. On integrating the equation (6.46) over Ω^*, we get

$$\int_{\Omega^*} \mathrm{div}_{\Omega}\tau \, d\mathcal{H}^2 = -\int_{\Omega^*} \frac{T_2}{\mathrm{I}_1} x_2 \, d\mathcal{H}^2. \tag{6.47}$$

By Gauss's theorem and the boundary condition (6.9), the first integral becomes

$$\int_{\Omega^*} \mathrm{div}_{\Omega}\tau \, d\mathcal{H}^2 = \int_{\partial\Omega^*} \tau \cdot n \, ds = \int_{\hat{\hat{b}}} \tau \cdot n \, ds = \int_{\hat{\hat{b}}} \tau_n ds,$$

where τ_n is the normal component of τ along the segment $\hat{\hat{b}}$, i.e., $\tau_n = \tau \cdot n$. By the mean value theorem, we get

$$\int_{\hat{\hat{b}}} \tau_n ds = \bar{\tau}_n \hat{b},$$

with $\bar{\tau}_n$ the mean value of τ_n over $\hat{\hat{b}}$. Moreover, since T_2 and I_1 are constant along the cylinder, we also get

$$\int_{\Omega^*} \frac{T_2}{\mathrm{I}_1} x_2 \, d\mathcal{H}^2 = \frac{T_2}{\mathrm{I}_1} \int_{\Omega^*} x_2 \, d\mathcal{H}^2 = \frac{T_2 \mathrm{S}_1^*}{\mathrm{I}_1},$$

where S_1^* is the static moment of inertia of Ω^* with respect to the x_1-axis. For $\hat{b}$ sufficiently small, we may assume that $\bar{\tau}_n$ well approximates the pointwise value of τ_n along $\hat{\hat{b}}$. In this case, we then write

$$\tau_n = -\frac{T_2 \mathrm{S}_1^*}{\mathrm{I}_1 \hat{b}}, \tag{6.48}$$

which is commonly called the **Jourawski formula**, after Dmitrii Ivanovich Zhuravskii (1821–1891), also transliterated as Jourawski from the Cyrillic.

While building a wood-based bridge 60 meters long with nine spans across the Verebya River for the Russian army, Zhuravskii realized that an evaluation of the load-bearing capacity of the structural elements based only on the Navier polynomial was insufficient for the reliability of the bridge, because the shear stresses had a nonnegligible effect on the structural elements used. From that observation, Zhuravskii set in motion the ideas that led to the formula (6.48). We generally write it simply as

$$\tau_n = -\frac{T\mathrm{S}}{\mathrm{I}\hat{b}},$$

leaving understood that the static moment S and the moment of inertia I have to be referred to the appropriate axes.

In general, in presence of T_1 and T_2 over Ω, we have

$$\tau_n = -\left(\frac{T_1 \mathrm{S}_2^*}{\mathrm{I}_2 \hat{b}} + \frac{T_2 \mathrm{S}_1^*}{\mathrm{I}_1 \hat{b}} \right). \tag{6.49}$$

The minus sign in equations (6.48) and (6.49) indicates that if the component τ_n is positive, and is directed outward Ω^*.

6.11 Shear Actions May Produce Torsional Effects: The Shear Center

Consider a cylinder with cross section as depicted in Figure 6.19. A shear force T acts on it, as indicated in the figure.

To determine an approximation of the shear stress distribution due to T in terms of the Jourawski formula, some steps appear necessary:

1. Determine the mass center and the relevant principal inertial frame.
2. Then start considering progressively portions Ω^* of the section. Among several possible choices, the most convenient ones are those for which $\hat{b}$ attains the smallest value. Other options would weaken the Jourawski approximation.
3. The choice of the progressive portions Ω^* determines one of the normals to the various straight boundaries $\hat{\hat{b}}$, i.e., the flows $\tau_n n$ that can be drawn by taking into account the algebraic sign of the static moment S^* and the direction of T. To draw it, one can imagine that Ω is filled by a liquid flowing out of Ω at a point along the direction where T is applied and in the same direction, by following the shortest path.
4. Calculate τ_n using the Jourawski formula as Ω^* varies. Take Ω_1^* as in Figure 6.20. Its static moment S_1^* with respect to the x_1-axis is

$$S_1^*(\Omega_1^*)_\tau = -s\xi_1\frac{a}{2}.$$

Then, since $T = T_2$, I_1 and s are constants; along the upper horizontal sector of Ω, the distribution of τ_n is linear. It is zero along the vertical segment at the point A where $\xi_1 = 0$ and is maximal at the corner. In the interval $[(a - \frac{s}{2}), (a + \frac{s}{2})]$, we essentially extrapolate the distribution. Beyond the corner, the value of the distribution is equal to the maximum reached on the horizontal part. When we

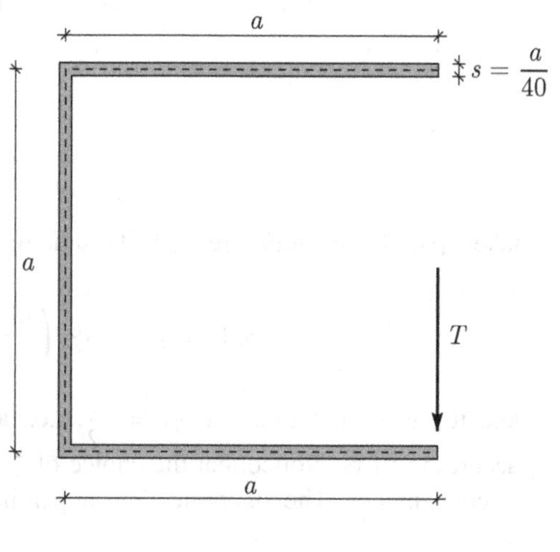

Fig. 6.19 A special simply connected cross section of a de Saint-Venant cylinder

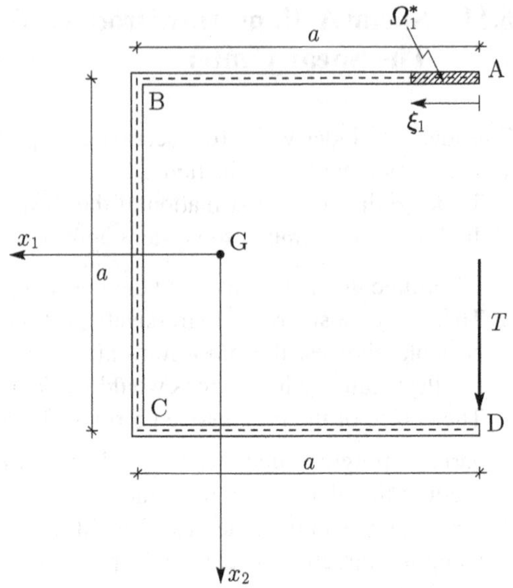

Fig. 6.20 The section in Figure 6.19 with the center of mass G, the principal inertial frame, and a choice of Ω_1^*, progressively growing with the local instrumental coordinate ξ_1

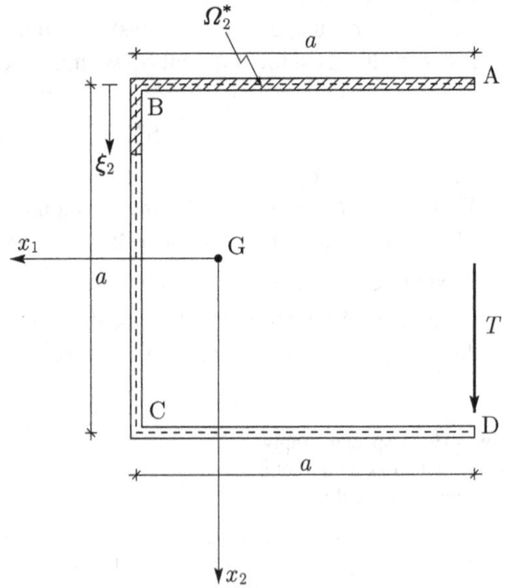

Fig. 6.21 The choice of Ω_2^*

take a part Ω_2^* as in Figure 6.21; its static moment is quadratic in ξ_2, namely

$$\mathrm{S}_1^*(\Omega_2^*)_\tau = -s\xi_2\left(\frac{a-\xi_2}{2}\right) - s\frac{a^2}{2},$$

and it has its maximum at $\xi_2 = \dfrac{a}{2}$, i.e., at $x_2 = 0$ for S_1^*, it changes sign according to x_2. Notice that the choice of ξ_1 and ξ_2 is dictated only by the sake of convenience. The algebraic signs appearing in $\mathrm{S}_1^*(\Omega_1^*)$ and $\mathrm{S}_1^*(\Omega_2^*)$ take into

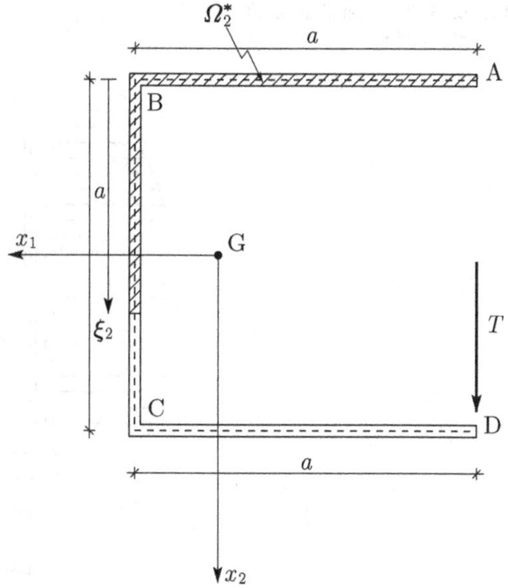

Fig. 6.22 Ω_2^* crosses the x_1-axis

account that Ω_1^* and Ω_2^* are selected in the sector of the coordinate frame where x_2 is negative.
When Ω_2^* crosses the x_1-axis, as in Figure 6.22, we have a positive contribution from the portion of Ω_2^* in the plane sector where x_2 is positive.
When we increase Ω^* by following the flow of shear stresses τ_n, the value at D is zero as a consequence of the boundary condition $\tau \cdot n = 0$.

The resulting diagram of the τ_n distribution is shown in Figure 6.23.

The total shear stress pertaining to the portion AB of Ω is a force q_1, with modulus given by the area of the diagram pertaining to that portion times the thickness s, and direction as represented in Figure 6.24. We may determine analogous forces along the portions BC and CD. Call them q_2 and q_3, respectively. By the symmetry of Ω, we have $q_1 = -q_3$. Moreover, the sum of the q forces gives T, so $q_2 = T$. However, the reduction of the force system q_1, q_2, and q_3 to any point of the plane x_1x_2, except the points on a straight line—call it r—implies the presence of an in-plane moment, i.e., a torsional effect over the section. Figure 6.25 summarizes the argument.

Exercise 12. *Show that the line r in Figure 6.25 is independent of the reduction point selected for the force system q_1, q_2, and q_3, obtained by the Jourawski formula from the scheme in Figure 6.19.*

The intersection point between r and the x_1-axis—call it C_T as in Figure 6.25—has a special status: *no shear force applied there generates torsional effects over the cylinder's cross section, irrespective of its orientation.*

Fig. 6.23 The distribution of the values of τ_n

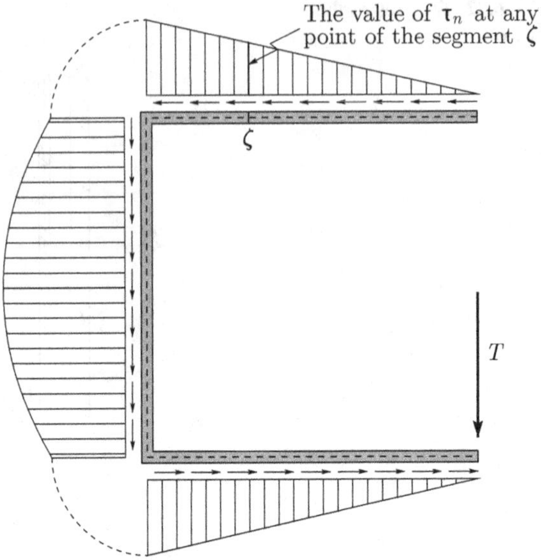

Fig. 6.24 The flow of τ_n

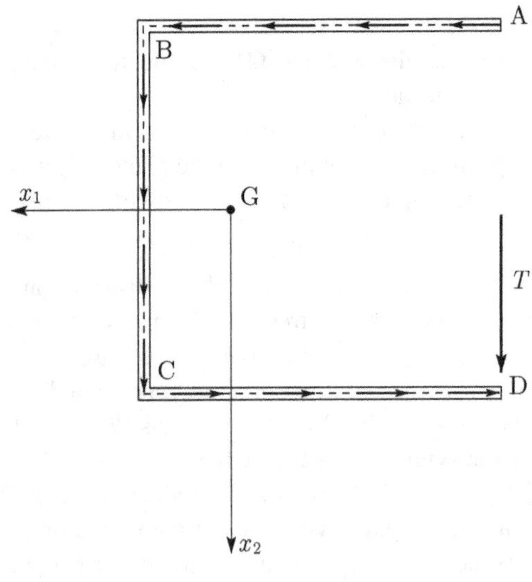

To prove such a statement, we have to recall first that T in Figure 6.19 is directed along the x_2-axis, which determines a principal inertial frame. Consider a force $\bar{T}$ acting along x_1, as depicted in Figure 6.26. By symmetry, the shortest paths followed by the shear stresses τ_n to "flow out" along the direction of $\bar{T}$ are those denoted by arrows in Figure 6.27.

Along the segment ζ in Figure 6.27, $\tau_n = 0$, by symmetry. We have then two choices of the portion of Ω for applying the Jourawski formula (see Fig. 6.28). In the first case (Fig. 6.28-a), the Jourawski formula furnishes a value pertaining to all

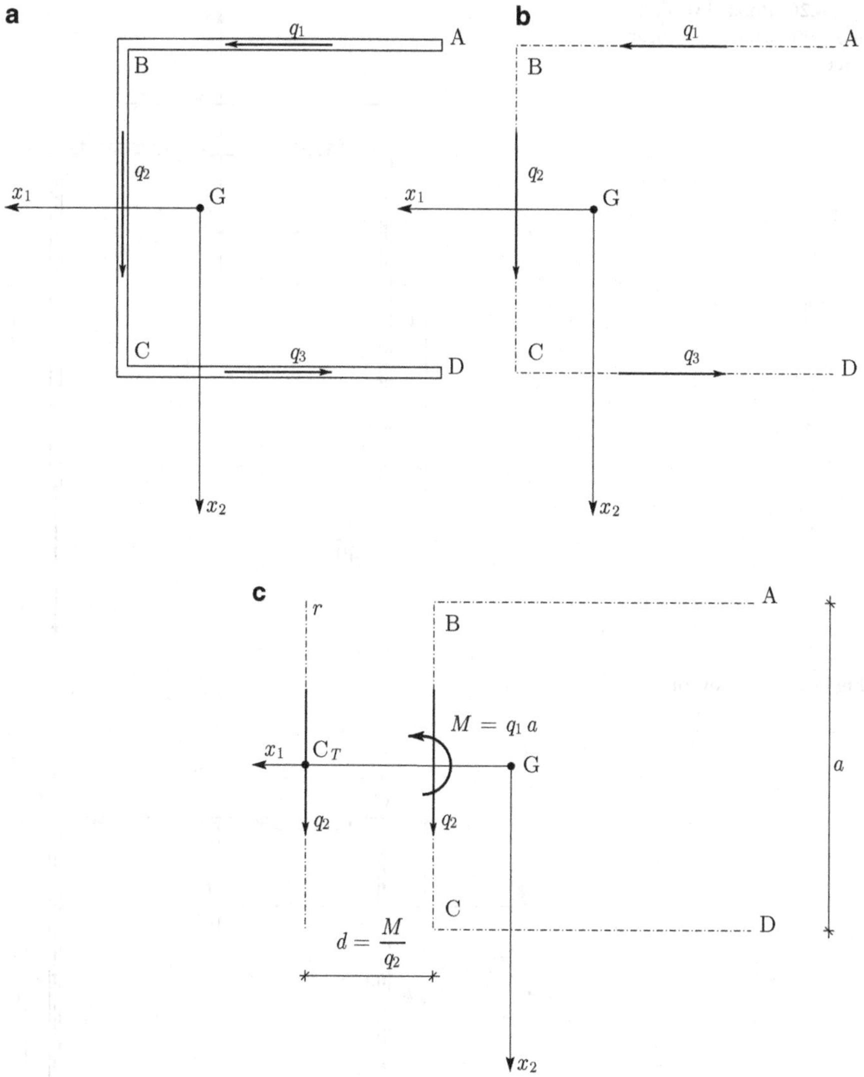

Fig. 6.25 (a) τ_n resultants within the section. (b) q_1, q_2, and q_3 as a planar system of forces. (c) Reduction of the force system to the point on the x_1-axis at the center of the vertical part of the section

points of the segment ζ_1. In the second case (Fig. 6.28-b), by symmetry, half of the value reached pertains to the segment denoted by ζ_1, while the other half pertains to that denoted by ζ_2. Analogous reasoning has to be followed when we adopt the options in Figure 6.29. In any case, the resulting distribution is the one represented in Figure 6.30.

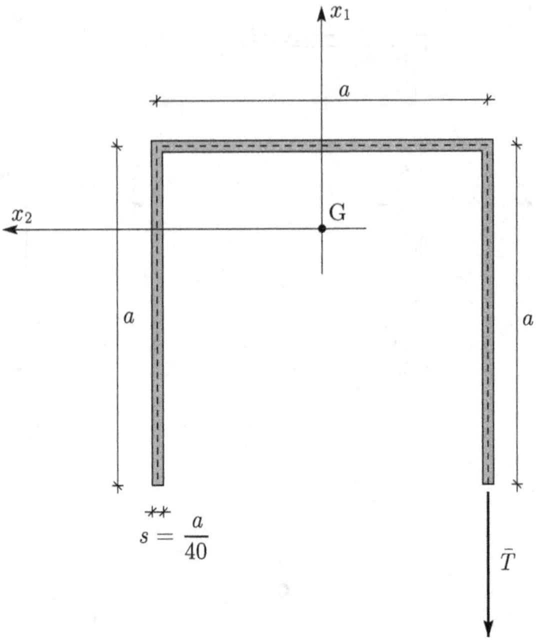

Fig. 6.26 Rotated section about 90° with a new shear force

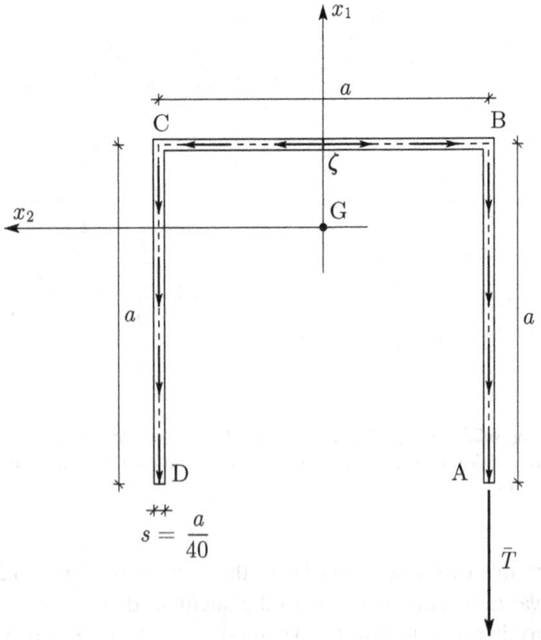

Fig. 6.27 The flow of τ_n

Fig. 6.28 Two possible choices of Ω^*

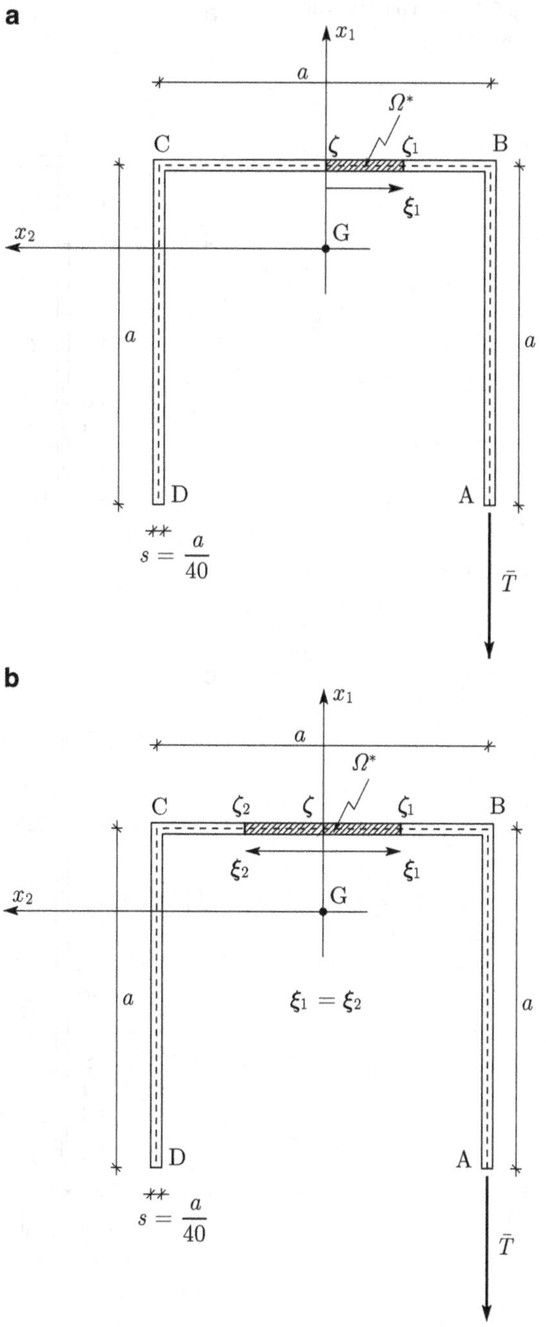

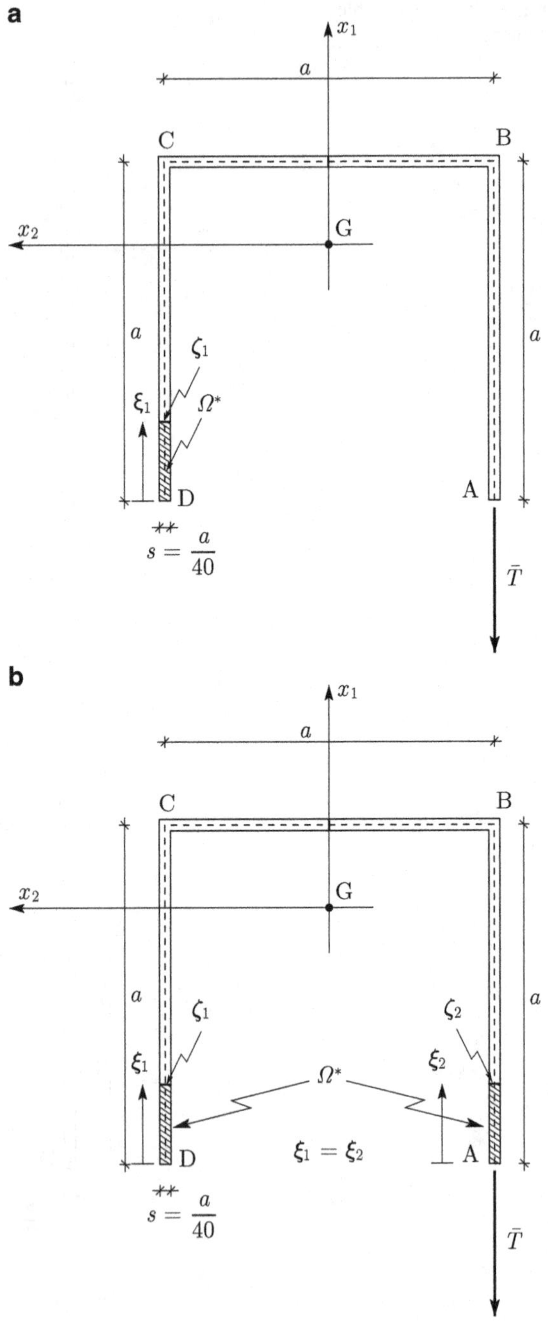

Fig. 6.29 Two other possible choices of Ω^*

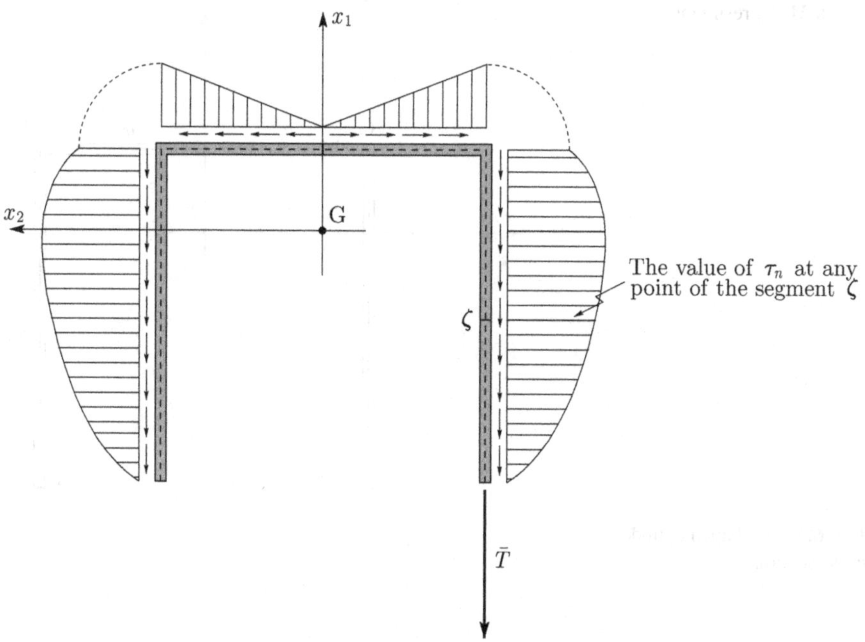

Fig. 6.30 The distribution of the values of τ_n, according to $\bar{T}$

Exercise 13. *Compute explicitly the distribution of the values of τ_n represented in Figure 6.30.*

The resulting total shear stresses emerging from the distribution in Figure 6.30 are those represented in Figure 6.31. By symmetry, we have $q_1 = q_2$. Equilibrium requires that $q_3 + q_4 = \bar{T}$. The straight line r analogous to that in Figure 6.25 coincides with the x_1-axis. The point C_T in Figure 6.25 is then the intersection of two nonparallel straight lines characterized by the property that a shear force applied at this point does not generate torsional effects over the section. Every force in the cylinder's cross-sectional plane applied at C_T can be decomposed along these two straight lines, and the distribution of the τ_n-values is the sum of the two distributions associated with the two components, each not generating torsional effects.

Exercise 14. *Determine the shear center of the section in Figure 6.32 in two cases: $a = 2l$ and $a = l$. Take $a =$* 40 cm, *$20s = a$, $T =$* 90 kN.

Suggestion. Find the mass center and the associated central principal inertial frame. Decompose T along the axes of this frame and proceed as in the discussion above. Could we follow the same procedure with respect to axes not determining a central principal inertial frame?

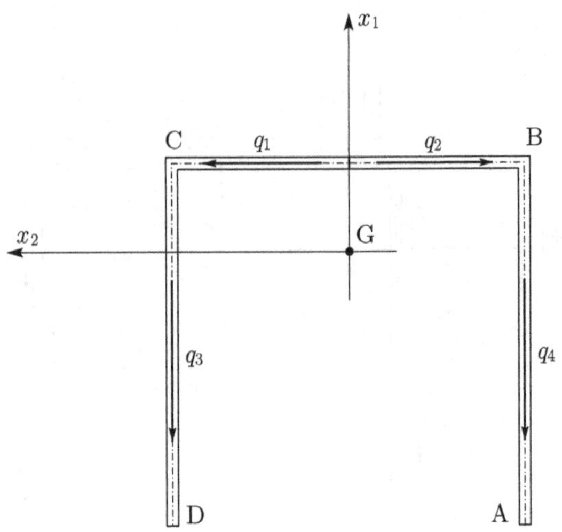

Fig. 6.31 τ_n resultants

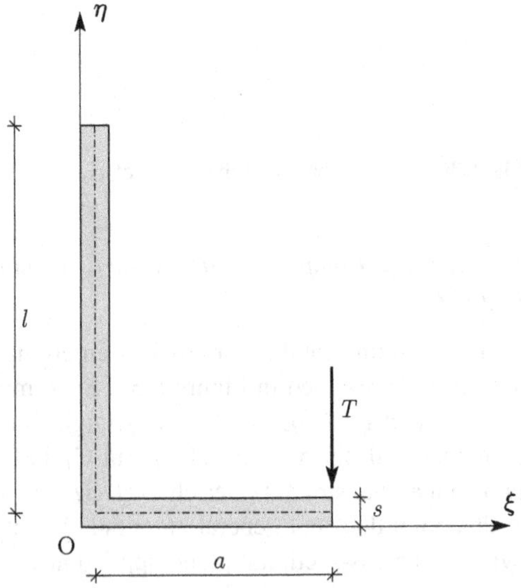

Fig. 6.32 L-shaped cylinder cross section

Exercise 15. *Prove or disprove that the position of the shear center is independent of the intensity of the applied in-plane force.*

Exercise 16. *Determine the τ_n distribution for the sections in Figure 6.33, according to the data reported there.*

A symmetry axis for the beam's cross section plays the role of the straight line r above when we have an applied shear force parallel to it.

Exercise 17. *Determine the shear center in all cases in Figure 6.33.*

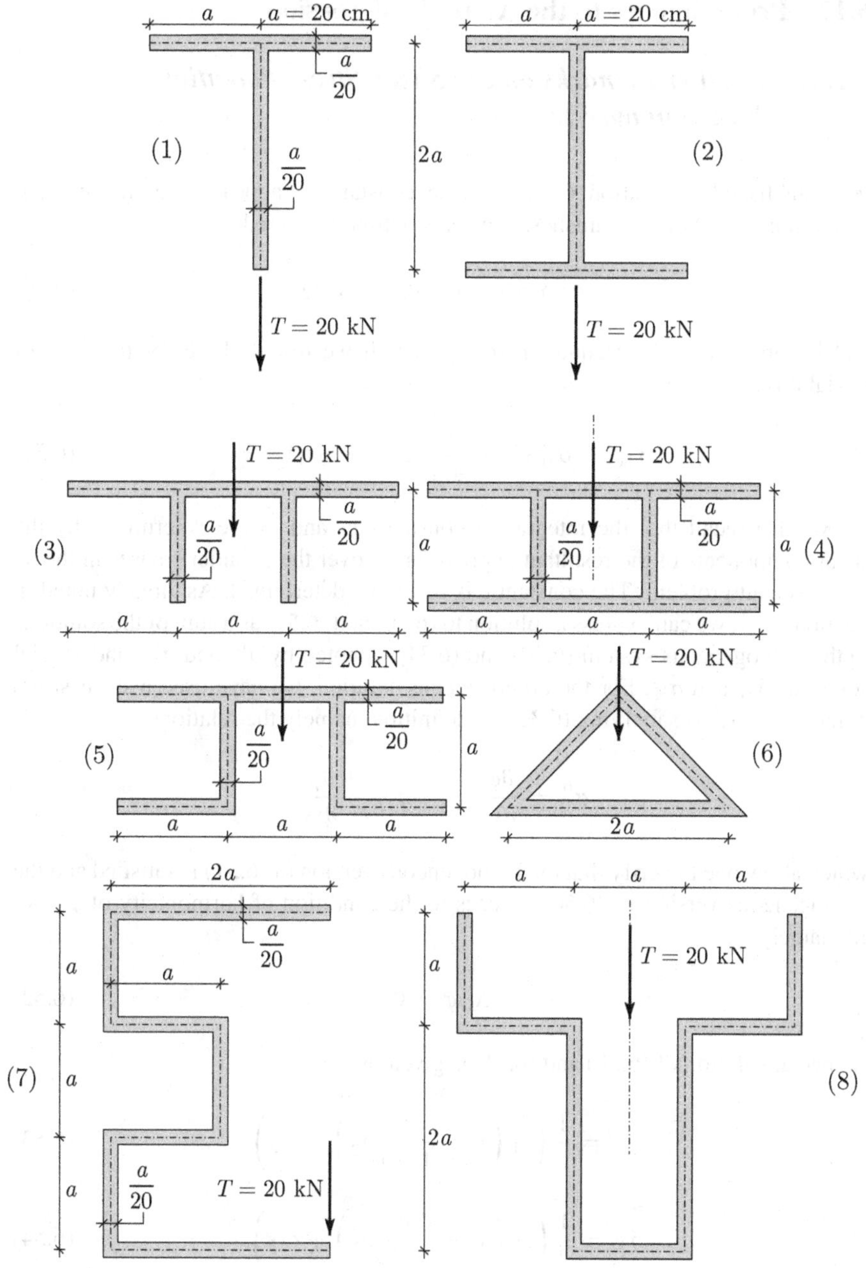

Fig. 6.33 Cylinder cross sections under shear forces

6.12 Preliminaries to the Analysis of Torsion

6.12.1 *Further Remarks on the Shear Stress Potential Function over Ω*

We established in Section 6.5.1 that the constant c_0 appearing in the explicit expression (6.23) of σ_{33} vanishes. Hence, equation (6.24) reduces to

$$\sigma_{31,1} + \sigma_{32,2} = c_1 x_1 + c_2 x_2 \tag{6.50}$$

and is supplemented by equation (6.29), which we rewrite here for the sake of readability:

$$\sigma_{32,1} - \sigma_{31,2} = \frac{\nu}{1+\nu}(c_1 x_2 - c_2 x_1) + \bar{c}. \tag{6.51}$$

We also recall that the integration constants c_1 and c_2 are determined by the shear components of the resultant applied forces over the cylinder's bases in the de Saint-Venant problem. The constant $\bar{c}$ is still to be determined. As already noted in Section 6.5.1, we can express a solution to (6.50) and (6.51) as a sum of the solutions to the homogeneous system (6.30) and (6.31), denoted by σ_{31}^0 and σ_{32}^0, and special solutions $\bar{\sigma}_{31}$ and $\bar{\sigma}_{32}$. For the homogeneous solution, we have used a shear stress potential $\varphi(x_1, x_2)$ satisfying (6.32) by definition, namely the relations

$$\sigma_{31}^0 = \frac{\partial\varphi}{\partial x_1}, \qquad \sigma_{32}^0 = \frac{\partial\varphi}{\partial x_2},$$

which allow one to verify that the homogeneous version of (6.51) is satisfied and the homogeneous version of (6.50) reduces to the condition of harmonicity of φ over Ω, namely

$$\Delta_\Omega \varphi = 0. \tag{6.52}$$

A special solution of (6.50) and (6.51) is given by

$$\bar{\sigma}_{31} = \frac{1}{2}\left(c_1\left(x_1^2 - \frac{\nu}{1+\nu}x_2^2\right) - \bar{c}x_2\right), \tag{6.53}$$

$$\bar{\sigma}_{32} = \frac{1}{2}\left(c_2\left(x_2^2 - \frac{\nu}{1+\nu}x_1^2\right) + \bar{c}x_1\right), \tag{6.54}$$

as we can verify by substitution in (6.50) and (6.51). Consequently, the boundary condition (6.34) becomes

$$\frac{d\varphi}{dn} = -\frac{c_1}{2}\left(x_1^2 - \frac{\nu}{1+\nu}x_2^2\right)n_1 - \frac{c_2}{2}\left(x_2^2 - \frac{\nu}{1+\nu}x_1^2\right)n_2 + \frac{\bar{c}}{2}\left(x_2 n_1 - x_1 n_2\right).$$

Such an expression may suggest that we imagine φ as the sum of three unknown functions $\varphi_1, \varphi_2, \varphi_t$, namely

$$\varphi(x_1, x_2) = -\frac{c_1}{2}\varphi_1(x_1, x_2) - \frac{c_2}{2}\varphi_2(x_1, x_2) + \frac{\bar{c}}{2}\varphi_t(x_1, x_2),$$

each term a harmonic function that satisfies the boundary conditions

$$\frac{d\varphi_1}{dn} = \left(x_1^2 - \frac{\nu}{1+\nu}x_2^2\right)n_1, \tag{6.55}$$

$$\frac{d\varphi_2}{dn} = \left(x_2^2 - \frac{\nu}{1+\nu}x_1^2\right)n_2, \tag{6.56}$$

$$\frac{d\varphi_t}{dn} = x_2 n_1 - x_1 n_2. \tag{6.57}$$

This choice allows us to express separately the contributions of the shear components T_{01} and T_{02} of the applied force resultants, since they are included in c_1 and c_2, the rest pertaining to $\frac{\bar{c}}{2}\varphi_t(x_1, x_2)$. On integrating (6.55) along $\partial\Omega$ and using Gauss's theorem, we obtain

$$\int_{\partial\Omega} \frac{d\varphi_1}{dn} ds = \int_{\partial\Omega} \left(x_1^2 - \frac{\nu}{1+\nu}x_2^2\right) n_1 ds = \int_{\Omega} \frac{\partial}{\partial x_1}\left(x_1^2 - \frac{\nu}{1+\nu}x_2^2\right) d\mathcal{H}^2$$
$$= 2\int_{\Omega} x_1 d\mathcal{H}^2 = 2\mathrm{S}_2 = 0,$$

with the last identity justified by the choice of the reference frame as a central principal inertial frame. In an analogous way, for (6.56) we get

$$\int_{\partial\Omega} \frac{d\varphi_2}{dn} ds = \int_{\partial\Omega} \left(x_2^2 - \frac{\nu}{1+\nu}x_1^2\right) n_2 ds = \int_{\Omega} \frac{\partial}{\partial x_2}\left(x_2^2 - \frac{\nu}{1+\nu}x_1^2\right) d\mathcal{H}^2$$
$$= 2\int_{\Omega} x_2 d\mathcal{H}^2 = 2\mathrm{S}_1 = 0.$$

The condition

$$\int_{\partial\Omega} \frac{d\varphi}{dn} ds = 0$$

generated by (6.52), as shown in Section 6.5.1, implies

$$\int_{\partial\Omega} \frac{d\varphi_t}{dn} ds = \int_{\partial\Omega} (x_2 n_1 - x_1 n_2) ds = 0.$$

6.12.2 The Prandtl Function

Instead of considering the shear stress potential function $\varphi(x_1, x_2)$, we can use the **Prandtl function** $f_p(x_1, x_2)$, named for Ludwig Prandtl (1875–1973) and defined to be a differentiable function such that

$$\sigma_{31}^0 = \frac{\partial f_p}{\partial x_2}, \qquad \sigma_{32}^0 = -\frac{\partial f_p}{\partial x_1}.$$

With this choice, equation (6.30), i.e., the homogeneous part of (6.50), reduces to

$$\Delta_\Omega f_p = 0.$$

In terms of f_p, the boundary condition reads

$$f_{p,2} n_1 - f_{p,1} n_2 = -(\bar{\sigma}_{31} n_1 + \bar{\sigma}_{32} n_2).$$

Consider Figure 6.34, showing a portion of Ω. The components of the normal n are $n_1 = \cos\alpha_1$ and $n_2 = \cos\alpha_2$. By the analysis of the triangles determined by the normal and the tangent to a point of $\partial\Omega$, where they are uniquely defined, we obtain

$$n_1 = \frac{dx_2}{ds}, \qquad n_2 = -\frac{dx_1}{ds},$$

where s is the arc length along $\partial\Omega$, a curve defined in parametric form by continuous and differentiable functions (it is an assumption) $x_1 = \tilde{x}_1(s)$ and $x_2 = \tilde{x}_2(s)$.

By substitution, we get

$$f_{p,1} \frac{dx_1}{ds} + f_{p,2} \frac{dx_2}{ds} = \frac{df_p}{ds} = \left(\bar{\sigma}_{32} \frac{dx_1}{ds} - \bar{\sigma}_{31} \frac{dx_2}{ds} \right).$$

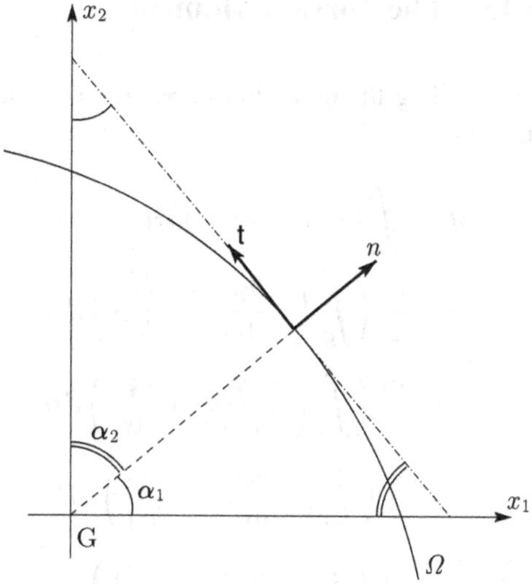

Fig. 6.34 A portion of Ω and the link between the normal and the tangent at a point

Integration of the previous expression leads to

$$f_p(\tilde{x}_1(s), \tilde{x}_2(s)) = \tilde{f}_p(s) = \int_{\partial\Omega} \left(\bar{\sigma}_{32} \frac{dx_1}{ds} - \bar{\sigma}_{31} \frac{dx_2}{ds} \right) ds$$

$$= -\frac{c_1}{2} \int_{\partial\Omega} \left(x_1^2 - \frac{\nu}{1+\nu} x_2^2 \right) dx_2 + \frac{c_2}{2} \int_{\partial\Omega} \left(x_2^2 - \frac{\nu}{1+\nu} x_1^2 \right) dx_1$$

$$+ \frac{\bar{c}}{2} \int_{\partial\Omega} (x_1 dx_1 + x_2 dx_2) .$$

Then $f_p(\tilde{x}_1(s), \tilde{x}_2(s)) = \tilde{f}_p(s)$ is the sum of three polynomials that can be explicitly calculated. Let us write

$$\tilde{f}_p(s) = -\frac{c_1}{2} \mathbb{P}(x_2) + \frac{c_2}{2} \mathbb{Q}(x_1) + \frac{\bar{c}}{4} (x_1^2 + x_2^2).$$

The boundary condition reduces in this way to the prescription of the Prandtl function along $\partial\Omega$, namely

$$\tilde{f}_p(s) = h(s),$$

with h a known function. The relation between the shear stress potential and the Prandtl function is given by the identities

$$\varphi_{,1} = f_{p,2}, \qquad \varphi_{,2} = -f_{p,1}.$$

6.13 The Torsion Moment

By recalling the definition in Section 6.3 and using the shear stress potential, we may write

$$
\begin{aligned}
M_t &= \int_\Omega (\sigma_{32}x_1 - \sigma_{31}x_2)d\mathcal{H}^2 \\
&= -\frac{c_1}{2}\left(\int_\Omega \left(x_1\frac{\partial\varphi_1}{\partial x_2} - x_2\frac{\partial\varphi_1}{\partial x_1}\right)d\mathcal{H}^2 + \int_\Omega \left(x_1^2x_2 - \frac{\nu}{1+\nu}x_2^3\right)d\mathcal{H}^2\right) \\
&\quad -\frac{c_2}{2}\left(\int_\Omega \left(x_1\frac{\partial\varphi_2}{\partial x_2} - x_2\frac{\partial\varphi_2}{\partial x_1}\right)d\mathcal{H}^2 + \int_\Omega \left(-x_2^2x_1 + \frac{\nu}{1+\nu}x_1^3\right)d\mathcal{H}^2\right) \\
&\quad +\frac{\bar{c}}{2}\left(\int_\Omega \left(x_1\frac{\partial\varphi_t}{\partial x_2} - x_2\frac{\partial\varphi_t}{\partial x_1}\right)d\mathcal{H}^2 + \int_\Omega (x_1^2 + x_2^2)\,d\mathcal{H}^2\right) \\
&= -\frac{c_1}{2}\left(\int_\Omega \left(x_1\frac{\partial\varphi_1}{\partial x_2} - x_2\frac{\partial\varphi_1}{\partial x_1}\right)d\mathcal{H}^2 + \mathrm{I}_{122} - \frac{\nu}{1+\nu}\mathrm{I}_{111}\right) \\
&\quad -\frac{c_2}{2}\left(\int_\Omega \left(x_1\frac{\partial\varphi_2}{\partial x_2} - x_2\frac{\partial\varphi_2}{\partial x_1}\right)d\mathcal{H}^2 - \mathrm{I}_{112} + \frac{\nu}{1+\nu}\mathrm{I}_{222}\right) \\
&\quad +\frac{\bar{c}}{2}\left(\int_\Omega \left(x_1\frac{\partial\varphi_t}{\partial x_2} - x_2\frac{\partial\varphi_t}{\partial x_1}\right)d\mathcal{H}^2 + \mathrm{I}_0\right),
\end{aligned}
$$

where

$$
\mathrm{I}_{122} := \int_\Omega x_1^2x_2\,d\mathcal{H}^2, \qquad \mathrm{I}_{111} := \int_\Omega x_2^3\,d\mathcal{H}^2,
$$

$$
\mathrm{I}_{112} := \int_\Omega x_1x_2^2\,d\mathcal{H}^2, \qquad \mathrm{I}_{222} := \int_\Omega x_1^3\,d\mathcal{H}^2
$$

are *third-rank inertia moments* and

$$
\mathrm{I}_0 := \int_\Omega (x_1^2 + x_2^2)\,d\mathcal{H}^2
$$

is the *polar inertia moment* with respect to the center of mass of Ω.

Contributions to the torsion moment by the shear components of the applied forces appear in the terms multiplied by c_1 and c_2 (recall that $c_1 = -\frac{T_{01}}{\mathrm{I}_2}$ and $c_2 = -\frac{T_{02}}{\mathrm{I}_1}$, as derived in Section 6.5.2). Consider the case $c_1 = 0$ and $c_2 = 0$:

$$
M_t = \frac{\bar{c}}{2}\left(\int_\Omega \left(x_1\frac{\partial\varphi_t}{\partial x_2} - x_2\frac{\partial\varphi_t}{\partial x_1}\right)d\mathcal{H}^2 + \mathrm{I}_0\right).
$$

By writing $\bar{a}$ for the vector with components $\bar{a}_1 = -x_2$ and $\bar{a}_2 = x_1$, we compute

$$\begin{aligned}
&\int_{\Omega}\left(x_1\frac{\partial\varphi_t}{\partial x_2} - x_2\frac{\partial\varphi_t}{\partial x_1}\right)d\mathcal{H}^2\\
&= \int_{\Omega}\left((x_1\varphi_t)_{,2} - (x_2\varphi_t)_{,1}\right)d\mathcal{H}^2 = \int_{\Omega}\operatorname{div}_{\Omega}(\varphi_t\bar{a})\,d\mathcal{H}^2\\
&= \int_{\partial\Omega}\varphi_t\bar{a}\cdot n\,ds = \int_{\partial\Omega}\varphi_t(x_1n_2 - x_2n_1)ds\\
&= -\int_{\partial\Omega}\varphi_t\frac{d\varphi_t}{dn}ds,
\end{aligned}$$

with the last equality guaranteed by the boundary condition (6.57) (by definition, $\frac{d\varphi_t}{dn} = D\varphi_t \cdot n$). Then using Gauss's theorem and taking into account that φ_t is harmonic, we get

$$\begin{aligned}
\int_{\partial\Omega}\varphi_t\frac{d\varphi_t}{dn}ds &= \int_{\partial\Omega}\varphi_t D\varphi_t\cdot n\,ds = \int_{\Omega}\operatorname{div}_{\Omega}(\varphi_t D\varphi_t)d\mathcal{H}^2\\
&= \int_{\Omega}(\varphi_t\Delta\varphi_t + D\varphi_t\cdot D\varphi_t)\,d\mathcal{H}^2 = \int_{\Omega}|D\varphi_t|^2 d\mathcal{H}^2,
\end{aligned}$$

so that

$$M_t = \frac{\bar{c}}{2}\left(\mathrm{I}_0 - \int_{\Omega}|D\varphi_t|^2 d\mathcal{H}^2\right) = \frac{\bar{c}}{2}\left(\mathrm{I}_0 - \int_{\Omega}(\varphi_{t,1}^2 + \varphi_{t,2}^2)\,d\mathcal{H}^2\right). \tag{6.58}$$

6.14 $\bar{c}$ and Torsional Curvature

From the relation (6.58), we realize that $\bar{c}$ has physical dimensions of a stress divided by a length, since the terms in parentheses have dimensions of length to the fourth power. A kinematic analysis allows us to give further meaning to $\bar{c}$. To this end, let us consider the (local) small rotation tensor $\omega := \mathrm{Skw}Du$ appearing in the natural additive decomposition $Du = \varepsilon + \omega$. By abandoning the distinction between covariant and contravariant components, as we have always done so far in this chapter, we write

$$\omega_{ik} = \frac{1}{2}(u_{i,k} - u_{k,i}).$$

Since in this chapter we are under the regularity conditions ensuring the validity of the Beltrami–Donati–Michell equations, the derivative of ω can be computed, and we can write

$$\omega_{ik,h} = \varepsilon_{ih,k} - \varepsilon_{kh,i}, \tag{6.59}$$

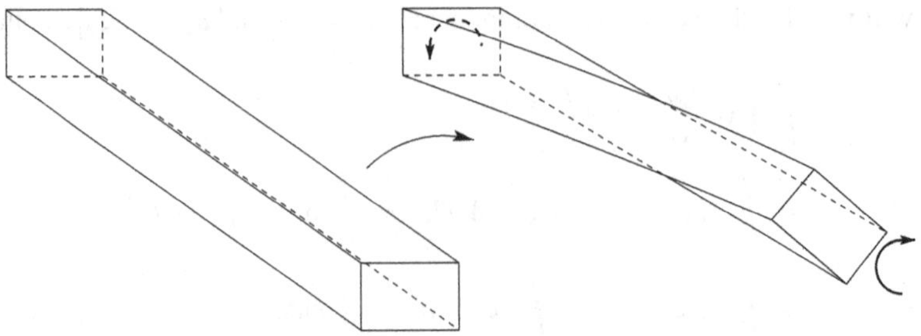

Fig. 6.35 Twist of a straight beam: $\omega_{21,3}$ describes the consequent curvature

as emerges from a direct computation:

$$\begin{aligned}\omega_{ik,h} &= \frac{1}{2}\left(u_{i,k} - u_{k,i}\right)_{,h} = \frac{1}{2}\left(u_{i,kh} - u_{k,ih}\right) \\ &= \frac{1}{2}\left(u_{i,hk} - u_{k,hi}\right) + \frac{1}{2}\left(u_{h,ik} - u_{h,ik}\right) \\ &= \frac{1}{2}\left(u_{i,h} + u_{h,i}\right)_{,k} - \frac{1}{2}\left(u_{k,h} + u_{h,k}\right)_{,i} \\ &= \varepsilon_{ih,k} - \varepsilon_{kh,i}.\end{aligned}$$

The component $\omega_{21,3}$ has a precise physical meaning: it denotes, in fact, the **torsional curvature** κ_t determined by the relative twist between two cylinder cross sections (Fig. 6.35 presents an example).

From the relation (6.59) and taking into account that the cylinder is composed of a linear-elastic isotropic material, we get

$$\omega_{21,3} = \varepsilon_{32,1} - \varepsilon_{31,2} = \frac{1}{2\mu}\left(\sigma_{32,1} - \sigma_{31,2}\right). \tag{6.60}$$

The specific assumptions $c_1 = 0$ and $c_2 = 0$ considered in the previous section and corresponding to torsion without shear imply from equation (6.51)

$$\omega_{21,3} = \frac{\bar{c}}{2\mu},$$

i.e., the torsional curvature in the present conditions, is a constant

$$\kappa_t = \omega_{21,3} = \frac{\bar{c}}{2\mu}, \tag{6.61}$$

so that

$$\bar{c} = 2\mu\kappa_t.$$

Consequently, we have

$$\sigma_{31} = \sigma_{31}^0 + \bar{\sigma}_{31} = \frac{\bar{c}}{2}\frac{\partial \varphi_t}{\partial x_1} - \frac{\bar{c}}{2}x_2 = \frac{\bar{c}}{2}\left(\varphi_{t,1} - x_2\right) = \mu\kappa_t\left(\varphi_{t,1} - x_2\right),$$

$$\sigma_{32} = \sigma_{32}^0 + \bar{\sigma}_{32} = \frac{\bar{c}}{2}\frac{\partial \varphi_t}{\partial x_2} + \frac{\bar{c}}{2}x_1 = \frac{\bar{c}}{2}\left(\varphi_{t,2} + x_1\right) = \mu\kappa_t\left(\varphi_{t,2} + x_1\right),$$

and

$$\varepsilon_{31} = \frac{\sigma_{31}}{2\mu} = \frac{\kappa_t}{2}\left(\varphi_{t,1} - x_2\right), \tag{6.62}$$

$$\varepsilon_{32} = \frac{\sigma_{32}}{2\mu} = \frac{\kappa_t}{2}\left(\varphi_{t,2} + x_1\right). \tag{6.63}$$

All these relations have a counterpart in terms of the Prandtl function f_p. In making them explicit, we find it convenient to adopt for f_p a decomposition analogous to that used for φ:

$$f_p(x_1,x_2) = -\frac{c_1}{2}f_{p1}(x_1,x_2) + \frac{c_2}{2}f_{p2}(x_1,x_2) + \frac{\bar{c}}{2}f_{pt}(x_1,x_2). \tag{6.64}$$

In the case treated here, namely $c_1 = 0$ and $c_2 = 0$, or better *pure torsion*, if we disregard the possible bending effects, since they do not contribute to the shear stresses, f_{pt} is harmonic over Ω with a Dirichlet boundary condition, namely

$$\Delta_\Omega f_{pt} = 0 \quad \text{in} \quad \Omega,$$

$$f_{pt}(x_1,x_2) = \frac{1}{2}\left(x_1^2 + x_2^2\right) \quad \text{along} \quad \partial\Omega,$$

as can be seen by inserting the assumption (6.64) into the boundary condition derived in Section 6.12.2.

Let us define ϕ by

$$\phi(x_1,x_2) := f_{pt}(x_1,x_2) - \frac{1}{2}\left(x_1^2 + x_2^2\right).$$

In terms of ϕ, the previous system becomes

$$\Delta_\Omega \phi(x_1,x_2) = -2 \quad \text{in} \quad \Omega, \tag{6.65}$$

$$\phi(x_1,x_2) = 0 \quad \text{along} \quad \partial\Omega, \tag{6.66}$$

and the shear stresses can be written as

$$\sigma_{31} = \mu\kappa_t\phi_{,2}, \qquad \sigma_{32} = -\mu\kappa_t\phi_{,1}. \tag{6.67}$$

The choice of introducing $\phi(x_1, x_2)$ is motivated by the possibility of linking ϕ directly with what we can call **torsional stiffness**, i.e., the scalar

$$\mu \mathsf{K}_t := \mu \left(\mathrm{I}_0 - \int_\Omega |D\varphi_t|^2 d\mathcal{H}^2 \right)$$

appearing in the expression (6.58), and linking M_t to the torsional curvature:

$$\kappa_t = \frac{M_t}{\mu \mathsf{K}_t}. \tag{6.68}$$

In fact, using (6.67), we can write

$$M_t = \int_\Omega (\sigma_{32}x_1 - \sigma_{31}x_2)\, d\mathcal{H}^2 = -\mu\kappa_t \int_\Omega (x_1\phi_{,1} + x_2\phi_{,2})\, d\mathcal{H}^2$$
$$= -\mu\kappa_t \int_\Omega p \cdot D\phi \, d\mathcal{H}^2,$$

with $p = (x_1, x_2)^{\mathrm{T}}$. We have also

$$\mathrm{div}(\phi\, p) = p \cdot D\phi + 2\phi,$$

as a result of the direct computation of the divergence, so that

$$\int_\Omega p \cdot D\phi \, d\mathcal{H}^2 = \int_\Omega (\mathrm{div}(\phi\, p) - 2\phi)\, d\mathcal{H}^2$$
$$= \int_{\partial\Omega} \phi p \, ds - 2\int_\Omega \phi \, d\mathcal{H}^2 = -2\int_\Omega \phi \, d\mathcal{H}^2,$$

the last identity justified by the boundary condition (6.66). Hence, we can write

$$M_t = 2\mu\kappa_t \int_\Omega \phi \, d\mathcal{H}^2.$$

By comparison with the relation (6.68), we obtain

$$\mathsf{K}_t = 2\int_\Omega \phi \, d\mathcal{H}^2.$$

In other words, the volume delimited by ϕ over Ω times μ is half the torsional stiffness. It is ϕ that is often called the Prandtl function. Here, we prefer to call f_p as the Prandtl function while we refer to ϕ as the **modified Prandtl function**. In any case, terminology a part, the treatment presented in this chapter is completely standard.

6.15 Maximality of $|\tau|^2$

If the modulus of the shear vector τ were maximal at a given point of Ω, we would have there

$$\Delta_\Omega |\tau|^2 < 0, \tag{6.69}$$

since $|\tau|$ is a function of two variables, namely x_1 and x_2. In terms of the modified Prandtl function, the left-hand side of the inequality (6.69) reads

$$\begin{aligned}\Delta_\Omega |\tau|^2 &= \Delta_\Omega \left(\sigma_{31}^2 + \sigma_{32}^2\right) = \mu^2 \kappa_t^2 \Delta_\Omega \left(\phi_{,1}^2 + \phi_{,2}^2\right) \\ &= \mu^2 \kappa_t^2 \Delta_\Omega \left(D\phi \cdot D\phi\right) = \mu^2 \kappa_t^2 (\phi_{,11}^2 + 2\phi_{,12}^2 + \phi_{,22}^2),\end{aligned}$$

due to the validity of equation (6.65). Consequently, the inequality (6.69) can be verified only over $\partial\Omega$, where we can extend ϕ. Hence, $|\tau|$ *is maximal at the boundary* of Ω.

As an example for visualizing the maximality of $|\tau|$, consider Ω to be a circle with radius $\bar{r}$. In this case, we can choose

$$\phi(x_1, x_2) = \hat{a}(x_1^2 + x_2^2 - \bar{r}^2),$$

with $\hat{a}$ a constant to be determined. We compute

$$\Delta_\Omega \phi = 4\hat{a}, \tag{6.70}$$

so that we should have

$$\hat{a} = -\frac{1}{2}$$

to allow the equation (6.70) to be consistent with (6.65). Then ϕ reduces to

$$\phi(x_1, x_2) = \frac{1}{2}\left(\bar{r}^2 - x_1^2 - x_2^2\right),$$

and we have

$$\sigma_{31} = -\mu\kappa_t x_2, \qquad \sigma_{32} = \mu\kappa_t x_1,$$

so that the distribution of τ is linear, as its components σ_{31} and σ_{32} are, and it is maximal at the boundary (see Fig. 6.36). For the constant K_t, we have

$$\mathsf{K}_t = \int_\Omega \left(\bar{r}^2 - x_1^2 - x_2^2\right) d\mathcal{H}^2 = \frac{\pi \bar{r}^4}{2} = \mathrm{I}_0,$$

since Ω is the circle in Figure 6.36.

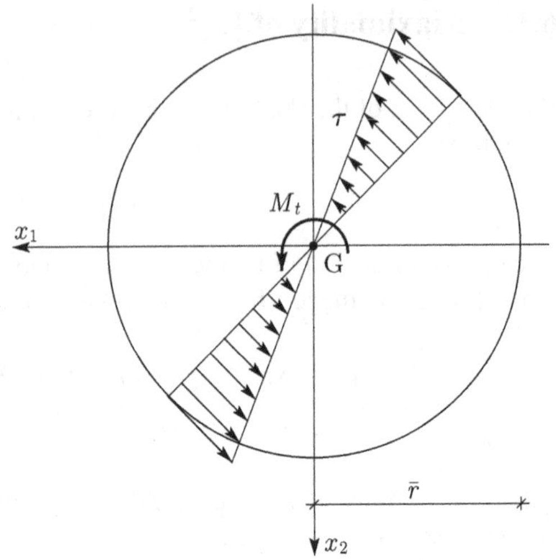

Fig. 6.36 The distribution of τ on a circular cylindrical cross section under pure torsion

6.16 A Kinematic Interpretation of φ_t in Pure Torsion

The ansatz on the structure of the stress in the de Saint-Venant problem and the constitutive structures considered imply

$$\varepsilon_{12} = 0.$$

Moreover, ε_{31} and ε_{32} turn out to be independent of x_3, since σ_{31} and σ_{32} are—recall (6.1). Consequently, the relations (6.59) furnish the following identities:

$$\omega_{21,1} = -\varepsilon_{11,2}, \qquad \omega_{21,2} = \varepsilon_{22,1},$$

$$\omega_{32,1} = \varepsilon_{31,2}, \qquad \omega_{32,2} = \varepsilon_{32,2} - \varepsilon_{22,3},$$

$$\omega_{23,3} = \varepsilon_{33,2}, \qquad \omega_{13,1} = \varepsilon_{11,3} - \varepsilon_{31,1},$$

$$\omega_{13,2} = -\varepsilon_{32,1}, \qquad \omega_{13,3} = -\varepsilon_{33,1},$$

in addition to (6.60). Consider the case of pure torsion: shear external actions, tractions, and bending moments are absent. Then $\sigma_{33} = 0$, and we get

$$\varepsilon_{11} = \varepsilon_{22} = \varepsilon_{33} = \varepsilon_{12} = 0,$$

while ε_{31} and ε_{32} are given by (6.62) and (6.63). The components of derivatives of ω in the frame adopted for the cylinder become

$$\omega_{21,1} = 0,$$

$$\omega_{21,2} = 0,$$

$$\omega_{32,1} = \frac{\kappa_t}{2}\left(\varphi_{t,12} - 1\right),$$

$$\omega_{32,2} = \frac{\kappa_t}{2}\varphi_{t,22},$$

$$\omega_{32,3} = 0,$$

$$\omega_{31,1} = \frac{\kappa_t}{2}\varphi_{t,11},$$

$$\omega_{31,2} = \frac{\kappa_t}{2}\left(\varphi_{t,12} + 1\right),$$

$$\omega_{31,3} = 0,$$

in addition to (6.61). On integrating (6.61), we obtain

$$\omega_{21} = \kappa_t x_3,$$

while from the other components of ω, on integrating the previous relations, we get

$$\omega_{31} = -\omega_{13} = \frac{\kappa_t}{2}\left(\varphi_{t,1} + x_2\right),$$

$$\omega_{32} = -\omega_{23} = \frac{\kappa_t}{2}\left(\varphi_{t,2} - x_1\right).$$

Since $Du = \varepsilon + \omega$ (with the identification of covariant and contravariant components adopted in this chapter), we then get

$$u_{1,1} = 0, \qquad u_{1,2} = -\kappa_t x_3, \qquad u_{1,3} = -\kappa_t x_2,$$

$$u_{2,1} = \kappa_t x_3, \qquad u_{2,2} = 0, \qquad u_{2,3} = \kappa_t x_1,$$

$$u_{3,1} = \kappa_t \varphi_{t,1}, \qquad u_{3,2} = \kappa_t \varphi_{t,2}, \qquad u_{3,3} = 0,$$

and by integration,

$$u_1 = -\kappa_t x_3 x_2, \qquad u_2 = \kappa_t x_1 x_3, \qquad u_3 = \kappa_t \varphi_t(x_1, x_2).$$

The relation between u_3 and φ_t shows that during pure torsion, there is warping of the cross section, and it is independent of x_3. The warping shape is completely described by the component of the potential stress function associated with the torsion.

6.17 A Geometric Property of ϕ

Consider the intersections of the graph of ϕ and planes parallel to Ω, namely the closed curves

$$\phi(x_1, x_2) = \text{const.} \tag{6.71}$$

Let ν be the tangent to one of the curves at a point where it can be uniquely defined. The normal to ν and the curve is n. If we calculate the normal and the tangential components of τ there, we get

$$\tau_n = \sigma_{31} n_1 + \sigma_{32} n_2 = \mu \kappa_t \left(\phi_{,2} \frac{dx_2}{ds} + \phi_{,1} \frac{dx_1}{ds} \right) = \mu \kappa_t \frac{d\phi}{ds} = 0,$$

where s is the arc length along the curve, and the last identity is justified by the relation (6.71) defining the curve itself, and

$$\tau_\nu = \sigma_{31} \nu_1 + \sigma_{32} \nu_2 = -\mu \kappa_t \left(\phi_{,2} \frac{dx_2}{dn} + \phi_{,1} \frac{dx_1}{dn} \right) = \mu \kappa_t \frac{d\phi}{dn}.$$

In other words, τ is purely tangential to curves $\phi(x_1, x_2) = \text{const.}$

In the same way, a similar property also holds for every piecewise smooth closed curve in Ω under conditions of pure torsion. In fact, let γ be any such curve and let $\mathcal{D}_\gamma \subseteq \Omega$ be the region surrounded by it. Using Gauss's theorem, we obtain

$$\int_\gamma \tau \cdot n \, ds = \int_{\mathcal{D}_\gamma} \text{div}_\Omega \tau \, d\mathcal{H}^2 = 0$$

since $\text{div}_\Omega \tau = 0$ for $\sigma_{33} = 0$.

6.18 The Bredt Formula

Consider a thin-walled beam doubly connected, a tube. Its prototype cross section Ω_γ consists of a piecewise smooth closed curve γ without intersections, thickened smoothly. Precisely, when γ is smooth, we write

$$\bar{\Omega}_\gamma := \left\{ x = y(s) + \alpha(s) n(s) \mid y(s) \in \gamma, \alpha(s) \in \left[-\frac{\hat{b}(s)}{2}, \frac{\hat{b}(s)}{2} \right], s \in \left[0, \bar{l}\right] \right\}, \tag{6.72}$$

where $\bar{l}$ is the length of γ, $\hat{b}(s)$ is the width at the point s, $n(s)$ is the outward unit normal to γ. In the presence of corners, i.e., when γ is just piecewise smooth, the

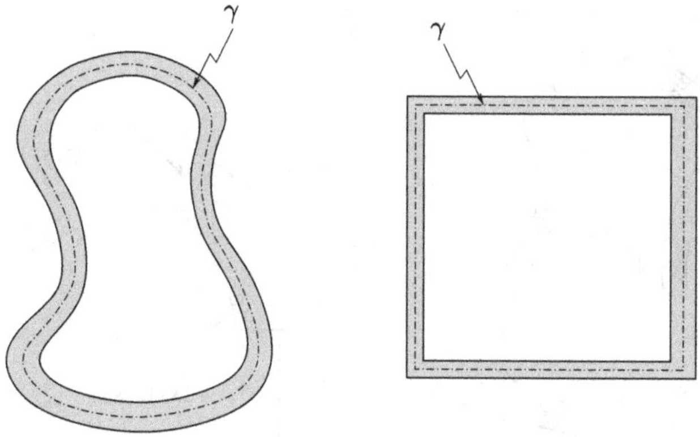

Fig. 6.37 Examples of Ω_γ

formal expression of Ω_γ is more intricate and involves the normal cones at the corners. Figure 6.37 shows two possible examples of acceptable regions Ω_γ.

In what follows in this section we shall smooth corners with the aim of furnishing an approximate evaluation of τ, which is analogous in spirit to what we have done for the Jourawski formula, i.e., a treatment based only on equilibrium reasoning. As a consequence of the previous results, we know that over Ω_γ, τ is tangential to the outer and inner boundaries. Under the condition of pure torsion, to ensure equilibrium, we must have

$$M_{0t} = M_t = \int_{\Omega_\gamma} (y(s) - y_0) \times \tau \, d\mathcal{H}^2 \tag{6.73}$$

with M_t the applied torsional moment and y_0 an arbitrary point in the plane containing Ω_γ. Moreover, the absence of applied shear forces implies that for any direction denoted by a unit vector $\hat{e}$,

$$\int_{\Omega_\gamma} \tau \cdot \hat{e} \, d\mathcal{H}^2 = 0. \tag{6.74}$$

Using the mean value theorem, we can write equation (6.74) as

$$\int_{\Omega_\gamma} \tau \cdot \hat{e} \, d\mathcal{H}^2 = \int_\gamma \bar{\tau}(s)\hat{b}(s) \cdot \hat{e} \, ds, \tag{6.75}$$

with $\bar{\tau}(s)$ the average of τ along the thickness of Ω_γ at s. Write $\bar{\tau}(s) = \bar{\bar{\tau}}(s)\nu(s)$, with $\bar{\bar{\tau}}(s)$ the modulus of $\bar{\tau}$ and ν the tangent to γ. Then equation (6.75) becomes

$$\int_\gamma \bar{\tau}(s)\hat{b}(s) \cdot \hat{e} \, ds = \int_\gamma \bar{\bar{\tau}}(s)\hat{b}(s)\nu(s) \cdot \hat{e} \, ds.$$

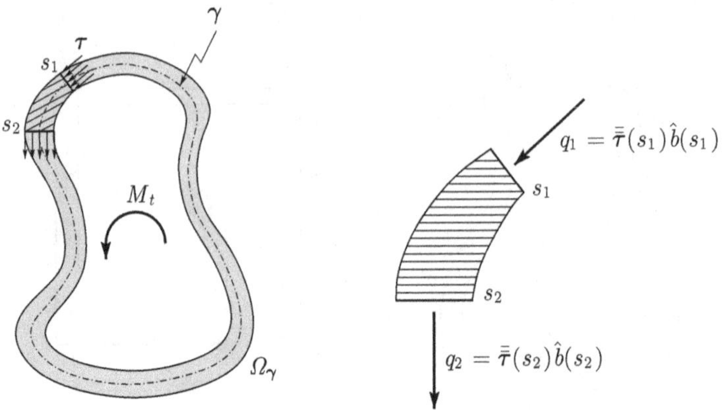

Fig. 6.38 A visualization of the constant value $q = \bar{\bar{\tau}}(s)\hat{b}(s)$. The portion of Ω_γ between s_1 and s_2 has no lateral flow of q. Its translational equilibrium can then be written as $q_1 = q_2$. In other words, at the segment at s_2, the two portions of Ω_γ meeting there exchange forces q_2 and $-q_2$ by the action–reaction principle: they are equal in modulus to q_1

The validity of the constraint (6.74), determined by the absence of shear external actions, implies that the product $\bar{\bar{\tau}}(s)\hat{b}(s)$ is constant. Let us denote such a constant by q, following tradition and paying attention not to confuse it with the rotation vector used in Chapters 1, 2, 3, and 4; ours is an abuse of notation. However, that vector does not play a role in this chapter, so that we have opted for the notation used. Figure 6.38 visualizes the proof of the previous statement.

As a consequence, we have

$$\int_\gamma \bar{\bar{\tau}}(s)\hat{b}(s)\nu(s)\cdot\hat{e}\,ds = q\int_0^{2\pi}\cos\vartheta\,d\vartheta = 0,$$

where ϑ is the angle between ν and $\hat{e}$. It ranges over $[0, 2\pi]$, because γ is closed and without self-intersections. It is exactly the arbitrariness of $\hat{e}$ that implies $q = $ const. Then from the balance (6.73), we get

$$|M_t| = \int_{\Omega_\gamma} |(y(s) - y_0)\times\tau|\,d\mathcal{H}^2 = q\int_\gamma |(y(s) - y_0)\times\nu|\,ds = 2|D_\gamma|q.$$

That the second integral in the previous expression is equal to twice the area $|D_\gamma|$ of the region D_γ delimited by γ follows by the geometric considerations depicted in Figure 6.39.

From the previous relation, we then get

$$\bar{\bar{\tau}}(s) = \frac{|M_t|}{2|D_\gamma|\hat{b}(s)},$$

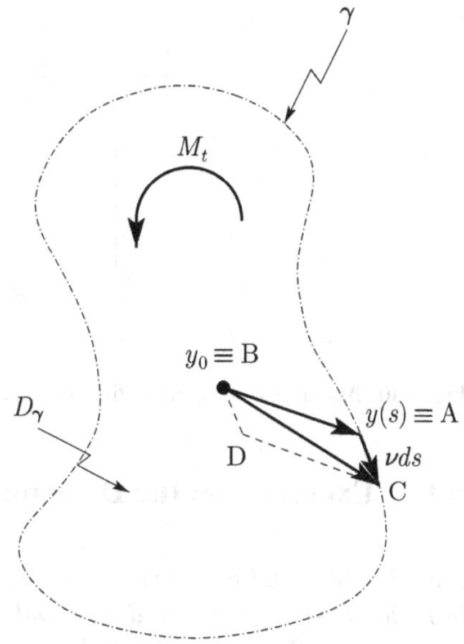

Fig. 6.39 The product $[(y(s) - y_0) \times \nu]\, ds$ at a point s of the curve is a vector with modulus equal to the area of the parallelogram ABDC, which is twice the area of the triangle ABC. By varying s, the triangles of type ABC span D_γ. Then the integral of $|(y(s) - y(0)) \times \nu(s)|$ over γ is the sum of all parallelograms of type ABCD, i.e., twice the area of D_γ

which is called the **Bredt formula**, after Rudolph Bredt (1842–1900). It is well approximated when $\hat{b}(s)$ is very small, so that we can "confuse" $\bar{\bar{\tau}}(s)$ with $|\tau|(s)$. The Bredt formula becomes particularly significant when $\hat{b}(s)$ is constant and very small. In that case, we write

$$|\tau| = \frac{|M_t|}{2|D_\gamma|\hat{b}}.$$

We write $\bar{\kappa}(s)$ for the curvature averaged at s along the width. In the spirit leading to the Bredt formula, we define the torsional angle ϑ of the thin-walled tube with cross section Ω_γ, imagining $\max_s \hat{b}(s)$ to be very small, as

$$\vartheta := \int_\gamma \bar{\kappa}_t(s) ds.$$

The assumed homogeneous isotropic linear-elastic constitutive structures imply

$$\vartheta = \int_\gamma \frac{|\tau(s)|}{\mu} ds = \frac{|M_t|}{2|D_\gamma|\mu} \int_\gamma \frac{1}{\hat{b}(s)} ds, \tag{6.76}$$

which is often called the **second Bredt formula**.

Exercise 18. *Prove the formula (6.76).*

Exercise 19. *Calculate the distribution of τ in the section in Figure 6.40.*

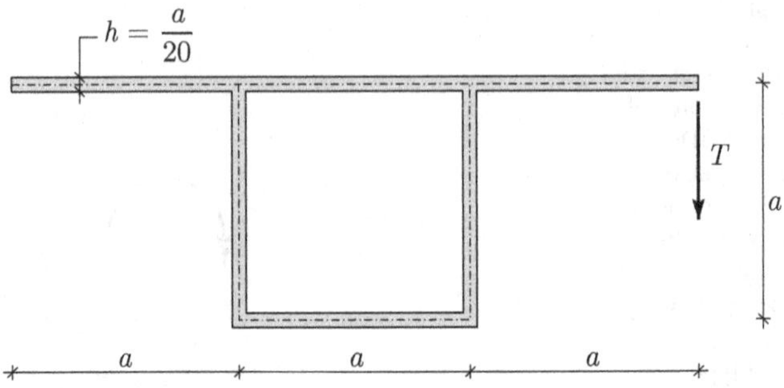

Fig. 6.40 A doubly connected section loaded by a shear force not applied at the shear center

6.19 Exercises on the Determination of Shear Stresses

Exercise 20. *Find the shear center* C_T *for the cylinder cross section represented in Figure 6.41, varying the dimensionless parameter* k *and for* $k = 1$, $k = 0$, *and* $k \to \infty$. *Lengths refer to the dashed line;* h *is constant.*

Remarks and solution. The barycenter G is located on the axis x_2 for symmetry reasons (Fig. 6.42). The cross section is ideally divided into four rectangles of length ka, ka, a, and $2a$, respectively, all having width h. For the pertinent areas A_1, A_2, A_3, and A_4, we obtain (A denotes the total area)

$$A_1 = A_2 = kah, \qquad A_3 = ha, \qquad A_4 = 2ha, \qquad A = \sum_{i=1}^{4} A_i = ah(3 + 2k).$$

In the frame $\{O, \xi, \eta\}$, the coordinates η_i of the centers of mass of the four rectangles are

$$\eta_1 = \eta_2 = a\left(1 - \frac{k}{2}\right), \qquad \eta_3 = \frac{a}{2}, \qquad \eta_4 = a.$$

Consequently, the η-coordinate of the section's barycenter is

$$\eta_G = \frac{\sum_{i=1}^{4} A_i \eta_i}{A} = \frac{a\left(\frac{5}{2} + 2k - k^2\right)}{(3 + 2k)}.$$

If $k \to 0$, $\eta_G \to \frac{5}{6}a$ and if $k = 1$, then $\eta_G = \frac{7}{10}a$. The principal inertial axes are denoted by x_1 and x_2 (Fig. 6.42). The moment of inertia pertaining to the x_2-axis is

$$I_2 = 2haka^2 + \frac{h(2a)^3}{12} = ha^3\left(2k + \frac{2}{3}\right).$$

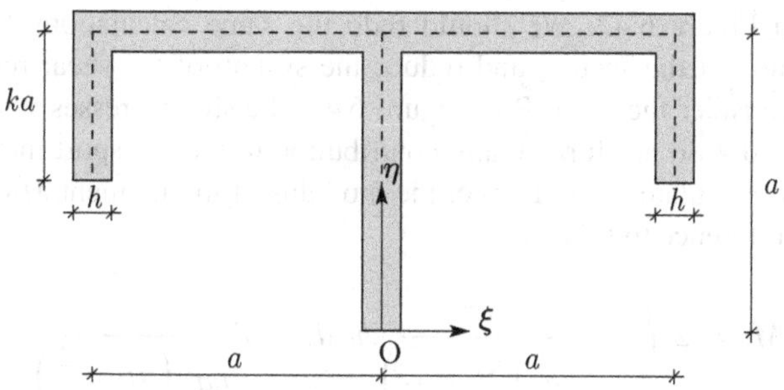

Fig. 6.41 Thin cross section consisting of straight parts

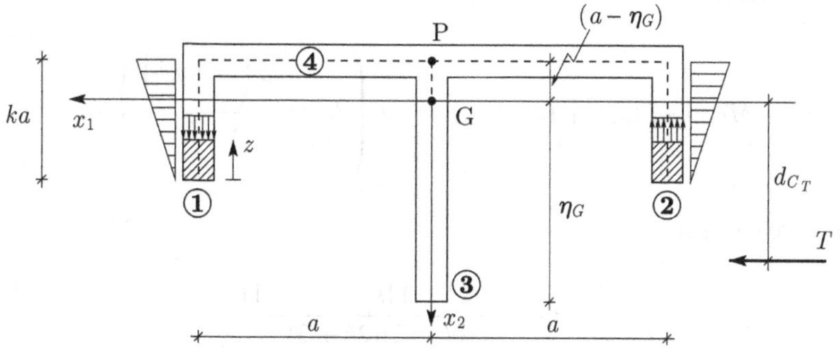

Fig. 6.42 Determination of the position of the shear center

In the calculation, the rectangle 3 has been neglected together with the moment of inertia of the rectangles 1 and 2 with respect to their barycenter axes. When $k \to 0$, we have $I_2 \to \frac{2}{3}ha^3$, and for $k = 1$, $I_2 = \frac{8}{3}ha^3$.

Since x_2 is a symmetry axis, the shear center belongs to it. To find its coordinate d_{C_T}, we apply a shear force T orthogonal to x_2 (Fig. 6.42) and determine the distribution of τ with respect to the rectangles 1 and 2 using the Jourawski formula. By fixing a local coordinate z as in Figure 6.42, $z \in [0, ka]$, for rectangle 1 we get

$$\tau(z) = \frac{TS_2^*(z)}{I_2 h} = T\frac{az}{ha^3\left(2k + \frac{2}{3}\right)}.$$

with τ directed as in Figure 6.42. For rectangle 2, we get

$$\tau(z) = \frac{TS_2^*(z)}{I_2 h} = -T\frac{az}{ha^3\left(2k + \frac{2}{3}\right)},$$

directed as in Figure 6.42. We should redo the same calculations for the other rectangular parts of the section and reduce the system of the shear resultants to a point. If we consider the point P in Figure 6.42, the shear stresses with respect to rectangles 3 and 4 do not furnish any contribution to the transport moment, so we do not have to calculate them. Hence, the modulus of the moment M_P of the shear stresses with reference to P is

$$M_P = 2\int_0^{ka} T\frac{az}{ha^3\left(2k+\dfrac{2}{3}\right)} ha\,dz = T\frac{ha^4k^2}{ha^3\left(2k+\dfrac{2}{3}\right)},$$

and it is oriented counterclockwise. Equilibrium requires

$$M_P = -T(d_{C_T} + (a-\eta_G)) = -T\left(d_{C_T} + \frac{a\left(\dfrac{1}{2}+k^2\right)}{3+2k}\right),$$

from which we get

$$d_{C_T} = -\frac{a(12k^3+11k^2+3k+1)}{(6k+2)(2k+3)}.$$

As $k \to 0$, we get $d_{C_T} \to -\dfrac{a}{6}$, and the shear center coincides with P; for $k = 1$, we obtain $d_{C_T} = -\dfrac{27a}{40}$ (Fig. 6.43). Finally, if $k \to \infty$, then $d_{C_T} \to -\infty$.

Exercise 21. *Compute the distribution of the shear stresses induced by the couple* M_t *for the two thin sections of Figure 6.44. Lengths refer to the dashed lines. Assume* $M_t = 5\,\text{kNm}$, $a = 5\,\text{mm}$ *for the tubular section and* $M_t = 2\,\text{kNm}$, $a = 10\,\text{mm}$ *for the open section.*

Remarks and solution. The first section is tubular, and we can use the Bredt formula with

$$|\Omega| = (100 + 25\pi)a^2$$

the area of the region delimited by the dashed line (Fig. 6.45). We obtain

$$|\tau| = \frac{M_t}{2|\Omega|a} \approx 112\ \text{N/mm}^2.$$

The second section is open and can be ideally stretched to an equivalent slender rectangle with sides a and $a(20 + 10\pi)$. Since $a \ll a(20 + 10\pi)$, we can neglect

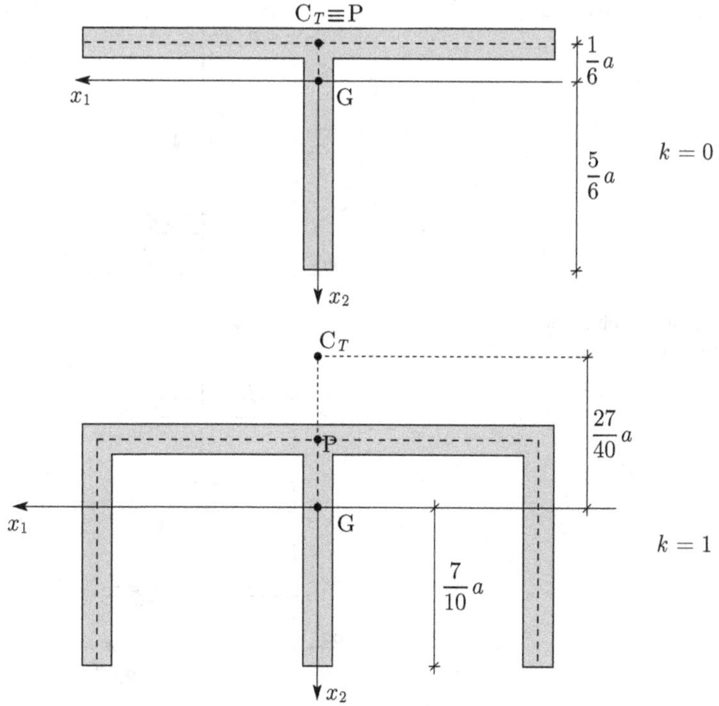

Fig. 6.43 Particular situations corresponding to $k = 0$ and $k = 1$

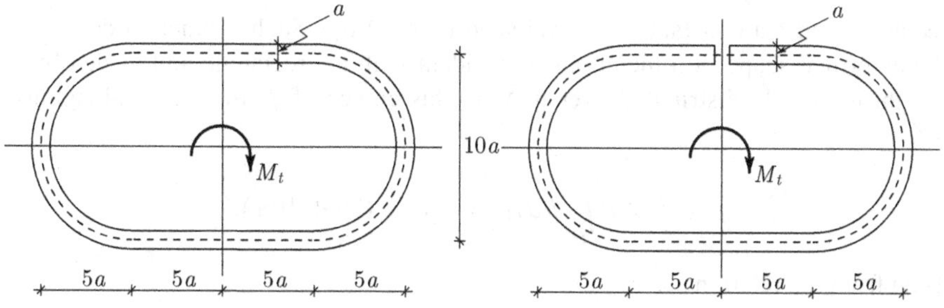

Fig. 6.44 Two thin sections subjected to pure torsion

what happens along the smallest sides, focusing attention on the stress distribution depicted in Figure 6.46. In this case, for the modified Prandtl function we can choose the expression

$$\phi = \frac{4\hat{f}}{a^2}(a - \xi)\xi,$$

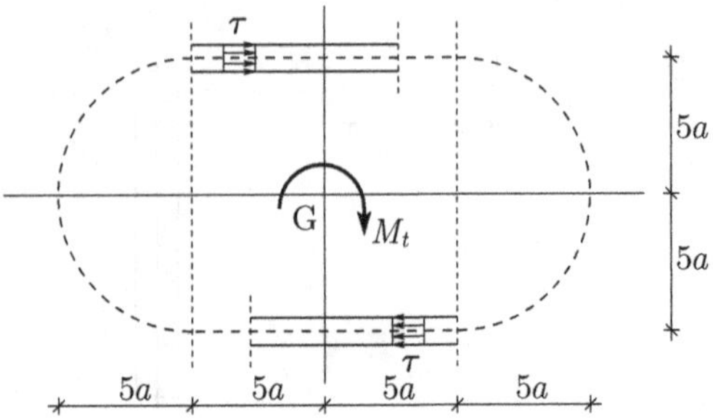

Fig. 6.45 Distribution of the shear stresses in the tubular section

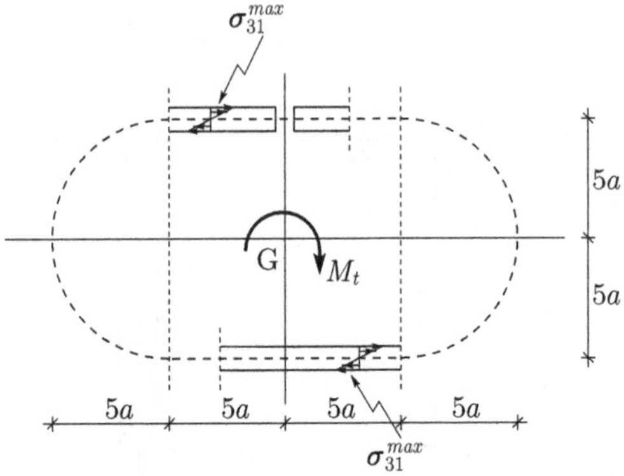

Fig. 6.46 Distribution of the shear stresses in the open section

where $\xi = x_2 + \dfrac{a}{2}$ with x_2 the principal inertia axis parallel to the short side of the rectangle and $\hat{f}$ a constant to be evaluated ($\xi \in [0, a]$). Such a function describes the deformed shape of a membrane coincident with the rectangle and loaded by a weight uniformly distributed over it. With this choice of ϕ, the torsional rigidity $\mu \mathsf{K}_t$ is

$$\mu \mathsf{K}_t = 2\mu \int_{\Omega} \phi \, d\mathcal{H}^2 = \frac{4}{3}\mu \hat{f} a^2 (20 + 10\pi).$$

Then from the relation

$$M_t = \mu \kappa_t \mathsf{K}_t,$$

we get

$$\hat{f} = \frac{3}{4} \frac{M_t}{\mu \kappa_t a^2 (20 + 10\pi)},$$

i.e.,

$$\phi = \frac{3M_t}{\mu \kappa_t a^4 (20 + 10\pi)} (a - \xi)\xi,$$

and consequently,

$$\sigma_{31} = \mu\kappa_t \frac{d\phi}{d\xi} = \frac{3M_t}{a^4(20+10\pi)}(a-2\xi).$$

The maximum modulus of the shear stress is reached at $\xi = 0$ and at $\xi = a$, where we get

$$\sigma_{31}^{max} = \frac{M_t}{\frac{1}{3}a^3(20+10\pi)},$$

so that since we have chosen to neglect the contribution of σ_{32} for $a \ll a(20+10\pi)$, we find that the maximal shear stress is approximately 117 N/mm^2.

The two sections experience similar values of the maximal shear stresses. The advantage in using the first section in technological applications rests in the higher value of M_t and the lower value of the thickness.

Exercise 22. *For the section in Figure 6.47, estimate the stress components σ_{31} and σ_{32} at the point* C *by the Jourawski formula. Assume $T = 50$ kN, $R = 100$ mm.*

Remarks and solution. The barycenter G is located on the x_2-axis due to the symmetry. The area is

$$A = \pi R^2 - \frac{\pi R^2}{16},$$

and in the frame $\{O, x_0, x_2\}$, the barycenter is at the point $(0, x_2(G))$ with

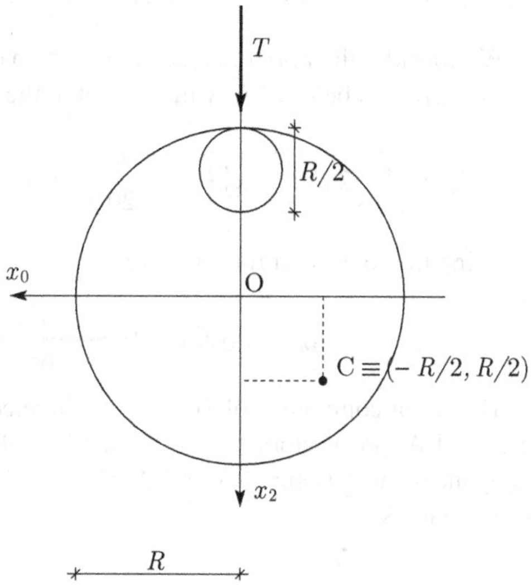

Fig. 6.47 Symmetric section with a circular hole subjected to a shear force acting along the symmetry axis

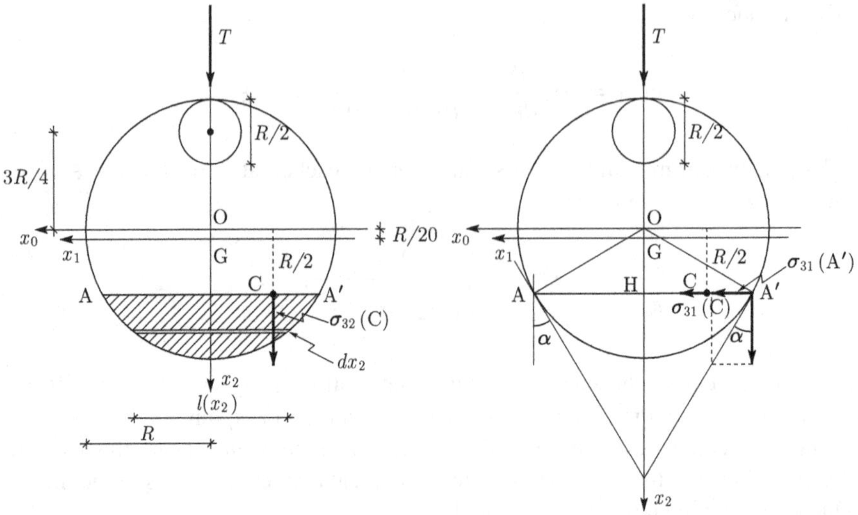

Fig. 6.48 Barycenter, principal axes of inertia, and determination of the components of the shear stresses σ_{31}(C) and σ_{32}(C)

$$x_2(\mathrm{G}) = \frac{R}{20}.$$

The principal inertial frame is represented in Figure 6.48. The moment of inertia with respect to the x_1-axis is

$$\mathrm{I}_1 = \left(\frac{\pi R^4}{4} + \pi R^2 \frac{R^2}{400}\right) - \left(\frac{\pi R^4}{1024} + \frac{\pi R^2}{16}\frac{(16)^2}{400}R^2\right) = \pi R^4 \frac{1083}{5120} \approx 0.2115\pi R^4.$$

We consider the horizontal segment $\overline{\mathrm{AA}'}$ passing through C. The static moment S_1^* of the region below $\overline{\mathrm{AA}'}$ with respect to the x_1-axis is

$$\mathrm{S}_1^* = \int_{R/2}^{R} 2\sqrt{R^2 - x_2^2}\left(x_2 - \frac{R}{20}\right) dx_2 = \frac{(63\sqrt{3} - 4\pi)}{240} R^3 \approx 0.4023R^3.$$

Using the Jourawski formula, we obtain

$$\tau_n = \sigma_{32}(\mathrm{C}) \approx 0.3496\frac{T}{R^2} \approx 1.748 \ \mathrm{N/mm}^2.$$

This shear component of the stress is directed downward. The total shear stresses at A and A′ are vectors tangent to the boundary of the section. Their horizontal components at the points A and A′ (Fig. 6.48) are directed inside the section and have modulus

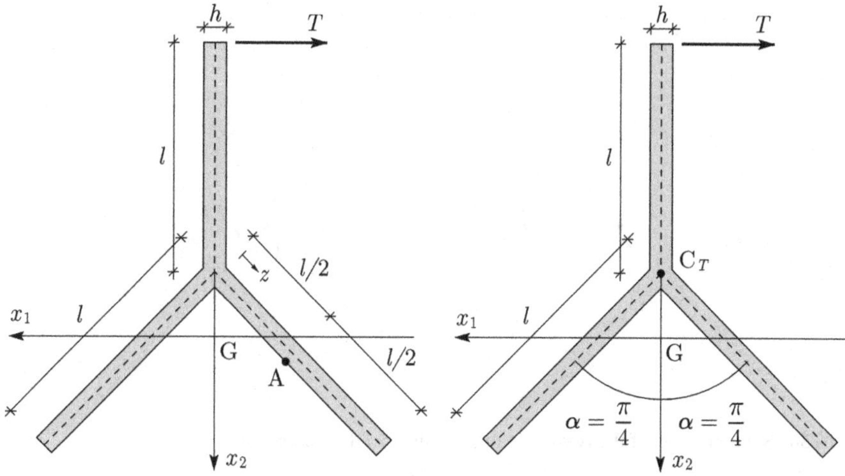

Fig. 6.49 Thin section with an eccentric shear force and position of the shear center C_T

$$\sigma_{31}(\text{A}) = \sigma_{31}(\text{A}') = \sigma_{32}(\text{C}) \tan\alpha, \qquad \tan\alpha = \frac{1}{\sqrt{3}}, \qquad (\alpha = \pi/6).$$

The σ_{31} component of the stress varies linearly along $\overline{\text{AA}'}$, so that at C, we have

$$\sigma_{31}(\text{C}) = \sigma_{31}(\text{A}')\frac{\overline{\text{HC}}}{\overline{\text{HA}'}} = \frac{\sigma_{32}(\text{C})}{\sqrt{3}}\frac{R}{2}\frac{2}{R\sqrt{3}} = \frac{\sigma_{32}(\text{C})}{3} \approx 0.583 \text{ N/mm}^2,$$

directed toward the point H (Fig. 6.48).

Exercise 23. *The section represented in Figure 6.49 is subjected to a horizontal shear force T. Find an approximate value of the shear stress at the point* A. *The thickness h is constant; lengths refer to the dashed line. The inclined sides are at* 45°. *Assume* $T = 10\,\text{kN}$, $l = 150\,\text{mm}$, $h = 15\,\text{mm}$.

Remarks and solution. By symmetry, the shear center C_T is at the point shown in Figure 6.49. The barycenter G is on the x_2-axis, but the determination of its precise position is not necessary for the solution of the exercise. The moment of inertia I_2 with respect to the x_2-axis is

$$I_2 = \frac{lh^3}{12} + 2\int_0^l h(z\sin\alpha)^2 dz = \frac{lh^3}{12} + 2h(\sin\alpha)^2\frac{l^3}{3}$$
$$= \frac{lh^3}{12} + \frac{hl^3}{3} \approx 16.917 \cdot 10^6 \text{ mm}^4 \qquad \left(\alpha = \frac{\pi}{4}\right).$$

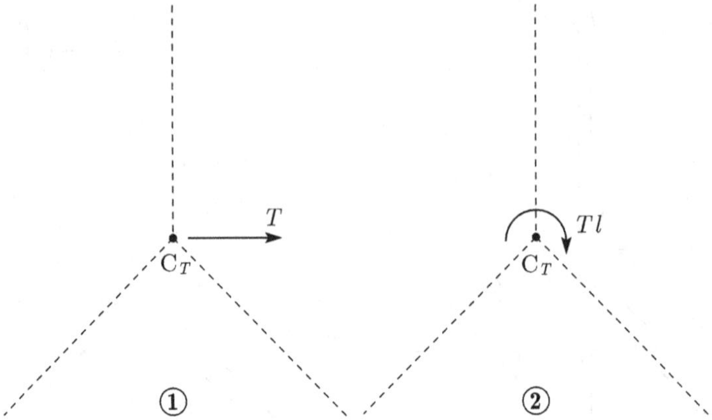

Fig. 6.50 Superposition of the two cases of pure shear and pure torsion

As shown in Figure 6.50, the action of T reduces to the superposition of a shear-without-torsion T applied at C_T and a torsion couple Tl, which can be subdivided into three equal parts, M_{ti}, $i = 1, 2, 3$, each given by

$$M_{ti} = \frac{M_t}{3} = \frac{Tl}{3},$$

acting on each rectangle of the figure, since they are all equal, so that they have the same torsional stiffness. By referring to the remarks on rectangles under torsion in Exercise 21 of this chapter, we write for the shear stress due to torsion at A (say $\tau_{tor}(\mathrm{A})$)

$$\tau_{tor}(\mathrm{A}) = \frac{M_{ti}}{\frac{lh^3}{3}} h = \frac{T}{h^2} \approx 44.44 \ \mathrm{N/mm^2}.$$

It is directed as shown in Figure 6.51.

As regards the shear stress at A induced by the shear action of T (we write $\tau_n(\mathrm{A})$ for it), using the Jourawski formula, we obtain

$$\tau_n(\mathrm{A}) = \frac{TS_2^*}{\mathrm{I}_2 h} = \frac{Thl^2 3\sqrt{2}}{16h\left(\frac{lh^3}{12} + \frac{hl^3}{3}\right)} \approx 3.527 \ \mathrm{N/mm^2},$$

which is directed as shown in Figure 6.52.

The linearized elastic setting allows superposition of effects, so that the resultant shear stress at A has modulus

$$\tau(\mathrm{A}) = |3.527 - 44.44| \approx 40.91 \ \mathrm{N/mm^2}$$

and is directed upward.

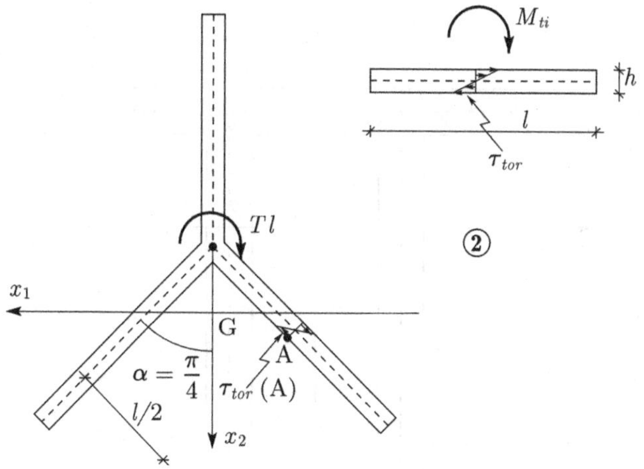

Fig. 6.51 Case of pure torsion

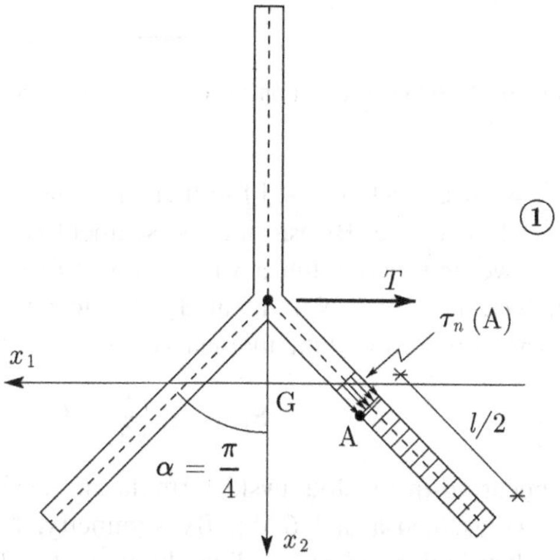

Fig. 6.52 Case of pure shear

Exercise 24. *The section in Figure 6.53 undergoes the action of a shear force T along the symmetry axis. Find the distribution of the shear stresses. Assume a =* 45 mm, r = 12 mm, c = 30 mm, h = 90 mm, T = 90 kN.

Remarks and solution. The shear center belongs to the x_2-axis. The area A, the position of the barycenter G with respect to the upper margin in Figure 6.53, and the principal moment of inertia I_1 are

$$A = 3888 \text{ mm}^2, \qquad d \approx 34.3 \text{ mm}, \qquad I_1 \approx 3975696 \text{ mm}^4.$$

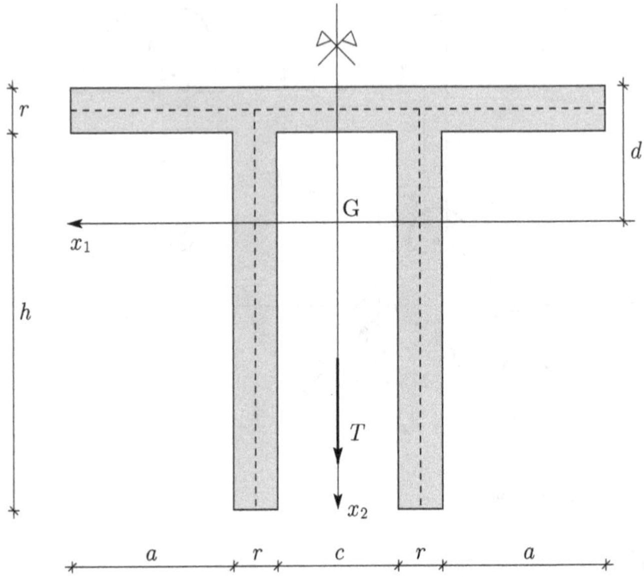

Fig. 6.53 Symmetric section subjected to pure shear

We use the Jourawski formula and consider a family of segments orthogonal to the dashed line. By varying the segment in the region between the points A and B_1, we find there a linear variation of τ from zero along the vertical segment at A to $\tau(B_1) \approx 28.8$ N/mm^2 at B_1, due to the numerical values of T, r, I_1, and the expression of the static moment

$$S_1^* = rz_1\left(\frac{r}{2} - d\right), \qquad z_1 \in [0, a],$$

appearing in the Jourawski formula and evaluated with respect to the coordinate z_1 (Figs. 6.54-a and 6.55). By symmetry, the value of τ at E is zero, and the τ distribution decreases linearly to that value of zero starting from $\tau(B_2) \approx$ 9.6 N/mm^2. In the left-hand portion of the section in Figure 6.54-c, we consider segments between F and H with the pertinent local coordinate $z_2 \in [0, h]$; τ begins at the value $\tau(F) \approx 46.2$ N/mm^2 at F and increases parabolically until the x_1-axis as the static moment of vertical portions is a quadratic function of the vertical abscissa z_2 as we vary the segment orthogonal to the dashed line as in Figure 6.54-c. The value 46.2 N/mm^2 is approximately the sum of the values reached in the linear diagrams on the horizontal side of the section when we prolong them linearly to the midline of the vertical portion of the section itself. Such a procedure is standard in all similar cases. When the segment defining the portion considered crosses the x_1-axis, the portion between the segment itself and the x_1-axis has static moment with algebraic sign opposite to that pertaining to the parts before the x_1-axis. Consequently, the τ distribution decreases to zero. We determine analogously the distribution of τ in the portions $\overline{ED_2}$ and $\overline{D_1C}$. It is endowed with

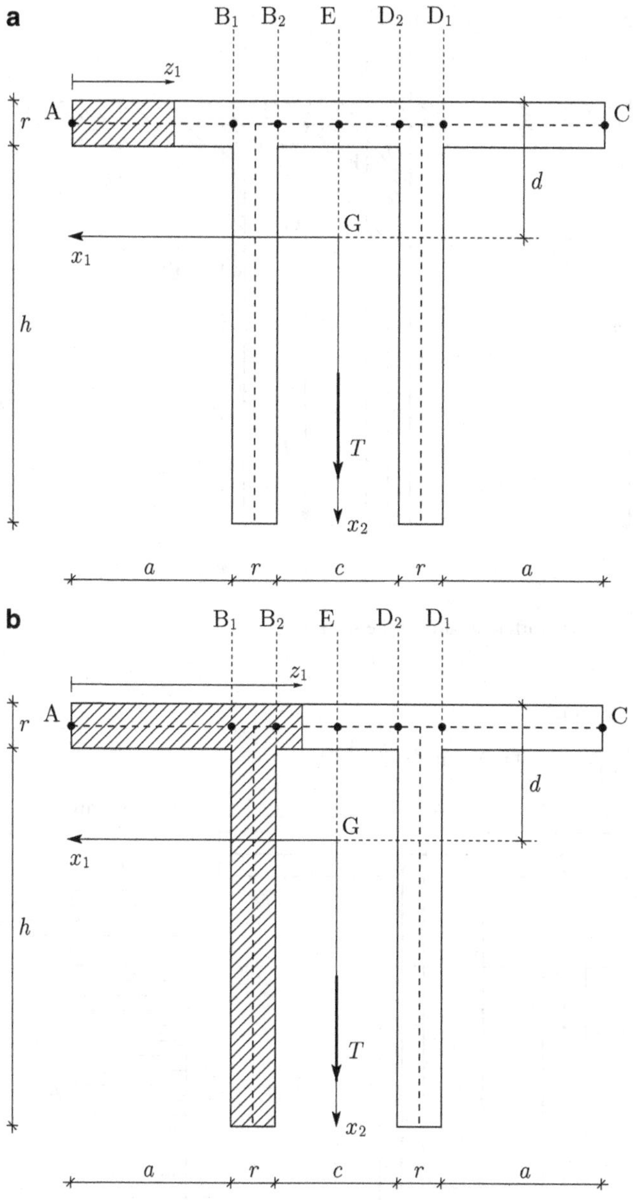
a
B1 B2 E D2 D1
z1
A
C
r
d
G
x1
h
T
x2
a r c r a
b
B1 B2 E D2 D1
z1
A
C
r
d
G
x1
h
T
x2
a r c r a

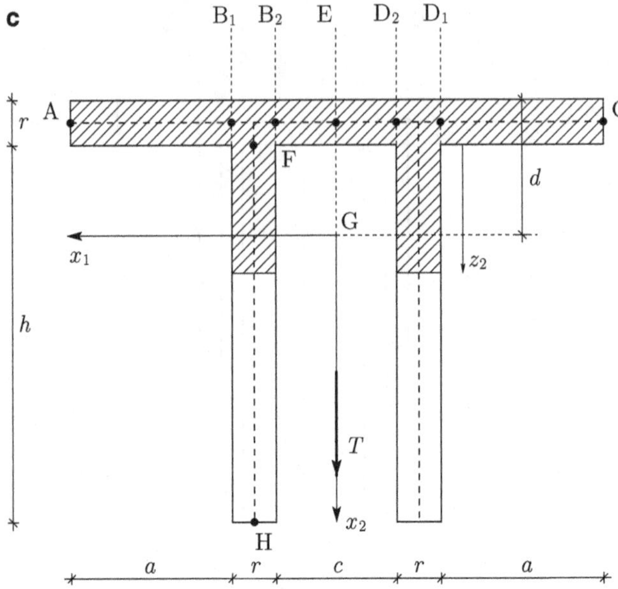

Fig. 6.54 Segments in different parts of the section

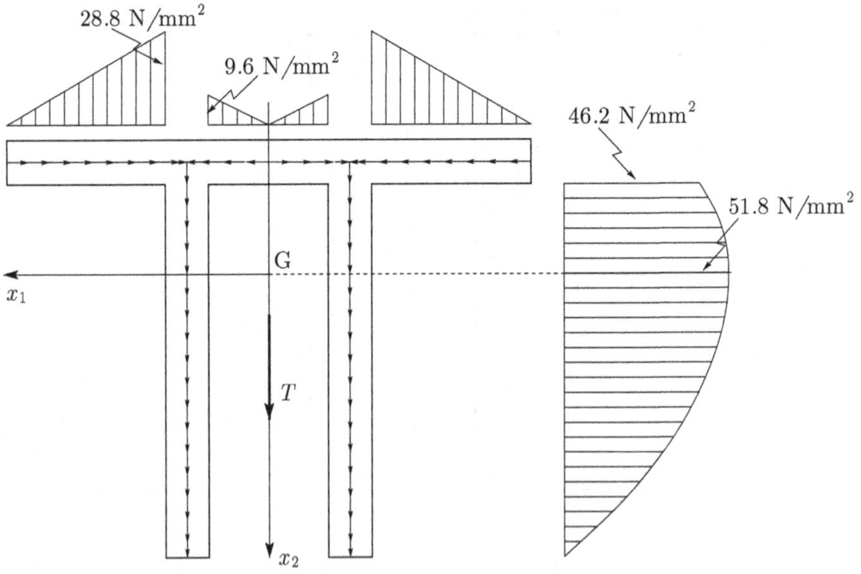

Fig. 6.55 Representation of the modulus of the shear stresses in the section

Fig. 6.56 Symmetric section under pure shear

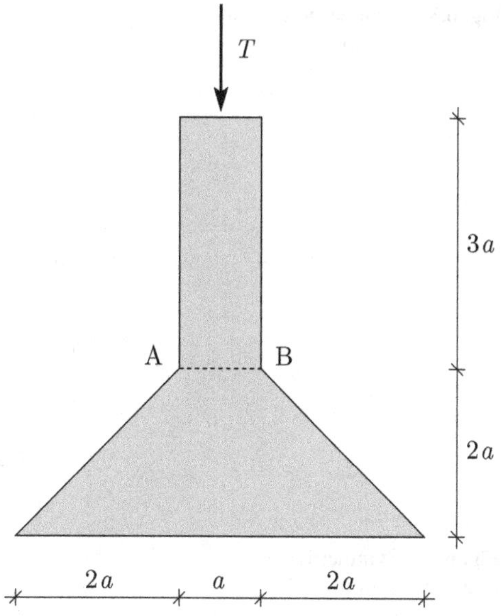

mirror symmetry (Fig. 6.55). Along the remaining vertical part of the section, the distribution of τ is once again quadratic and is equal to that pertaining to the other vertical part (Fig. 6.55).

6.20 Further Exercises on Shear Stresses

Exercise 25. *Find the value of* τ *along the segment* $\overline{\text{AB}}$ *of the section in Figure 6.56. Assume* $a = 10\,\text{cm}$, $T = 300\,\text{kN}$.

Solution. $\tau = \dfrac{252}{869}\dfrac{T}{a^2} \approx 8.7\ \text{N/mm}^2$, directed downward.

Exercise 26. *A torsional couple* M_t *acts on the section in Figure 6.57. Find the distribution of* τ *over the section. Lengths refer to the midline. Suppose that* $h = 10\,\text{mm}$, $a_1 = 10h$, $a_2 = 15h$, $M_t = 2\,\text{kNm}$.

Exercise 27. *Find the position of the shear center* C_T *for the section in Figure 6.58. Lengths refer to the dashed line.*

Exercise 28. *A shear force* T *acts on the section in Figure 6.59. Determine the horizontal segment where the average value of the vertical component of the shear stress,* τ_n*, is maximal.*

Exercise 29. *A shear force* T *acts on the rectangular section in Figure 6.60. Calculate* τ_n *over the segment* $\overline{\text{AB}}$ *as a function of* T *and* l.

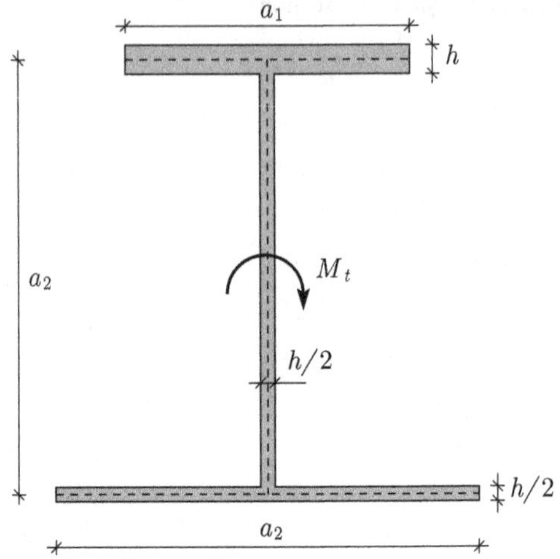

Fig. 6.57 Symmetric section under pure torsion

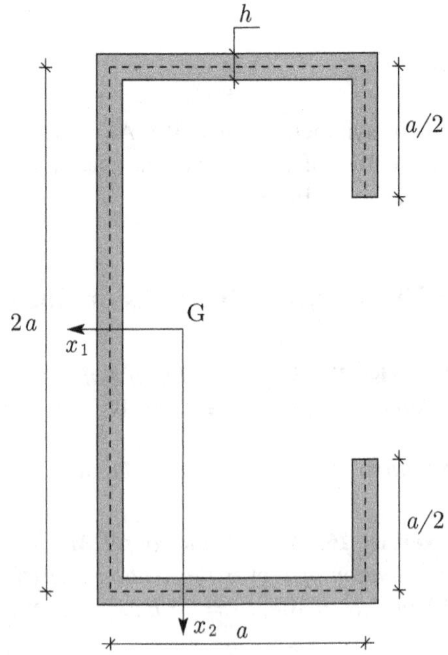

Fig. 6.58 Symmetric C-shaped section

Exercise 30. *For the T-shaped section in Figure 6.61, calculate the shear stress* τ_n *at the point* B. *Assume* $a = 8\,\text{cm}$, $T = 10\,\text{kN}$.

Exercise 31. *For the thin section in Figure 6.62, find the position of the segment* $\overline{\text{AB}}$ *where the value of* τ_n *is maximal according to the Jourawski formula. Assume* $h = 30a$, $a = 10\,\text{mm}$.

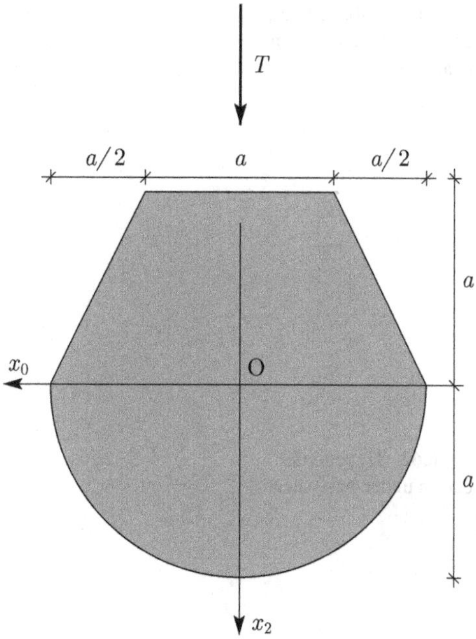

Fig. 6.59 Symmetric section under pure shear

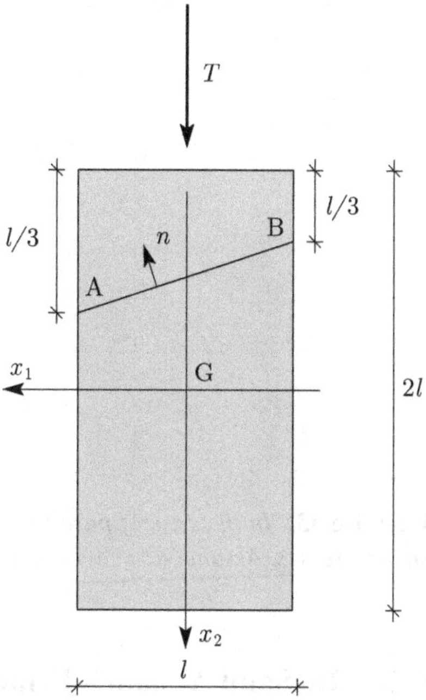

Fig. 6.60 Rectangular section under pure shear

Exercise 32. *Find the position of the shear center* C_T *for the thin section in Figure 6.63. Lengths refer to the dashed line. Assume* $h = R/10$.

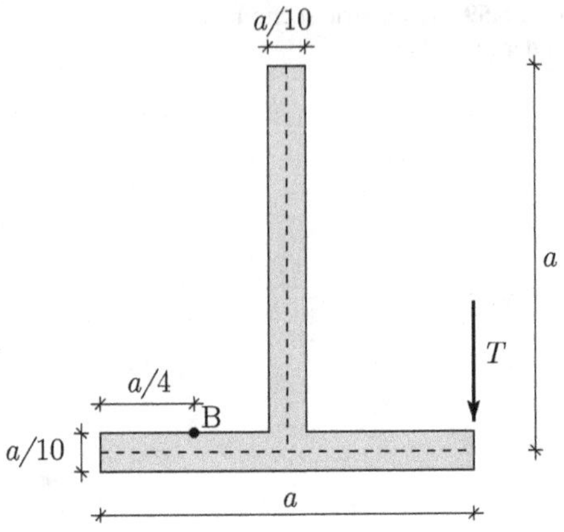

Fig. 6.61 T-shaped cross section under an eccentric shear force

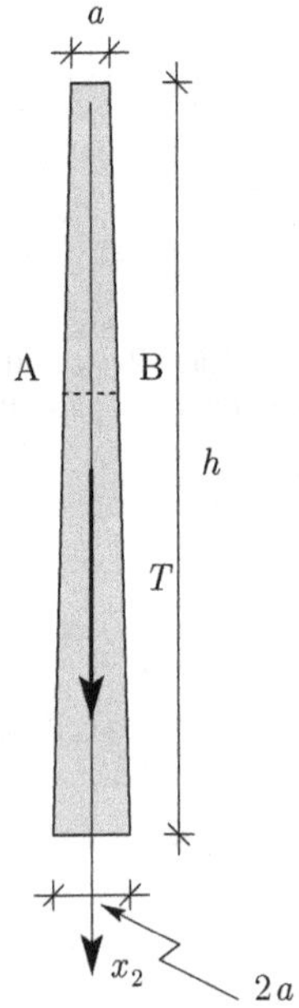

Fig. 6.62 Trapezoidal section under pure shear

Exercise 33. *In the case of pure torsion in Figure 6.64, find the distribution of the shear stresses. Assume* $a = 50\,\text{mm}$, $h = 12\,\text{mm}$, $M_t = 2700\,\text{Nm}$.

6.21 De Saint-Venant's Principle and Toupin's Theorem

In the Navier polynomial we find just the resultants of the applied actions. In other words, different load distributions having the same resultants produce the same σ_{33}. Also, in discussing the determination of the shear stresses σ_{31} and σ_{32}, we have

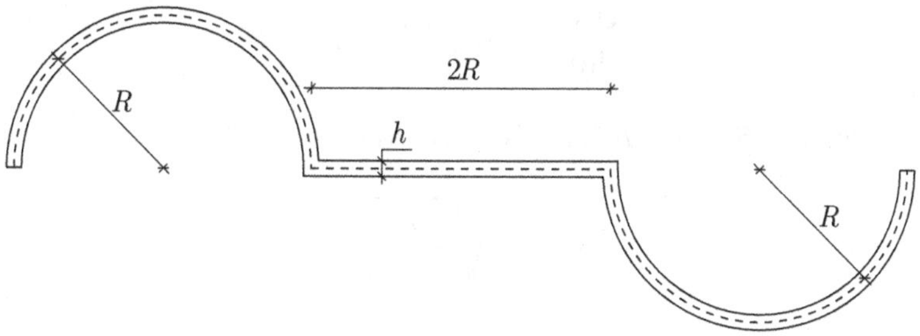

Fig. 6.63 Thin section with open medium line

Fig. 6.64 Thin tubular section in pure torsion

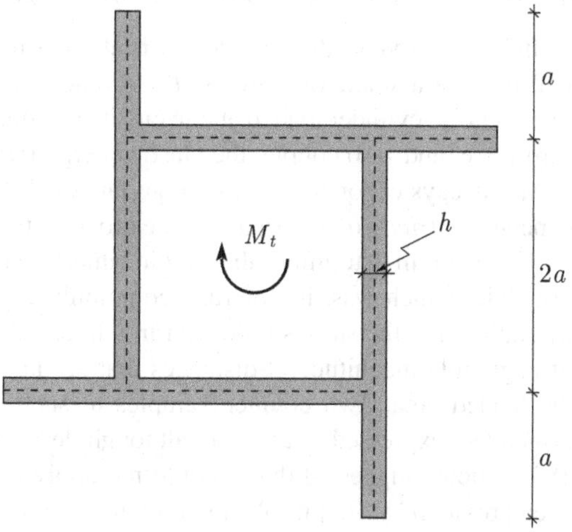

found constants written once again in terms of load resultants. These circumstances have their origin in what de Saint-Venant used as a principle—his analysis was in terms of displacements rather than stresses as we have done in the previous section—formalized by Richard Toupin in the following theorem, for which we weaken the assumption of isotropy adopted so far in this chapter and consider the generic linear-elastic constitutive structures

$$\sigma_{ij} = \mathbb{C}_{ijhk}\varepsilon_{hk},$$

denoting by μ_M and μ_m the *maximum* and *minimum eigenvalues* of $\mathbb{C}$, respectively.

Theorem 34 (Richard Toupin, 1965, a rigorous version of de Saint-Venant's principle). *Let a cylinder of arbitrary length and cross section be loaded only on one end by an arbitrary system of self-equilibrated forces. Then the stored elastic energy* $\mathsf{U}(s)$ *in the cylinder beyond a distance s from the loaded end bears a ratio to the total stored energy* $\mathsf{U}(0)$ *that always satisfies the inequality*

$$\frac{\mathsf{U}(s)}{\mathsf{U}(0)} \leq \exp\left(-\frac{s-l}{s_c(l)}\right),$$

where the characteristic decay length $s_c(l)$ is given by

$$s_c(l) = \sqrt{\frac{\mu^*}{\rho\omega_0^2(l)}},$$

with $\mu^* := \dfrac{\mu_M^2}{\mu_m}$, ρ the mass density, $\omega_0(l)$ the smallest characteristic frequency of free vibration of a cylindrical segment with length $l > 0$.

In Toupin's view, the parameter l is at one's disposal to choose in a manner that will provide a small value for $s_c(l)$. In other words, the theorem affirms that in a linear-elastic cylinder loaded at one end by a system of forces statically equivalent to zero force, and zero couple, the elastic energy (which is stored within the material) always decays exponentially as we go far from the loaded end, with a decay length depending *strictly* on the shape of the cross section.

The theorem determines the way in which we have to interpret de Saint-Venant's principle, which was, in contrast, commonly expressed even in classical treatises by stating that the strains produced in a linear-elastic cylinder loaded as above are of negligible magnitude at distances that are large compared with the diameter of the loaded area. Two counterexamples to such a statement of de Saint-Venant's principle—expressed in this way although de Saint-Venant himself in his 1853 and 1855 articles suggested that it could not apply to any cross-sectional shape—have been provided by Toupin. We mention just the first counterexample, which involves a cylinder with a cross section as depicted in Figure 6.65. Consider two points P and P′ in a cross section distant from the cylinder's bases loaded by two torsional moments M_t and $-M_t$. With the assignment of such torques, even if they are small, for sufficiently small thickness s, we can expect, in fact, that the strain at P is much larger than that at P′.

A formal proof of Toupin's theorem requires some analytical preliminaries, which are given below.

6.21.1 Inequalities

Let v and w be two vector fields defined over a region $\mathcal{B}$ of Euclidean space, where they are square integrable. For these fields, we can prove the following inequality:

$$\int_{\mathcal{B}} \langle \mathsf{v}, \mathsf{w} \rangle d\mu \leq \left(\int_{\mathcal{B}} |\mathsf{v}|^2 d\mu \int_{\mathcal{B}} |\mathsf{w}|^2 d\mu \right)^{\frac{1}{2}}, \tag{6.77}$$

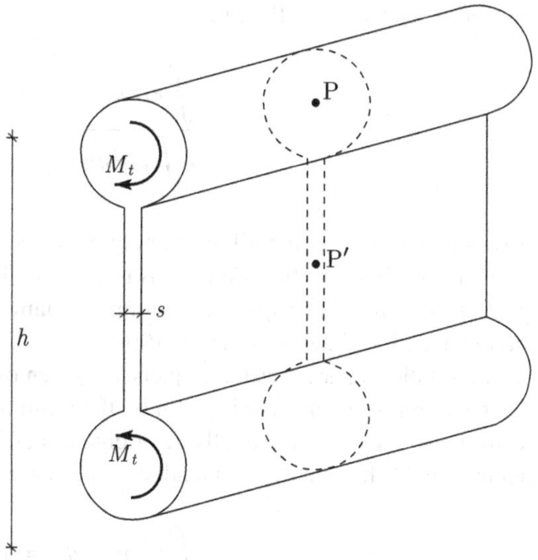

Fig. 6.65 Toupin's counterexample to the common statement of de Saint-Venant's principle before his 1965 theorem

which is called the *Schwarz inequality*.[3] The proof can be found in typical textbooks on functional analysis.

For all positive numbers α, r, and s, we have

$$\sqrt{rs} \le \frac{1}{2}\left(\alpha r + \frac{s}{\alpha}\right). \tag{6.78}$$

It is the *geometric–arithmetic mean inequality*. By combining the inequalities (6.77) and (6.78), we get

$$\int_{\mathcal{B}} \langle \mathsf{v}, \mathsf{w} \rangle d\mu \le \frac{1}{2}\left(\alpha \int_{\mathcal{B}} |\mathsf{v}|^2 d\mu + \frac{1}{\alpha}\int_{\mathcal{B}} |\mathsf{w}|^2 d\mu\right)$$

for all $\alpha > 0$.

6.21.2 The Rayleigh Principle

For a linear-elastic body with (stored) elastic energy density

$$e(\varepsilon) = \frac{1}{2}(\mathbb{C}\varepsilon)\cdot\varepsilon,$$

[3]The inequality was first suggested by Augustin-Louis Cauchy (1789–1857); then it was expressed in integral form by Viktor Bunyakovsky (1804–1889) and appeared in Hermann Amandus Schwarz's (1843–1921) treatise on mathematical analysis.

the stationary values of the ratio

$$\frac{\displaystyle\int_{\mathcal{B}} e(\varepsilon)d\mu}{\displaystyle\frac{1}{2}\int_{\mathcal{B}} |u|^2 d\mu} = \rho\omega^2,$$

provided that ε is compatible, namely $\varepsilon = \operatorname{Sym} Du$, furnish the characteristic frequencies of its free vibrations. This is the so-called *Rayleigh principle*, also called the *Rayleigh–Ritz principle*, after John William Strutt (1842–1919), third Baron Rayleigh, and Walther Ritz (1878–1909).

The smallest characteristic frequency is given by the minimum of the ratio above. The next largest eigenvalue is given by the minimum of the quotient over the set of vector fields u orthogonal to all vector eigenfunctions $v^{(\alpha)}$ corresponding to the first eigenvalue. Orthogonality is intended in the sense that

$$\int_{\mathcal{B}} \langle u, v^{(\alpha)} \rangle d\mu = 0.$$

Since in this chapter we are exploiting systematically the identification of $\mathbb{R}^3$ with its dual, the previous condition can also be written as

$$\int_{\mathcal{B}} u \cdot v^{(\alpha)} d\mu = 0.$$

In the case treated here, the first eigenvalue provided by the ratio above is zero, and the corresponding vector eigenfunctions describe six linearly independent rigid-body motions in three-dimensional Euclidean space. And a vector field u is orthogonal to them when

$$\int_{\mathcal{B}} u\, d\mu = 0 \quad \text{and} \quad \int_{\mathcal{B}} (x - x_0) \times u\, d\mu = 0. \tag{6.79}$$

Consequently, the first nonzero eigenvalue, denoted by ω_0, is given by

$$\rho\omega_0^2 = \min_{u \in \mathcal{C}} \frac{\displaystyle\int_{\mathcal{B}} e(\varepsilon)d\mu}{\displaystyle\frac{1}{2}\int_{\mathcal{B}} |u|^2 d\mu},$$

where $\mathcal{C}$ is the space of continuously differentiable vector fields over $\mathcal{B}$ satisfying the conditions (6.79).

6.21.3 Proof of Toupin's Theorem

Proof. Let $\mathcal{B}_s$ be the part of the cylinder beyond distance s from the loaded end; Ω_s denotes the *cross section at s*. Since bulk loads are not considered in de Saint-Venant problem, the integral balances over $\mathcal{B}_s$ can be written as

$$\int_{\partial\mathcal{B}_s} \mathsf{t}\, d\mathcal{H}^2 = 0, \qquad \int_{\partial\mathcal{B}_s} (x - x_0) \times \mathsf{t}\, d\mathcal{H}^2 = 0.$$

Consequently, since the lateral boundary is free of loads, so is the extreme end of the cylinder included in $\mathcal{B}_s$, the above balances become

$$\int_{\Omega_s} \mathsf{t}\, d\mathcal{H}^2 = 0, \qquad \int_{\Omega_s} (x - x_0) \times \mathsf{t}\, d\mathcal{H}^2 = 0. \tag{6.80}$$

Also, the elastic energy of $\mathcal{B}_s$—call it $\mathsf{U}(s)$—reduces to half of the work of t over Ω_s, namely ($\sigma \in \mathrm{Sym}$)

$$\begin{aligned} \mathsf{U}(s) &= \frac{1}{2}\int_{\mathcal{B}_s} (\mathbb{C}\varepsilon)\cdot\varepsilon\, d\mu \\ &= \frac{1}{2}\int_{\mathcal{B}_s} \sigma\cdot Du\, d\mu = \frac{1}{2}\int_{\mathcal{B}_s} \left(\operatorname{div}(\sigma^{\mathrm{T}}u) - u\cdot\operatorname{div}\sigma\right) d\mu \\ &= \frac{1}{2}\int_{\partial\mathcal{B}_s} \sigma n\cdot u\, d\mathcal{H}^2 = \frac{1}{2}\int_{\Omega_s} \mathsf{t}\cdot u\, d\mathcal{H}^2. \end{aligned} \tag{6.81}$$

The validity of the balances (6.80) implies also that the previous expression remains invariant if we superpose on u an arbitrary rigid displacement, i.e., if we substitute u with

$$\bar{u} = u + c + W(x - x_0),$$

with $c \in \mathbb{R}^3$, $W \in \mathrm{Skw}(\mathbb{R}^3, \mathbb{R}^3)$, x_0 an arbitrary fixed point. The balances (6.80) imply, in fact,

$$\int_{\Omega_s} \mathsf{t}\cdot(c + W(x - x_0))\, d\mathcal{H}^2 = 0,$$

since c and W are constant and $\mathsf{t}\cdot W(x - x_0) = q\cdot(x - x_0)\times\mathsf{t}$, with q the axial vector of W. Consequently, we can write

$$\mathsf{U}(s) = \frac{1}{2}\int_{\Omega_s} \mathsf{t}\cdot\bar{u}\, d\mathcal{H}^2.$$

Then we estimate $\mathsf{U}(s)$ using both the Schwarz and geometric-arithmetic mean inequalities, and we obtain

$$\mathsf{U}(s) \leq \frac{1}{4}\left(\alpha \int_{\Omega_s} |\mathsf{t}|^2 d\mathcal{H}^2 + \frac{1}{\alpha}\int_{\Omega_s} |\bar{u}|^2 d\mathcal{H}^2\right) \tag{6.82}$$

for $\alpha > 0$. However, since the eigenvalues of $\mathbb{C}^2 := \mathbb{C}^{\mathrm{T}}\mathbb{C}$, with the superscript T indicating major transposition,[4] are the squares of those pertaining to $\mathbb{C}$, we obtain

$$|\mathsf{t}|^2 = (\mathbb{C}\varepsilon)n \cdot (\mathbb{C}\varepsilon)n \leq \mu_M^2 |\varepsilon|^2. \tag{6.83}$$

However, we have also

$$\mu_m |\varepsilon|^2 \leq (\mathbb{C}\varepsilon)\cdot\varepsilon = 2e(\varepsilon) \leq \mu_M |\varepsilon|^2. \tag{6.84}$$

By multiplying the inequalities (6.84) by the ratio $\dfrac{\mu_M^2}{\mu_m}$, we get

$$\mu_M^2 |\varepsilon|^2 \leq 2\frac{\mu_M^2}{\mu_m} e(\varepsilon),$$

i.e.,

$$|\mathsf{t}|^2 \leq 2\mu^* e(\varepsilon),$$

which implies

$$\mathsf{U}(s) \leq \frac{1}{4}\left(2\alpha\mu^* \int_{\Omega_s} e(\varepsilon) d\mathcal{H}^2 + \frac{1}{\alpha}\int_{\Omega_s} |\bar{u}|^2 d\mathcal{H}^2\right).$$

Consider a slice of the cylinder included between Ω_s and Ω_{s+l}, with $l > 0$ some positive distance from Ω_s on the opposite side of Ω_0, where the external loads are applied. Denote by $\mathcal{B}_{s,l}$ such a part of the cylinder and define $Q(s,l)$ as

$$Q(s,l) := \frac{1}{l}\int_s^{s+l} \mathsf{U}(\zeta) d\zeta. \tag{6.85}$$

By integrating the inequality (6.82) from s to $s + l$, we obtain

$$lQ(s,l) \leq \frac{1}{4}\left(2\alpha\mu^* \int_{\mathcal{B}_{s,l}} e(\varepsilon) d\mu + \frac{1}{\alpha}\int_{\mathcal{B}_{s,l}} |\bar{u}|^2 d\mu\right). \tag{6.86}$$

[4] $\mathbb{C}^{\mathrm{T}}$ is such that $\mathbb{C}A \cdot B = A \cdot \mathbb{C}^{\mathrm{T}} B$, for any pair of second-rank symmetric tensors A and B over $\mathbb{R}^3$.

The next step is to find an estimate for the last integral in the previous inequality. For this, a lemma is useful.

Lemma 35. *For u a vector field integrable over $\mathcal{B} \subset \mathcal{E}^3$, we can always choose $c \in \mathbb{R}^3$ and $W \in \mathrm{Skw}(\mathbb{R}^3, \mathbb{R}^3)$ such that*

$$\bar{u} = u + c + W(x - x_0)$$

satisfies the integral constraints

$$\int_{\mathcal{B}} \bar{u}\, d\mu = 0, \qquad \int_{\mathcal{B}} (x - x_0) \times \bar{u}\, d\mu = 0. \tag{6.87}$$

Proof. Without loss of generality, we can choose the arbitrary point x_0 to be coincident with the center of mass of $\mathcal{B}$. We then compute

$$\int_{\mathcal{B}} \bar{u}\, d\mu = \int_{\mathcal{B}} (c + W(x - x_0) + u) d\mu = c\, \mathrm{vol}(\mathcal{B}) + \int_{\mathcal{B}} u\, d\mu,$$

since $\int_{\mathcal{B}} (x - x_0) d\mu$ is the static moment of $\mathcal{B}$, and it vanishes, since it is calculated with respect to axes crossing the center of mass. Consequently, to satisfy the first relation (6.87), it suffices to choose

$$c = -\frac{1}{\mathrm{vol}(\mathcal{B})} \int_{\mathcal{B}} u\, d\mu.$$

The second integral in (6.87) leads to

$$\begin{aligned} \int_{\mathcal{B}} (x - x_0) \times \bar{u}\, d\mu &= -c \times \int_{\mathcal{B}} (x - x_0) d\mu \\ &+ \int_{\mathcal{B}} (x - x_0) \times W(x - x_0) d\mu + \int_{\mathcal{B}} (x - x_0) \times u\, d\mu, \end{aligned} \tag{6.88}$$

so that the first term on the right-hand side vanishes, for it includes the static moment calculated with respect to axes crossing the mass center of $\mathcal{B}$. Consider the integrand in the second term and write p for $x - x_0$. In components, we get

$$(p \times Wp) := \mathsf{e}_{ijk} p_k W_{jl} p_l,$$

with e_{ijk} the Ricci tensor. However, with q the axial vector of the skew-symmetric tensor W, we can write

$$W_{jl} = \mathsf{e}_{jlm} q_m,$$

so that

$$
\begin{aligned}
\mathsf{e}_{ijk} p_k W_{jl} p_l &= \mathsf{e}_{ijk} p_k \mathsf{e}_{jlm} q_m p_l \\
&= -\mathsf{e}_{ikj} \mathsf{e}_{jlm} q_m p_l p_k = -(\delta_{il}\delta_{km} - \delta_{kl}\delta_{im}) q_m p_l p_k \\
&= p_i p_m q_m - p_k p_k \delta_{im} q_m.
\end{aligned}
$$

In other words, we can write

$$
p \times Wp = \mathrm{Dev}(p \otimes p) q,
$$

where

$$
\mathrm{Dev}(p \otimes p) := p \otimes p - \mathrm{tr}(p \otimes p) I,
$$

with I the second-rank unit tensor. On inserting such a result into equation (6.88), the second integral constraint (6.87) can be written as

$$
\mathcal{I} q = \hat{V}, \tag{6.89}
$$

where

$$
\hat{V} := \int_{\mathcal{B}} p \times u \, d\mu
$$

and

$$
\mathcal{I} := -\int_{\mathcal{B}} \mathrm{Dev}(p \otimes p) d\mu
$$

is the tensor of inertia of $\mathcal{B}$, and as such is positive definite. Then the algebraic equation (6.89) always admits a solution for arbitrary $\hat{V}$. The existence of such a solution ensures the validity of equations (6.87) and consequently, the orthogonality of $\bar{u}$ (constructed by adding to u an appropriate rigid displacement) to every rigid displacement. □

Due to the property established in the lemma, by the Rayleigh principle, we can write

$$
\frac{1}{2} \int_{\mathcal{B}_{s,l}} |\bar{u}|^2 d\mu \le \frac{1}{\rho \omega_0^2(l)} \int_{\mathcal{B}_{s,l}} e(\bar{\varepsilon}) d\mu = \frac{1}{\rho \omega_0^2(l)} \int_{\mathcal{B}_{s,l}} e(\varepsilon) d\mu,
$$

where $\bar{\varepsilon} := \mathrm{Sym}(D\bar{u}) = \mathrm{Sym}(Du)$ by the definition of $\bar{u}$. Consequently, the inequality (6.86) becomes

$$
Q(s,l) \le \frac{1}{2l} \left(\alpha \mu^* + \frac{1}{\alpha \rho \omega_0^2(l)} \right) \int_{\mathcal{B}_{s,l}} e(\varepsilon) d\mu. \tag{6.90}
$$

Moreover, since equation (6.81) shows that the energy $\mathsf{U}(s)$ of a generic portion of the cylinder of type $\mathcal{B}_s$ is given only by the traction work over Ω_s, we can write

$$\frac{1}{l}\int_{\mathcal{B}_{s,l}} e(\varepsilon)d\mu = \frac{1}{l}\left(\mathsf{U}(s) - \mathsf{U}(s+l)\right) = -\frac{dQ(s,l)}{ds},$$

so that the inequality (6.90) can be written as

$$s_c(l,\alpha)\frac{dQ(s,l)}{ds} + Q(s,l) \leq 0, \tag{6.91}$$

where

$$s_c(l,\alpha) = \frac{1}{2}\left(\alpha\mu^* + \frac{1}{\alpha\rho\omega_0^2(l)}\right).$$

The function $s_c(l,\cdot)$ has a minimum $s_c(l)$ at

$$\alpha = \frac{1}{\sqrt{\mu^*\rho\omega_0^2(l)}}, \tag{6.92}$$

given by

$$s_c(l) = \sqrt{\frac{\mu^*}{\rho\omega_0^2(l)}}.$$

By choosing α as in the formula (6.92) and integrating the inequality (6.91) between $s_1 \geq 0$ and $s_2 \geq s_1$, we get

$$\frac{Q(s_2,l)}{Q(s_1,l)} \leq \exp\left(-\frac{s_2 - s_1}{s_c(l)}\right). \tag{6.93}$$

The positive definiteness of the constitutive elastic tensor $\mathbb{C}$ implies from equation (6.81) that $\mathsf{U}(\cdot)$ is a nonincreasing function of s. Then since $Q(s,l)$ is by definition (6.85) the mean value of the function $\mathsf{U}(\cdot)$ between s and $s+l$, we can write

$$\mathsf{U}(s+l) \leq Q(s,l) \leq \mathsf{U}(s),$$

so that the inequality (6.93) implies

$$\frac{\mathsf{U}(s_2+l)}{\mathsf{U}(s_1)} \leq \exp\left(-\frac{s_2 - s_1}{s_c(l)}\right).$$

By choosing $s_1 = 0$ and $s_2 = s - l$, we obtain the inequality in the statement of Toupin's theorem. □

The idea leading to Toupin's theorem has been extended variously. We limit here the discussion to the original case, for that allows us to deepen our comprehension of the results about the de Saint-Venant problem. Such results will be useful later—among other things—when we shall go beyond the traditional format of continuum mechanics as described so far to construct coarse one-dimensional models of beams, rods, columns, etc., for obvious computational advantages.

Chapter 7
Critical Conditions: Yield Criteria

We have so far focused attention on simple elastic materials,[1] being conscious of our limited view, dictated by our original didactic purposes. Real materials are, in contrast, often not simple, and when a material's response to external actions is elastic, such behavior is not indefinite; rather, it has a finite range, starting from states that we can consider relaxed—the body is left alone without experiencing external actions, and the decay of memory process is presumed to be ended—even in some approximate sense. The notions of elastic range and relaxed state can be defined rigorously even in a very abstract way in terms of the theory of dynamical systems. Having in mind that this book is meant essentially for "beginners", we do not go into detail. We just mention a well-known example that clarifies the situation: the traction test of a cylindrical steel specimen. Figure 7.1 shows the typical recorded stress–strain relationship.

The behavior appears to be linear up to a certain stress value, indicated by σ_Y in the figure, where there is a phase transition, and increasing strain does not produce stress increments (fluctuations apart); also the strain in that region is irreversible, as can be seen by removing the load. In metals, such a behavior is the macroscopic appearance of single and/or collective motions of dislocations through crystal grains: the basic origin of the strain irreversibility. Beyond a certain value of the strain (elongation in the specific example discussed here), the stress–strain relationship becomes nonlinearly nondecreasing (*hardening* regime) until the stress reaches another value beyond which the material is unstable (*softening* regime), and the test can be continued only by controlling the deformation. When we "reverse" the load, we find similar behavior, and the limit stress value for the linear-elastic regime can be reached at $-\sigma_Y$ (*traction–compression symmetry*) or even at a different value. Such a test reveals prominent aspects of what is called elastic–plastic behavior.

[1]More precisely, we have been referred to hyperelastic materials, since we have always considered the (stored) elastic energy.

P.M. Mariano, L. Galano, *Fundamentals of the Mechanics of Solids*,
DOI 10.1007/978-1-4939-3133-0_7

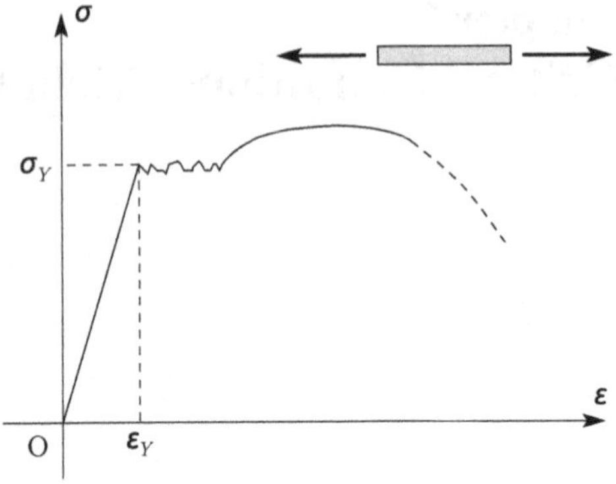

Fig. 7.1 Stress–strain relationship typically recorded in a traction test on a steel specimen

In the case of traction–compression symmetric behavior, with reference to the one-dimensional case in Figure 7.1, we say that the material behaves in a linear-elastic way when

$$|\sigma| \le \sigma_Y,$$

i.e., when

$$f(\sigma) := |\sigma| - \sigma_Y \le 0. \tag{7.1}$$

The constraint (7.1) defines the critical condition but does not furnish by itself information on the postcritical behavior. In other words, from the inequality (7.1) we are not able to deduce whether the material behavior is elastic–plastic or brittle or something else when the stress increases beyond σ_Y.

In three dimensions, the possible expressions of $f(\cdot)$—what we commonly call the **yield function**—are not so simple as that appearing in the **yield criterion** (7.1).

There are many proposals of criteria available in the current literature. Each is associated with a certain specific physical mechanism. The choice then depends on what is suggested by the circumstance under analysis and what we believe to be its prominent aspects.

We list in this chapter some criteria, some historically prominent, some particularly popular. They are all in reference to a point within a body; in this sense, they are *local*.

7.1 Tresca's Criterion

With **Tresca's criterion**, we assume that at a point, a critical stress state is reached when in one of the planes crossing it, the tangential stress reaches a certain critical value k_T determined by, e.g., a traction test as described above.

To formalize this, consider the eigenvalues of the pertinent stress at a point—call them $\sigma_I, \sigma_{II}, \sigma_{III}$. We say that the stress state in the material is admissible at the point considered when

$$\frac{1}{2}\max\left(|\sigma_I - \sigma_{II}|, |\sigma_{II} - \sigma_{III}|, |\sigma_{III} - \sigma_I|\right) - k_T \leq 0.$$

Equivalently, we can define

$$f_T(\sigma) := \left[(\sigma_I - \sigma_{II})^2 - 4k_T^2\right]\left[(\sigma_{II} - \sigma_{III})^2 - 4k_T^2\right]\left[(\sigma_{III} - \sigma_I)^2 - 4k_T^2\right], \tag{7.2}$$

so that the yield criterion (the admissibility condition for the stress) reads

$$f_T(\sigma) \leq 0. \tag{7.3}$$

The domain described by the inequality (7.3) in the space $(\sigma_I, \sigma_{II}, \sigma_{III})$ is a straight prism with hexagonal basis, wrapped around the straight line inclined at 45° with respect to all three axes σ_I, σ_{II}, and σ_{III} as in Figure 7.2. In the plane $\sigma_{III} = 0$, the domain is represented in Figure 7.3. The constant k_T is estimated by uniaxial tension–compression tests. In the case of symmetric material behavior, using the function (7.2), we obtain $k_T = \frac{\sigma_Y}{2}$, since the sole nonzero eigenvalue of

$$\begin{pmatrix} \sigma_Y & 0 & 0 \\ 0 & 0 & 0 \\ 0 & 0 & 0 \end{pmatrix}$$

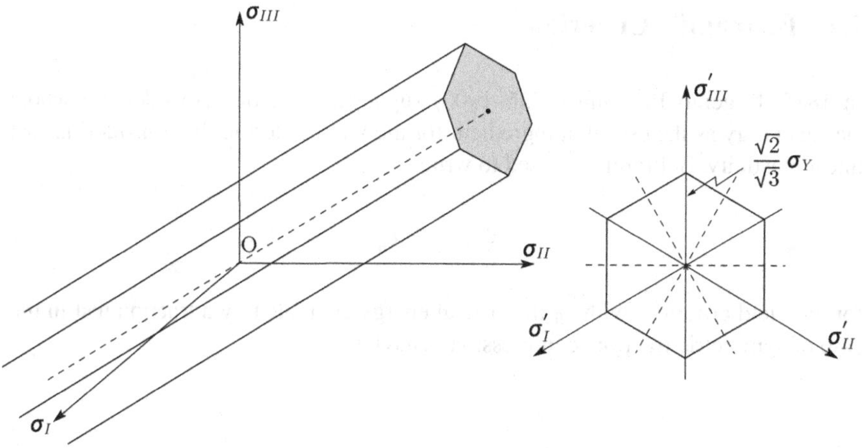

Fig. 7.2 The admissibility domain defined by the yield criterion (7.3) in the space $(\sigma_I, \sigma_{II}, \sigma_{III})$ and deviatoric section

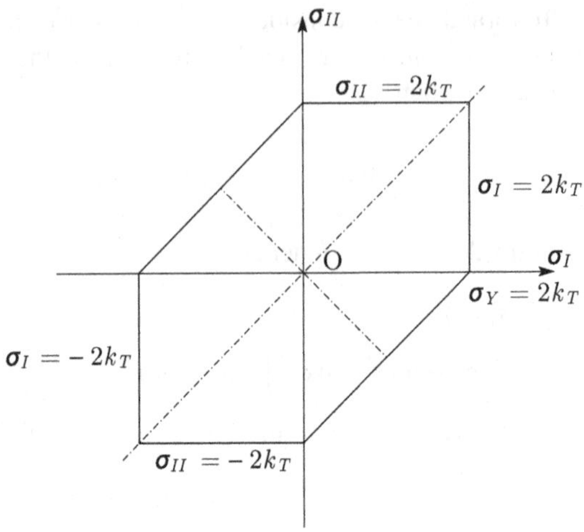

Fig. 7.3 Tresca's admissibility domain in the plane $\sigma_{III} = 0$

is just σ_Y. Henry Tresca (1814–1885) introduced this criterion in 1864 and discussed it in three additional papers published in 1867, 1868, and 1872. Adhémard Jean-Claude Barré de Saint-Venant (1797–1886) reanalyzed it in 1870 and 1872. The criterion is now named after Tresca. For the de Saint-Venant problem, the Tresca criterion reduces to

$$\sqrt{\sigma_{33}^2 + 4\tau^2} \leq \sigma_Y,$$

in which τ denotes here the modulus of the shear stress.

7.2 Beltrami's Criterion

In 1885, Eugenio Beltrami (1836–1900) suggested that one consider the whole elastic energy as the essential ingredient for the yield criterion. By considering just linear elasticity, Beltrami proposed to write

$$\frac{1}{2}\sigma \cdot \varepsilon \leq k_B$$

for the yield criterion, with k_B the critical energy estimated by a traction test in the case of symmetric traction–compression behavior.

7.3 Huber–von Mises–Hencky Criterion

From experimental data on metals, Titus Maksymilian Huber (1872–1950) observed that in case of hydrostatic pressure at a point, the material never reaches the limit σ_Y measured in the uniaxial traction test. Huber's observation suggested to Richard Edler von Mises (1883–1953) in 1913 that a relevant yield criterion should be written in terms of the deviatoric part σ^d of the Cauchy stress, defined analogously to the deviatoric part of ε—the tensor ε^d appearing in Chapter 1—by

$$\sigma^d = \sigma - \frac{1}{3}(\operatorname{tr}\sigma)I,$$

with I is the second-rank unit tensor. As in the case of ε^d, we immediately compute $\operatorname{tr}\sigma^d = 0$. Later, in 1924, with reference to linear elasticity, Heinrich Hencky (1885–1951) expressed the criterion based on Huber's and von Mises's ideas in terms of the deviatoric part of the elastic energy density. The first step is the following decomposition in terms of deviatoric stress and strain:

$$\frac{1}{2}\sigma \cdot \varepsilon = \frac{1}{2}\left(\sigma^d + \frac{1}{3}(\operatorname{tr}\sigma)I\right) \cdot \left(\varepsilon^d + \frac{1}{3}(\operatorname{tr}\varepsilon)I\right) = \frac{1}{2}\sigma^d \cdot \varepsilon^d + \frac{1}{6}(\operatorname{tr}\sigma)(\operatorname{tr}\varepsilon).$$

The last equality follows by taking into account that $\sigma^d \cdot I = \operatorname{tr}\sigma^d = 0$ and $\varepsilon^d \cdot I = \operatorname{tr}\varepsilon^d = 0$.

We define the product

$$e^d := \frac{1}{2}\sigma^d \cdot \varepsilon^d$$

as the **deviatoric elastic** (stored) **energy density**, denoting it by e^d.

When the material in the linear-elastic phase is isotropic, the previous expression becomes

$$e^d = \frac{1}{4\mu}\left(\sigma_{ij} - \frac{\lambda}{3\lambda + 2\mu}(\operatorname{tr}\sigma)\delta_{ij}\right)\sigma_{ij} - \frac{(\operatorname{tr}\sigma)^2}{6(3\lambda + 2\mu)} = \frac{1}{12\mu}\left(3\sigma_{ij}\sigma_{ij} - (\operatorname{tr}\sigma)^2\right).$$

In terms of the deviatoric component of the stress, we simply write

$$e^d = \frac{1}{4\mu}\sigma^d \cdot \sigma^d = -\frac{J_2}{2\mu},$$

where J_2 is the *second invariant* of σ^d, i.e., the sum of the determinants of the minors of σ^d around its principal diagonal. **Huber–von Mises–Hencky's yield criterion** is expressed by the inequality

$$e^d \leq \frac{k_{HMH}^2}{2\mu}, \tag{7.4}$$

where k_{HMH} is a constant estimated by a traction–compression uniaxial test. The ratio of its square with 2μ is the relevant deviatoric energy.

Exercise 1. *Compute explicitly the deviatoric part of the stress with components*

$$\begin{pmatrix} \sigma_Y & 0 & 0 \\ 0 & 0 & 0 \\ 0 & 0 & 0 \end{pmatrix}$$

and show that

$$k_{HMH} = \frac{\sigma_Y}{\sqrt{3}}.$$

In the linear isotropic elastic case, the inequality (7.4) reads

$$|J_2| \leq k_{HMH}^2,$$

so that the yield function can be chosen as

$$f_{HMH}(\sigma) := J_2 + k_{HMH}^2,$$

or alternatively,

$$f_{HMH}(\sigma) := \sqrt{|J_2|} - k_{HMH}.$$

In terms of the principal stresses—the eigenvalues of σ—for the first choice of $f(\sigma)$ above, the admissibility condition

$$f_{HMH}(\sigma) \leq 0$$

reads

$$(\sigma_I - \sigma_{II})^2 + (\sigma_{II} - \sigma_{III})^2 + (\sigma_{III} - \sigma_I)^2 - 6k_{HMH}^2 \leq 0.$$

Such an inequality describes a cylinder with circular base in the space $(\sigma_I, \sigma_{II}, \sigma_{III})$, wrapped around the straight line inclined at 45° with respect to all three axes σ_I, σ_{II}, and σ_{III} (see Fig. 7.4). In the plane $\sigma_{III} = 0$, the admissibility domain is in Fig. 7.5. For the de Saint-Venant problem, the Huber–von Mises–Hencky criterion reduces to

$$\sqrt{\sigma_{33}^2 + 3\tau^2} \leq \sigma_Y.$$

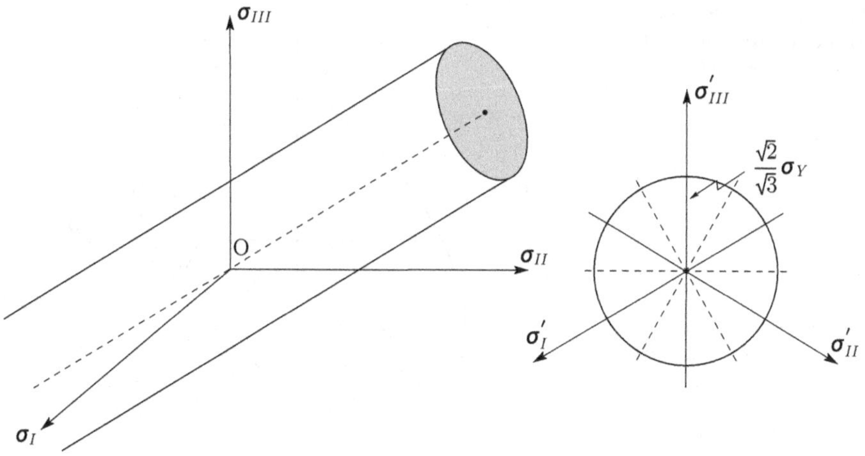

Fig. 7.4 Admissibility region determined by the Huber–von Mises–Hencky criterion in the space $(\sigma_I, \sigma_{II}, \sigma_{III})$ and deviatoric section

7.4 The Drucker–Prager Criterion

The Huber–von Mises–Hencky criterion excludes participation of the hydrostatic component of the stress at a point in the definition of critical states. The choice appears appropriate for metals, in general, but it seems to be unsatisfactory for granular materials (where the hydrostatic pressure governs compactification), classes of rocks, even concrete. For these cases, Daniel C. Drucker (1918–2001) and William Prager (1903–1980) proposed in 1952 a modification of the Huber–von Mises–Hencky criterion that reads

$$f_{DP}(\sigma) \leq 0,$$

where

$$f_{DP}(\sigma) := \sqrt{|J_2|} - k_{DP1} - k_{DP2} \operatorname{tr} \sigma, \tag{7.5}$$

with k_{DP1} and k_{DP2} appropriate constants. In the case of asymmetric material behavior (traction–compression), with yield stress σ_{Yt} under a traction test and σ_{Yc} in compression conditions, k_{DP1} and k_{DP2} have the following form:

$$k_{DP1} = \frac{2}{\sqrt{3}} \frac{(\sigma_{Yc}\sigma_{Yt})}{(\sigma_{Yc} + \sigma_{Yt})}, \qquad k_{DP2} = \frac{1}{\sqrt{3}} \frac{(\sigma_{Yt} - \sigma_{Yc})}{(\sigma_{Yc} + \sigma_{Yt})}.$$

These two expressions follow by writing first the critical condition $\sqrt{|J_2|} = k_{DP1} + k_{DP2} \operatorname{tr} \sigma$ in the case of uniaxial traction and then the same expression

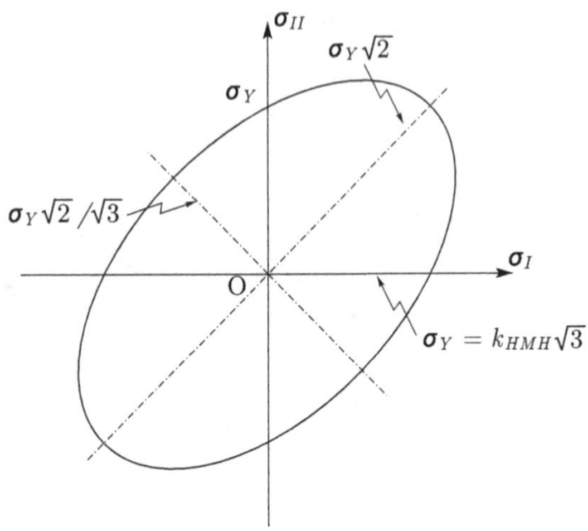

Fig. 7.5 The Huber–von Mises–Hencky admissibility domain in the plane $\sigma_{III} = 0$

in compression uniaxial conditions. In the space of the triplet $(\sigma_I, \sigma_{II}, \sigma_{III})$, the admissibility criterion defines a cone with circular basis. Further extensions of the Drucker–Prager criterion are available in the current literature.

7.5 Hill's Criterion

In his scientific career, Rodney Hill (1921–2011) proposed various yield criteria, first with the aim of accounting for material anisotropies. Here, we recall just briefly the simplest of his proposals:

$$f_H(\sigma) := \sigma^d_{max} - k_H,$$

where σ^d_{max} is the maximum absolute eigenvalue of σ^d, and k_H has the same meaning but with reference to the critical conditions in a uniaxial traction test, i.e., it is the maximum eigenvalue of the deviatoric stress component. In the space $(\sigma_I, \sigma_{II}, \sigma_{III})$, the admissibility condition

$$f_H(\sigma) \leq 0$$

describes a region similar to that pertaining to Tresca's criterion. Figure 7.6 presents a comparison among the Tresca, Huber–von Mises–Hencky, and Hill criteria in the case $\sigma_{III} = 0$. Tresca's condition appears to be more stringent in favor of safety.

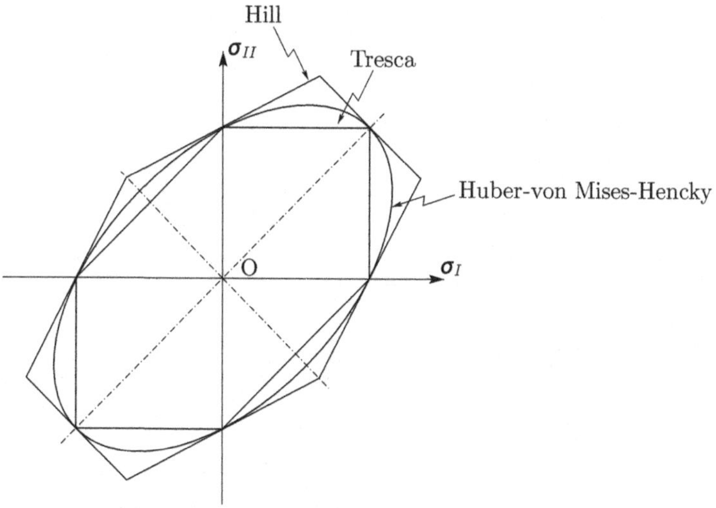

Fig. 7.6 Comparison among three yield criteria in the plane $\sigma_{III} = 0$

7.6 Objectivity of the Yield Function

Under classical changes in observers, the stress σ becomes $Q\sigma Q^{\mathrm{T}}$, with Q in $SO(3)$. This issue has been already discussed in previous chapters. The action of $Q \in SO(3)$ alters the frame(s) in the ambient space but does not change the material properties. In other words, if σ satisfies the admissibility condition (the yield criterion), then $Q\sigma Q^{\mathrm{T}}$ must also be admissible. The circumstance is verified when we require that the yield function f be objective, i.e.,

$$f(\sigma) = f(Q\sigma Q^{\mathrm{T}})$$

for *any* $Q \in SO(3)$.

This requirement implies that f has to depend only on the invariants of σ, namely $\operatorname{tr}\sigma$, $\det\sigma$, and the second invariant, i.e., $\frac{1}{2}\left((\operatorname{tr}\sigma)^2 - \operatorname{tr}\sigma^2\right)$. Let us call I_2 the second invariant of σ. It is connected with J_2, the second invariant of σ^d, by the relation $J_2 = I_2 - \frac{1}{3}(\operatorname{tr}\sigma)^2$, so that the Huber–von Mises–Hencky and the Drucker–Prager criteria fall under the objectivity conditions.

Chapter 8
Rod Models

The analysis of the equilibrium of three-dimensional slender cylinders is rather complicated as it emerges, at least to us, from the discussion of the de Saint-Venant problem. The difficulty increases when we consider framed structures. The task becomes formidable in the presence of large strains and dynamics. The presence of a dominant dimension suggests, however, that one explore the possibility of constructing coarse models in which the sole independent space variable is the arc length along the dominant axis—it can be generically a curve. We can construct different models of these cylinders, or rods, by including or excluding essential peculiar aspects of their mechanical behavior. Often, ad hoc assumptions appear in technical theories of rods, leading to the neglect of certain kinematic terms regarded as small. In the nonlinear case, they often lead to equations no easier to analyze than those obtained from the three-dimensional theory avoiding the introduction of further ad hoc assumptions. In this last case, we can derive a rod model in two ways:

1. through some approximation procedure from the three-dimensional nonlinear elasticity (top-to-bottom procedure);
2. by considering a one-dimensional continuum (the curve determining the dominant axis of the cylinder) and assigning to each point information on the rod's cross section.

Both approaches to constructing one-dimensional-in-space-variables rod models can be followed even when slenderness is not so pronounced or the body is relatively thick. In these cases, the utility of the resulting equations may be seriously impaired, although in the case of approximation procedures, convergence would be ensured.

In this chapter, we shall focus our attention on the second way of deriving rod models. We call the result a *directed description of rods*.

Before going into details, we warn the reader that we use the term *rod* in a broad sense. For us, it is a generic term for (in alphabetical order) *arch, bar, beam, column, ring, shaft*, etc.

P.M. Mariano, L. Galano, *Fundamentals of the Mechanics of Solids*,
DOI 10.1007/978-1-4939-3133-0_8

8.1 Basis Conceptual Aspects of the Director-Based Description of Rods: Reference Places

Consider

- (*i*) a smooth curve $s \longmapsto \varphi_0(s) \in \mathcal{E}^3$, $s \in [0, l]$;
- (*ii*) three unit vector fields $s \longmapsto \hat{e}_\Gamma(s) = \hat{e}_\Gamma(\varphi_0(s)) \in S^2$, $\Gamma = 1, 2, 3$, orthogonal to each other, $\hat{e}_3 = \hat{e}_1 \times \hat{e}_2$ at every s, and $\langle \hat{e}_3(s), \varphi_0'(s) \rangle_{\mathbb{R}^3} > 0$, where the prime denotes in this chapter differentiation with respect to s.

We may take in $\mathcal{E}^3$ a fixed orthonormal basis e_1, e_2, e_3 such that

$$\hat{e}_\Gamma(s) = Q_0(s)e_\Gamma,$$

where $Q_0(s)$ is the value of a differentiable map $s \longmapsto Q_0(s) \in SO(3)$, $s \in [0, l]$. The reference shape of a tangible rod is then the set

$$\mathcal{B} := \left\{ x \in \mathcal{E}^3 | x = \varphi_0(s) + \sum_{\alpha=1}^{2} \xi^\alpha \hat{e}_\alpha(s), \ \ (\xi^1, \xi^2, s) \in \mathcal{X} \times (0, l) \right\},$$

where $\mathcal{X}$ is a two-dimensional compact open set in $\mathbb{R}^2$, with piecewise smooth boundary $\partial\mathcal{X}$. We shall assume that $\left| \frac{d\varphi_0}{ds} \right| = 1$ for every s, so that l is the length of the rod. We can consider $\varphi_0(\cdot)$ the line of centroids, i.e., a line crossing at every s the mass center of the rod section spanned by the vectors $\hat{e}_1(s)$ and $\hat{e}_2(s)$ as ξ^1 and ξ^2 in vary over $\mathcal{X}$. For the reference place, we can also write

$$\mathcal{B} := \left\{ x \in \mathcal{E}^3 | \ x = \varphi_0(s) + \sum_{\alpha=1}^{2} \xi^\alpha Q_0(s) e_\alpha, \ \ (\xi^1, \xi^2, s) \in \mathcal{X} \times (0, l) \right\}.$$

We shall then identify geometrically the physical rod with the closure of $\mathcal{B}$.

The vector $\hat{e}_3(s)$ orients at s the transversal section spanned by $\hat{e}_1(s)$ and $\hat{e}_2(s)$. When $\hat{e}_3(s) = \varphi_0'(s)$, such a section is properly the cross section at s. The description of the rod's geometry in terms of a curve and vector fields over it and above all, the role these vectors will play later are the reasons for calling the resulting scheme director-based (or directed). Figure 8.1 reproduces graphically what we have discussed above.

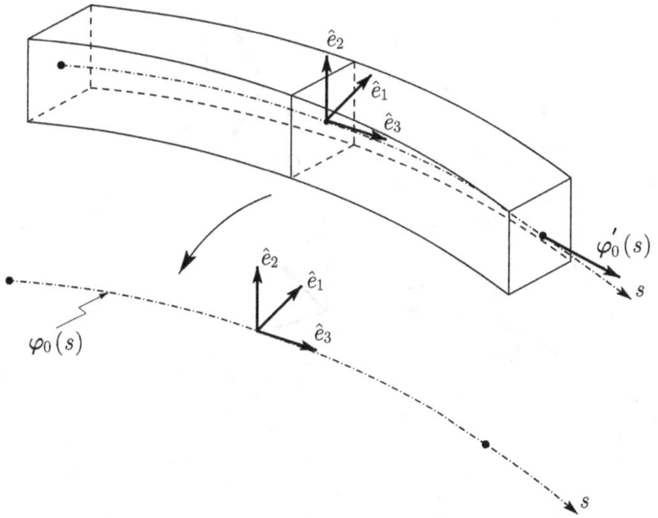

Fig. 8.1 From a three-dimensional shape to a one-dimensional model with directors

8.2 Deformation and Motions

– We assume that during the deformation, the transversal sections (those spanned by $\hat{e}_1$ and $\hat{e}_2$ in the reference place) behave rigidly (Bernoulli's hypothesis).

The assumption excludes warping of the rod's cross section. It allows us to identify the rod's shape after deformation with the set

$$\mathcal{B}_a := \left\{ y \in \tilde{\mathcal{E}}^3 \,|\, y = \bar{y} + \sum_{\alpha=1}^{2} \xi^\alpha \mathsf{d}_\alpha(s) \, , \right.$$

$$\left. \bar{y} := \varphi(\varphi_0(s)), \ \mathsf{d}_\alpha(s) = Q(s)\hat{e}_\alpha(s), \ \ (\xi^1, \xi^2, s) \in \mathcal{X} \times (0, l) \right\} ;$$

φ is a one-to-one differentiable orientation-preserving function.[1] The curve $s \longmapsto \varphi(\varphi_0(s))$, the centroid locus of the deformed rod, can be considered to be parameterized by the arc length $\bar{s}(s)$, given by the map

$$s \longmapsto \bar{s}(s) := \int_0^s \left| \varphi'(z) \right| dz.$$

[1]**Warning:** φ must not be confused with the stress potential function in Chapter 6.

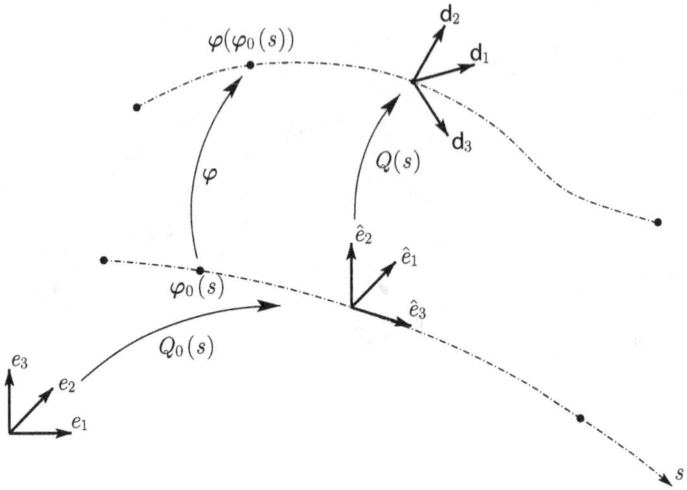

Fig. 8.2 Kinematics in the director-based rod model

Moreover, we still impose

$$\left\langle \mathsf{d}_3(s), \varphi'(s) \right\rangle_{\tilde{\mathbb{R}}^3} > 0$$

as a counterpart of the analogous condition in the reference shape. Figure 8.2 shows schematically the kinematics of the director-based model just described.

– Unless otherwise stated, we shall refer to a straight reference shape of the rod without too much loss of generality, since we are discussing large strains. Consequently, for us, the actual shape $\mathcal{B}_a$ is then

$$\mathcal{B}_a := \left\{ y \in \tilde{\mathcal{E}}^3 | \, y = \bar{y} + \sum_{\alpha=1}^{2} \xi^\alpha \mathsf{d}_\alpha(s) \, , \right.$$

$$\left. \bar{y} := \varphi(s), \; \mathsf{d}_\alpha(s) = Q(s) e_\alpha, \;\; (\xi^1, \xi^2, s) \in \mathcal{X} \times (0, l) \right\} .$$

The current shape of the rod is determined by the two maps $s \longmapsto \varphi(s) \in \tilde{\mathcal{E}}^3$ and $s \longmapsto Q(s) \in SO(3)$. In other words, we determine the deformed line of the centroids (the centers of mass) and add at each of its points information on the current arrangement of the pertinent transversal section. We are then reducing a rod to a one-dimensional continuum with every point endowed not only with the standard three translational degrees of freedom of a point in three-dimensional space, but also with *additional* rotational degrees of freedom. In this way, the material element placed at a point in the one-dimensional continuum is considered a (finite-size) rigid body able to rotate independently of its neighbors; we picture the rod as a deck of infinitely many two-dimensional cards, each one coinciding with a rod cross section. For this reason, such a one-dimensional scheme in space gives rise to a continuum going beyond the traditional format described so far, where a material element is described just by a geometric point and, as such, has only three degrees of freedom.

- We shall tacitly assume that the rod does not self-intersect, deforming in the sense that we assume the existence of a well-defined smooth inverse mapping $\bar{s} \longmapsto s$, with $\bar{s}$ the arc length over the curve φ, the deformed rod axis. Then we shall call the derivative

$$\lambda(s) := \frac{d\bar{s}(s)}{ds}$$

the *stretch* developed in going from $\mathcal{B}$ to $\mathcal{B}_a$.

Each point x in $\mathcal{B}$ is parameterized by ξ^1, ξ^2, and s as is every y in $\mathcal{B}_a$ at a fixed instant t. Then we can evaluate the deformation gradient F by computing the derivatives with respect to ξ^1, ξ^2, and s. The result is

$$F = \mathsf{d}_1 \otimes e_1 + \mathsf{d}_2 \otimes e_2 + \left(\varphi_{,s} + \sum_{\alpha=1}^{2} \xi^\alpha \mathsf{d}_{\alpha,s} \right) \otimes e_3,$$

where ", s" as a subscript denotes the derivative with respect to s. In particular, we set

$$F_0(s) = \left. F(\xi^1, \xi^2, s) \right|_{\xi^1=\xi^2=0},$$

for the deformation gradient over the line, so that

$$F_0(s) = \mathsf{d}_1 \otimes e_1 + \mathsf{d}_2 \otimes e_2 + \varphi_{,s} \otimes e_3.$$

Time comes into play for describing motions, i.e., mappings

$$(s,t) \longmapsto y := \varphi(s,t) + \sum_{\alpha=1}^{2} \xi^\alpha \mathsf{d}_\alpha(s,t) \tag{8.1}$$

assumed to be twice piecewise differentiable in time.

8.3 Derivatives

Let us begin with the relation

$$\mathsf{d}_\alpha(s,t) = Q(s,t) e_\alpha,$$

which defines d_α, and assume that d_α is twice differentiable. We then compute

$$\mathsf{d}_{\alpha,s} = Q_{,s} e_\alpha = Q_{,s} Q^{\mathrm{T}} \mathsf{d}_\alpha.$$

Actually, the tensor field $s \longmapsto Q_{,s}Q^{\mathrm{T}}$ at each instant t is skew-symmetric (for the proof, we recall that we just need to compute the derivative with respect to s of the defining relation $Q(s)Q^{\mathrm{T}}(s) = I$, with I the second-rank identity), and we write

$$\mathsf{d}_{\alpha,s} = \Omega \mathsf{d}_{\alpha} = \omega \times \mathsf{d}_{\alpha}, \tag{8.2}$$

where $\Omega := Q_{,s}Q^{\mathrm{T}}$, and $\omega(s,t)$ is the corresponding axial vector. An analogous expression holds for the time derivative of d_{α}, namely

$$\dot{\mathsf{d}}_{\alpha} = \dot{Q}e_{\alpha} = \dot{Q}Q^{\mathrm{T}}\mathsf{d}_{\alpha},$$

so that since $W := \dot{Q}Q^{\mathrm{T}}$ is skew-symmetric, we write also

$$\dot{\mathsf{d}}_{\alpha} = W\mathsf{d}_{\alpha} = w \times \mathsf{d}_{\alpha}, \tag{8.3}$$

where w is the axial vector of W. We have also

$$\frac{\partial}{\partial t}\mathsf{d}_{\alpha,s} = \dot{\omega} \times \mathsf{d}_{\alpha} + \omega \times \dot{\mathsf{d}}_{\alpha} = \dot{\omega} \times \mathsf{d}_{\alpha} + \omega \times w \times \mathsf{d}_{\alpha}$$

and

$$\frac{\partial}{\partial s}\dot{\mathsf{d}}_{\alpha} = w_{,s} \times \mathsf{d}_{\alpha} + w \times \mathsf{d}_{\alpha,s} = w_{,s} \times \mathsf{d}_{\alpha} + w \times \omega \times \mathsf{d}_{\alpha}.$$

Since by Schwartz's theorem, the two mixed derivatives equal each other, we get

$$(\dot{\omega} - w \times \omega) \times \mathsf{d}_{\alpha} = (w_{,s} + w \times \omega) \times \mathsf{d}_{\alpha},$$

or more concisely,

$$w_{,s} \times \mathsf{d}_{\alpha} = (\overset{\triangledown}{\omega} - w \times \omega) \times \mathsf{d}_{\alpha},$$

where

$$\overset{\triangledown}{\omega} := \dot{\omega} - w \times \omega$$

is the rate of change of ω relative to an observer moving with the spatial frame $\{\mathsf{d}_{\Gamma}\}$.

Write now ζ for the vector

$$\zeta := \frac{\partial \varphi}{\partial s} - \mathsf{d}_3,$$

which measures how much the normal to the plane containing what is a cross section at s in the reference place deviates from the tangent to the deformed centroid axis at the same s. Then ζ *accounts for shear*. We compute

$$
\begin{aligned}
\overset{\nabla}{\zeta} :=& \frac{\partial}{\partial t}\left(\frac{\partial \varphi}{\partial s} - \mathsf{d}_3\right) - w \times \left(\frac{\partial \varphi}{\partial s} - \mathsf{d}_3\right) \\
=& \frac{\partial \dot{\varphi}}{\partial s} - \dot{\mathsf{d}}_3 - w \times \frac{\partial \varphi}{\partial s} + w \times \mathsf{d}_3 \\
=& \frac{\partial \dot{\varphi}}{\partial s} - w \times \mathsf{d}_3 - w \times \frac{\partial \varphi}{\partial s} + w \times \mathsf{d}_3 \\
=& \frac{\partial \dot{\varphi}}{\partial s} - w \times \frac{\partial \varphi}{\partial s} = \overset{\nabla}{\overline{\frac{\partial \varphi}{\partial s}}},
\end{aligned}
\tag{8.4}
$$

where we have used the relation (8.3).

8.4 Sectional Stress Power

Consider a cross section at s in the reference place of a three-dimensional rod. The covector

$$\hat{\tau}(x) := P(x)e_3$$

is the contact action exchanged between the material element at $y(x)$ in the deformed shape and the part of the rod having in the reference shape normal $-e_3$ at the same s. Figure 8.3 illustrates the scenario.

We call the integral over the section of the power density $\hat{\tau} \cdot \dot{y}$, namely

$$\mathcal{P}_{\mathcal{X}} := \int_{\mathcal{X}} \hat{\tau} \cdot \dot{y}\, d\xi^1 d\xi^2,$$

the **sectional stress power**.

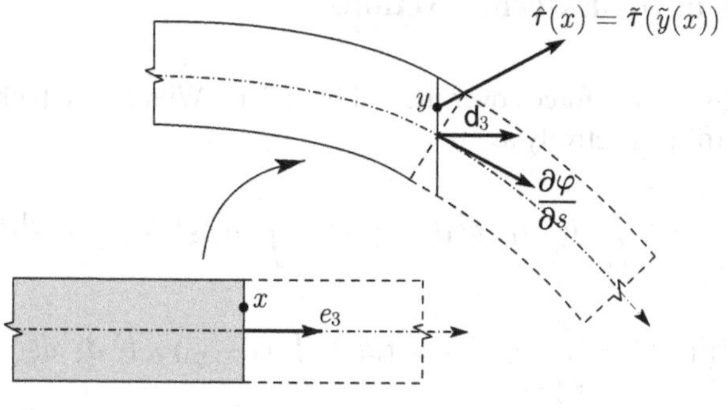

Fig. 8.3 Sketch of the mapping $x \longmapsto \hat{\tau}(x)$

Using (8.1), we obtain

$$\mathcal{P}_{\mathcal{X}} = \int_{\mathcal{X}} \hat{\tau} \cdot \dot{y}\, d\xi^1 d\xi^2 = \int_{\mathcal{X}} \hat{\tau} \cdot \left(\dot{\varphi}(s,t) + \sum_{\alpha=1}^{2} \xi^\alpha \dot{\mathsf{d}}_\alpha(s,t) \right) d\xi^1 d\xi^2$$

$$= \int_{\mathcal{X}} \hat{\tau} \cdot \dot{\varphi}(s,t)\, d\xi^1 d\xi^2 + \int_{\mathcal{X}} \hat{\tau} \cdot \sum_{\alpha=1}^{2} \xi^\alpha w \times \mathsf{d}_\alpha\, d\xi^1 d\xi^2$$

$$= \dot{\varphi} \cdot \int_{\mathcal{X}} \hat{\tau}\, d\xi^1 d\xi^2 + w \cdot \int_{\mathcal{X}} \sum_{\alpha=1}^{2} \xi^\alpha \mathsf{d}_\alpha \times \hat{\tau}\, d\xi^1 d\xi^2$$

$$= \dot{\varphi} \cdot \mathsf{n} + w \cdot \mathsf{m},$$

where

$$\mathsf{n} := \int_{\mathcal{X}} \hat{\tau}\, d\xi^1 d\xi^2 \quad \text{and} \quad \mathsf{m} := \int_{\mathcal{X}} \sum_{\alpha=1}^{2} \xi^\alpha \mathsf{d}_\alpha \times \hat{\tau}\, d\xi^1 d\xi^2.$$

Since in the one-dimensional scheme with directors developed so far, we assign at every point of the centroid line information on the whole cross section there, $\mathsf{n}(s)$ is the contact traction at s in that scheme, and $\mathsf{m}(s)$ is a contact couple, which develops power in the rotational degrees of freedom assigned to every s, in addition to what holds in the standard continuum scheme. If we take a portion of the rod between s_1 and s_2, with $s_2 > s_1$, the difference

$$\mathsf{n}(s_2,t) \cdot \dot{\varphi}(s_2,t) + \mathsf{m}(s_2,t) \cdot w(s_2,t) - \mathsf{n}(s_1,t) \cdot \dot{\varphi}(s_1,t) - \mathsf{m}(s_1,t) \cdot w(s_1,t)$$

is the external power of the contact actions exerted by the rest of the rod over the part between s_1 and s_2.

8.5 Power of the External Actions

Let f denote distributed forces over the rod boundary. With $b^\ddagger$ the bulk actions, we define $\bar{\mathsf{n}}^\ddagger$ and $\bar{\mathsf{m}}^\ddagger$ respectively as

$$\bar{\mathsf{n}}^\ddagger(s,t) := \int_{\partial\mathcal{X}} \mathsf{f}(\xi^1(l), \xi^2(l), s, t)\, dl + \int_{\mathcal{X}} b^\ddagger(\xi^1, \xi^2, s, t)\, d\xi^1 d\xi^2,$$

$$\bar{\mathsf{m}}^\ddagger(s,t) := \int_{\partial\mathcal{X}} (y - y_0) \times \mathsf{f}\, dl + \int_{\mathcal{X}} (y - y_0) \times b^\ddagger\, d\xi^1 d\xi^2,$$

where l is the arclength along $\partial\mathcal{X}$, y_0 is a fixed point in space, and for the sake of notational conciseness, we have omitted in the definition of $\bar{\mathsf{m}}^{\ddagger}$ to make explicit the dependence on ξ^1, ξ^2, s, and the time t. Due to the presence of $b^{\ddagger}$, we can decompose $\bar{\mathsf{n}}^{\ddagger}$ and $\bar{\mathsf{m}}^{\ddagger}$ additively into inertial and noninertial factors, since $b^{\ddagger}$ can be so decomposed by definition. Consequently, we shall write

$$\bar{\mathsf{n}}^{\ddagger} = \bar{\mathsf{n}} + \bar{\mathsf{n}}^{in}, \qquad \bar{\mathsf{m}}^{\ddagger} = \bar{\mathsf{m}} + \bar{\mathsf{m}}^{in},$$

where the superscript "in" denotes the inertial components defined by

$$\bar{\mathsf{n}}^{in}(s,t) := \int_{\mathcal{X}} b^{in}(\xi^1, \xi^2, s, t)\, d\xi^1 d\xi^2 \tag{8.5}$$

and

$$\bar{\mathsf{m}}^{in}(s,t) := \int_{\mathcal{X}} (y - y_0) \times b^{in}\, d\xi^1 d\xi^2. \tag{8.6}$$

By $\hat{\mathsf{f}}$ and $\hat{\mathsf{m}}$ we shall denote external noninertial forces and couples applied to finite numbers $\bar{k}_s \geq 0$ and $\bar{\bar{k}}_s \geq 0$, respectively, of discrete points of the centroid line. These points can be the orthogonal projections over the centroid line of points over the boundary of the rod where localized loads are applied. Also, the isolated applied forces $\hat{\mathsf{f}}$ can be the resultants over a specific section of external actions that can even have inertial nature as in the case of shocks. They generate discontinuities in the distributions of $\bar{\mathsf{n}}^{\ddagger}$ and $\bar{\mathsf{m}}^{\ddagger}$, which are necessary to ensure balance at those points. The power of external actions over a portion of the rod between s_1 and s_2, with $s_2 > s_1$, under (at least square) integrability assumptions for $\bar{\mathsf{n}}^{\ddagger}$, $\bar{\mathsf{m}}^{\ddagger}$, and the velocities, can then be written as

$$\int_{s_1}^{s_2} (\bar{\mathsf{n}}^{\ddagger}(s,t) \cdot \dot{\varphi}(s,t) + \bar{\mathsf{m}}^{\ddagger}(s,t) \cdot w(s,t))\, ds$$

$$+ \sum_{k=1}^{\bar{k}_s} (\hat{\mathsf{f}}(s_k,t) \cdot \dot{\varphi}(s_k,t)) + \sum_{r=1}^{\bar{\bar{k}}_s} (\hat{\mathsf{m}}(s_r,t) \cdot w(s_r,t)).$$

8.6 Changes in Observers

Classical changes in observers in the physical space imply for $\dot{\varphi}(s,t)$ the same transformation $v \longmapsto v^*$ as that presented in Chapter 2, so that we can write

$$\dot{\varphi}^*(s,t) = \dot{\varphi}(s,t) + c(t) + q(t) \times (\varphi(s,t) - y_0). \tag{8.7}$$

We have, in fact, two observers rototranslating one with respect to the other. Each of them may evaluate the velocity $\dot{\varphi}(s,t)$ of a given point $\varphi(s,t)$ of the centroid line and the pullback $\dot{\varphi}^*(s,t)$ from the other observer of the velocity measured in the rototranslational frame. Such a transformation has also an effect on w. In fact, with w, the *rod's cross-sectional rotational velocity* evaluated by a given observer, the pullback w^* in the relevant frame of the velocity measured by the second (rototranslating) observer reads

$$w^* = w + q. \tag{8.8}$$

In fact, with d'_α the image of the vector d_α in the frame defining the second observer, we have

$$\mathsf{d}'_\alpha = Q(t)\mathsf{d}_\alpha$$

with $Q(t) \in SO(3)$ the orthogonal time-dependent transformation linking the two observers. The derivative with respect to time reads

$$\dot{\mathsf{d}}'_\alpha = \dot{Q}\mathsf{d}_\alpha + Q\dot{\mathsf{d}}_\alpha.$$

Then, by defining $\dot{\mathsf{d}}^*_\alpha$ as

$$\dot{\mathsf{d}}^*_\alpha = Q^{\mathrm{T}}\dot{\mathsf{d}}'_\alpha$$

and using the relation (8.3), we obtain

$$\dot{\mathsf{d}}^*_\alpha = Q^{\mathrm{T}}\dot{Q}\mathsf{d}_\alpha + \dot{\mathsf{d}}_\alpha = q \times \mathsf{d}_\alpha + w \times \mathsf{d}_\alpha = (q + w) \times \mathsf{d}_\alpha,$$

which justifies, by comparison with equation (8.3), the choice (8.8).

8.7 External Power Invariance for Director-Based Rod Models

We write $\mathcal{P}^{ext}_{(s_1,s_2)}(\dot{\varphi}, w)$ for the power on the part of the rod between s_1 and s_2, with $s_2 > s_1$, in the one-dimensional, director-based scheme. As we explained in the previous section, it reads

$$\begin{aligned}
\mathcal{P}^{ext}_{(s_1,s_2)}(\dot{\varphi}, w) :=\ & \mathsf{n}(s_2,t)\cdot\dot{\varphi}(s_2,t) - \mathsf{n}(s_1,t)\cdot\dot{\varphi}(s_1,t) \\
& + \mathsf{m}(s_2,t)\cdot w(s_2,t) - \mathsf{m}(s_1,t)\cdot w(s_1,t) \\
& + \int_{s_1}^{s_2} (\bar{\mathsf{n}}^{\ddagger}(s,t)\cdot\dot{\varphi}(s,t) + \bar{\mathsf{m}}^{\ddagger}(s,t)\cdot w(s,t))\, ds \\
& + \sum_{k=1}^{\bar{k}_s} \hat{\mathsf{f}}(s_k,t)\cdot\dot{\varphi}(s_k,t) + \sum_{r=1}^{\bar{\bar{k}}_s} \hat{\mathsf{m}}(s_r,t)\cdot w(s_r,t).
\end{aligned}$$

In accord with the path described in Chapter 3, we impose now an invariance requirement.

Axiom 1 (Power invariance for the director-based rod model). The external power $\mathcal{P}^{ext}_{(s_1,s_2)}(\dot{\varphi}, w)$ is invariant under isometry-based changes in observers, i.e.,

$$\mathcal{P}^{ext}_{(s_1,s_2)}(\dot{\varphi}, w) = \mathcal{P}^{ext}_{(s_1,s_2)}(\dot{\varphi}^*, w^*)$$

for every choice of c and q, and for every interval (s_1, s_2).

Since $\mathcal{P}^{ext}_{(s_1,s_2)}$ is linear in the velocities, the axiom implies

$$\mathcal{P}^{ext}_{(s_1,s_2)}(c + q \times (\varphi - y_0), q) = 0,$$

for every choice of c, q, and the interval (s_1, s_2) considered, i.e.,

$$\begin{aligned}
&c \cdot \left(\mathsf{n}(s_2,t) - \mathsf{n}(s_1,t) + \int_{s_1}^{s_2} \bar{\mathsf{n}}^{\ddagger}(s,t) ds + \sum_{k=1}^{\bar{k}_s} \hat{\mathsf{f}}(s_k,t) \right) \\
&+ q \cdot \Bigg(\mathsf{m}(s_2,t) - \mathsf{m}(s_1,t) + \int_{s_1}^{s_2} \bar{\mathsf{m}}^{\ddagger}(s,t)\, ds \\
&+ (\varphi(s_2,t) - y_0) \times \mathsf{n}(s_2,t) - (\varphi(s_1,t) - y_0) \times \mathsf{n}(s_1,t) \\
&+ \sum_{k=1}^{\bar{k}_s} (\varphi(s_k,t) - y_0) \times \hat{\mathsf{f}}(s_k,t) + \sum_{r=1}^{\bar{\bar{k}}_s} \hat{\mathsf{m}}(s_r,t) \\
&+ \int_{s_1}^{s_2} (\varphi(s,t) - y_0) \times \bar{\mathsf{n}}^{\ddagger}(s,t)\, ds \Bigg) = 0.
\end{aligned}$$

The arbitrariness of c and q implies the global balances

$$\mathsf{n}(s_2,t) - \mathsf{n}(s_1,t) + \int_{s_1}^{s_2} \bar{\mathsf{n}}^{\ddagger}(s,t)\, ds + \sum_{k=1}^{\bar{k}_s} \hat{\mathsf{f}}(s_k,t) = 0 \tag{8.9}$$

and

$$\begin{aligned}
&\mathsf{m}(s_2,t) - \mathsf{m}(s_1,t) \\
&+ (\varphi(s_2,t) - y_0) \times \mathsf{n}(s_2,t) - (\varphi(s_1,t) - y_0) \times \mathsf{n}(s_1,t) \\
&+ \int_{s_1}^{s_2} \bar{\mathsf{m}}^{\ddagger}(s,t)\, ds + \sum_{k=1}^{\bar{k}_s} (\varphi(s_k,t) - y_0) \times \hat{\mathsf{f}}(s_k,t) \\
&+ \sum_{r=1}^{\bar{\bar{k}}_s} \hat{\mathsf{m}}(s_r,t) + \int_{s_1}^{s_2} (\varphi(s,t) - y_0) \times \bar{\mathsf{n}}^{\ddagger}(s,t)\, ds = 0.
\end{aligned} \tag{8.10}$$

Take $s_1 \leq \bar{s} \leq s_2$ and imagine that $\mathsf{n}(\cdot, t)$ has at $\bar{s}$ a bounded discontinuity. For α a nonnegative number, we denote by n^+ and n^- the limits

$$\mathsf{n}^{\pm}(\bar{s}, t) := \lim_{\alpha \to 0} \mathsf{n}(\bar{s} \pm \alpha, t).$$

The jump $[\mathsf{n}]$ of n is defined by

$$[\mathsf{n}] := \mathsf{n}^+ - \mathsf{n}^-,$$

and the average $\langle \mathsf{n} \rangle$ by

$$\langle \mathsf{n} \rangle := \frac{(\mathsf{n}^+ + \mathsf{n}^-)}{2}.$$

Analogous definitions hold for m and the other fields defined over the centroid line involved here. Discontinuity points for $\mathsf{n}(\cdot, t)$ are those where external isolated actions $\hat{\mathsf{f}}$ are applied. We then have

$$\mathsf{n}(s_2, t) - \mathsf{n}(s_1, t) = \int_{s_1}^{s_2} \frac{\partial \mathsf{n}(s, t)}{\partial s} ds + \sum_{k=1}^{\bar{k}_s} [\mathsf{n}](s_k, t).$$

By inserting the result in the integral balance (8.9), we obtain

$$\int_{s_1}^{s_2} \left(\frac{\partial \mathsf{n}(s, t)}{\partial s} + \bar{\mathsf{n}}^{\ddagger}(s, t) \right) ds + \sum_{k=1}^{\bar{k}_s} \left([\mathsf{n}] + \hat{\mathsf{f}} \right) (s_k, t) = 0.$$

The arbitrariness of the interval of integration (i.e., of the part considered in the rod) implies that of $\bar{k}_s \geq 0$, so that we must have

$$\frac{\partial \mathsf{n}}{\partial s} + \bar{\mathsf{n}}^{\ddagger} = 0 \tag{8.11}$$

along the centroid line, and

$$[\mathsf{n}] + \hat{\mathsf{f}} = 0 \tag{8.12}$$

at the points of the same line where isolated external actions occur. When $\hat{\mathsf{f}} = 0$, the balance (8.12) requires that n be continuous.

Assume now that $\mathsf{m}(\cdot, t)$ is continuous and continuously differentiable over the centroid line except at $\bar{\bar{k}}_s \geq 0$ points, where it has bounded discontinuities. Such points are those where the (purely noninertial) isolated couples $\hat{\mathsf{m}}$ are applied. As in the case of n, we then have

$$\mathsf{m}(s_2, t) - \mathsf{m}(s_1, t) = \int_{s_1}^{s_2} \frac{\partial \mathsf{m}(s, t)}{\partial s} ds + \sum_{r=1}^{\bar{\bar{k}}_s} [\mathsf{m}](s_r, t).$$

The integral balance (8.10) then becomes

$$\int_{s_1}^{s_2} \left(\frac{\partial \mathsf{m}}{\partial s} + \frac{\partial \varphi}{\partial s} \times \mathsf{n} + \bar{\mathsf{m}}^{\ddagger} \right) ds + \sum_{r=1}^{\bar{\bar{k}}_s} ([\mathsf{m}] + \hat{\mathsf{m}})(s_r, t)$$
$$+ \int_{s_1}^{s_2} (\varphi(s,t) - y_0) \times \left(\frac{\partial \mathsf{n}}{\partial s} + \bar{\mathsf{n}}^{\ddagger} \right) ds$$
$$+ \sum_{k=1}^{\bar{k}_s} (\varphi(s,t) - y_0) \times ([\mathsf{n}] + \hat{\mathsf{f}})(s_k, t) = 0.$$

The local balances (8.11) and (8.12) allow us to write the integral balance of couples as

$$\int_{s_1}^{s_2} \left(\frac{\partial \mathsf{m}}{\partial s} + \frac{\partial \varphi}{\partial s} \times \mathsf{n} + \bar{\mathsf{m}}^{\ddagger} \right) ds + \sum_{r=1}^{\bar{\bar{k}}_s} ([\mathsf{m}] + \hat{\mathsf{m}})(s_r, t) = 0.$$

The arbitrariness of the integration interval implies that of $\bar{\bar{k}}_s$ and the validity of the local balances

$$\frac{\partial \mathsf{m}}{\partial s} + \frac{\partial \varphi}{\partial s} \times \mathsf{n} + \bar{\mathsf{m}}^{\ddagger} = 0 \tag{8.13}$$

along the centroid line, and

$$[\mathsf{m}] + \hat{\mathsf{m}} = 0 \tag{8.14}$$

at the isolated points where external couples are applied.

8.8 Identification of the Inertial Terms

8.8.1 Kinetic Energy Averaged over the Cross Section

On averaging the kinetic energy density over the generic cross section at s, we obtain

$$\frac{1}{2}\int_{\mathcal{X}} \rho |\dot{y}|^2 d\xi^1 d\xi^2 = \frac{1}{2}\int_{\mathcal{X}} \rho \left(\dot{\varphi} + \sum_{\alpha=1}^{2} \xi^{\alpha} \dot{\mathsf{d}}_{\alpha} \right) \cdot \left(\dot{\varphi} + \sum_{\beta=1}^{2} \xi^{\beta} \dot{\mathsf{d}}_{\beta} \right) d\xi^1 d\xi^2$$
$$= \frac{1}{2} |\dot{\varphi}(s,t)|^2 \int_{\mathcal{X}} \rho(\xi^1, \xi^2, s)\, d\xi^1 d\xi^2$$
$$+ \dot{\varphi}(s,t) \cdot \sum_{\alpha=1}^{2} w(s,t) \times \mathsf{d}_{\alpha}(s,t) \int_{\mathcal{X}} \rho \xi^{\alpha} d\xi^1 d\xi^2$$
$$+ \frac{1}{2} \sum_{\alpha=1}^{2} \sum_{\beta=1}^{2} (w(s,t) \times \mathsf{d}_{\alpha}(s,t)) \cdot (w(s,t) \times \mathsf{d}_{\beta}(s,t)) \int_{\mathcal{X}} \rho \xi^{\alpha} \xi^{\beta} d\xi^1 d\xi^2,$$

where we have exploited the natural identification of $\mathbb{R}^3$ with its dual and the relation (8.3) in the second term on the right-hand side. The integrals

$$\int_{\mathcal{X}} \rho\xi^1 d\xi^1 d\xi^2 \quad \text{and} \quad \int_{\mathcal{X}} \rho\xi^2 d\xi^1 d\xi^2$$

correspond respectively to the static moments S_2 and S_1 of the cross section at s with respect to d_1 and d_2. Since the frame considered is in reference to the cross sectional center of mass, we have $S_2 = S_1 = 0$. Then we get

$$\dot{\varphi}(s,t) \cdot \sum_{\alpha=1}^{2} w(s,t) \times \mathsf{d}_\alpha(s,t) \int_{\mathcal{X}} \rho\xi^\alpha d\xi^1 d\xi^2 = 0.$$

Given two nonparallel vectors $a, b \in \mathbb{R}^3$, the Lagrange formula for the triple vector product prescribes

$$a \times b \times a = (a \cdot a)b - (a \cdot b)a = (|a|^2 I - a \otimes a)b.$$

Then we compute

$$\frac{1}{2}\int_{\mathcal{X}} \sum_{\alpha=1}^{2}\sum_{\beta=1}^{2} \rho\xi^{\alpha^2} (w(s,t) \times \mathsf{d}_\beta) \cdot (w(s,t) \times \mathsf{d}_\alpha)\, d\xi^1 d\xi^2$$

$$= \frac{1}{2}\int_{\mathcal{X}} \sum_{\alpha=1}^{2}\sum_{\beta=1}^{2} \xi^{\alpha^2} (\mathsf{d}_\alpha \times w \times \mathsf{d}_\beta) \cdot w\, d\xi^1 d\xi^2$$

$$= \frac{1}{2}\left(\int_{\mathcal{X}} \left(\sum_{\alpha=1}^{2} \xi^{\alpha^2} I - \sum_{\alpha=1}^{2}\sum_{\beta=1}^{2} \xi^\alpha \xi^\beta \mathsf{d}_\alpha \otimes \mathsf{d}_\beta\right) \rho d\xi^1 d\xi^2\right) w \cdot w,$$

where I is the second-rank unit tensor. By defining the scalar $\rho_{\mathcal{X}}$ and the second-rank tensor $I_{\mathcal{X}}$ respectively by

$$\rho_{\mathcal{X}}(s) := \int_{\mathcal{X}} \rho(\xi^1, \xi^2, s)\, d\xi^1 d\xi^2,$$

and

$$I_{\mathcal{X}}(s,t) := \int_{\mathcal{X}} \left(\sum_{\alpha=1}^{2} \xi^{\alpha^2} I - \sum_{\alpha=1}^{2}\sum_{\beta=1}^{2} \xi^\alpha \xi^\beta \mathsf{d}_\alpha \otimes \mathsf{d}_\beta\right) \rho d\xi^1 d\xi^2,$$

the **kinetic energy** at the instant t **of the cross section** at s in the reference configuration reads

$$\frac{1}{2}\int_{\mathcal{X}} \rho|\dot{y}|^2 d\xi^1 d\xi^2 = \frac{1}{2}\rho_{\mathcal{X}}|\dot{\varphi}|^2 + \frac{1}{2} I_{\mathcal{X}} w \cdot w.$$

We shall write $kin_{\mathcal{X}}$ for the right-hand side of the above expression.

8.8.2 The Inertia Balance and the Identification

We assume that for every portion of the rod between s_1 and s_2, the inertial components of $\bar{\mathsf{n}}^{\ddagger}$ and $\bar{\mathsf{m}}^{\ddagger}$, namely $\bar{\mathsf{n}}^{in}$ and $\bar{\mathsf{m}}^{in}$, satisfy the integral balance

$$\frac{d}{dt}\int_{s_1}^{s_2} kin_{\mathcal{X}}\, ds + \frac{1}{2}\int_{s_1}^{s_2} \dot{I}_{\mathcal{X}} w\cdot w\, ds + \int_{s_1}^{s_2} \bar{\mathsf{n}}^{in}\cdot\dot{\varphi}\, ds + \int_{s_1}^{s_2} \bar{\mathsf{m}}^{in}\cdot w\, ds = 0 \qquad (8.15)$$

for every choice of the velocity fields $\dot{\varphi}$ and w. On considering the explicit expression of $kin_{\mathcal{X}}$ and computing the time derivative, the balance (8.15) becomes

$$\int_{s_1}^{s_2} (\rho_{\mathcal{X}}\ddot{\varphi} + \bar{\mathsf{n}}^{in})\cdot\dot{\varphi}\, ds + \int_{s_1}^{s_2} (I_{\mathcal{X}}\dot{w} + \dot{I}_{\mathcal{X}} w + \bar{\mathsf{m}}^{in})\cdot w\, ds = 0.$$

Then, the arbitrariness of the velocity implies

$$\bar{\mathsf{n}}^{in} = -\rho_{\mathcal{X}}\ddot{\varphi},$$

$$\bar{\mathsf{m}}^{in} = -I_{\mathcal{X}}\dot{w} - \dot{I}_{\mathcal{X}} w.$$

Using the relation (8.3), the time derivative of $I_{\mathcal{X}}$ reads

$$\begin{aligned}\dot{I}_{\mathcal{X}} &= -\int_{\mathcal{X}}\sum_{\alpha=1}^{2}\sum_{\beta=1}^{2}\xi^{\alpha}\xi^{\beta}(\dot{\mathsf{d}}_{\alpha}\otimes\mathsf{d}_{\beta} + \mathsf{d}_{\alpha}\otimes\dot{\mathsf{d}}_{\beta})\,\rho d\xi^1 d\xi^2\\ &= -\int_{\mathcal{X}}\sum_{\alpha=1}^{2}\sum_{\beta=1}^{2}\xi^{\alpha}\xi^{\beta}\left((w\times\mathsf{d}_{\alpha})\otimes\mathsf{d}_{\beta} + \mathsf{d}_{\alpha}\otimes(w\times\mathsf{d}_{\beta})\right)\rho d\xi^1 d\xi^2.\end{aligned}$$

However, since

$$\left((w\times\mathsf{d}_{\alpha})\otimes\mathsf{d}_{\beta}\right)w = (\mathsf{d}_{\beta}\cdot w)(w\times\mathsf{d}_{\alpha}),$$

$$\left(\mathsf{d}_{\alpha}\otimes(w\times\mathsf{d}_{\beta})\right)w = \left(w\cdot(w\times\mathsf{d}_{\beta})\right)\mathsf{d}_{\alpha},$$

and obviously,

$$w\times Iw = w\times w = 0,$$

we get

$$\begin{aligned}\dot{I}_{\mathcal{X}} w &= -w\times\left(\int_{\mathcal{X}}\sum_{\alpha=1}^{2}\sum_{\beta=1}^{2}\xi^{\alpha}\xi^{\beta}\mathsf{d}_{\alpha}\otimes\mathsf{d}_{\beta}\,\rho d\xi^1 d\xi^2\right)w\\ &= w\times\left(\int_{\mathcal{X}}\left(\sum_{\alpha=1}^{2}\xi^{\alpha^2} I - \sum_{\alpha=1}^{2}\sum_{\beta=1}^{2}\xi^{\alpha}\xi^{\beta}\mathsf{d}_{\alpha}\otimes\mathsf{d}_{\beta}\right)\rho d\xi^1 d\xi^2\right)w.\end{aligned} \qquad (8.16)$$

Define now the **moment of momentum** $\mathsf{H}(s,t)$ **pertaining** at the instant t **to the rod's cross section** at s as

$$\mathsf{H}(s,t) := \int_{\mathcal{X}} \rho(y-\varphi) \times \dot{y}\, d\xi^1 d\xi^2.$$

From the definition (8.1), we have

$$(y-\varphi(s,t)) = \sum_{\alpha=1}^{2} \xi^\alpha \mathsf{d}_\alpha$$

and

$$\sum_{\alpha=1}^{2} \xi^\alpha \dot{\mathsf{d}}_\alpha = \sum_{\alpha=1}^{2} \xi^\alpha w \times \mathsf{d}_\alpha = w \times (y-\varphi(s,t)).$$

Consequently, H can be written as

$$\begin{aligned}\mathsf{H}(s,t) &= \int_{\mathcal{X}} \rho(y-\varphi) \times \dot{\varphi}\, d\xi^1 d\xi^2 + \int_{\mathcal{X}} \rho(y-\varphi) \times \sum_{\alpha=1}^{2} \xi^\alpha \dot{\mathsf{d}}_\alpha d\xi^1 d\xi^2 \\ &= \left(\int_{\mathcal{X}} \rho \sum_{\alpha=1}^{2} \xi^\alpha d\xi^1 d\xi^2\right) \mathsf{d}_\alpha \times \dot{\varphi} + \int_{\mathcal{X}} \rho \sum_{\alpha=1}^{2} \xi^\alpha \mathsf{d}_\alpha \times w \times \sum_{\beta=1}^{2} \xi^\beta \mathsf{d}_\beta d\xi^1 d\xi^2 \\ &= \left(\int_{\mathcal{X}} \left(\sum_{\alpha=1}^{2} \xi^{\alpha^2} I - \sum_{\alpha=1}^{2}\sum_{\beta=1}^{2} \xi^\alpha \xi^\beta \mathsf{d}_\alpha \otimes \mathsf{d}_\beta\right) \rho\, d\xi^1 d\xi^2\right) w,\end{aligned}$$

where we have taken into account (*i*) that ξ^1 and ξ^2 are with reference to the cross-sectional center of mass, so that

$$\mathrm{S}_\alpha = \int_{\mathcal{X}} \rho \xi \alpha\, d\xi^1 d\xi^2 = 0, \quad \alpha = 1, 2,$$

as we have already mentioned, and (*ii*) the Lagrange formula for the triple vector product. On inserting the result into the identity (8.16), we then obtain

$$\dot{I}_{\mathcal{X}} w = w \times \mathsf{H},$$

so that

$$\bar{\mathsf{m}}^{in} = -I_{\mathcal{X}} \dot{w} - w \times \mathsf{H}.$$

8.8.3 A Further Justification of the Inertia Balance

We can justify the use of (8.15) by recalling the definition (8.5) and (8.6). Since

$$b^{in} = -\rho\ddot{y} = -\frac{d}{dt}(\rho\dot{y}) \tag{8.17}$$

(see Chapter 3), we can rewrite the definition (8.5) of $\bar{\mathsf{n}}^{in}$ as

$$\bar{\mathsf{n}}^{in}(s,t) = -\frac{d}{dt}\hat{\mathsf{L}}(s,t),$$

where

$$\hat{\mathsf{L}}(s,t) := \int_{\mathcal{X}} \rho(\xi^1,\xi^2,s)\dot{y}(\xi^1,\xi^2,s,t)\,d\xi^1 d\xi^2.$$

Using the definition (8.1), we then have

$$\hat{\mathsf{L}}(s,t) = \int_{\mathcal{X}} \rho(\xi^1,\xi^2,s)\dot{\varphi}(s,t)\,d\xi^1 d\xi^2 + \int_{\mathcal{X}} \sum_{\alpha=1}^{2} \xi^\alpha \dot{\mathsf{d}}_\alpha \rho(\xi^1,\xi^2,s) d\xi^1 d\xi^2$$

$$= \rho_{\mathcal{X}}(s)\dot{\varphi}(s,t) + \sum_{\alpha=1}^{2} \dot{\mathsf{d}}_\alpha \mathsf{S}_\alpha = \rho_{\mathcal{X}}(s)\dot{\varphi}(s,t)$$

In fact, as above, the static momenta vanish, since ξ^1 and ξ^2 are with reference to the cross-sectional center of mass. Substitution into the above expression of $\bar{\mathsf{n}}^{in}$ determines $\bar{\mathsf{n}}^{in} = -\rho_{\mathcal{X}}(s)\ddot{\varphi}$, as derived in the previous section. Moreover, since $b^{in} = -\rho\ddot{y}$, we can write the relation (8.6) as

$$\bar{\mathsf{m}}^{in} = -\frac{d}{dt}\mathsf{H}$$

with H as defined in a previous section, where the results show that

$$\mathsf{H} = I_{\mathcal{X}} w,$$

so that

$$\dot{\mathsf{H}} = I_{\mathcal{X}}\dot{w} + \dot{I}_{\mathcal{X}} w = I_{\mathcal{X}}\dot{w} + w \times \mathsf{H},$$

a relation already obtained from the assumption (8.15).

The balance (8.15) is a version in the present setting of the integral balance (3.17).

8.9 Summary

In summary, the balance equations for the director-based description of rods presented in this chapter read

$$\frac{\partial \mathsf{n}}{\partial s} + \bar{\mathsf{n}} = \rho_{\mathcal{X}} \ddot{\varphi},$$

$$\frac{\partial \mathsf{m}}{\partial s} + \frac{\partial \varphi}{\partial s} \times \mathsf{n} + \bar{\mathsf{m}} = I_{\mathcal{X}} \dot{w} + w \times \mathsf{H}$$

along the rod axis and

$$[\mathsf{n}] + \hat{\mathsf{f}} = 0,$$

$$[\mathsf{m}] + \hat{\mathsf{m}} = 0,$$

at the points where concentrated external forces $\hat{\mathsf{f}}$ and couples $\hat{\mathsf{m}}$ are applied. Explicitly, we write

$$\mathsf{n} = T_1 \mathsf{d}_1 + T_2 \mathsf{d}_2 + N \mathsf{d}_3,$$

where N is the amplitude of the action normal to the cross section; T_1 and T_2 are the shear forces in the frame defined over the cross section itself by d_1 and d_2. Analogously, for m we write

$$\mathsf{m} = M_1 \mathsf{d}_1 + M_2 \mathsf{d}_2 + M_t \mathsf{d}_3,$$

where M_1 and M_2 are the bending moments around the pertinent d_α, while M_t is the torsion moment. We have adopted the same symbols—we refer to T_1, T_2, N, M_1, M_2, M_t—of Chapter 6 for the components of the contact actions over the rod's cross section, because in the conditions of the de Saint-Venant problem, they reduce precisely to those defined in Chapter 6.

For the applied distributed noninertial actions, we write also

$$\bar{\mathsf{n}} = \bar{q}_1 \mathsf{d}_1 + \bar{q}_2 \mathsf{d}_2 + \bar{p} \mathsf{d}_3$$

and

$$\bar{\mathsf{m}} = \bar{m}_1 \mathsf{d}_1 + \bar{m}_2 \mathsf{d}_2 + \bar{m}_t \mathsf{d}_3.$$

8.10 A Special Case

Consider a straight rod remaining straight after deformation. This case restricts the treatment to rigid changes of place and approximately to the small-strain regime. Neglect inertia. Select also the triple $\{e_\Gamma\}$, $\Gamma = 1, 2, 3$, to be orthogonal at each s. In this case, the balance equations read in components

$$\frac{dN}{ds} + \bar{p} = 0,$$
$$\frac{dT_1}{ds} + \bar{q}_1 = 0,$$
$$\frac{dT_2}{ds} + \bar{q}_2 = 0,$$
$$\frac{dM_1}{ds} - T_2 + \bar{m}_1 = 0,$$
$$\frac{dM_2}{ds} - T_1 + \bar{m}_2 = 0,$$
$$\frac{dM_t}{ds} + \bar{m}_t = 0.$$

We can easily find a solution to such a system when the rod is constrained and the constraints (*i*) are equal to the number of the degrees of freedom and (*ii*) are well posed in the sense explained in Chapter 1.

In the case of piecewise straight rods, the previous equations apply on the straight portions, providing the addition of corner continuity conditions except in the case in which concentrated forces or couples are applied, so that a bounded discontinuity is necessary for n and m, respectively, in order to ensure locally the balance of the actions, as shown in the previous section.

We provide examples in the following sections, where we apply the previous notions to the analysis of the statics of framed structures. We also include in these structures rods with a natural curved reference shape, a case requiring a refined analysis.

8.11 Starting from a Curved Reference Shape

Let us consider the line of centroids to be an arbitrary smooth curve in the reference shape. Such a curve is determined by the function $s \longmapsto \varphi_0(s)$ introduced in Section 8.1. Write $\{e_\Gamma\}$, $\Gamma = 1, 2, 3$, for the orthogonal fixed basis in the reference space. As above, we can select a differentiable function

$$s \longmapsto Q_0(s) \in SO(3)$$

such that

$$\hat{e}_\Gamma(s) = Q_0(s) e_\Gamma .$$

After deformation, we have

$$\mathsf{d}_\Gamma(s) = Q(s)\hat{e}_\Gamma(s).$$

Then we compute

$$\mathsf{d}_{\Gamma,s} = Q_{,s}\hat{e}_\Gamma + Q\hat{e}_{\Gamma,s} = Q_{,s}Q^{\mathrm{T}}\mathsf{d}_\Gamma + QQ_{0,s}e_\Gamma = \Omega\mathsf{d}_\Gamma + QQ_{0,s}Q_0^{\mathrm{T}}Q_0 e_\Gamma$$
$$= \Omega\mathsf{d}_\Gamma + Q\Omega_0\hat{e}_\Gamma = \Omega\mathsf{d}_\Gamma + Q\Omega_0 Q^{\mathrm{T}}\mathsf{d}_\Gamma,$$

where $\Omega_0 := Q_{0,s}Q_0^{\mathrm{T}}$ is a skew-symmetric tensor. The second-rank tensor

$$\Omega + Q\Omega_0 Q^{\mathrm{T}} \tag{8.18}$$

is skew-symmetric, since Ω and Ω_0 are, and

$$(Q\Omega_0 Q^{\mathrm{T}})^{\mathrm{T}} = Q\Omega_0^{\mathrm{T}} Q^{\mathrm{T}} = -Q\Omega_0 Q^{\mathrm{T}}.$$

Consequently, with $\bar{\omega}$ the axial vector of the tensor (8.18), we have just to substitute ω with the appropriate $\bar{\omega}$ once a deformed reference shape has been selected. Imagine that we parameterize the curved reference shape by arc length $\tilde{s}$ given by a map

$$s \longmapsto \tilde{s}(s) := \int_0^s \left|\varphi_0'(z)\right| dz.$$

For a function $f(\tilde{s}(s))$, we have

$$\frac{df}{d\tilde{s}} = \frac{1}{\tilde{\lambda}}\frac{df}{ds},$$

where $\tilde{\lambda}$ is the stretch from the straight configuration to the reference curved one, and it is never zero by assumption, i.e.,

$$\tilde{\lambda}(s) = \frac{d\tilde{s}(s)}{ds}.$$

What we have developed so far applies also to the cases in which $s \longmapsto \varphi_0(s)$ is piecewise smooth with the sole proviso that we must pay attention to the balance conditions at corners, as will appear in some exercises discussed in the next sections of this chapter.

8.12 Statically Determined (Isostatic) Framed Structures

Consider a supported rod of length l, loaded by a distributed force of density $\bar{q}$ in two-dimensional space, where we fix an orthogonal frame $\{\mathrm{A}, y, s\}$, as in Figure 8.4. Let us compute the reactive forces at A and B, those in Figure 8.5, imagining the rod to be rigid.

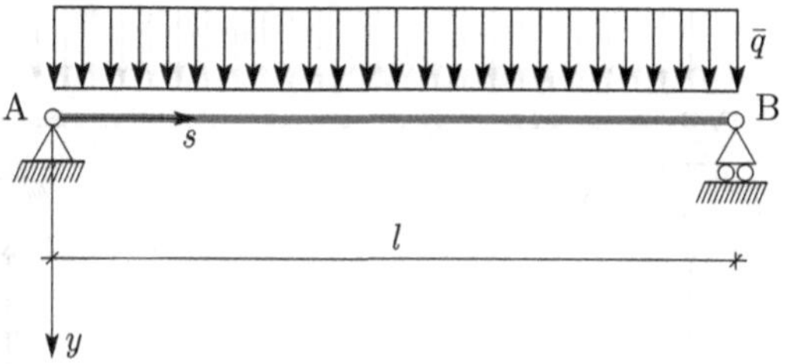

Fig. 8.4 A supported rod in two-dimensional space

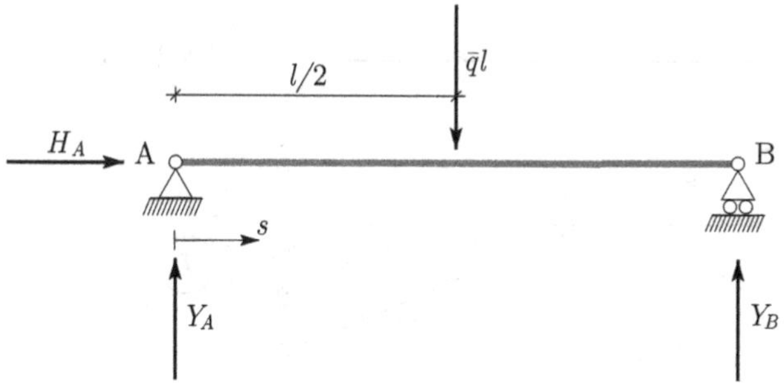

Fig. 8.5 The scheme of reactive forces

To calculate the reactive forces, we can substitute the distributed load with its resultant $\bar{q}l$, applied at the barycenter of the load diagram. By equalizing to zero the sum of forces and that of couples evaluated with respect to the point A, we get

$$H_A = 0, \qquad Y_A + Y_B = \bar{q}l, \qquad Y_B l = \frac{\bar{q}l^2}{2}.$$

Consequently, we obtain

$$Y_A = \frac{\bar{q}l}{2}, \qquad Y_B = \frac{\bar{q}l}{2}.$$

In the present two-dimensional setting, the balance equations read

$$\frac{dN}{ds} = 0, \qquad \frac{dT}{ds} + \bar{q} = 0, \qquad \frac{dM}{ds} - T = 0,$$

where here, T stands for T_2 in the balances summarized in Section 8.10, M for M_1, $\bar{p} = \bar{m}_1 = 0$, $\bar{q}_2 = \bar{q}$. At the point A, we have

$$N_A = H_A = 0, \qquad T_A = Y_A = \frac{\bar{q}l}{2}, \qquad M_A = 0.$$

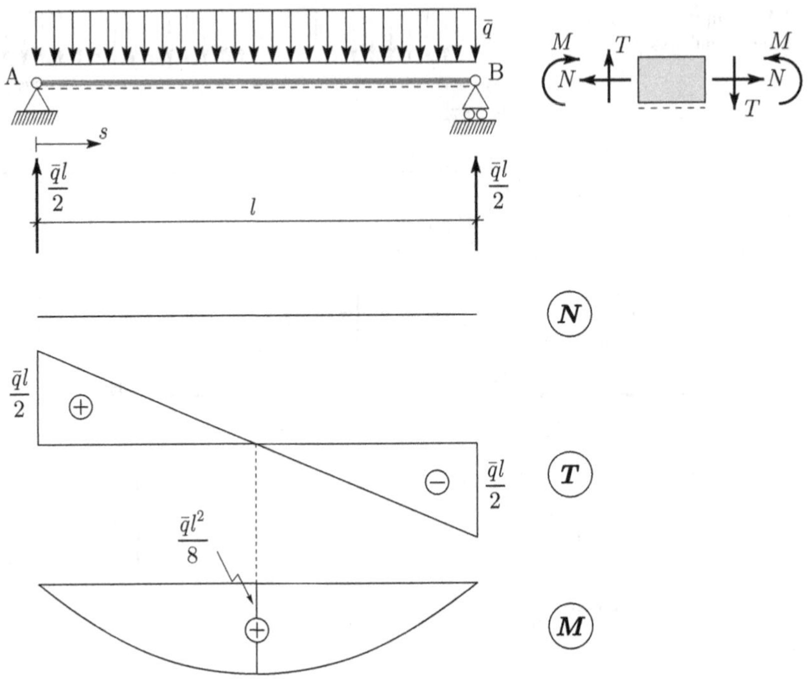

Fig. 8.6 Distributions of the action characteristics. The scheme on the right-hand side (figure top) indicates the convention on positive signs used here

On integrating the balance equations with the previous conditions, we get

$$N(s) = 0, \qquad T(s) = \frac{\bar{q}l}{2} - \bar{q}s, \qquad M(s) = \frac{\bar{q}l}{2}s - \frac{\bar{q}s^2}{2}$$

(see also Figure 8.6).

The same result can be obtained via integral balances evaluated over subsequent parts of the rod. To this end, it is convenient to start from one of the extreme points of the rod: let us say A, as indicated by an eye—showing the point of view—in Figure 8.7. Consider then the generic part from A to a point with abscissa s. The horizontal component of the balance of forces over that part reads

$$N(s) = 0$$

and does not vary as s varies. The vertical component of that balance is given by

$$T(s) - Y_A + \bar{q}s = 0,$$

according to the agreement on positive signs indicated in Figure 8.6. Since we are watching from the left side, the positive algebraic sign are those on the left side in the rectangle in the figure. Then we write

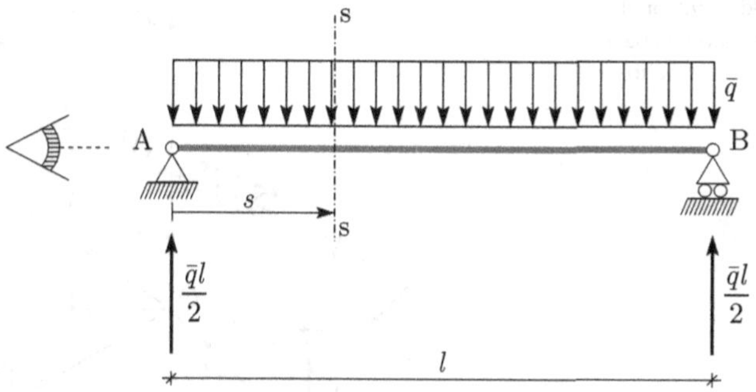

Fig. 8.7 The choice of subsequent parts over the rod, moving the abscissa s from the point A

$$T(s) = \frac{\bar{q}l}{2} - \bar{q}s.$$

The same convention on algebraic signs holds also for the balance of couples, calculated with respect to the extreme right-hand point with abscissa s, which is

$$M(s) - Y_A s + \bar{q}s\frac{s}{2} = 0,$$

i.e.,

$$M(s) = \frac{\bar{q}l}{2}s - \frac{\bar{q}s^2}{2}.$$

Further examples are present in the subsequent sections.

8.13 Remarks on the Statics of 1-Dimensional Rigid Bodies in 3-Dimensional Space

A general expression of what we have done in the previous section opens a path for the analysis of more intricate structures.

Consider a one-dimensional rigid body that occupies a portion of a piecewise smooth line in $\mathcal{E}^3$ and take an orthogonal frame $\{O, e_1, e_2, e_3\}$. The vector position of a point P is then $x_P = (P - O)$ (Fig. 8.8).

As we have described so far, consider isolated forces $\hat{\mathsf{f}}_{(i)}$ and couples $\hat{\mathsf{m}}_{(j)}$, applied respectively to $\bar{k}_s \geq 0$ and $\bar{\bar{k}}_s \geq 0$ points. In the frame selected, $\hat{\mathsf{f}}_{(i)}$ and $\hat{\mathsf{m}}_{(j)}$ have components $\hat{\mathsf{f}}_{(i)1}$, $\hat{\mathsf{f}}_{(i)2}$, $\hat{\mathsf{f}}_{(i)3}$, and $\hat{\mathsf{m}}_{(j)1}$, $\hat{\mathsf{m}}_{(j)2}$, $\hat{\mathsf{m}}_{(j)3}$, respectively. Assume also that distributed forces $\bar{\mathsf{f}}_{(j)}(s)$ and couples $\bar{\mathsf{m}}_{(p)}(s)$ act on $\bar{r} \geq 0$ and $\bar{\bar{r}} \geq 0$ intervals of the body. Figure 8.8 depicts the situation.

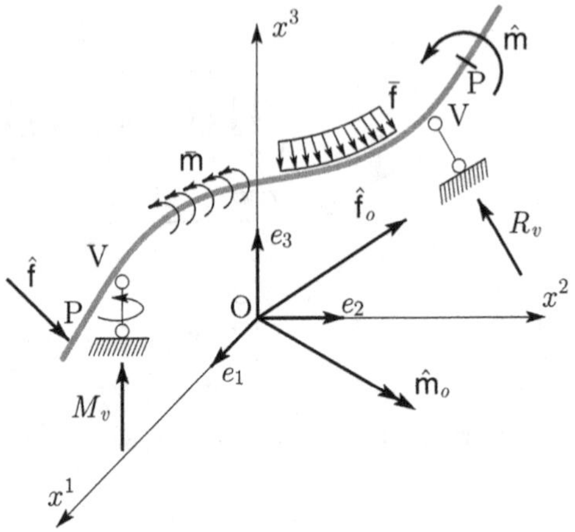

Fig. 8.8 One-dimensional body in 3-dimensional space: external and reactive actions

With the aim of evaluating the equilibrium in the reference configuration, we *reduce* the system of external actions to the origin of the frame considered, i.e., a force $\hat{f}_o$ and a couple $\hat{m}_o$ applied at O such that

$$\hat{f}_o = \sum_{i=1}^{\bar{k}_s} \hat{f}_{(i)} + \sum_{j=1}^{\bar{r}} \int_{s_{(j)1}}^{s_{(j)2}} \bar{f}_{(j)}(s)\, ds,$$

$$\hat{m}_o = \sum_{j=1}^{\bar{\bar{k}}_s} \hat{m}_{(j)} + \sum_{p=1}^{\bar{\bar{r}}} \int_{s_{(p)1}}^{s_{(p)2}} \bar{m}_{(p)}(s)\, ds + \sum_{i=1}^{\bar{k}_s} x_{P(i)} \times \hat{f}_{(i)}$$

$$+ \sum_{j=1}^{\bar{r}} \int_{s_{(j)1}}^{s_{(j)2}} x_{P(j)} \times \bar{f}_{(j)}(s) ds,$$

where $s_{(\cdot)1} < s_{(\cdot)2}$.

The body considered is subjected to $n_t \geq 0$ translational and $n_r \geq 0$ rotational simple constraints, each one reacting with a force R_v or a couple M_v. The balance equations in the reference frame then read

$$\hat{f}_o + \sum_{i=1}^{n_t} R_{v(i)} = 0, \tag{8.19}$$

$$\hat{m}_o + \sum_{i=1}^{n_r} M_{v(i)} + \sum_{i=1}^{n_t} x_{V(i)} \times R_{v(i)} = 0. \tag{8.20}$$

Write $\boldsymbol{r}$ for the list $(R_{v(1)}, \ldots, R_{v(n_t)}, M_{v(1)}, \ldots, M_{v(n_r)})^{\mathrm{T}}$, taken as a vector column and including the components of all reactions, and $\boldsymbol{b}$ for the list

$(-\hat{\mathsf{f}}_{o1}, -\hat{\mathsf{f}}_{o2}, -\hat{\mathsf{f}}_{o3}, -\hat{\mathsf{m}}_{o1}, -\hat{\mathsf{m}}_{o2}, -\hat{\mathsf{m}}_{o3})^{\mathrm{T}}$. In terms of $\boldsymbol{r}$ and $\boldsymbol{b}$, we can write the balance equations as the algebraic system

$$\boldsymbol{B}\boldsymbol{r} = \boldsymbol{b}. \tag{8.21}$$

With $n_v = n_t + n_r$, the matrix $\boldsymbol{B}$ has dimension $6 \times n_v$.

- If the system (8.21) has only one solution, we say that the body under consideration is **statically determinate**: the balance equations are sufficient to determine the unique set of reactive actions able to equilibrate the external ones.
- If the system (8.21) has no solution, we say that the body is **hypostatic**: the reactive actions are insufficient to balance the external ones.
- If the system (8.21) has infinitely many solutions, we say that the body is **statically indeterminate** or **hyperstatic**: the number and disposition of the constraints allow infinitely many choices of the reactive actions that balance the external ones. The degree of infinity—let us denote it by $\hat{h}$—is called the **degree of hyperstaticity**.

When we have six constraints ($n_v = 6$, $\boldsymbol{B}$ is a square matrix), if $\det \boldsymbol{B} \neq 0$, the algebraic system (8.21) has only one solution for every $\boldsymbol{b}$. In this case, the body is **isostatic**. If $\det \boldsymbol{B} = 0$, i.e., $\operatorname{rank}(\boldsymbol{B}) = r_B < 6$, we have two possibilities: $\operatorname{rank}(\boldsymbol{B}) = \operatorname{rank}(\hat{\mathbf{B}})$, with $\hat{\mathbf{B}}$ the matrix obtained by including in the matrix at the left-hand side of equation (8.21) the column on the right-hand side of the same equation, with ∞^{6-r_B} solutions and $\hat{h} = 6 - r_B$ degree of hyperstaticity, and $\operatorname{rank}(\boldsymbol{B}) \neq \operatorname{rank}(\hat{\mathbf{B}})$, the hypostatic state.

Since the body considered here is rigid, the work of the external actions must be zero. Rigidity implies in fact that every admissible motion must be a time-dependent isometry, so that (see Chapter 3) the power of the external actions must vanish, and by time-integration, also the **work** performed by the external actions denoted by L^{ext} vanishes; indeed, it is evaluated on a rigid displacement. The external actions include the reactive ones, so that the identity

$$L^{ext}(u_R) = 0,$$

with u_R a rigid displacement, can be rewritten as

$$L_a^{ext}(u_R) + L_r^{ext}(u_R) = 0, \tag{8.22}$$

where L_a^{ext} is the work performed by the applied external actions and L_r^{ext} is the work performed by the reactive actions in the possible constraint failures. By taking into account the relations (1.5) and (8.21), we can rewrite the identity (8.22) in terms of the vectors $\boldsymbol{r}, \boldsymbol{b}, \boldsymbol{p}, \overline{\boldsymbol{p}}$ and the matrices $\boldsymbol{A}$ and $\boldsymbol{B}$. A consequence is the identity

$$\boldsymbol{B} = \boldsymbol{A}^{\mathrm{T}},$$

which is commonly called **static–dynamic duality**. A consequence is that a rigid body is kinematically isodeterminate if and only if it is isostatic.

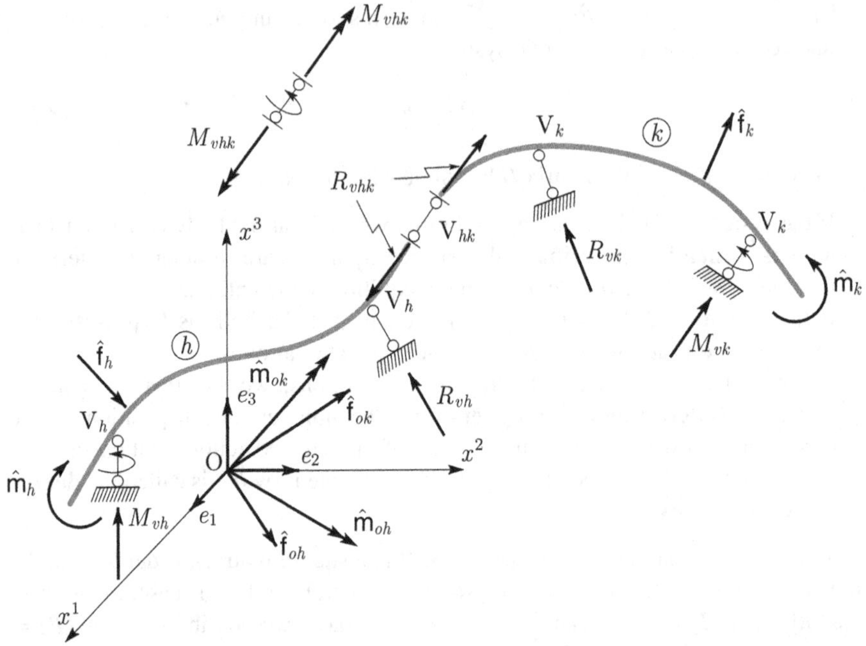

Fig. 8.9 System of 1-dimensional rigid bodies in 3-dimensional space: external and reactive actions

- When $\hat{h} > 0$, there are $\hat{h}$ reactive forces and/or couples, denoted by $X_1, X_2, \ldots, X_{\hat{h}}$, that cannot be determined using only the balance equations; $X_1, X_2, \ldots, X_{\hat{h}}$ are called **hyperstatic unknowns**. We call $\hat{h}$ as the **degree of hyperstaticity**. When a structure has a degree of hyperstaticity equal to $\hat{h}$, we say in short that the structure is *$\hat{h}$-times hyperstatic*, as we say *$\hat{l}$-times labile* for the degree of lability $\hat{l}$
- The degree of lability $\hat{l}$ and the degree of hyperstaticity $\hat{h}$ are related by

$$6 - n_v = \hat{l} - \hat{h}, \tag{8.23}$$

  when the body has no ringlike portions; in this case, the function defining the rod's axis is one-to-one everywhere.

When we consider a system of K interconnected rigid bodies (Fig. 8.9 presents an example), the balance equations can be summarized once again into an algebraic system of the type $\boldsymbol{B}\boldsymbol{r} = \boldsymbol{b}$, but now the matrix $\boldsymbol{B}$ is $6K \times n_v$. The relation (8.23) becomes

$$6K - n_v = \hat{l} - \hat{h}. \tag{8.24}$$

- If $6K > n_v$, we have $\hat{l} > 0$: the structure is labile in its entirety, although portions of it could have some degree of hyperstaticity.
- If $6K = n_v$ and $\mathrm{rank}(\boldsymbol{B}) = 6K$, we have $\hat{l} = \hat{h} = 0$: the structure is isostatic.

- If $6K = n_v$ and rank$(\boldsymbol{B}) < 6K$, we have $\hat{l} = \hat{h} \neq 0$: the structure has a degree of lability equal to that of hyperstaticity. In other words, there are superabundant constraints with respect to the degrees of freedom, but they are disposed in a way allowing some rigid kinematics.
- If $6K < n_v$, we have $\hat{h} > 0$: the structure is hyperstatic, although it could have some degrees of lability.

The condition $6K = n_v$ is necessary but not sufficient to ensure isostaticity; the condition $\hat{l} = \hat{h} = 0$ is sufficient.

The previous definitions hold in a two-dimensional setting once we substitute 6 with 3.

Consider now a rigid rod with axis determining a closed circuit in three-dimensional space. If we cut the circuit at a point, we have twelve action characteristics equally divided between the two sides of the cut as a consequence of the action–reaction principle. Besides the continuity conditions at the cut, in the absence of external actions applied there, we do not have sufficient balance equations to determine the action characteristics at the cut. Consequently, the presence of the closed circuit can be interpreted as the introduction of a constraint with multiplicity equal to six. If we are in a two-dimensional space, a closed circuit can be interpreted as the introduction of a constraint with multiplicity equal to three (see Fig. 8.10). Write n_c for the number of circuits in the structure. The relation (8.24) becomes

$$6K - (n_v + 6n_c) = \hat{l} - \hat{h}. \tag{8.25}$$

In the two-dimensional setting, we substitute 6 with 3 (see Fig. 8.10).

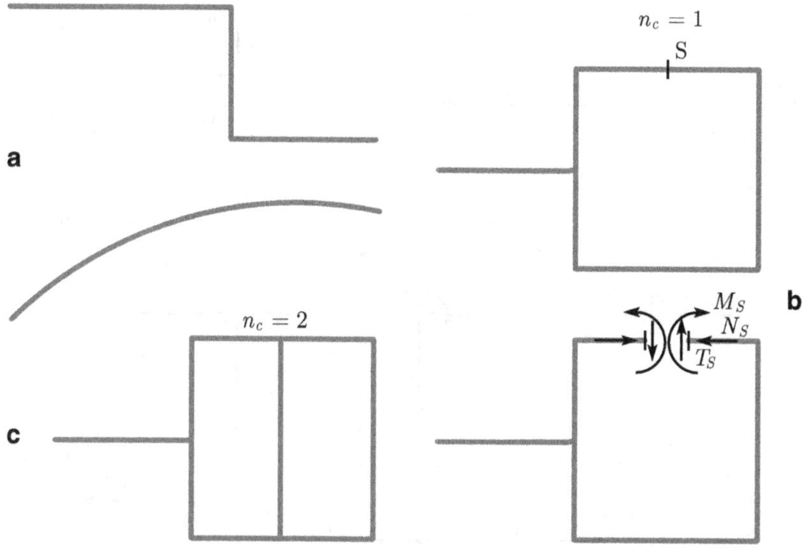

Fig. 8.10 Examples of rods with axes described by one-to-one functions (a) and those admitting closed circuits (b), (c)

8.14 Exercises

Exercise 2. *Calculate the reactive actions on the rigid rod in Figure 8.11.*

Remarks and solution. The structure is isostatic because it is constrained by three pendulums with axes not intersecting at the same point, so that a rotation center does not exist. By taking into account the notation in Figure 8.11, the balance equations read

$$H_A + H_C - \hat{\mathsf{f}} = 0, \quad Y_B - \hat{\mathsf{f}} = 0, \quad -\hat{\mathsf{m}} + \hat{\mathsf{f}}l - Y_B 2l - H_C l = 0,$$

where momenta have been calculated with respect to the point A, whence

$$H_A = 2\hat{\mathsf{f}} + \frac{\hat{\mathsf{m}}}{l}, \quad H_C = -\left(\hat{\mathsf{f}} + \frac{\hat{\mathsf{m}}}{l}\right), \quad Y_B = \hat{\mathsf{f}}.$$

The orientation assumed for the reaction H_C in Figure 8.11 is not correct: the negative sign indicates that the effective orientation of H_C is opposite to that initially assumed (Fig. 8.12).

Exercise 3. *Find the reactive actions for the cantilever rigid rod in Figure 8.13.*

Remarks and solution. The rod is subjected to a distributed triangular load with maximal value $\bar{q}_0$, statically equivalent to a vertical concentrated force $\bar{q}_0 l/2$ applied

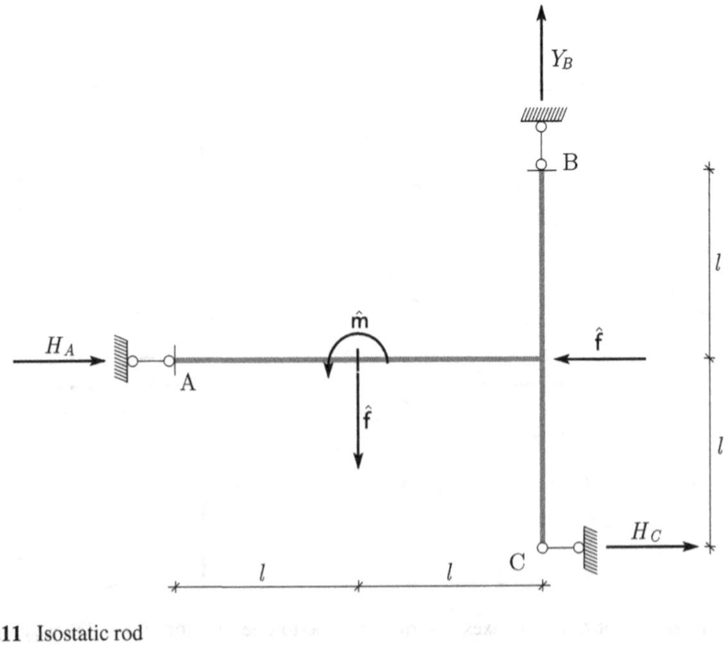

Fig. 8.11 Isostatic rod

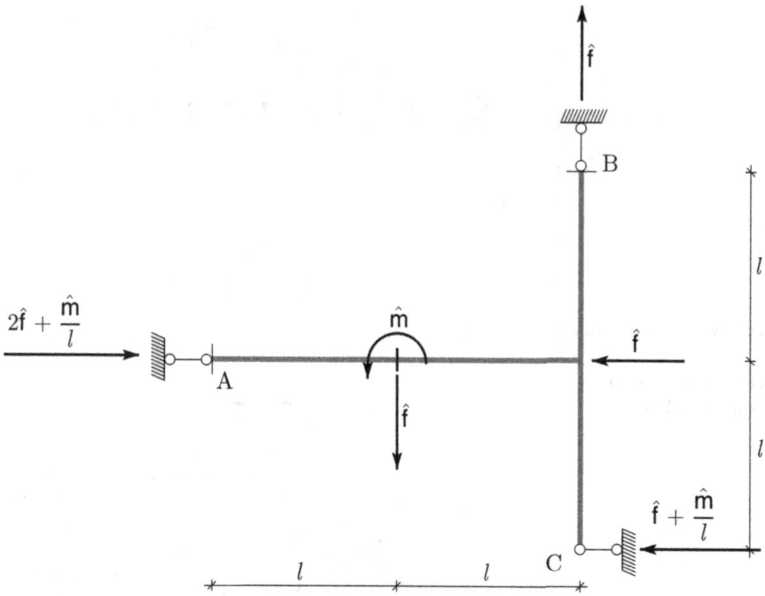

Fig. 8.12 Values of the reactive actions

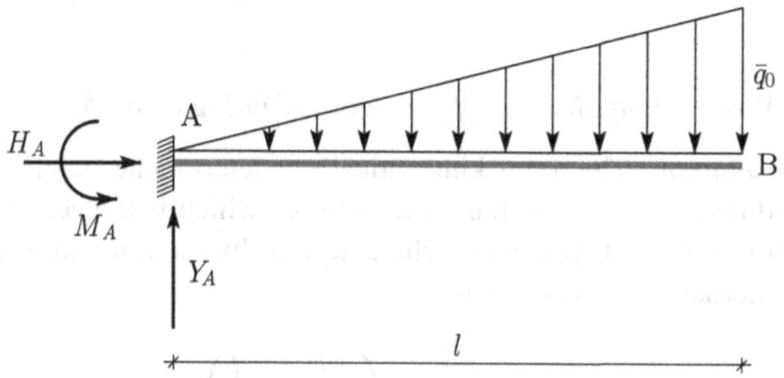

Fig. 8.13 Cantilever rigid rod with a triangular distributed load

at distance of $l/3$ from the point B. By evaluating momenta with respect to the point A and adopting the notation in Figure 8.13, the balance equations read

$$H_A = 0, \qquad Y_A - \frac{\bar{q}_0 l}{2} = 0, \qquad M_A - \frac{\bar{q}_0 l}{2}\frac{2}{3}l = 0,$$

so we get (Fig. 8.14)

$$H_A = 0, \qquad Y_A = \frac{\bar{q}_0 l}{2}, \qquad M_A = \frac{\bar{q}_0 l^2}{3}.$$

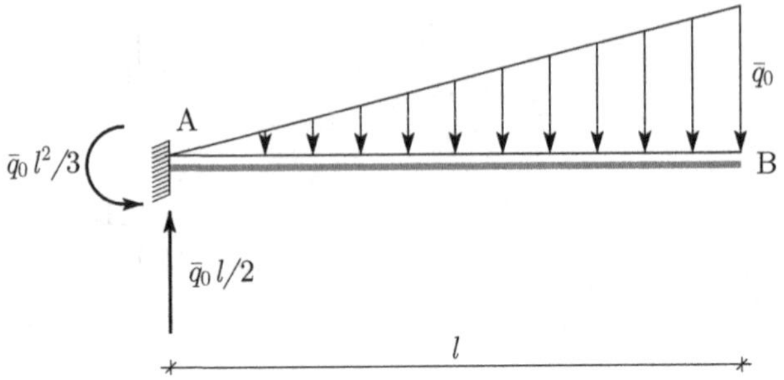

Fig. 8.14 Values of the reactive actions

Fig. 8.15 Labile rod in equilibrium conditions

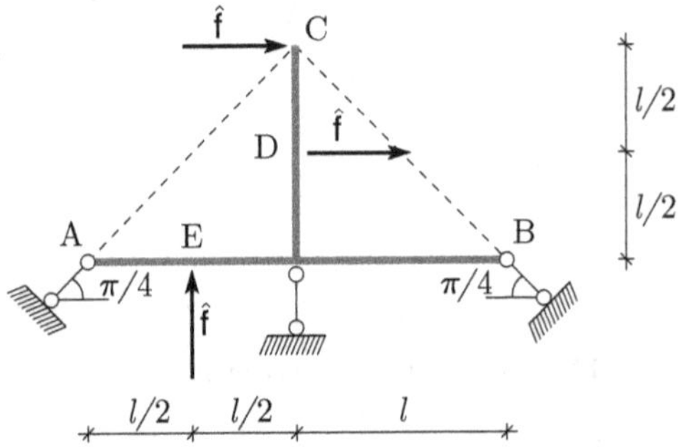

Exercise 4. *Verify the equilibrium of the rigid rod in Figure 8.15.*

Remarks and solution. The rod is kinematically indeterminate (labile): the axes of the three pendulums intersect at the same point C, which is the center of rotation. Given an arbitrary rigid displacement characterized by the rotation ω about C, the work of the external forces is given by

$$L^{ext} = \hat{\mathfrak{f}} u_D - \hat{\mathfrak{f}} u_E = \hat{\mathfrak{f}}\left(\omega\frac{l}{2} - \omega\frac{l}{2}\right) = 0.$$

The displacements of the points D and E appear in the diagrams of Figure 8.16. Although the rod is labile, the specific load condition ensures equilibrium.

Exercise 5. *Calculate the reactive forces on the structure in Figure 8.17, made of rigid rods.*

Remarks and solution. The structure consists of three rods with nine simple constraints (five external constraints and four internal constraints) independent of each other, so it is isostatic. We compute the reactions of the external constraints using the balance equations for the system, considered as a unique rigid body, and two additional balances corresponding to the equilibrium of momenta with respect to the internal hinges in B and D. On adopting the notation in Figure 8.17, such balances read

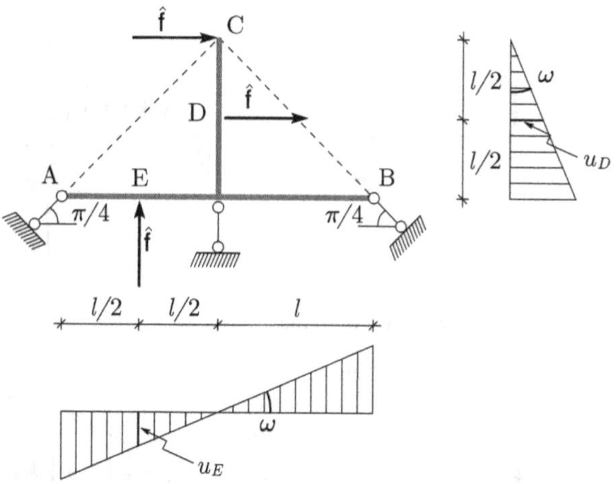

Fig. 8.16 Diagrams of the rigid linearized displacement of the rod

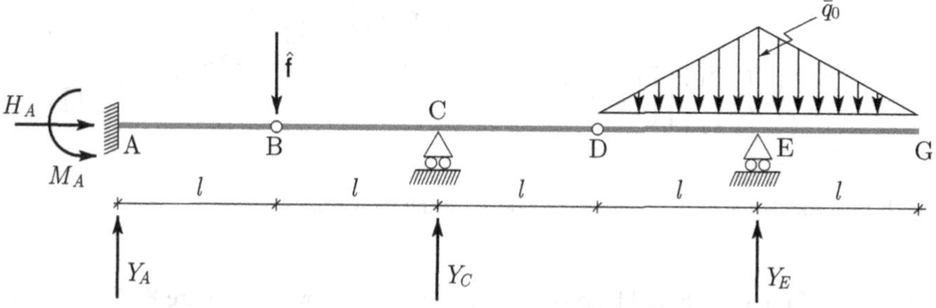

Fig. 8.17 Isostatic rod with internal hinges

$$H_A = 0, \quad Y_A + Y_C + Y_E - \hat{\mathrm{f}} - \bar{q}_0 l = 0, \quad M_A - 4lY_A - 2lY_C + 3l\hat{\mathrm{f}} = 0,$$

$$-lY_A + M_A = 0, \quad lY_E - \bar{q}_0 l^2 = 0.$$

Then we obtain (Fig. 8.18)

$$H_A = 0, \quad Y_A = \hat{\mathrm{f}}, \quad Y_C = 0, \quad M_A = \hat{\mathrm{f}}l, \quad Y_E = \bar{q}_0 l.$$

Equilibrium in the vertical direction for the part $\overline{\mathrm{DG}}$ implies $Y_D = 0$, while that for the part $\overline{\mathrm{BD}}$ requires $Y_B = 0$, because $Y_C = Y_D = 0$ and the external force $\hat{\mathrm{f}}$ is totally sustained by the part $\overline{\mathrm{AB}}$. Finally, the internal horizontal reactions H_B and H_D vanish, because there are no horizontal forces acting on the system. For the reaction M_A, we act in terms of (virtual) work (Fig. 8.19) to verify the result already obtained. We eliminate the simple rotational constraint at A (so a hinge has to be inserted) so that the system may admit a linearized rigid displacement. With these new conditions, the external work is given by

$$L^{ext} = -M_A\omega + \hat{\mathrm{f}}u_B = 0.$$

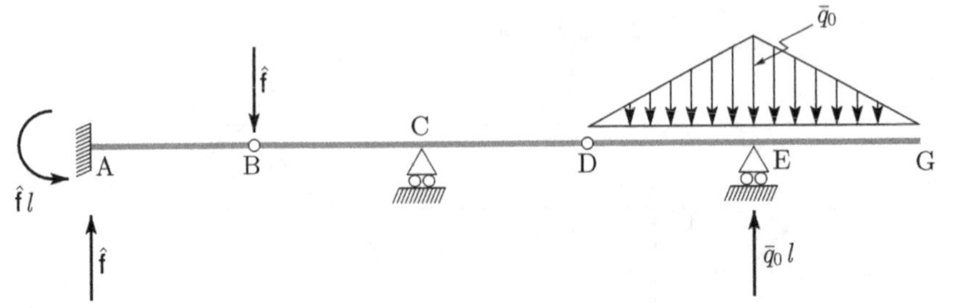

Fig. 8.18 Reactions and active forces over the structure

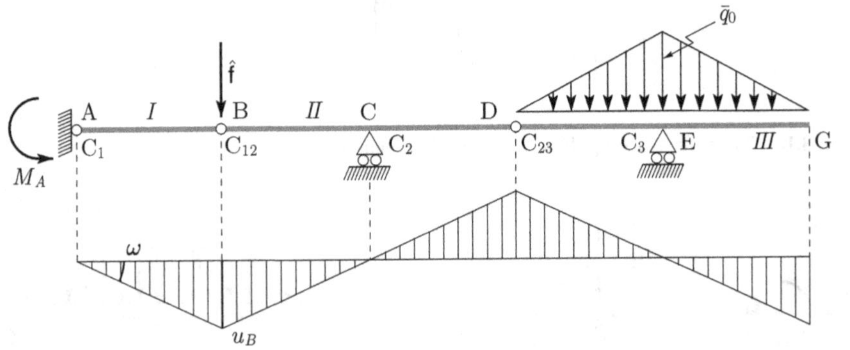

Fig. 8.19 Calculation of the external reaction M_A using the virtual work

The work of the distributed load is equal to zero. As shown in Figure 8.19, we have $u_B = l\omega$ and

$$L^{ext} = (-M_A + \hat{f}l)\omega = 0,$$

which gives $M_A = \hat{f}l$, a value already known.

Exercise 6. *Calculate the reactive forces exerted on the structure in Figure 8.20 by the constraints imposed by the environment.*

Remarks and solution. There are four rods and twelve simple constraints (the internal hinge in C is equivalent to four simple constraints) that do not allow rigid displacements (they are well posed in this sense).[2] First, we write the balance equations of the whole structure, considering the point D as a pole for momenta

$$H_D + \hat{f} = 0, \qquad Y_A + Y_D - \hat{f} = 0, \qquad M_A - 2lY_A - \hat{f}l = 0.$$

---

[2]In fact, the system can be considered to be made of two rigid bodies, and the rotation centers C_1, C_2, and C_{12} are not collinear.

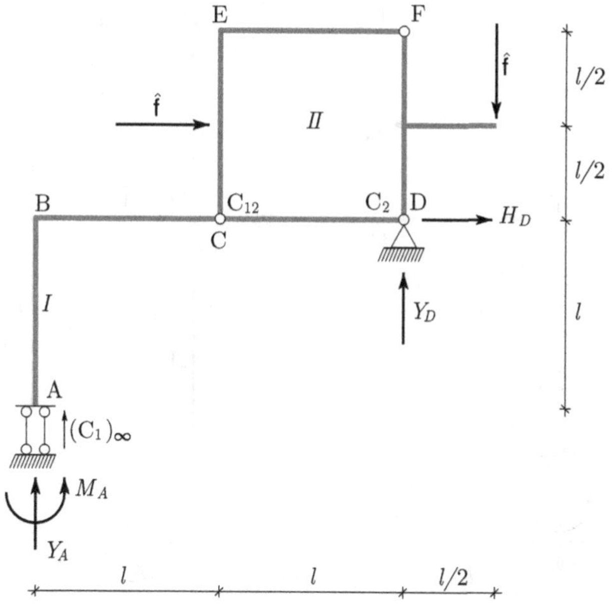

Fig. 8.20 Isostatic structure

There are four unknowns and three equations. However, if the structure is in an equilibrium state, every part of it has to be in the same state (Euler's principle). Hence we may impose the balance of momenta of the part $\overline{\text{ABC}}$ with respect to C:

$$M_A - lY_A = 0,$$

so that we obtain

$$H_D = -\hat{\mathsf{f}}, \qquad M_A = -\hat{\mathsf{f}}l, \qquad Y_A = -\hat{\mathsf{f}}, \qquad Y_D = 2\hat{\mathsf{f}}.$$

The reactions with negative signs have opposite orientation to that initially assumed (Fig. 8.21).

Exercise 7. *Find the action characteristics in the structure in Figure 8.22.*

Remarks and solution. The structure is isostatic. The balance of forces in the horizontal direction implies $H_E = 0$. Moreover, the balance of couples evaluated with respect to the point B reads

$$-Y_E 3l + Q\frac{l}{3} = 0 \quad \Longrightarrow \quad Y_E = \frac{Q}{9} \quad \left(Q = \frac{\bar{q}_0 l}{2}\right).$$

Finally, the balance of forces in the vertical direction implies $Y_B = \dfrac{10}{9}Q$.

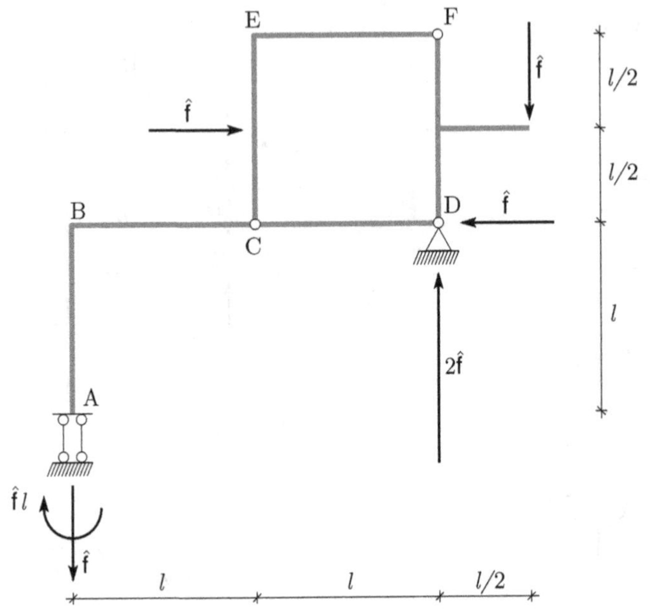

Fig. 8.21 Structure with the reactions of the external constraints

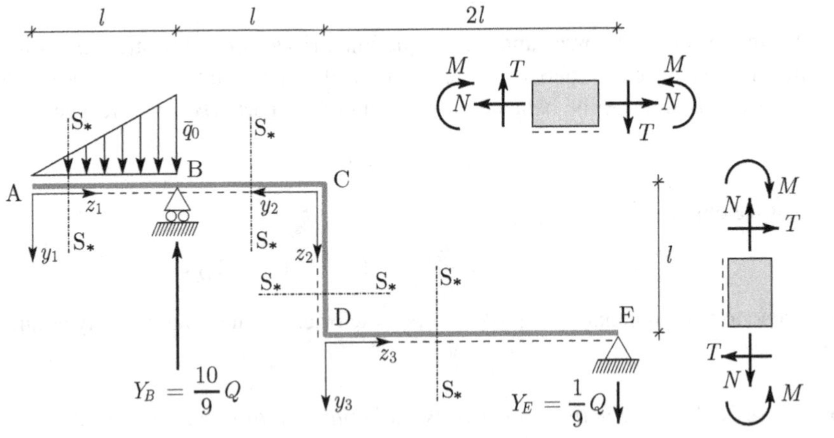

Fig. 8.22 Simple supported rod with two angular deviations of the axis. The schemes depicted as rectangles indicate the positive signs used here

Consider three different reference frames denoted by $\{y_1, z_1\}$, $\{y_2, z_2\}$ and $\{y_3, z_3\}$ for the parts $\overline{\text{AC}}$, $\overline{\text{CD}}$, and $\overline{\text{DE}}$, respectively. Take a generic section S_* in four different portions of the rod, namely $\overline{\text{AB}}$, $\overline{\text{BC}}$, $\overline{\text{CD}}$, and $\overline{\text{DE}}$. In $\overline{\text{AB}}$, consider the balance of the portion $\overline{\text{AS}_*}$, with S_* at z_1, in which the value of the load is $\bar{q}(z_1) = \bar{q}_0 z_1 / l$. The balance equations for $\overline{\text{AS}_*}$ can be written as

$$\begin{cases} N(z_1) = 0 \\ T(z_1) + \dfrac{\bar{q}_0 z_1}{l}\dfrac{z_1}{2} = 0 \\ M(z_1) + \dfrac{\bar{q}_0 z_1}{l}\dfrac{z_1}{2}\dfrac{z_1}{3} = 0 \end{cases} \implies \begin{cases} N(z_1) = 0 \\ T(z_1) = -\dfrac{\bar{q}_0 z_1^2}{2l} \\ M(z_1) = -\dfrac{\bar{q}_0 z_1^3}{6l} \end{cases} \qquad z_1 \in [0, l].$$

By choosing S_* in $\overline{BC}$, the balance equations of the portion $\overline{AS_*}$ can be written as

$$\begin{cases} N(z_1) = 0 \\ T(z_1) + Q - \dfrac{10}{9}Q = 0 \\ M(z_1) - Q\left(\dfrac{2}{3}l - z_1\right) - \dfrac{10}{9}Q(z_1 - l) = 0 \end{cases} \implies$$

$$\begin{cases} N(z_1) = 0 \\ T(z_1) = \dfrac{Q}{9} \\ M(z_1) = \dfrac{Q}{9}(z_1 - 4l) \end{cases} \qquad z_1 \in [l, 2l].$$

By choosing S_* in the third part $\overline{CD}$, the balance equations of the portion $\overline{S_*DE}$ read

$$\begin{cases} N(z_2) - \dfrac{Q}{9} = 0 \\ T(z_2) = 0 \\ M(z_2) + \dfrac{Q}{9}2l = 0 \end{cases} \implies \begin{cases} N(z_2) = \dfrac{Q}{9} \\ T(z_2) = 0 \\ M(z_2) = -2l\dfrac{Q}{9} \end{cases} \qquad z_2 \in [0, l].$$

Finally, in the part $\overline{DE}$, the balance equations for the portion $\widehat{S_*E}$ are

$$\begin{cases} N(z_3) = 0 \\ T(z_3) - \dfrac{Q}{9} = 0 \\ M(z_3) + \dfrac{Q}{9}(2l - z_3) = 0 \end{cases} \implies \begin{cases} N(z_3) = 0 \\ T(z_3) = \dfrac{Q}{9} \\ M(z_3) = \dfrac{Q}{9}(z_3 - 2l) \end{cases} \qquad z_3 \in [0, 2l].$$

The diagrams of the action characteristics are depicted in Figure 8.23. At the points C and D shear and axial forces balance each other, due to the presence of a right angle.

Exercise 8. *Calculate the action characteristics in the system in Figure 8.24.*

Remarks and solution. The structure is kinematically indeterminate: the external constraints allow a rigid horizontal translation. However, for the particular load condition the equilibrium is possible. The balance of forces along the vertical direction and the one of the couples evaluated with respect to the point A are

$$Y_A + Y_B = 0, \qquad -\hat{m} + 2lY_B = 0,$$

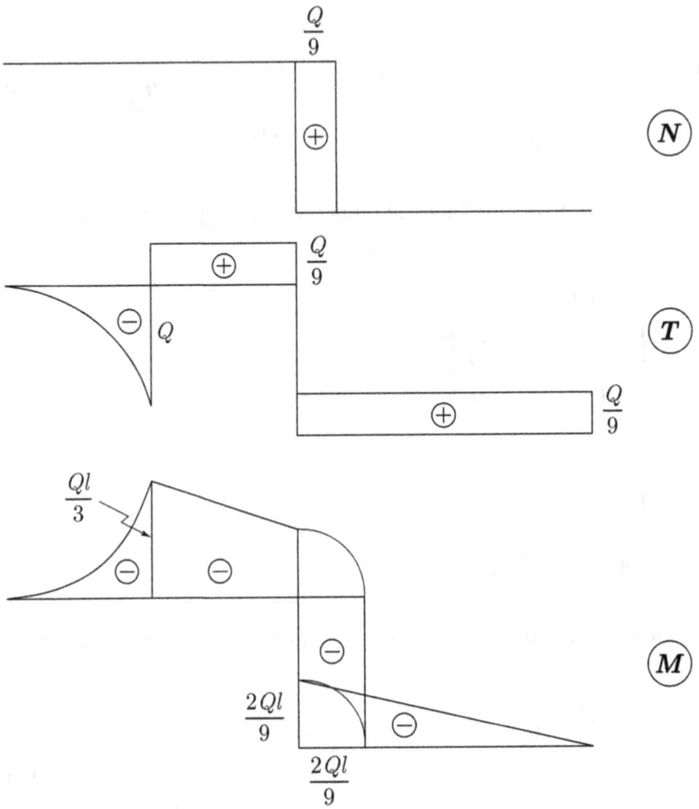

Fig. 8.23 Diagrams of the action characteristics of the rod

Fig. 8.24 Structure with horizontal lability

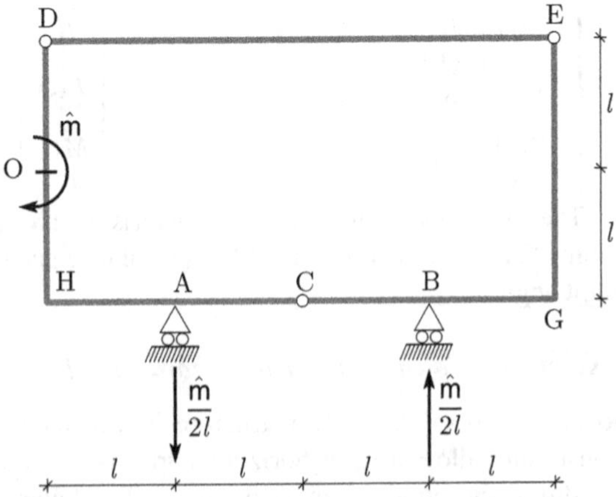

where Y_A and Y_B are respectively the reactions of the supports at A and B, and they are equal to

$$Y_A = -\frac{\hat{\mathfrak{m}}}{2l}, \qquad Y_B = \frac{\hat{\mathfrak{m}}}{2l}.$$

To calculate the six reactions of the internal hinges, we separate the three rods $\overline{CD}$, $\overline{CE}$ and $\overline{DE}$ and write the pertinent balance equations. The result is shown in Figure 8.25. Table 8.1 collects the equations of the action characteristics with reference to the local frames in Figure 8.25. Their diagrams are in Figure 8.26.

Exercise 9. *Calculate the action characteristics in the system analyzed in Exercise 6.*

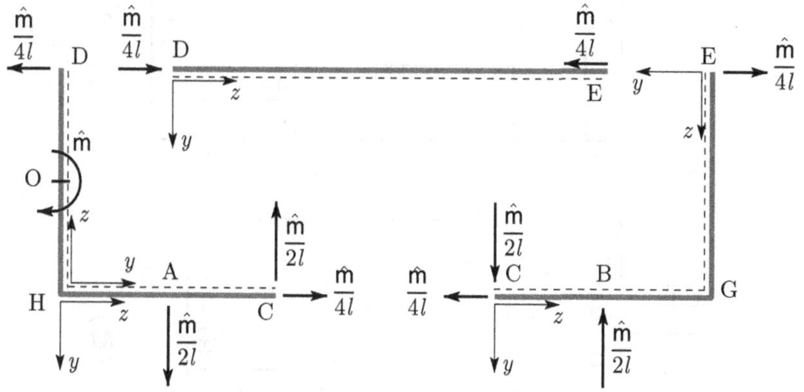

Fig. 8.25 Reactive forces and local frames

Table 8.1 Action characteristics

Part	$N(z)$	$T(z)$	$M(z)$
$\overline{CB}$, $z \in [0, l]$	$\frac{\hat{\mathfrak{m}}}{4l}$	$-\frac{\hat{\mathfrak{m}}}{2l}$	$\frac{\hat{\mathfrak{m}}}{2l}z$
$\overline{BG}$, $z \in [l, 2l]$	$\frac{\hat{\mathfrak{m}}}{4l}$	0	$\frac{\hat{\mathfrak{m}}}{2}$
$\overline{EG}$, $z \in [0, 2l]$	0	$\frac{\hat{\mathfrak{m}}}{4l}$	$\frac{\hat{\mathfrak{m}}}{4l}z$
$\overline{DE}$, $z \in [0, 4l]$	$-\frac{\hat{\mathfrak{m}}}{4l}$	0	0
$\overline{HO}$, $z \in [0, l]$	0	$-\frac{\hat{\mathfrak{m}}}{4l}$	$-\frac{\hat{\mathfrak{m}}}{2} - \frac{\hat{\mathfrak{m}}}{4l}z$
$\overline{OD}$, $z \in [l, 2l]$	0	$-\frac{\hat{\mathfrak{m}}}{4l}$	$\frac{\hat{\mathfrak{m}}}{4l}(2l - z)$
$\overline{HA}$, $z \in [0, l]$	$\frac{\hat{\mathfrak{m}}}{4l}$	0	$-\frac{\hat{\mathfrak{m}}}{2}$
$\overline{AC}$, $z \in [l, 2l]$	$\frac{\hat{\mathfrak{m}}}{4l}$	$-\frac{\hat{\mathfrak{m}}}{2l}$	$\frac{\hat{\mathfrak{m}}}{2l}(z - 2l)$

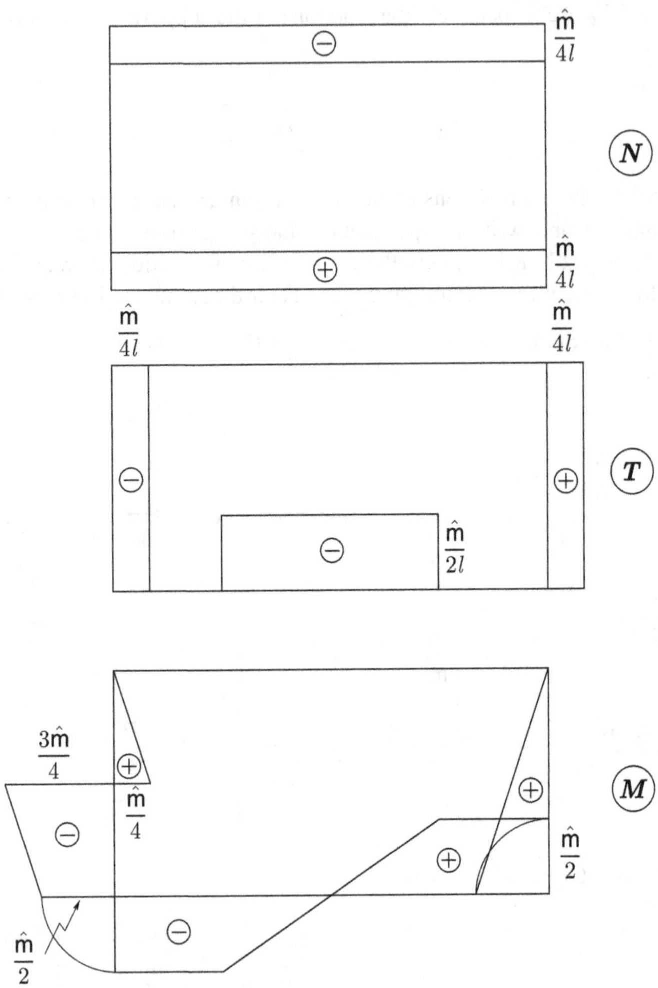

Fig. 8.26 Diagrams of the action characteristics

Remarks and solution. The external reactive forces have been already determined in Exercise 6 of this chapter. We open the circuit $\overline{\text{CDEG}}$ at E and consider the two internal forces Y_E and H_E (unknown reactive forces of the hinge), as in Figure 8.27. Then we write the balance of the couples for the portion $\overline{\text{EDC}}$ with respect to C and for $\overline{\text{EHG}}$ with respect to G (see Fig. 8.27),

$$lY_E - H_E l - \frac{\hat{\mathrm{f}}l}{2} = 0, \qquad H_E l - \frac{\hat{\mathrm{f}}l}{2} = 0,$$

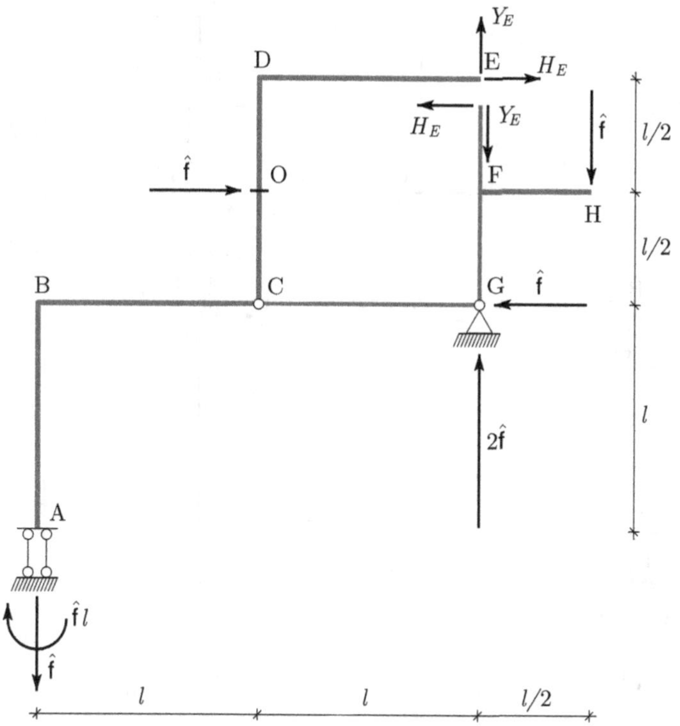

Fig. 8.27 Calculation of two internal reactive forces

obtaining (Fig. 8.28)

$$Y_E = \hat{f}, \qquad H_E = \frac{\hat{f}}{2}.$$

On adopting the frames of reference in Figure 8.28, we determine the diagrams of $N(z)$, $T(z)$, and $M(z)$ (Fig. 8.29).

Exercise 10. *Find the action characteristics in the truss system in Figure 8.30.*

Remarks and solution. The *truss system* (so called because it consists of rods connected *only* by hinges *and* the external actions are just forces applied over the hinges) is isostatic. The external reactive forces have been depicted in Figure 8.30, and they can be determined by the balance equations of the system, considered as a single rigid body:

$$H_E + \frac{\hat{f}_2}{2} = 0, \qquad \hat{f}_1 + \frac{\sqrt{3}}{2}\hat{f}_2 - Y_A - Y_E = 0, \qquad 2lY_E - l\frac{\sqrt{3}}{2}\hat{f}_2 = 0,$$

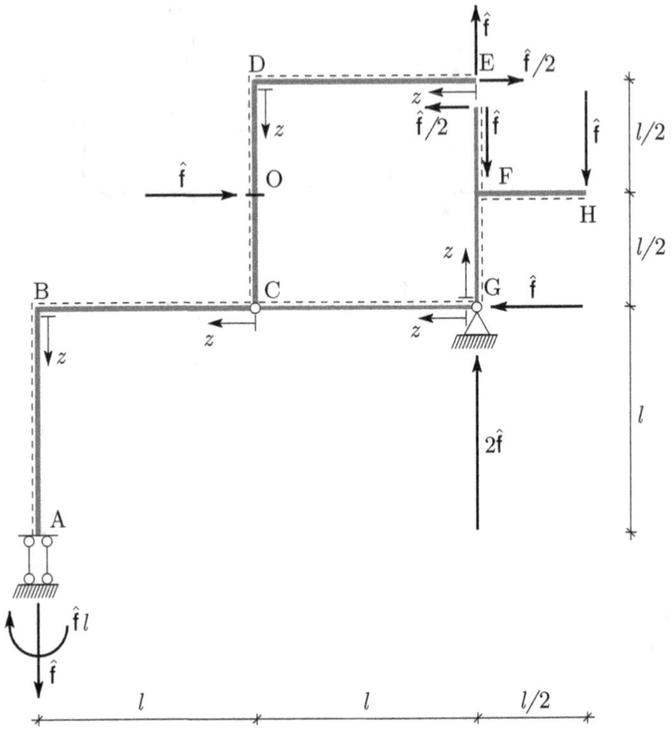

Fig. 8.28 Values of the internal reactive forces at E

which furnish (Fig. 8.31)

$$H_E = -\frac{\hat{f}_2}{2}, \qquad Y_A = \hat{f}_1 + \frac{\sqrt{3}}{4}\hat{f}_2, \qquad Y_E = \frac{\sqrt{3}}{4}\hat{f}_2.$$

In a truss system, the only nonzero action characteristic is the normal force N, which is constant in each rod. The rods subjected to a tensile action are called *ties*; those compressed are called *struts*. A way to determine the normal actions in these systems is the method of the *equilibrium of nodes*.

Consider first the node B with all pertinent known and unknown forces and write their balance equation (Fig. 8.32):

$$N_{BC} = 0, \qquad N_{AB} + \hat{f}_1 = 0,$$

where N_{BC} and N_{AB} are the normal actions over the rods $\overline{BC}$ and $\overline{AB}$, respectively. We then obtain

$$N_{BC} = 0, \qquad N_{AB} = -\hat{f}_1,$$

which means that the rod $\overline{AB}$ is a strut.

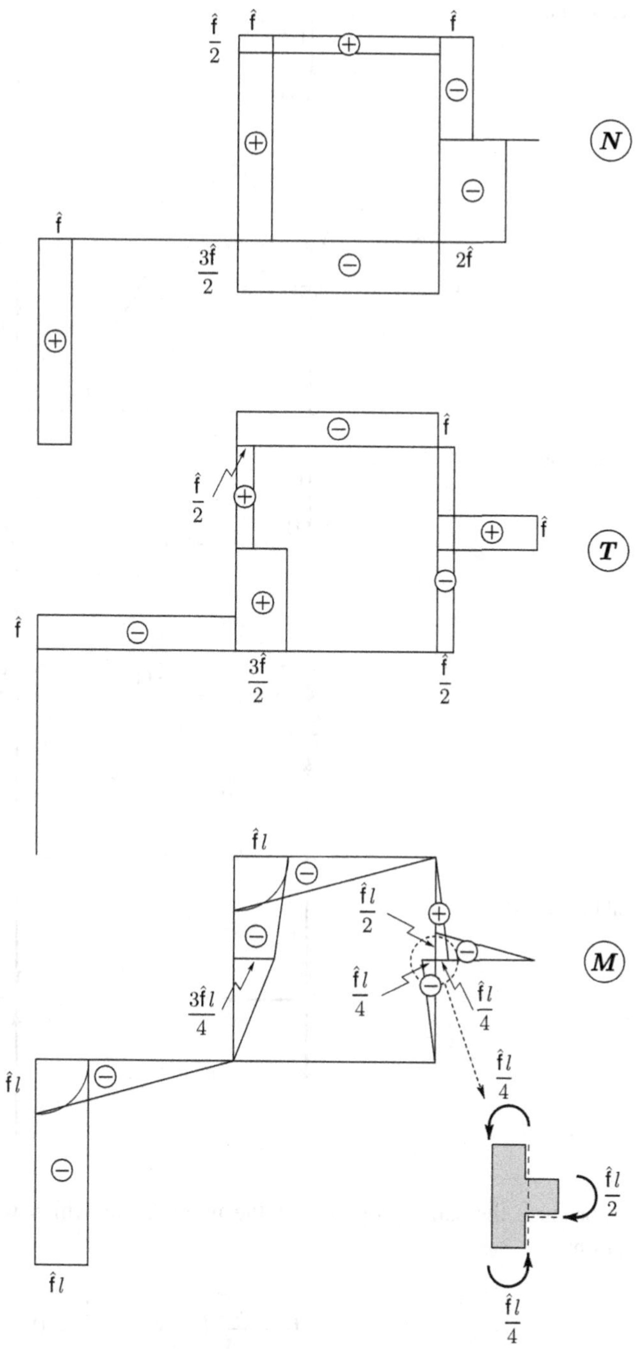

Fig. 8.29 Diagrams of the action characteristics

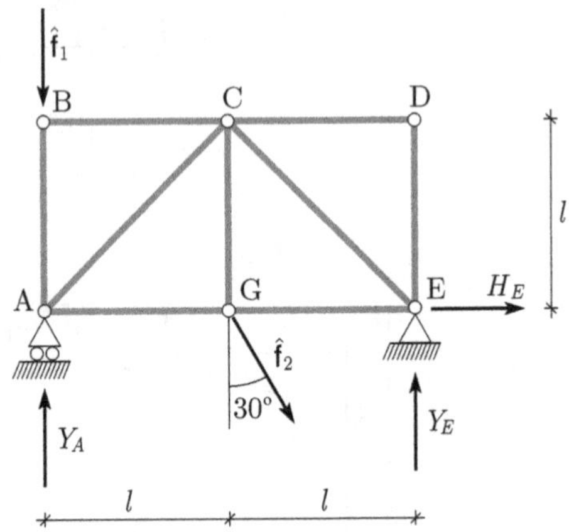

Fig. 8.30 Truss system

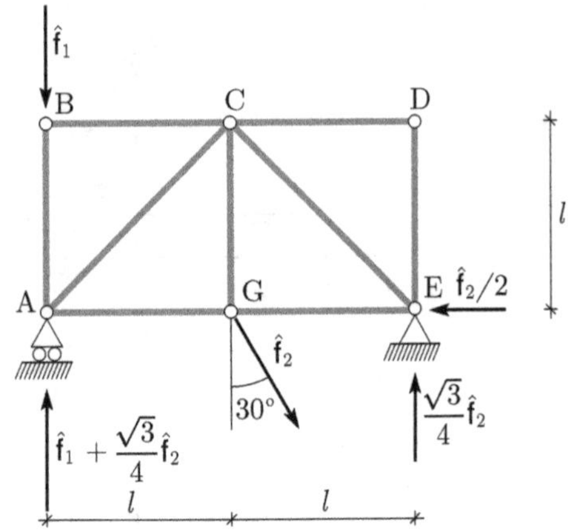

Fig. 8.31 External reactive forces

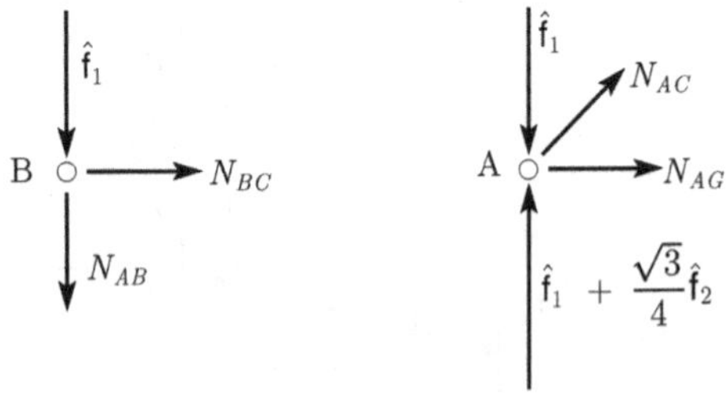

Fig. 8.32 Equilibrium of two nodes

We then reproduce the same program for the node A, for which we have the balance equations

$$N_{AG} + N_{AC}\frac{\sqrt{2}}{2} = 0, \qquad \hat{f}_1 - \hat{f}_1 - \frac{\sqrt{3}}{4}\hat{f}_2 - N_{AC}\frac{\sqrt{2}}{2} = 0,$$

Table 8.2 Action characteristics

Rod	Type of action	Value
$\overline{AB}$	Strut	$-\hat{f}_1$
$\overline{BC}$	Unloaded	0
$\overline{CD}$	Unloaded	0
$\overline{DE}$	Unloaded	0
$\overline{EG}$	Strut	$\hat{f}_2\left(\frac{\sqrt{12}}{8}-\frac{1}{2}\right)$
$\overline{AG}$	Tie	$\hat{f}_2\frac{\sqrt{12}}{8}$
$\overline{AC}$	Strut	$-\hat{f}_2\frac{\sqrt{6}}{4}$
$\overline{GC}$	Tie	$\hat{f}_2\frac{\sqrt{3}}{2}$
$\overline{CE}$	Strut	$-\hat{f}_2\frac{\sqrt{6}}{4}$

Fig. 8.33 Normal actions

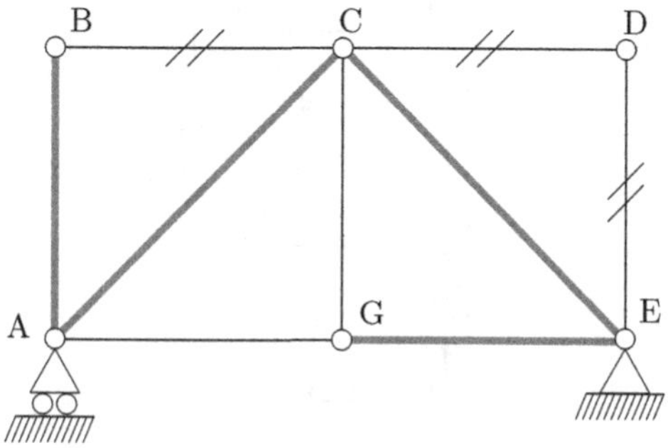

which give

$$N_{AG} = \frac{\sqrt{12}}{8}\hat{f}_2, \qquad N_{AC} = -\frac{\sqrt{6}}{4}\hat{f}_2,$$

so that the rod $\overline{AG}$ is a tie and the rod $\overline{AC}$ is a strut. The indices to N denote the rod. By following this procedure and considering other nodes, we calculate the unknown normal actions in each rod. Table 8.2 collects these values (the ties have a positive normal action; the struts have negative values). We show a synthetic view of the result in Figure 8.33: the ties are shown by thin lines, while the struts are represented by thick lines. The method can be generally applied to all isostatic truss structures.

To verify the results, we can use an alternative method based on so-called *Ritter's sections*.

Imagine that we divide the structure into two distinct parts by ideally cutting three nonparallel rods, with N unknown. Two examples are shown in Figure 8.34. A section of this type is called a *Ritter's section*. Consider the section 1-1 in Figure 8.34 and the left part of the system (Fig. 8.35). The pertinent balance equations—they are the balance of couples with respect to the point A, the balance of couples with respect to C, and the balance of forces in the vertical direction—read

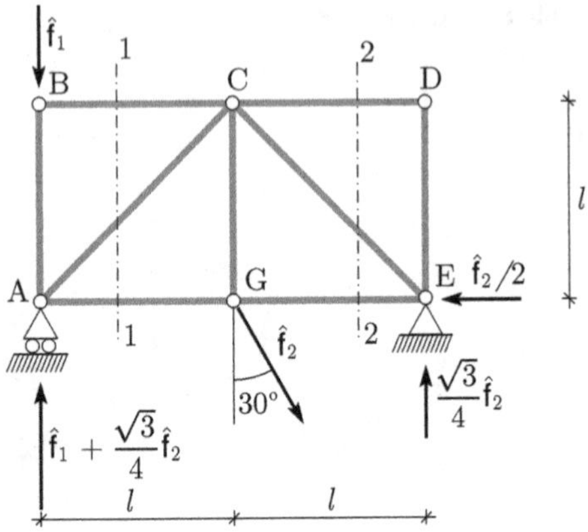

Fig. 8.34 Examples of Ritter's sections

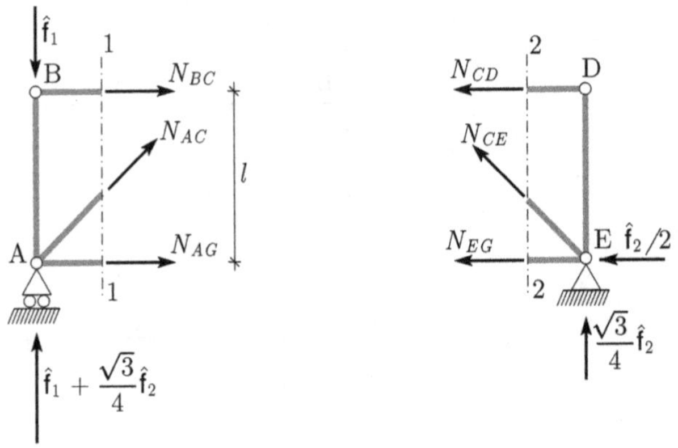

Fig. 8.35 Ritter's section method

$$N_{BC} = 0, \quad lN_{AG} + \hat{f}_1 l - \hat{f}_1 l - \frac{\sqrt{3}}{4}\hat{f}_2 l = 0, \quad N_{AC}\frac{\sqrt{2}}{2} - \hat{f}_1 + \hat{f}_1 + \frac{\sqrt{3}}{4}\hat{f}_2 = 0,$$

i.e.,

$$N_{BC} = 0, \qquad N_{AG} = \hat{f}_2\frac{\sqrt{3}}{4}, \qquad N_{AC} = -\hat{f}_2\frac{\sqrt{6}}{4}.$$

Hence, the beam $\overline{AG}$ is a tie, while $\overline{AC}$ is a strut. By considering different cuts, it is possible to evaluate the normal actions in all the rods (Fig. 8.35).

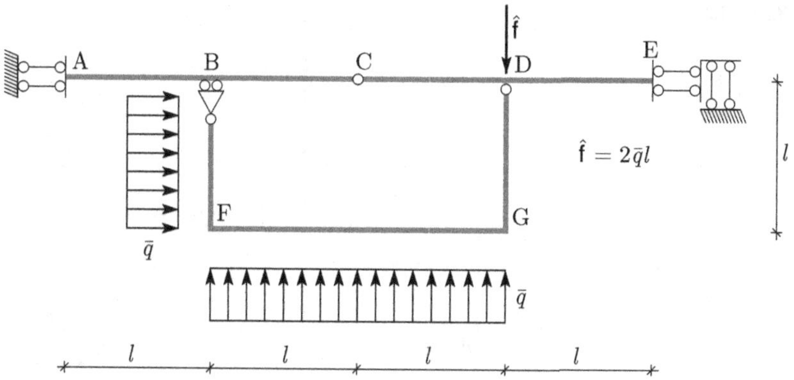

Fig. 8.36 Kinematically indeterminate system, with a particular load condition ensuring equilibrium

Fig. 8.37 Spatial rod

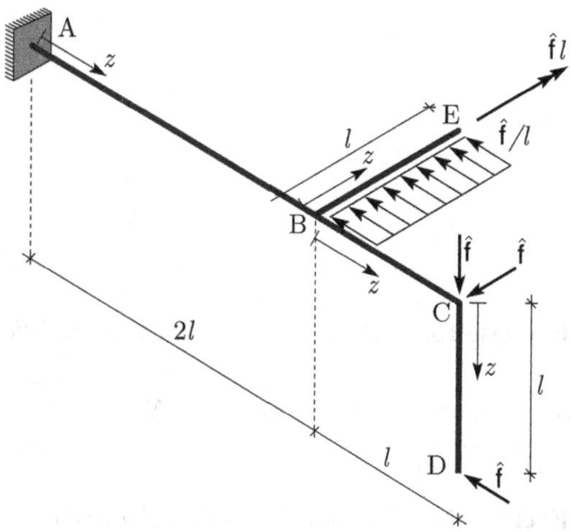

8.15 Further exercises on isostatic framed structures

Exercise 11. *Calculate the action characteristics in the system in Figure 8.36.*

Some elements of the solution: $M(\mathrm{A}) = \dfrac{3\bar{q}l^2}{4}$, oriented counterclockwise; $M(\mathrm{F}) = \dfrac{\bar{q}l^2}{2}$.

Exercise 12. *Calculate the action characteristics in the structure in Figure 8.37.*

Some elements of the solution: $N(\mathrm{A}) = -2\hat{\mathrm{f}}$, $M_t(\mathrm{A}) = 0$.

Fig. 8.38 Helix

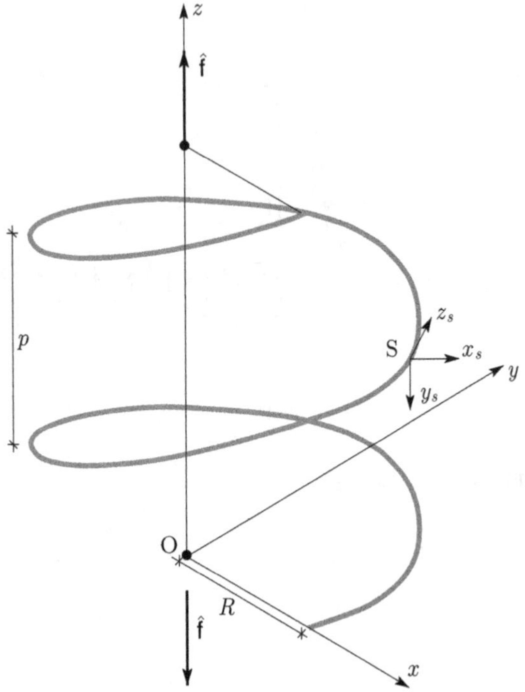

Exercise 13. *Calculate the action characteristics in the structure in Figure 8.38.*

Some elements of the solution: $N = \dfrac{2p\hat{\mathsf{f}}}{c}$, $M_t = \dfrac{\omega R^2\hat{\mathsf{f}}}{c}$, $\omega = 4\pi$, $c = \sqrt{\omega^2 R^2 + 4p^2}$.

Exercise 14. *Calculate the action characteristics in the system in Figure 8.39.*

Some elements of the solution: The structure is twice labile; $N_{AD} = \hat{\mathsf{f}}$, $M(\mathrm{A}) = \dfrac{\hat{\mathsf{f}}l}{\sqrt{2}}$.

Exercise 15. *Calculate the action characteristics in the structure in Figure 8.40.*

Some elements of the solution: The structure is isostatic; $Y_A = \hat{\mathsf{f}}$, and it is oriented upward; $M(\mathrm{D}) = \dfrac{5\hat{\mathsf{f}}l}{2}$.

Exercise 16. *Find the values of the bending moments in the sections* A *and* B *of the structure in Figure 8.41. Consider* $\hat{\mathsf{f}} = \bar{q}l$ *and* $\hat{\mathsf{m}} = \bar{q}l^2/2$.

Solution. $M(\mathrm{A}) = 15\bar{q}l^2/8$, $M(\mathrm{B}) = 2\bar{q}l^2$.

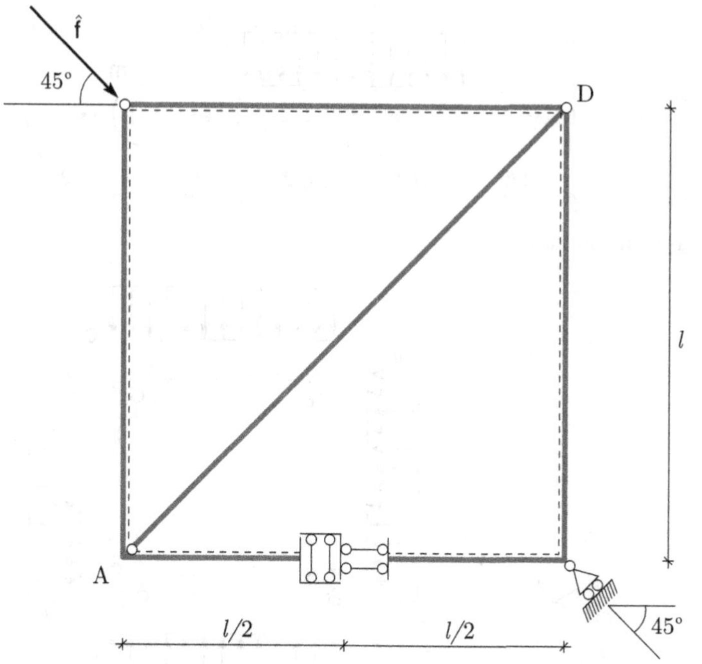

Fig. 8.39 Kinematically indeterminate structure with a particular load condition ensuring equilibrium

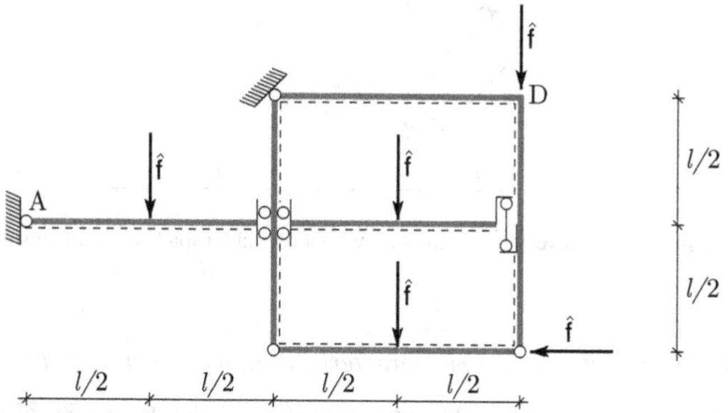

Fig. 8.40 Isostatic structure

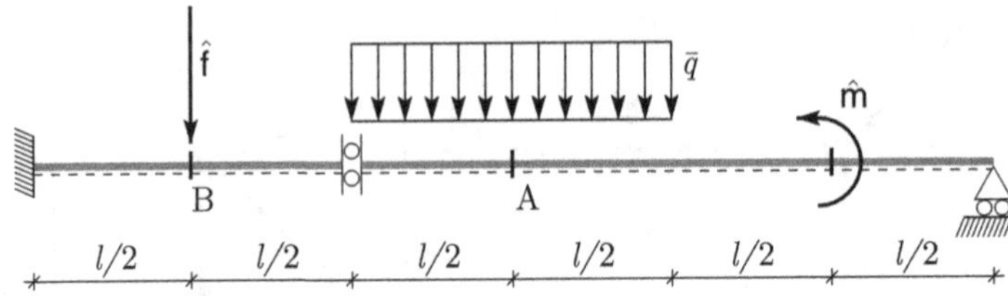

Fig. 8.41 Isostatic structure

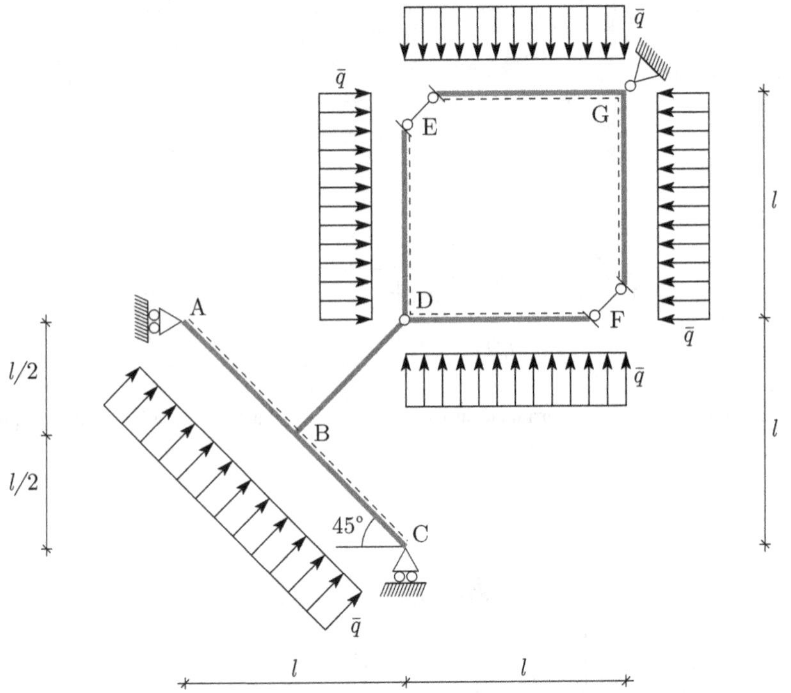

Fig. 8.42 Kinematically indeterminate structure with a particular load condition ensuring equilibrium

Exercise 17. *Calculate the action characteristics in the structure in Figure 8.42.*

Some elements of the solution: The structure is twice labile; $Y_C = \bar{q}l$ is directed downward; $H_G = 0$.

8.16 Weak Balances: The Inner Power

In deriving the inertia terms in the balance equations, we have assumed the velocity to be continuous along the centroid line, i.e., $\dot{\varphi}$ and w are continuous. The assumption implies that

$$\mathsf{n}(s_2,t)\cdot\dot{\varphi}(s_2,t)-\mathsf{n}(s_1,t)\cdot\dot{\varphi}(s_1,t)$$
$$=\int_{s_1}^{s_2}\left(\frac{\partial\mathsf{n}}{\partial s}\cdot\dot{\varphi}-\mathsf{n}\cdot\frac{\partial\dot{\varphi}}{\partial s}\right)ds+\sum_{k=1}^{\bar{k}_s}[\mathsf{n}(s_k,t)]\cdot\dot{\varphi}(s_k,t)$$

and

$$\mathsf{m}(s_2,t)\cdot w(s_2,t)-\mathsf{m}(s_1,t)\cdot w(s_1,t)$$
$$=\int_{s_1}^{s_2}\left(\frac{\partial\mathsf{m}}{\partial s}\cdot w-\mathsf{m}\cdot\frac{\partial w}{\partial s}\right)ds+\sum_{r=1}^{\bar{\bar{k}}_s}[\mathsf{m}(s_r,t)]\cdot w(s_r,t).$$

By taking into account the previous expressions and using the balance equations, we can manipulate the expression of $\mathcal{P}^{ext}_{(s_1,s_2)}(\dot{\varphi},w)$, obtaining

$$\mathcal{P}^{ext}_{(s_1,s_2)}(\dot{\varphi},w)=\int_{s_1}^{s_2}\left(\mathsf{n}\cdot\left(\frac{\partial\dot{\varphi}}{\partial s}-w\times\frac{\partial\varphi}{\partial s}\right)+\mathsf{m}\cdot\frac{\partial w}{\partial s}\right)ds.$$

We call the right-hand-side integral the **inner power** *in the rod*, denoting it by $\mathcal{P}^{inn}_{(s_1,s_2)}$. We leave the proof to the reader as an exercise.

The corotational derivative (8.4) implies then

$$\mathcal{P}^{inn}_{(s_1,s_2)}=\int_{s_1}^{s_2}\left(\mathsf{n}\cdot\overset{\nabla}{\zeta}+\mathsf{m}\cdot\frac{\partial w}{\partial s}\right)ds.$$

Recall that we have determined the expression for the external power after evaluating the sectional power, i.e., the average over a generic cross section of the power developed by the tension over the section itself. Thus, we find it reasonable to imagine that $\mathcal{P}^{inn}_{(s_1,s_2)}$ equals the inner power in the part between s_1 and s_2 of the three-dimensional rod. In other words, we should have

$$\int_{\mathcal{X}}\int_{s_1}^{s_2}P\cdot\dot{F}\,ds\,d\xi^1d\xi^2=\int_{s_1}^{s_2}\left(\mathsf{n}\cdot\overset{\nabla}{\zeta}+\mathsf{m}\cdot\frac{\partial w}{\partial s}\right)ds. \tag{8.26}$$

Since (see Section 8.2)

$$F=\sum_{\alpha=1}^{2}\mathsf{d}_\alpha\otimes e_\alpha+\left(\frac{\partial\varphi}{\partial s}+\sum_{\alpha=1}^{2}\xi^\alpha\mathsf{d}_{\alpha,s}\right)\otimes e_3$$
$$=\sum_{\alpha=1}^{2}\mathsf{d}_\alpha\otimes e_\alpha+\left(\frac{\partial\varphi}{\partial s}+\omega\times(y-\varphi)\right)\otimes e_3,$$

because

$$\sum_{\alpha=1}^{2} \xi^{\alpha} \mathsf{d}_{\alpha,s} = \omega \times \sum_{\alpha=1}^{2} \xi^{\alpha} \mathsf{d}_{\alpha} = \omega \times (y - \varphi),$$

we get

$$\begin{aligned}
\dot{F} &= \sum_{\alpha=1}^{2} \dot{\mathsf{d}}_{\alpha} \otimes e_{\alpha} + \left(\frac{\partial \dot{\varphi}}{\partial s} + \dot{\omega} \times (y - \varphi) + \omega \times (\dot{y} - \dot{\varphi}) \right) \otimes e_3 \\
&= \sum_{\alpha=1}^{2} (w \times \mathsf{d}_{\alpha}) \otimes e_{\alpha} + \left(\frac{\partial \dot{\varphi}}{\partial s} + \dot{\omega} \times (y - \varphi) + \omega \times w \times (y - \varphi) \right) \otimes e_3.
\end{aligned}$$

In the basis e_1, e_2, e_3, we have also

$$P = \mathsf{t}_1 \otimes e_1 + \mathsf{t}_2 \otimes e_2 + \hat{\tau} \otimes e_3,$$

with $\hat{\tau} := Pe_3$, as introduced in Section 8.4. Then we compute

$$\begin{aligned}
P \cdot \dot{F} &= \hat{\tau} \cdot \frac{\partial \dot{\varphi}}{\partial s} + \dot{\omega} \cdot ((y - \varphi) \times \hat{\tau}) \\
&\quad + \hat{\tau} \cdot (\omega \times w \times (y - \varphi)) + \sum_{\alpha=1}^{2} w \cdot (\mathsf{d}_{\alpha} \times \mathsf{t}_{\alpha}).
\end{aligned} \tag{8.27}$$

A digression is now necessary to express the last term in the previous expression in a convenient way. The local balance of couples $PF^* \in \mathrm{Sym}(\tilde{\mathbb{R}}^3, \tilde{\mathbb{R}}^3)$ implies

$$\sum_{\alpha=1}^{2} \frac{\partial y}{\partial \xi^{\alpha}} \times \mathsf{t}_{\alpha} + \frac{\partial y}{\partial s} \times \hat{\tau} = 0. \tag{8.28}$$

The proof of this statement follows by writing the product FP^* in terms of the explicit expression of F and P above. We have, in fact,

$$\begin{aligned}
FP^* &= (\mathsf{d}_1 \otimes e_1)(e_1 \otimes \mathsf{t}_1) + (\mathsf{d}_2 \otimes e_2)(e_2 \otimes \mathsf{t}_2) \\
&\quad + \left(\left(\frac{\partial \varphi}{\partial s} + \sum_{\alpha=1}^{2} \xi^{\alpha} \mathsf{d}_{\alpha,s} \right) \otimes e_3 \right) (e_3 \otimes \hat{\tau}) \\
&= \mathsf{d}_1 \otimes \mathsf{t}_1 + \mathsf{d}_2 \otimes \mathsf{t}_2 + \left(\frac{\partial \varphi}{\partial s} + \sum_{\alpha=1}^{2} \xi^{\alpha} \mathsf{d}_{\alpha,s} \right) \otimes \hat{\tau} \\
&= \frac{\partial y}{\partial \xi^1} \otimes \mathsf{t}_1 + \frac{\partial y}{\partial \xi^2} \otimes \mathsf{t}_2 + \frac{\partial y}{\partial s} \otimes \hat{\tau}.
\end{aligned} \tag{8.29}$$

The left-hand side of (8.28) is the axial vector of the skew-symmetric part of the second-rank tensor on the right-hand side of equation (8.29)—prove such a statement as an exercise. Consequently, since PF^* is symmetric, its skew-symmetric part is zero, and so is the pertinent axial vector.[3]

---

[3]The relation (8.28) allows us to derive in another way the local balance of couples along the rod. In the coordinates ξ^1, ξ^2, s, used so far, we have

$$\mathrm{Div}P = \sum_{\alpha=1}^{2} \frac{\partial \mathsf{t}_\alpha}{\partial \xi^\alpha} + \frac{\partial \hat{\tau}}{\partial s}.$$

The local balance of forces $\mathrm{Div}P + b = \rho \ddot{y}$ then implies

$$\frac{\partial \hat{\tau}}{\partial s} = -\sum_{\alpha=1}^{2} \frac{\partial \mathsf{t}_\alpha}{\partial \xi^\alpha} - b + \rho \ddot{y}.$$

From the definition of m, we have also

$$\frac{\partial \mathsf{m}}{\partial s} = \frac{\partial}{\partial s} \int_{\mathcal{X}} \sum_{\alpha=1}^{2} \xi^\alpha \mathsf{d}_\alpha \times \hat{\tau}\, d\xi^1 d\xi^2 = \frac{\partial}{\partial s} \int_{\mathcal{X}} (y - \varphi) \times \hat{\tau}\, d\xi^1 d\xi^2$$

$$= \int_{\mathcal{X}} \frac{\partial y}{\partial s} \times \hat{\tau}\, d\xi^1 d\xi^2 - \frac{\partial \varphi}{\partial s} \times \int_{\mathcal{X}} \hat{\tau}\, d\xi^1 d\xi^2 + \int_{\mathcal{X}} (y - \varphi) \times \frac{\partial \hat{\tau}}{\partial s} d\xi^1 d\xi^2$$

$$= \int_{\mathcal{X}} \frac{\partial y}{\partial s} \times \hat{\tau}\, d\xi^1 d\xi^2 - \frac{\partial \varphi}{\partial s} \times \mathsf{n} - \int_{\mathcal{X}} (y - \varphi) \times \sum_{\alpha=1}^{2} \frac{\partial \mathsf{t}_\alpha}{\partial \xi^\alpha} d\xi^1 d\xi^2 - \int_{\mathcal{X}} (y - \varphi) \times b\, d\xi^1 d\xi^2$$

$$+ \int_{\mathcal{X}} (y - \varphi) \times \rho \ddot{y}\, d\xi^1 d\xi^2.$$

However, we have also

$$- \int_{\mathcal{X}} (y - \varphi) \times \sum_{\alpha=1}^{2} \frac{\partial \mathsf{t}_\alpha}{\partial \xi^\alpha} d\xi^1 d\xi^2 = \int_{\mathcal{X}} \sum_{\alpha=1}^{2} \frac{\partial}{\partial \xi^\alpha} ((y - \varphi) \times \mathsf{t}_\alpha)\, d\xi^1 d\xi^2$$

$$+ \int_{\mathcal{X}} \sum_{\alpha=1}^{2} \frac{\partial y}{\partial \xi^\alpha} \times \mathsf{t}_\alpha\, d\xi^1 d\xi^2 = - \int_{\partial\mathcal{X}} \sum_{\alpha=1}^{2} (y - \varphi) \times \mathsf{t}_\alpha \nu_\alpha\, dl + \int_{\mathcal{X}} \sum_{\alpha=1}^{2} \frac{\partial y}{\partial \xi^\alpha} \times \mathsf{t}_\alpha d\xi^1 d\xi^2,$$

where ν_α is the αth component of the normal to $\partial\mathcal{X}$ in the plane containing it (recall that t_α is a vector; ν_α is a scalar).

Consequently, we obtain

$$\frac{\partial \mathsf{m}}{\partial s} = \left(\int_{\partial\mathcal{X}} \sum_{\alpha=1}^{2} \frac{\partial y}{\partial \xi^\alpha} \times \mathsf{t}_\alpha + \frac{\partial y}{\partial s} \times \hat{\tau} \right) d\xi^1 d\xi^2 - \bar{\mathsf{m}} - \frac{\partial \varphi}{\partial s} \times \mathsf{n} + \int_{\mathcal{X}} (y - \varphi) \times \rho \ddot{y}\, d\xi^1 d\xi^2$$

$$= \left(\int_{\mathcal{X}} \sum_{\alpha=1}^{2} \frac{\partial y}{\partial \xi^\alpha} \times \mathsf{t}_\alpha + \frac{\partial y}{\partial s} \times \hat{\tau} \right) d\xi^1 d\xi^2 - \bar{\mathsf{m}} - \frac{\partial \varphi}{\partial s} \times \mathsf{n} + \dot{H},$$

where

Using equation (8.28), we then compute

$$
\begin{aligned}
w \cdot \sum_{\alpha=1}^{2} \mathsf{d}_\alpha \times \mathsf{t}_\alpha &= w \cdot \sum_{\alpha=1}^{2} \frac{\partial (y-\varphi)}{\partial \xi^\alpha} \times \mathsf{t}_\alpha \\
&= w \cdot \sum_{\alpha=1}^{2} \frac{\partial y}{\partial \xi^\alpha} \times \mathsf{t}_\alpha = -w \cdot \frac{\partial y}{\partial s} \times \hat{\tau} \\
&= -\hat{\tau} \cdot \left(w \times \frac{\partial \varphi}{\partial s} + w \times \sum_{\alpha=1}^{2} \xi^\alpha \mathsf{d}_{\alpha,s} \right) \\
&= -\hat{\tau} \cdot \left(w \times \frac{\partial \varphi}{\partial s} + w \times \omega \times \sum_{\alpha=1}^{2} \xi^\alpha \mathsf{d}_\alpha \right) \\
&= -\hat{\tau} \cdot \left(w \times \frac{\partial \varphi}{\partial s} + w \times (\omega \times (y-\varphi)) \right).
\end{aligned}
\tag{8.30}
$$

Write p for $(y-\varphi)$. Using the Lagrange formula for the triple vector product, we obtain

$$
\omega \times (w \times p) - w \times (\omega \times p) = (w \otimes \omega - \omega \otimes w)p = (\omega \times w) \times p. \tag{8.31}
$$

On inserting the result (8.30) in the expression (8.27) and using the identity (8.31), after integration we obtain

$$
\begin{aligned}
&\int_{\mathcal{X}} \int_{s_1}^{s_2} P \cdot \dot{F} ds\, d\xi^1 d\xi^2 \\
&= \int_{s_1}^{s_2} \left(\mathsf{n} \cdot \left(\frac{\partial \dot{\varphi}}{\partial s} - w \times \frac{\partial \varphi}{\partial s} \right) + \mathsf{m} \cdot (\dot{\omega} - w \times \omega) \right) ds \\
&= \int_{s_1}^{s_2} \left(\mathsf{n} \cdot \overline{\frac{\partial \varphi}{\partial s}}^{\nabla} + \mathsf{m} \cdot \overset{\nabla}{\omega} \right) ds.
\end{aligned}
\tag{8.32}
$$

$$
\bar{\mathsf{m}} = \sum_{\alpha=1}^{2} \int_{\partial \mathcal{X}} (y-\varphi) \times \mathsf{t}_\alpha \nu_\alpha ds + \int_{\mathcal{X}} (y-\varphi) \times b\, d\xi^1 d\xi^2 .
$$

The result (8.27) implies then the local balance of couples

$$
\frac{\partial \mathsf{m}}{\partial s} + \frac{\partial \varphi}{\partial s} \times \mathsf{n} + \bar{\mathsf{m}} = \dot{H}.
$$

On the other hand, the identity $w_{,s} \times \mathsf{d}_\alpha = (\overset{\nabla}{\omega} - w \times \omega) \times \mathsf{d}_\alpha)$, derived in Section 8.3, allows us to write

$$\frac{\partial w}{\partial s} \cdot \mathsf{m} = \frac{\partial w}{\partial s} \cdot \int_{\mathcal{X}} \sum_{\alpha=1}^{2} \xi^\alpha \mathsf{d}_\alpha \times \hat{\tau}\, d\xi^1 d\xi^2$$

$$= \int_{\mathcal{X}} \hat{\tau} \cdot \left(\frac{\partial w}{\partial s} \times \sum_{\alpha=1}^{2} \xi^\alpha \mathsf{d}_\alpha \right) d\xi^1 d\xi^2$$

$$= \int_{\mathcal{X}} \hat{\tau} \cdot \left(\overset{\nabla}{\omega} - w \times \omega \times \sum_{\alpha=1}^{2} \xi^\alpha \mathsf{d}_\alpha \right) d\xi^1 d\xi^2$$

$$= \int_{\mathcal{X}} \left(\sum_{\alpha=1}^{2} \xi^\alpha \mathsf{d}_\alpha \times \hat{\tau} \right) \cdot \overset{\nabla}{\omega}\, d\xi^1 d\xi^2 - \int_{\mathcal{X}} \left(\sum_{\alpha=1}^{2} \xi^\alpha \mathsf{d}_\alpha \times \hat{\tau} \right) \cdot (w \times \omega)\, d\xi^1 d\xi^2$$

$$= \mathsf{m} \cdot \overset{\nabla}{\omega} - \mathsf{m} \cdot (w \times \omega).$$

When we introduce this result in the right-hand side of equation (8.26) and compare the result with equation (8.32), we find

$$\mathsf{m} \cdot (w \times \omega) = 0.$$

Then for the **inner power** *along the rod*, we write

$$\mathcal{P}^{inn}_{(s_1, s_2)} = \int_{s_1}^{s_2} (\mathsf{n} \cdot \overset{\nabla}{\zeta} + \mathsf{m} \cdot \overset{\nabla}{\omega})\, ds. \tag{8.33}$$

Then the equality to the external power reads explicitly

$$\begin{aligned}
&\mathsf{n}(s_2, t) \cdot \dot{\varphi}(s_2, t) - \mathsf{n}(s_1, t) \cdot \dot{\varphi}(s_1, t) \\
&+ \mathsf{m}(s_2, t) \cdot w(s_2, t) - \mathsf{m}(s_1, t) \cdot w(s_1, t) \\
&+ \int_{s_1}^{s_2} \bar{\mathsf{n}}^{\ddagger}(s, t) \cdot \dot{\varphi}(s, t)\, ds + \int_{s_1}^{s_2} \bar{\mathsf{m}}^{\ddagger}(s, t) \cdot w(s, t)\, ds \\
&+ \sum_{k=1}^{\bar{k}_s} \hat{\mathsf{f}}(s_k, t) \cdot \dot{\varphi}(s_k, t) + \sum_{r=1}^{\bar{\bar{k}}_s} \hat{\mathsf{m}}(s_r, t) \cdot w(s_r, t) \\
&= \int_{s_1}^{s_2} \mathsf{n}(s, t) \cdot \overset{\nabla}{\zeta}(s, t)\, ds + \int_{s_1}^{s_2} \mathsf{m}(s, t) \cdot \omega(s, t)\, ds.
\end{aligned} \tag{8.34}$$

Such an identity is the *weak form of the balance equations for the rod* in the scheme adopted in this chapter.

8.17 Constitutive Restrictions

The expression (8.32) of the internal power suggests that we adopt a mechanical dissipation inequality written in terms of the corotational time derivative. Let $\psi(\xi^1, \xi^2, s, t)$ be the free energy at time t of a material element placed at (ξ^1, ξ^2, s) in the three-dimensional (original) rod. We define the **sectional free energy** $\psi_r(s)$ as the average of ψ over the cross section at s:

$$\psi_r(s) := \int_{\mathcal{X}} \psi(\xi^1, \xi^2, s, t)\, d\xi^1 d\xi^2.$$

We then write the mechanical dissipation inequality for a generic part of the rod between s_1 and s_2 as

$$\overline{\int_{s_1}^{s_2} \psi_r(s)\, ds}^{\nabla} - \mathcal{P}^{ext}_{(s_1,s_2)}(\dot{\varphi}, w) \leq 0$$

for every choice of s_1 and s_2, with $s_1 \leq s_2$, and the velocity fields. Then equation (8.33) allows us to write

$$\overline{\int_{s_1}^{s_2} \psi_r(s)\, ds}^{\nabla} - \int_{s_1}^{s_2} (\mathsf{n}(s,t) \cdot \overset{\nabla}{\zeta}(s,t) + \mathsf{m}(s,t) \cdot \overset{\nabla}{\omega}(s,t))\, ds \leq 0. \tag{8.35}$$

Here ζ and ω are strain measures, collecting what we have implied in the kinematic choice above: elongation and shear (both in ζ), bending and twist (both in ω).

We consider here only the case of **elastic rods**, those for which we assume the following constitutive structures:

$$\psi_r(s,t) = \tilde{\psi}_r(s, \zeta(s,t), \omega(s,t)),$$
$$\mathsf{n}(s,t) = \tilde{\mathsf{n}}(s, \zeta(s,t), \omega(s,t)),$$
$$\mathsf{m}(s,t) = \tilde{\mathsf{m}}(s, \zeta(s,t), \omega(s,t)).$$

By inserting them in (8.35) and computing the corotational time derivative of ψ_r, we get

$$\int_{s_1}^{s_2} \left(\left(\frac{\partial \psi_r}{\partial \zeta} - \mathsf{n} \right) \cdot \overset{\nabla}{\zeta} + \left(\frac{\partial \psi_r}{\partial \omega} - \mathsf{m} \right) \cdot \overset{\nabla}{\omega} \right) ds \leq 0.$$

The assumed arbitrariness of $\overset{\nabla}{\zeta}$ and $\overset{\nabla}{\omega}$, once ζ and ω are fixed, implies

$$\mathsf{n} = \frac{\partial \psi_r(s, \zeta, \omega)}{\partial \zeta}, \qquad \mathsf{m} = \frac{\partial \psi_r(s, \zeta, \omega)}{\partial \omega}.$$

With respect to the frame d_1, d_2, d_3 at s—a moving frame, in fact, the reason motivating us to write the mechanical dissipation in terms of corotational derivatives—the vectors ζ and ω are expressed by

$$\zeta = \hat{\gamma}_1 \mathsf{d}_1 + \hat{\gamma}_2 \mathsf{d}_2 + \hat{\varepsilon} \mathsf{d}_3,$$

where $\hat{\gamma}_1$ and $\hat{\gamma}_2$ are the shears along d_1 and d_2, and $\hat{\varepsilon}$ is the elongation (stretch) in the d_3 direction, and

$$\omega = \hat{\kappa}_1 \mathsf{d}_1 + \hat{\kappa}_2 \mathsf{d}_2 + \hat{\kappa}_t \mathsf{d}_3,$$

with $\hat{\kappa}_1$, $\hat{\kappa}_2$, and $\hat{\kappa}_t$ the curvatures along the relevant directions. In components related to the frame d_1, d_2, d_3, we then have

$$T_1 = \frac{\partial \psi_r}{\partial \hat{\gamma}_1}, \qquad T_2 = \frac{\partial \psi_r}{\partial \hat{\gamma}_2}, \qquad N = \frac{\partial \psi_r}{\partial \hat{\varepsilon}},$$

$$M_1 = \frac{\partial \psi_r}{\partial \hat{\kappa}_1}, \qquad M_2 = \frac{\partial \psi_r}{\partial \hat{\kappa}_2}, \qquad M_t = \frac{\partial \psi_r}{\partial \hat{\kappa}_t}.$$

8.18 Rotated Stress Vectors

For computational purposes, it can be useful to project the interaction vectors n and m into the local frame in the reference shape, since it is known and fixed once and for all, unless material mutations occur in the rod, changing its structure, which is, however, a topic not treated here.

At every s, the orthogonal frame $\{e_\Gamma\}$, $\Gamma = 1, 2, 3$, is related to $\{\mathsf{d}_\Gamma\}$ by a rotation $Q \in SO(3)$. We then define

$$\mathsf{n}_R := Q^{\mathrm{T}} \mathsf{n}, \qquad \mathsf{m}_R := Q^{\mathrm{T}} \mathsf{m},$$

$$\bar{\mathsf{n}}_R := Q^{\mathrm{T}} \bar{\mathsf{n}}, \qquad \bar{\mathsf{m}}_R := Q^{\mathrm{T}} \bar{\mathsf{m}},$$

$$\hat{\mathsf{n}}_R := Q^{\mathrm{T}} \hat{\mathsf{n}}, \qquad \hat{\mathsf{m}}_R := Q^{\mathrm{T}} \hat{\mathsf{m}},$$

and compute

$$\frac{\partial \mathsf{n}_R}{\partial s} = Q^{\mathrm{T}} \frac{\partial \mathsf{n}}{\partial s} - \omega \times \mathsf{n}_R, \tag{8.36}$$

$$\frac{\partial \mathsf{m}_R}{\partial s} = Q^{\mathrm{T}} \frac{\partial \mathsf{m}}{\partial s} - \omega \times \mathsf{m}_R. \tag{8.37}$$

Moreover, for the velocities $\dot{\varphi}(s,t)$ and $w(s,t)$, we define their rotated counterparts as

$$\dot{\varphi}_R := Q^{\mathrm{T}} \dot{\varphi}, \qquad w_R := Q^{\mathrm{T}} w,$$

and we obtain

$$Q^{\mathrm{T}}\frac{\partial \dot{\varphi}}{\partial t} = \frac{\partial \dot{\varphi}_R}{\partial t} + w \times \dot{\varphi}_R \tag{8.38}$$

and

$$Q^{\mathrm{T}}\mathrm{I}_{\mathcal{X}}\dot{w} = \mathrm{I}_{\mathcal{X}}\dot{w}_R, \qquad Q^{\mathrm{T}}(w \times \mathrm{I}_{\mathcal{X}}w) = w_R \times \mathrm{I}_{\mathcal{X}}w_R, \tag{8.39}$$

where we have used the identity $Q^{\mathrm{T}}\mathrm{I}_{\mathcal{X}}Q = \mathrm{I}_{\mathcal{X}}$, since $\mathrm{I}_{\mathcal{X}}$ collects the moments of inertia evaluated in a frame rotating with the section, so that the values in the frame $\{\mathsf{d}_\Gamma\}$ are the same as those relative to the frame $\{e_\Gamma\}$. Moreover, we have

$$Q^{\mathrm{T}}\left(\frac{\partial \varphi}{\partial s} \times \mathsf{n}\right) = \left(\frac{\partial \varphi}{\partial s}\right)_R \times \mathsf{n}_R, \tag{8.40}$$

where

$$\left(\frac{\partial \varphi}{\partial s}\right)_R := Q^{\mathrm{T}}\left(\frac{\partial \varphi}{\partial s}\right)$$

and

$$Q^{\mathrm{T}}(w \times H) = w_R \times H_R \tag{8.41}$$

with $H_R := \mathrm{I}_{\mathcal{X}} \times w_R$.

Exercise 18. *Prove the relations (8.36) through (8.41).*

On multiplying the balance equations by Q^{T} from the left, we get

$$\rho_{\mathcal{X}}\left(\frac{\partial \dot{\varphi}_R}{\partial t} + w \times \dot{\varphi}_R\right) = \frac{\partial \mathsf{n}_R}{\partial s} + \omega \times \mathsf{n}_R + \bar{\mathsf{n}}_R,$$

$$\mathrm{I}_{\mathcal{X}}\dot{w}_R - w_R \times \mathrm{I}_{\mathcal{X}}w_R = \frac{\partial \mathsf{m}_R}{\partial s} + \omega \times \mathsf{m}_R + \left(\frac{\partial \varphi}{\partial s}\right)_R \times \mathsf{n}_R + \bar{\mathsf{m}}_R,$$

$$[\mathsf{n}_R] + \hat{\mathsf{n}}_R = 0,$$

$$[\mathsf{m}_R] + \hat{\mathsf{m}}_R = 0.$$

Moreover, as regards ζ and ω, we define

$$\zeta_R := Q^{\mathrm{T}}\zeta, \qquad \omega_R := Q^{\mathrm{T}}\omega.$$

We have then

$$Q\dot{\zeta}_R = Q\dot{Q}^{\mathrm{T}}\zeta + \dot{\zeta} = \dot{\zeta} - Q^{\mathrm{T}}\dot{Q}\zeta = \dot{\zeta} - w \times \zeta = \overset{\nabla}{\zeta},$$

and analogously,

$$Q\dot{\omega}_R = \overset{\nabla}{\omega} .$$

Consequently, we compute

$$\mathsf{n}\cdot \overset{\nabla}{\zeta} + \mathsf{m}\cdot \overset{\nabla}{\omega} = QQ^{\mathrm{T}}\mathsf{n}\cdot \overset{\nabla}{\zeta} + QQ^{\mathrm{T}}\mathsf{m}\cdot \overset{\nabla}{\omega}$$

$$= Q^{\mathrm{T}}\mathsf{n}\cdot Q^{\mathrm{T}} \overset{\nabla}{\zeta} + Q^{\mathrm{T}}\mathsf{m}\cdot Q^{\mathrm{T}} \overset{\nabla}{\omega} = \mathsf{n}_R\cdot\dot{\zeta}_R + \mathsf{m}_R\cdot\dot{\omega}_R,$$

and we have

$$\int_{s_2}^{s_2}\left(\mathsf{n}\cdot \overset{\nabla}{\zeta} + \mathsf{m}\cdot \overset{\nabla}{\omega}\right) ds = \int_{s_2}^{s_2}\left(\mathsf{n}_R\cdot\dot{\zeta}_R + \mathsf{m}_R\cdot\dot{\omega}_R\right) ds.$$

The last integral is the inner power expressed in terms of the rotated stress vectors, an expression that suggests that we write the mechanical dissipation inequality in a form reminiscent of what we have done so far for Cauchy's bodies:

$$\frac{d}{dt}\int_{s_2}^{s_2}\psi_R(s,t)\,ds - \int_{s_2}^{s_2}\left(\mathsf{n}_R\cdot\dot{\zeta}_R + \mathsf{m}_R\cdot\dot{\omega}_R\right) ds \le 0 \tag{8.42}$$

for any choice of the rates involved. In this case, by reducing the analysis to the elastic behavior, we can assume that

$$\psi_R(s,t) = \tilde{\psi}_R(s,\zeta_R,\omega_R),$$
$$\mathsf{n}_R(s,t) = \tilde{\mathsf{n}}_R(s,\zeta_R,\omega_R),$$
$$\mathsf{m}_R(s,t) = \tilde{\mathsf{m}}_R(s,\zeta_R,\omega_R),$$

so that the arbitrariness of $\dot{\zeta}_R$ and $\dot{\omega}_R$ implies

$$\mathsf{n}_R = \frac{\partial\tilde{\psi}_R(s,\zeta_R,\omega_R)}{\partial\zeta_R}, \qquad \mathsf{m}_R = \frac{\partial\tilde{\psi}_R(s,\zeta_R,\omega_R)}{\partial\omega_R}.$$

If ψ_R admits second derivatives, we obtain

$$\left\{\begin{matrix}\dot{\mathsf{n}}_R\\ \dot{\mathsf{m}}_R\end{matrix}\right\} = \boldsymbol{C}(s,\zeta_R,\omega_R)\left\{\begin{matrix}\dot{\zeta}_R\\ \dot{\omega}_R\end{matrix}\right\},$$

with

$$\boldsymbol{C}(s,\zeta_R,\omega_R) = \begin{pmatrix} \dfrac{\partial^2\psi_R}{\zeta_R\zeta_R} & \dfrac{\partial^2\psi_R}{\zeta_R\omega_R} \\[2ex] \dfrac{\partial^2\psi_R}{\omega_R\zeta_R} & \dfrac{\partial^2\psi_R}{\omega_R\omega_R} \end{pmatrix}.$$

The tensor C links the time rate of the strain vectors to the rate of their stress counterparts. For this reason, we call $C(s, \zeta_R, \omega_R)$ the **tangential stiffness**. It defines the tangent hyperplane to the stress–strain surface. Regarding the reference place, $C(s, \zeta_R, \omega_R)$ is the stiffness matrix in the infinitesimal strain regime. This circumstance suggests to us in the frame $\{e_\Gamma\}$ a specific form of $C(s, \zeta_R, \omega_R)$ on the basis of the results emerging in the analysis of the de Saint-Venant problem, namely

$$C = \mathrm{diag}(GA_{t1}, GA_{t2}, EA, E\mathrm{I}_1, E\mathrm{I}_2, G\mathsf{K}_t),$$

where diag(., .., .) denotes the diagonal matrix with entries in the list; A_{t1} and A_{t2} are the shear areas related to e_1 and e_2 respectively; A is the cross-sectional area; I_1 and I_2 are the moments of inertia associated with e_1 and e_2 respectively; K_t is the torsional moment of inertia.

By the action of Q, for every differentiable vector field $(s, t) \longmapsto h(s, t)$, we have

$$Q\frac{\partial}{\partial t}Q^{\mathrm{T}}h = Q\dot{Q}^{\mathrm{T}}h + \dot{h} = \dot{h} - w \times h = \overset{\nabla}{h}.$$

Then we compute

$$\begin{Bmatrix} \overset{\nabla}{\mathsf{n}} \\ \overset{\nabla}{\mathsf{m}} \end{Bmatrix} = QCQ^{\mathrm{T}} \begin{Bmatrix} \overset{\nabla}{\zeta} \\ \overset{\nabla}{\omega} \end{Bmatrix}.$$

8.19 A Special Case: In-Plane Deformations in 3-Dimensional Space

Let us consider finite rotations about e_1. We take tensors $Q \in SO(3)$ with matrix form

$$Q = \begin{pmatrix} 1 & 0 & 0 \\ 0 & \cos\hat{\vartheta} & \sin\hat{\vartheta} \\ 0 & -\sin\hat{\vartheta} & \cos\hat{\vartheta} \end{pmatrix}.$$

We write $\mathsf{w}(s, t)$ and $\mathsf{v}(s, t)$ for what we call *axial* and *transversal* displacements at s and t. They are the values of two functions such that

$$\frac{\partial\varphi}{\partial s} = \mathsf{v}' e_2 + (1 + \mathsf{w}')e_3,$$

where the apex denotes differentiation with respect to s. By computing the rotated strain measure ζ_R, we then find

$$\zeta_{R3} = \mathsf{v}' \sin\hat{\vartheta} + (1 + \mathsf{w}') \cos\hat{\vartheta} - 1,$$

$$\zeta_{R2} = \mathsf{v}' \cos\hat{\vartheta} - (1 + \mathsf{w}') \sin\hat{\vartheta}.$$

When the rotation is small, i.e.,

$$\cos\hat{\vartheta} \approx 1 \quad \text{and} \quad \sin\hat{\vartheta} \approx \hat{\vartheta},$$

the previous expressions reduce to

$$\zeta_{R3} \approx \mathsf{w}' + \mathsf{v}'\hat{\vartheta},$$

$$\zeta_{R2} \approx \mathsf{v}' - (1 + \mathsf{w}')\hat{\vartheta}.$$

In the small-strain regime, when the rotations are small $\mathsf{v}'\hat{\vartheta}$ and $\mathsf{w}'\hat{\vartheta}$ are of higher order with respect to v' and w', so that we may write

$$\zeta_{R3} \approx \mathsf{w}', \qquad \zeta_{R2} \approx \mathsf{v}' - \hat{\vartheta}. \tag{8.43}$$

If in the second estimate we consider opposite rotations, i.e., $\vartheta = -\hat{\vartheta}$, with ϑ counterclockwise when $\hat{\vartheta}$ is clockwise, we write

$$\zeta_{R2} \approx \mathsf{v}' + \vartheta.$$

These expressions can be obtained by direct computation or by linearizing the scheme developed so far in the previous sections. In the next section, we adopt the first option. Our choice is motivated essentially by the desire for simplicity in presenting the material. We thereby pave the way for simple applications presented later.

8.20 Timoshenko's Rod

We consider a straight rod undergoing axial, shear, and bending planar deformations in the small-strain regime. Smallness is intended in the sense above. The displacement of a point x in a generic cross section in the plane spanned by e_2 and e_3 is then given by (see Fig. 8.43, where $\vartheta > 0$ if the rotation is counterclockwise)

$$u_2 = \mathsf{v}(s),$$

$$u_3 = \mathsf{w}(s) + \vartheta(s)\xi,$$

and the relevant components of the two-dimensional reduction of the small-strain tensor are

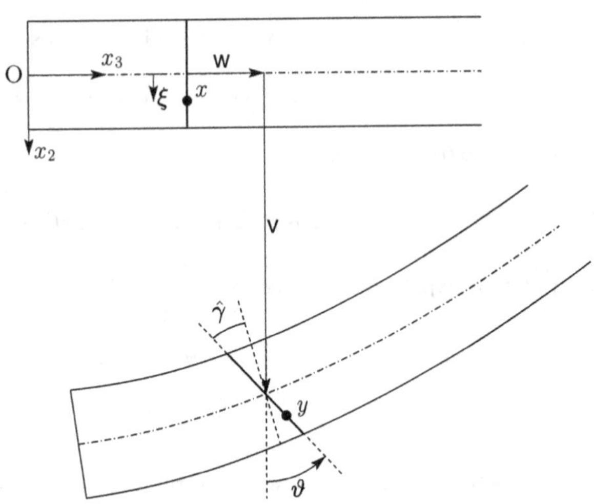

Fig. 8.43 Two-dimensional rod: kinematics without warping

$$
\begin{aligned}
\varepsilon_{22} &= \frac{\partial u_2}{\partial \xi} = 0, \\
\varepsilon_{23} &= \frac{1}{2}\left(\frac{\partial u_2}{\partial s} + \frac{\partial u_3}{\partial \xi}\right) = \frac{1}{2}(\mathsf{v}' + \vartheta), \\
\varepsilon_{33} &= \frac{\partial u_3}{\partial s} = \mathsf{w}' + \vartheta'\xi.
\end{aligned}
$$

The assumed absence of warping implies that we should consider the cross section (recall that it has no thickness) to be a rigid body, exactly as we have done so far.

By comparison with the relation (8.43), as strain measures we can distinguish the **elongation** (stretch)

$$\hat{\varepsilon} = \mathsf{w}', \tag{8.44}$$

the **shear**

$$\hat{\gamma} = \mathsf{v}' + \vartheta, \tag{8.45}$$

and the rod **curvature**

$$\hat{\kappa} = \vartheta'. \tag{8.46}$$

The relations (8.44) through (8.46) are the *compatibility conditions* linking the strain measures $\hat{\varepsilon}$, $\hat{\gamma}$, and $\hat{\kappa}$ to the degrees of freedom in the plane, those represented by the horizontal and vertical displacements w and v, and the rotation ϑ.

By writing T for the shear stress T_2 introduced in Chapter 6 and analogously M for M_1, $\bar{q}$ for $\bar{q}_2$, and $\bar{m}$ for $\bar{m}_1$, in the absence of inertial effects, the balance equations read in this case

$$N' + \bar{p} = 0, \tag{8.47}$$

$$T' + \bar{q} = 0, \tag{8.48}$$

$$M' - T + \bar{m} = 0. \tag{8.49}$$

The first equation is independent of the other two, which are, in contrast, coupled.

The remaining ingredient is the set of constitutive equations. To assign them, we can take advantage of de Saint-Venant's theory, so that we write

$$\hat{\varepsilon} = \frac{N}{EA}, \tag{8.50}$$

$$\hat{\gamma} = \frac{T}{GA^*}, \tag{8.51}$$

$$\hat{\kappa} = \frac{M}{EI}, \tag{8.52}$$

where E and G are the Young's and shear moduli respectively, A is the area of the generic cross section, A^* is the shear area ($A^* = A_{t2}$), I the cross-sectional inertial moment with respect to the axis orienting the plane ($\mathrm{I} = \mathrm{I}_1$). By inserting these constitutive relations and the compatibility conditions into the balance equations (8.47) through (8.49), we eventually obtain

$$(EA\mathrm{w}')' + \bar{p} = 0, \tag{8.53}$$

$$(GA^*(\mathrm{v}' + \vartheta))' + \bar{q} = 0, \tag{8.54}$$

$$(E\mathrm{I}\vartheta')' - GA^*(\mathrm{v}' + \vartheta) + \bar{m} = 0, \tag{8.55}$$

which constitute a scheme referred to as **Timoshenko's rod**, for Stepan Prokofevič Timošenko (1878–1972). The extension of the scheme to the case in which isolated external forces and/or couples are present is straightforward, along the guidelines already presented in this chapter.

Exercise 19. *Develop a version of Timoshenko's rod model in the presence of isolated applied external forces and couples.*

8.21 Weak Form of the Balance Equations

Consider the balance equations (8.47) through (8.49) for a given rod and multiply them by w_*, v_*, and ϑ_*, respectively. The asterisk in the index position indicates that the relevant fields are not necessarily solutions of the equations (8.53) through (8.55). They just satisfy the compatibility conditions (8.44) through (8.46) and those imposed by the constraints. In this sense, they are *virtual*. On integrating over the whole rod and using the compatibility conditions, we eventually write

$$\begin{aligned}&\int_0^l (\bar{p}\mathsf{w}_* + \bar{q}\mathsf{v}_* + \bar{m}\vartheta_*)\,ds + N(l)\mathsf{w}_*(l) - N(0)\mathsf{w}_*(0)\\ &+ T(l)\mathsf{v}_*(l) - T(0)\mathsf{v}_*(0) + M(l)\vartheta_*(l) - M(0)\vartheta_*(0)\\ &= \int_0^l (N\hat{\varepsilon}_* + T\hat{\gamma}_* + M\hat{\kappa}_*)\,ds,\end{aligned} \tag{8.56}$$

where $\hat{\varepsilon}_*$, $\hat{\gamma}_*$, and $\hat{\kappa}_*$ are the strain measures associated with w_*, v_*, and ϑ_*, respectively.

Conversely, consider *virtual* contact actions N_*, T_*, and M_*, so called because we imagine that they satisfy the balance equations (8.47) through (8.49) but *not necessarily* the constitutive relations (8.50) through (8.52) with the *real* elongation, shear, and curvature determined by the applied forces and boundary conditions. If we multiply the compatibility conditions by N_*, T_*, and M_*, integrate over the rod, and use the balance equations, we derive an equation differing from (8.56) *only* in the presence of the asterisk as a subscript for $\bar{p}$, $\bar{q}$, $\bar{m}$, N, T, and M rather than w, v, ϑ, $\hat{\varepsilon}$, $\hat{\gamma}$, and $\hat{\kappa}$.

Exercise 20. *Extend the relation (8.56) to the presence of applied forces and couples at specific points or imposed displacements or rotations at other isolated points. Then compare the result with the specific expressions used in the exercises presented later.*

Moreover, when we have a system of rods joined variously to each other, the relation (8.56) has to be extended to the entire system as a sum of the contributions of the various rods, including both distributed and concentrated forces and couples, applied displacements, and local structural failures at the constraints joining different rods or along the rod themselves. In general, we write

$$L^{ext} = L^{inn}, \tag{8.57}$$

where L^{ext} is the **work of all external actions** over a system and

$$L^{inn} = \sum_{k=1}^{K} \int_0^{l_k} (N_k\hat{\varepsilon}_k + T_k\hat{\gamma}_k + M_k\hat{\kappa}_k)\,ds$$

is the overall **inner** (or **internal**) **work**, where K is the total number of rods constituting the system. The expression can be extended to include localized structural "failures," as we shall see later. Depending on the viewpoint adopted, we can consider virtual the actions (external forces and couples and the triple N, T, M) or the kinematics (displacements, rotations, strain measures). We have three ways of interpreting the equation (8.57):

1. The work performed by every balanced system of actions over a framed structure in every displacement system compatible with the strain equals the internal work developed in the strain.
2. Given displacements, rotations, and strains, if the equation (8.57) holds for *every* system of equilibrated actions, then the strains are compatible with the displacements and rotations.
3. Given a system of actions, if equation (8.57) holds for *every* system of compatible displacements, rotations, and relevant strains, then the system of actions is balanced.

This last interpretation of equation (8.57) allows us to construct a procedure—called the **force method**—to analyze the statics of hyperstatic framed structures. However, before going into the details of this method—not the only possible one, but the one presented here—some remarks on the constitutive relations will useful for the subsequent developments.

8.22 Remarks on the Constitutive Equations

The relations (8.50) through (8.52) hold in the linear-elastic setting. We can extend our viewpoint while still focusing our attention on the small-strain regime and considering thermal variations and inelastic strain. Although we have not developed here a general treatment of thermoelasticity and (at least some aspects of) inelastic behavior, we mention here the expressions for some relevant constitutive relations valid in the *small-strain regime*, restricted to the two-dimensional setting. These relations will appear in some of the exercises presented later. In the presence of thermal variations, we have

$$\hat{\varepsilon} = \frac{N}{EA} + \hat{\varepsilon}_{\theta} \tag{8.58}$$

with

$$\hat{\varepsilon}_{\theta} = \alpha \delta\theta, \tag{8.59}$$

the **temperature-induced elongation** given by the **temperature variation** $\delta\theta$ multiplied by the **thermal dilatation coefficient** α,

$$\hat{\kappa} = \frac{M}{EI} + \hat{\kappa}_{\theta}, \tag{8.60}$$

with

$$\hat{\kappa}_\theta = 2\alpha\frac{\delta\theta}{h} \tag{8.61}$$

the **temperature-induced curvature**, an expression holding under the hypothesis that the (two-dimensional) rod of thickness h under investigation undergoes positive and negative temperature variations of the same magnitude at the extreme sides of the section (those sides at distance h) and from section to section.

We do not consider here thermal effects on the shear strain, which, however, can have an inelastic component $\hat{\gamma}_{in}$, a case in which we shall write

$$\hat{\gamma} = \frac{T}{GA^*} + \hat{\gamma}_{in}.$$

This type of inelastic effect can appear also for the elongation and the curvature. In including thermoelastic effects, we shall write

$$\hat{\varepsilon} = \frac{N}{EA} + \alpha\delta\theta + \hat{\varepsilon}_{in}$$

and

$$\hat{\kappa} = \frac{M}{EI} + 2\alpha\frac{\delta\theta}{h} + \hat{\kappa}_{in}.$$

The additive decomposition of the strain measures into elastic and inelastic components appears acceptable in the small-strain regime. In the presence of finite strains, we should account for multiplicative decomposition, considering appropriate additive factors only for the strain rates. However, this is an issue not tackled here.

8.23 The Force Method

Consider a planar structure consisting of K rods, the kth of them l_k long, endowed with interrod joints and constrained with respect to the external environment. If all rods are rigid, the structure we consider will have a degree of hyperstaticity $\hat{h} > 0$. In addition, the structure could also be labile ($\hat{l} > 0$) and subjected to a load condition ensuring equilibrium. To find the $\hat{h} > 0$ superabundant constraint reactions, we assume that the rods in the structure are linear elastic.

The *force method* develops along the following steps.

1. Select the so-called **principal structure**, removing $\hat{h}$ simple constraints (internal and/or external), denoting by X_i the *reaction of the* ith *removed constraint*. When such a constraint is external, X_i is the corresponding external reaction (force or

couple); when it is internal, X_i is an internal reaction or an action characteristic. The resulting reduced structure must have the same degree of lability of the original structure. In particular, if for the real structure, $\hat{l} = 0$, the principal structure must be isostatic. There are infinitely many possible choices of the principal structure. Experience leads us toward the most convenient choice with the consciousness that the evaluation of the equilibrium state is independent of the choice of the principal structure.

2. Determine the distribution of N, T, M on the principal structure with the original loads, obtaining the so-called **0-system**. Write for the resulting action characteristics N_0, T_0, M_0 just to remind ourselves of their pertinence to the principal system. The calculation does not require recourse to the constitutive structures.
3. Consider the principal structure $\hat{h}$ times *without* the external loads, applying in the ith step the X_i hyperstatic unknown alone, assumed to have unitary value, where the ith simple constraint has been eliminated. In this way, we construct the so-called **1-system, 2-system,** ..., **$\hat{h}$-system**. Compute then the pertinent distributions of N, T, M for all these systems and write for them N_i, T_i, M_i, $i = 1, 2, \ldots, \hat{h}$.
4. Since we are working here in the linear setting, we have superposition of effects. Hence, the real N, T, M distributions are given by

$$N = N_0 + \sum_{i=1}^{\hat{h}} X_i N_i, \qquad T = T_0 + \sum_{i=1}^{\hat{h}} X_i T_i, \qquad M = M_0 + \sum_{i=1}^{\hat{h}} X_i M_i.$$

5. For every step in item 3, write the external power performed by the forces and/or couples of the ith system on the real displacements of the points where they are applied and equalize it to the inner power involving N_i, T_i, M_i and the real compatible strains.
6. Eventually, we have a system of $\hat{h}$ algebraic equations with $\hat{h}$ unknowns. The ith equation in such a system is

$$\begin{aligned} &1 \cdot \eta_i + \sum_{p=1}^{n_f} R_{ip}\bar{\eta}_p + \sum_{q=1}^{n_e} R_{iq}\frac{R_q}{k_q} \\ &= \sum_{k=1}^{K} \int_0^{l_k} (N_{ik}\hat{\varepsilon}_k + T_{ik}\hat{\gamma}_k + M_{ik}\hat{\kappa}_k) ds + \sum_{p=1}^{m_f} C_{ip}\delta\bar{\eta}_p + \sum_{q=1}^{m_e} C_{iq}\frac{C_q}{k_q}. \end{aligned} \tag{8.62}$$

   - η_i represents the real displacement or the rotation of the point where the ith simple constraint has been removed: $\eta_i = 0$ if the constraint is ideal, $\eta_i = \bar{\eta}_i$ if the constraint has an inelastic failure equal to $\bar{\eta}_i$, $\eta_i = -\frac{X_i}{k_i}$ if the constraint is an elastic spring of stiffness k_i.

- $\bar{\eta}_p$ are the possible inelastic failures of the external constraints, and R_{ip} the corresponding reactions in the ith system.
- $\dfrac{R_q}{k_q}$ are the possible displacements or rotations of the points of application of the external elastic constraints with stiffness k_q; R_{iq} are the corresponding reactions in the ith system. We have

$$R_q = R_{0q} + \sum_{i=1}^{\hat{h}} X_i R_{iq},$$

with R_{0q} the pertinent reaction in the 0-system.
- $\delta\bar{\eta}_p$ are the possible inelastic failures of the internal constraints, and C_{ip} the corresponding reactions in the ith system (generally, C_{ip} are action characteristics).
- $\dfrac{C_q}{k_q}$ are the possible relative displacements or rotations of the sections in which the internal elastic constraints with stiffness k_q are applied; C_{iq} are the corresponding reactions in the ith system (generally, C_{iq} are action characteristics). We have

$$C_q = C_{0q} + \sum_{i=1}^{\hat{h}} X_i C_{iq},$$

with C_{0q} the pertinent reaction in the 0-system.
- In the case we consider of only linear thermoelastic behavior, the real compatible strains are

$$\hat{\varepsilon}_k = \frac{N_{0k} + \sum_{j=1}^{\hat{h}} X_j N_{jk}}{E_k A_k} + \alpha_k \delta\theta_k,$$

$$\hat{\gamma}_k = \frac{T_{0k} + \sum_{j=1}^{\hat{h}} X_j T_{jk}}{G_k A_k^*},$$

$$\hat{\kappa}_k = \frac{M_{0k} + \sum_{j=1}^{\hat{h}} X_j M_{jk}}{E_k I_k} + 2\alpha_k \frac{\delta\theta_k}{h_k}.$$

The relation (8.62) can be rewritten as

$$1 \cdot \eta_i = \eta_{i0} + X_1\eta_{i1} + \cdots + X_{\hat{h}}\eta_{i\hat{h}} + \eta_{if} + \eta_{ie} + \eta_{it}, \quad i = 1, 2, \ldots, \hat{h}, \qquad (8.63)$$

where

$$\eta_{i0} = \sum_{k=1}^{K} \int_0^{l_k} \left(N_{ik} \frac{N_{0k}}{E_k A_k} + T_{ik} \frac{T_{0k}}{G_k A_k^*} + M_{ik} \frac{M_{0k}}{E_k \mathrm{I}_k} \right) ds,$$

$$\eta_{ij} = \sum_{k=1}^{K} \int_0^{l_k} \left(N_{ik} \frac{N_{jk}}{E_k A_k} + T_{ik} \frac{T_{jk}}{G_k A_k^*} + M_{ik} \frac{M_{jk}}{E_k \mathrm{I}_k} \right) ds,$$

$$\eta_{if} = \sum_{p=1}^{m_f} C_{ip} \delta \bar{\eta}_p - \sum_{p=1}^{n_f} R_{ip} \bar{\eta}_p,$$

$$\eta_{ie} = \sum_{q=1}^{m_e} C_{iq} \frac{C_q}{k_q} - \sum_{q=1}^{n_e} R_{iq} \frac{R_q}{k_q},$$

$$\eta_{it} = \sum_{k=1}^{K} \int_0^{l_k} \left(N_{ik} \alpha_k \delta\theta_k + M_{ik} 2\alpha_k \frac{\delta\theta_k}{h_k} \right) ds.$$

Commonly, we call equations of the type (8.63) for a given framed structure the **Müller–Breslau equations**. It is evident that $\eta_{ij} = \eta_{ji}$ and $\eta_{ii} > 0$, where the last inequality does not imply summation over repeated indices.

An example clarifies the method. Consider the structure in Figure 8.44. If the rod is rigid, we will have three degrees of freedom and five rigid constraints, well posed in the sense that they do not allow any rigid displacement. Hence, the rod is twice hyperstatic, according to the definition already stated. A choice for the principal structure is that of Figure 8.45. With this choice, the hyperstatic unknowns X_1 and X_2 are the vertical reactions in the carriages at A and B, respectively.

Then the distributions of N_0, T_0, M_0 along the rod's axis are those illustrated in Figure 8.46.

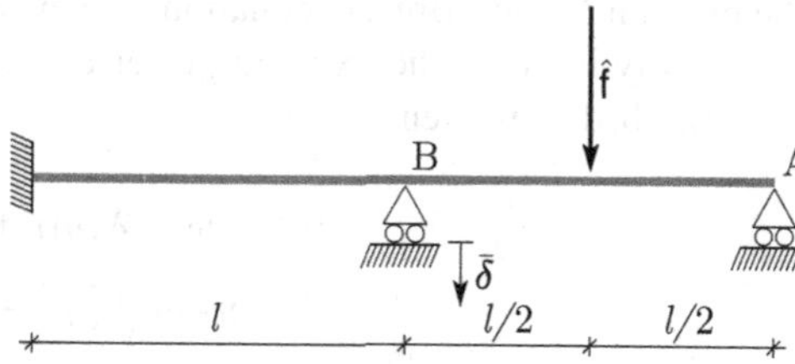

Fig. 8.44 Structure with two superabundant constraints with respect to the statically determined state, and a constraint failure $\bar{\delta}$

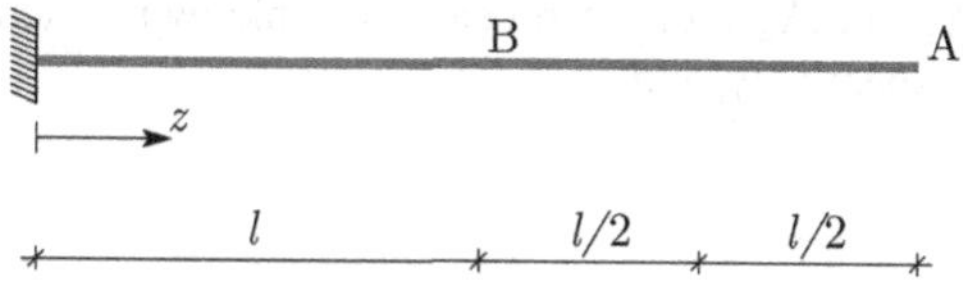

Fig. 8.45 Principal structure of the scheme in Figure 8.44

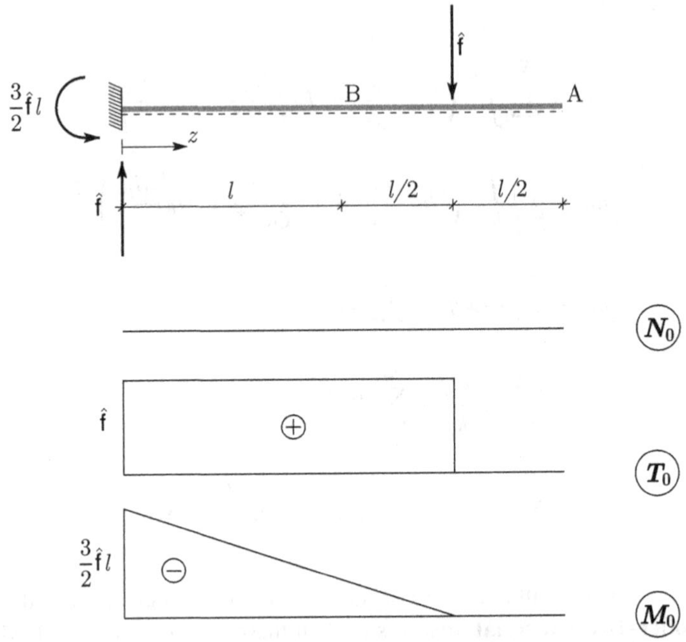

Fig. 8.46 Distributions of N_0, T_0, M_0 over the principal structure subjected to the original external load

As a subsequent step, we consider the same structure, but the load is now unitary and imposed at the point A ($X_1 = 1$). The result of the analysis is shown in Figure 8.47. An analogous analysis has to be done by applying a unitary force at the point B ($X_2 = 1$), where we have eliminated a constraint, as was done for A. The result is shown in Figure 8.48.

Notice that the *real* displacement of the point A is zero, due to the constraint, while that pertaining to B is equal to $\bar{\delta}$ as a consequence of the constraint failure. The relevant Müller–Breslau equations—they are just a way of expressing formally the equality between the external power and the internal one along the procedure just described—are then

$$0 = \eta_{10} + X_1\eta_{11} + X_2\eta_{12},$$

$$\bar{\delta} = \eta_{20} + X_1\eta_{21} + X_2\eta_{22};$$

X_1 and X_2 are, we repeat, the unknown reactions of the constraints at A and B, respectively, and

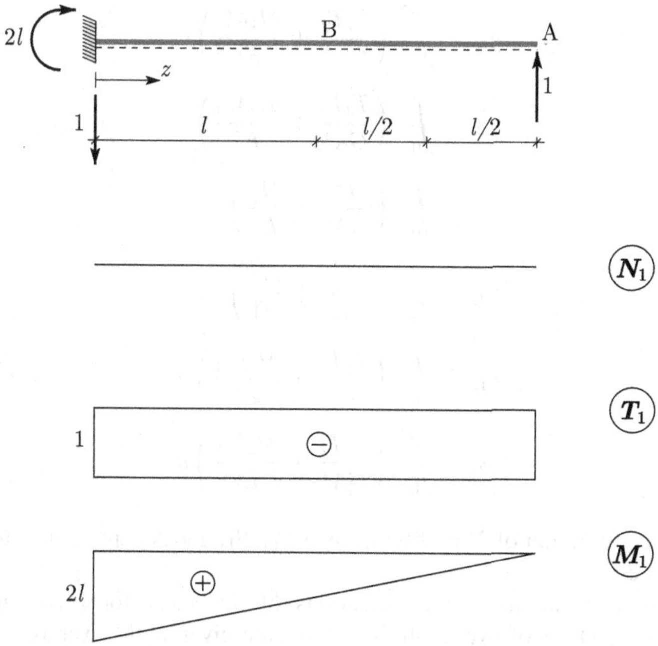

Fig. 8.47 Distributions of N_1, T_1, M_1 over the 1-system

Fig. 8.48 Distributions of N_2, T_2, M_2 over the 2-system

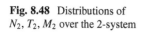

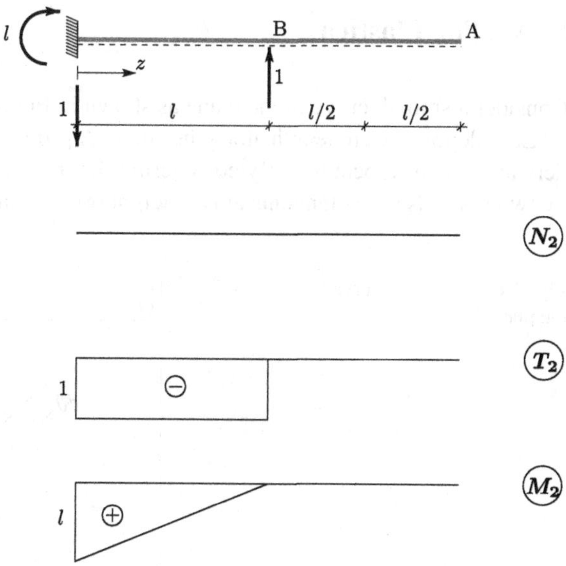

$$\eta_{10} = \int_0^{2l} \left(\frac{T_1 T_0}{GA^*} + \frac{M_1 M_0}{EI} \right) dz,$$

$$\eta_{20} = \int_0^{2l} \left(\frac{T_2 T_0}{GA^*} + \frac{M_2 M_0}{EI} \right) dz,$$

$$\eta_{11} = \int_0^{2l} \left(\frac{T_1^2}{GA^*} + \frac{M_1^2}{EI} \right) dz,$$

$$\eta_{22} = \int_0^{2l} \left(\frac{T_2^2}{GA^*} + \frac{M_2^2}{EI} \right) dz,$$

$$\eta_{12} = \int_0^{2l} \left(\frac{T_1 T_2}{GA^*} + \frac{M_1 M_2}{EI} \right) dz,$$

$$\eta_{21} = \int_0^{2l} \left(\frac{T_2 T_1}{GA^*} + \frac{M_2 M_1}{EI} \right) dz,$$

where the contribution of N is absent, since N_0, N_1, and N_2 are identically zero in this special case.

Often in applications, the contributions of the shear forces are neglected. Examples of analysis of hyperstatic structures are given in the exercises discussed later in this chapter.

8.24 The Elastica

Consider a smooth curve in the plane as shown in Figure 8.49.

Let s denote the arc length along the curve. A prime will denote in this section the derivative with respect to z. By considering the triangle in Figure 8.49, determined by two normals to the tangents at two neighboring points at distance ds, we get

Fig. 8.49 A smooth curve in the plane

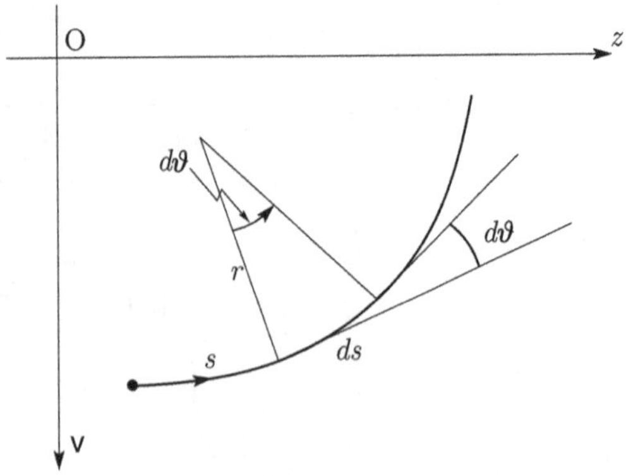

$$\frac{1}{r} = \frac{d\vartheta}{ds},$$

so that in accordance with the relation (8.46), we have for the curvature $\hat{\kappa}$ the identity

$$\hat{\kappa} = \frac{d\vartheta}{ds} = \frac{1}{r}.$$

On the other hand, from the geometric scheme in Figure 8.49, we get

$$\tan\vartheta = -\frac{d\mathrm{v}}{dz},$$

and the assumed smoothness allows us to write, after differentiation with respect to z,

$$(1 + \tan^2\vartheta)\frac{d\vartheta}{ds}\frac{ds}{dz} = -\frac{d^2\mathrm{v}}{dz^2}. \qquad (8.64)$$

Moreover, since

$$ds^2 = (dz^2 + d\mathrm{v}^2) = (1 + \mathrm{v}'^2)dz^2,$$

we may rewrite (8.64) as

$$(1 + \mathrm{v}'^2)\hat{\kappa}\sqrt{(1 + \mathrm{v}'^2)} = -\mathrm{v}'',$$

which is

$$\hat{\kappa} = -\frac{\mathrm{v}''}{(1 + \mathrm{v}'^2)^{\frac{3}{2}}}. \qquad (8.65)$$

When $|\mathrm{v}'| \ll 1$, we may accept the approximation

$$\hat{\kappa} \approx -\mathrm{v}'',$$

i.e.,

$$\mathrm{v}' = -\vartheta. \qquad (8.66)$$

If we look at the kinematics described in deriving the scheme of Timoshenko's rod, we realize that the identity (8.66) corresponds to the absence of shear strain:

$$\hat{\gamma} = \mathrm{v}' + \vartheta = 0.$$

In other words, we have described the kinematics of an inextensible rod, which experiences only bending. This is what we commonly call the **Bernoulli rod**, after the 1691 pioneering work of Jacob Bernoulli (1654–1705), although we should perhaps call it the **Bernoulli–Euler rod**, due to Euler's 1744 analysis of the equilibrium configurations of such a one-dimensional elastic body in terms of the minimizers of the functional

$$\int_0^l \frac{1}{r^2} ds$$

with $\frac{1}{r}$ determined by the nonlinear relation (8.65). Also, Euler introduced the analysis of bifurcated bend states, a problem that we shall discuss later. For the moment, we restrict attention to the approximation (8.66), and using (8.52), we write

$$EI\mathsf{v}'' + M = 0. \tag{8.67}$$

Once M is known, using equation (8.67), we can determine the deflection $\mathsf{v}(z)$ of the inextensible and unshearable rod considered here. The distribution of the bending moment along the rod can be easily determined when the rod itself is statically determinate. Additional difficulties arise when the rod is hyperstatic. In the case in which M can be differentiated twice and we do not have distributed couples and thermal effects, and the rod is straight in the reference place, we can use the balances (8.48) and (8.49) to write

$$M'' = -\bar{q},$$

i.e.,

$$(EI\mathsf{v}'')'' - \bar{q} = 0. \tag{8.68}$$

On integrating equation (8.67) for statically determinate rods or equation (8.68) for hyperstatic rods, we determine the deflection $\mathsf{v}(z)$ up to the appropriate integration constants, which follow from the boundary conditions.

We can also describe inextensible and unshearable rods by beginning with the rod representation in terms of a one-dimensional continuum, as developed so far. To this end, consider a deformation such that

$$\begin{aligned} \mathsf{d}_3 &= \cos\vartheta e_1 + \sin\vartheta e_3, \\ \mathsf{d}_1 &= -\sin\vartheta e_1 + \cos\vartheta e_3, \\ \mathsf{d}_2 &= e_2, \end{aligned}$$

where e_1, e_2, e_3 represent the orthogonal reference frame introduced in Section 8.2, and $\vartheta(\zeta)$ represents the bending angle constrained by the boundary conditions

$$\vartheta(0) = 0, \qquad \vartheta(1) = 0,$$

assuming that the rod has unitary length. The strain measure $\zeta = \varphi' - \mathsf{d}_3$ has components

$$\zeta = \hat{\varepsilon}\mathsf{d}_3 + \hat{\gamma}\mathsf{d}_1. \tag{8.69}$$

For the contact actions n and m, we also have

$$\mathsf{n} = N\mathsf{d}_3 + T\mathsf{d}_1, \qquad \mathsf{m} = M\mathsf{d}_2.$$

According to the constitutive restrictions discussed in Section 8.17, we consider N, T, M invertible functions of $\hat{\varepsilon}$, $\hat{\gamma}$, ϑ' and impose also the conditions

$$\varphi(0) = 0, \qquad \varphi'(0) = 0, \qquad \mathsf{n}(1) = \hat{\lambda}e_1, \qquad \hat{\lambda} > 0. \tag{8.70}$$

In terms of e_1 and e_2, the vector n reads

$$\mathsf{n} = N(\cos\vartheta e_1 + \sin\vartheta e_2) + T(-\sin\vartheta e_1 + \cos\vartheta e_2),$$

so that

$$(\mathsf{n} \cdot e_1)e_1 = N\cos\vartheta e_1 - T\sin\vartheta e_1.$$

Let us assume the absence of distributed and concentrated external actions, so that the balance of forces reads $\mathsf{n}' = 0$, that is, n is equal to a constant given by the boundary condition (8.70), which implies

$$N = \hat{\lambda}\cos\vartheta, \qquad T = -\hat{\lambda}\sin\vartheta.$$

Consequently, since from (8.69) we have

$$\varphi' = (1 + \hat{\varepsilon})\mathsf{d}_3 + \hat{\gamma}\mathsf{d}_1,$$

we compute

$$\varphi' \times \mathsf{n} = \hat{\lambda}\hat{\gamma}\cos\vartheta\mathsf{d}_3 \times \mathsf{d}_1 - \hat{\lambda}(1+\hat{\varepsilon})\sin\vartheta\mathsf{d}_1 \times \mathsf{d}_3 = -\hat{\lambda}((1+\hat{\varepsilon})\sin\vartheta + \hat{\gamma}\cos\vartheta)\mathsf{d}_2.$$

The assumed absence of distributed and concentrated momenta allows us to express the balance of couples $\mathsf{m}' + \varphi' \times \mathsf{n} = 0$ as

$$(M(\hat{\varepsilon}, \hat{\gamma}, \vartheta)')' - \hat{\lambda}((1 + \hat{\varepsilon})\sin\vartheta + \hat{\gamma}\cos\vartheta) = 0,$$

and by adopting the constitutive relation

$$M = EI\hat{\kappa} = -EI\vartheta',$$

where we assume constant the product EI, we can reduce the balance of couples to the semilinear differential equation

$$EI\vartheta'' + \hat{\lambda}((1+\hat{\varepsilon})\sin\vartheta + \hat{\gamma}\cos\vartheta) = 0.$$

In the absence of elongation and shear strain, i.e., when $\hat{\varepsilon} = 0$ and $\hat{\gamma} = 0$, the previous balance becomes

$$EI\vartheta'' + \hat{\lambda}\sin\vartheta = 0, \tag{8.71}$$

which is what we call **elastica**. It was Euler who first classified the solutions of such an equation.

8.25 Exercises on Isostatic and Hyperstatic Framed Structures

Exercise 21. *Find $\bar{q}$, $\delta\theta$, and ϑ_0 such that the vertical displacement η_C of the point C in the structure in Figure 8.50 vanishes. Write α for the thermal dilatation coefficient of the material. Take $k = 2EI/l$ in the rotational spring, with EI a constant. Neglect shear and axial deformabilities.*

Remarks and solution. First suppose that the elastic spring has infinite stiffness (Fig. 8.51). In this state, the structure is isostatic. In fact, there are two rods and six simple constraints ($\hat{l} = \hat{h}$): the rod $\overline{\text{BCD}}$ is fixed by the joint at B, and the rod $\overline{\text{AD}}$ does not admit rigid displacements ($\hat{l} = \hat{h} = 0$). To calculate the vertical displacement at C, we follow a virtual-work approach in terms of real deformations on the original structure (Fig. 8.50) and virtual action characteristics evaluated by

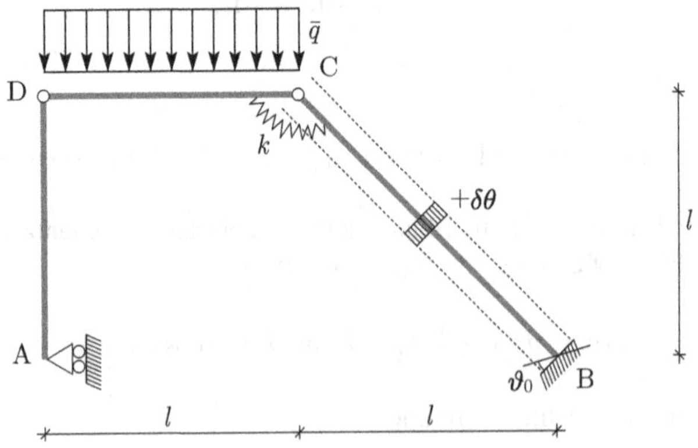

Fig. 8.50 A structure with a constraint failure, thermal variations, and distributed load

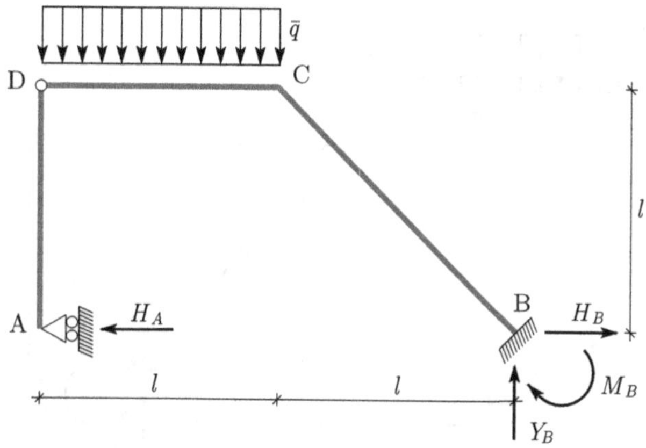

Fig. 8.51 Reactive forces

applying a unitary load $\hat{\mathfrak{f}}^* = 1$ at C on the structure without original load and spring, as in Figure 8.52-b. With this last loading condition alone, we denote by an asterisk the pertinent action characteristics.

The reactive forces exerted by the external constraints are displayed in Figure 8.52. The distributions of M, M^*, N^*, and the local abscissas z appear in Figure 8.53; the corresponding equations in the part $\overline{\text{CB}}$ are collected in Table 8.3. The external virtual work is then

$$L^{ext} = 1\eta_C + \vartheta_0 l.$$

In fact, if we suppose that the vertical displacement of the point C is directed downward, the first term in the external virtual work is positive; the second term is the work done by the reactive couple at B of the virtual system in the inelastic rotation ϑ_0 of the constraint. Both quantities are oriented clockwise, so the work is positive. The internal virtual work is given by

$$L^{inn} = \int_0^{l\sqrt{2}} \left(\frac{z}{\sqrt{2}}\right)\left(\frac{\bar{q}l^2}{2} + \frac{\bar{q}lz}{\sqrt{2}}\right)\frac{dz}{EI} + \int_0^{l\sqrt{2}} \left(-\frac{1}{\sqrt{2}}\right)\alpha\delta\theta \, dz.$$

The elastic spring does not contribute to the internal work, because the bending moment $M^*(\text{C})$ is zero. The contribution of $\delta\theta$ in the part $\overline{\text{CB}}$ is negative, because N^* is negative in this part. Hence, the identity $L^{ext} = L^{inn}$ can be written explicitly as follows:

$$\eta_C = -\vartheta_0 l - \alpha\delta\theta l + \frac{7\sqrt{2}}{12}\frac{\bar{q}l^4}{EI}.$$

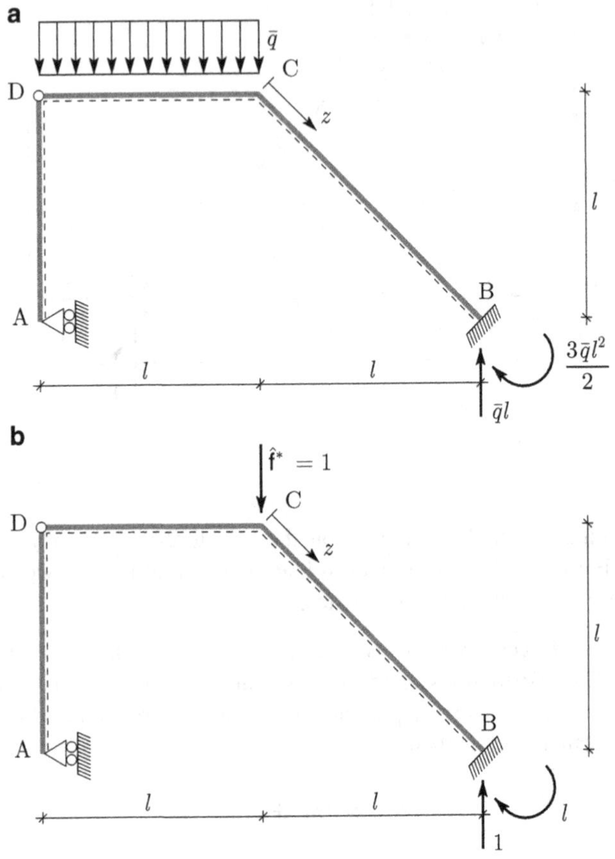

Fig. 8.52 (a) Effective system, and (b) virtual system

Finally, $\eta_C = 0$ furnishes

$$\bar{q} = (\vartheta_0 + \alpha\delta\theta)\frac{12}{7\sqrt{2}}\frac{EI}{l^3}.$$

Exercise 22. *For the truss system presented in Figure 8.54, calculate the displacement u_A of the point* A *in the direction of the applied force* $\hat{\mathsf{f}}$. *The axial stiffness EA of all rods is constant. The force* $\hat{\mathsf{f}}$ *in inclined at* 45°. *Find the limits of u_A for $k \to 0$ and $k \to \infty$.*

Remarks and solution. The structure is isostatic, because it is composed of simple triangles and is constrained to the ground by three independent pendulums (the elastic constraint in the top right-hand part of the structure is equivalent to a simple

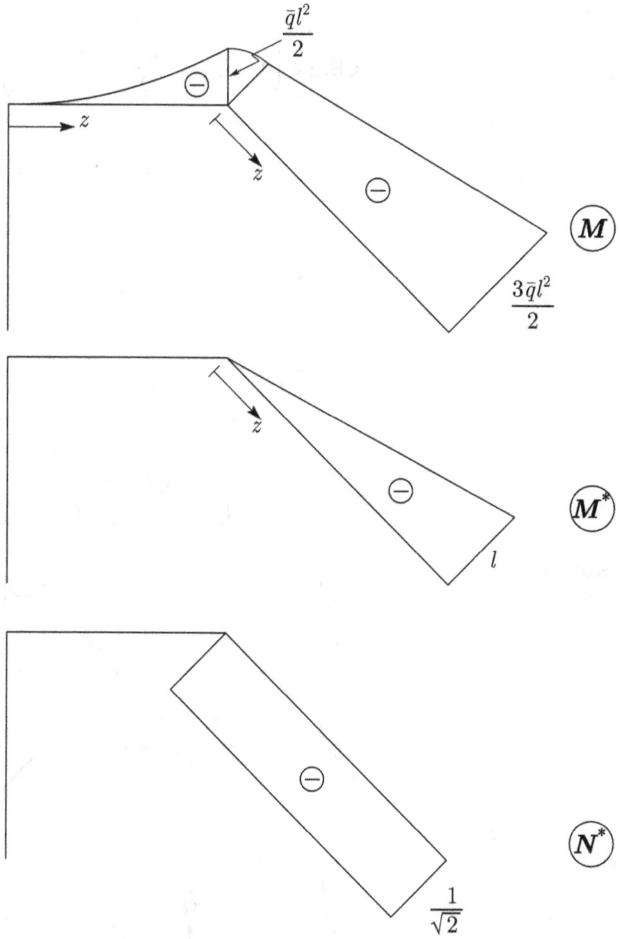

Fig. 8.53 Action characteristics of the two systems employed for calculation

pendulum when $k \to \infty$). The external reactive forces can be determined by considering the structure as a sole body (Fig. 8.55; moments refer to the point 4):

$$H_5 + \frac{\hat{\mathsf{f}}}{\sqrt{2}} = 0, \qquad Y_1 + Y_8 - \frac{\hat{\mathsf{f}}}{\sqrt{2}} = 0, \qquad Y_1 l = 0,$$

which give (Fig. 8.56)

$$Y_1 = 0, \qquad Y_8 = \frac{\hat{\mathsf{f}}}{\sqrt{2}}, \qquad H_5 = -\frac{\hat{\mathsf{f}}}{\sqrt{2}}.$$

Table 8.3 Action characteristics

	$M(z)$	$M^*(z)$	$N^*(z)$
$\overline{CB}, z \in \left[0, l\sqrt{2}\right]$	$-\frac{\bar{q}l^2}{2} - \frac{\bar{q}lz}{\sqrt{2}}$	$-\frac{z}{\sqrt{2}}$	$-\frac{1}{\sqrt{2}}$

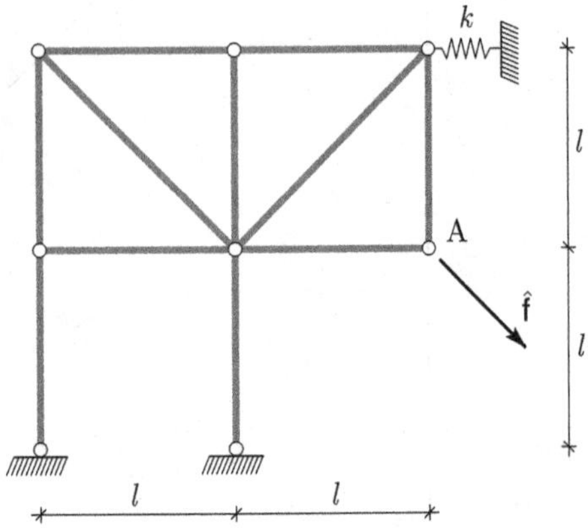

Fig. 8.54 Isostatic truss system with elastic spring

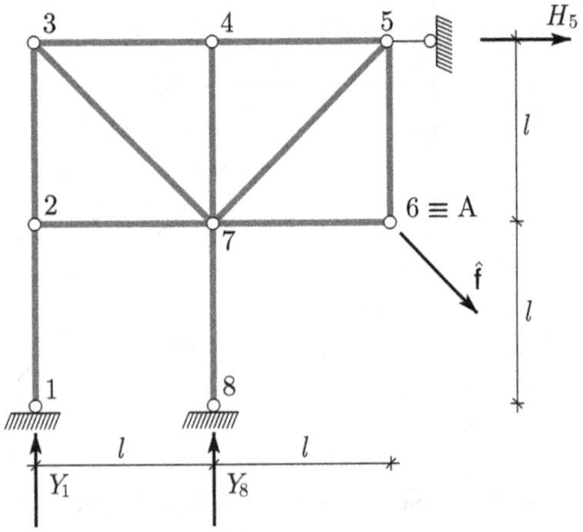

Fig. 8.55 External reactive forces in the structure

To compute the axial forces in the rods, we consider the equilibrium of the nodes. We begin from node 1, continuing with the analyses of nodes 2, 3, and 4. We obtain

$$N_{21} = N_{23} = N_{34} = N_{27} = N_{37} = N_{45} = N_{47} = 0.$$

From the equilibrium of the nodes 7 and 6 (Fig. 8.57), we get

$$N_{76} = \frac{\hat{f}}{\sqrt{2}}, \qquad N_{57} = -\hat{f}, \qquad N_{56} = \frac{\hat{f}}{\sqrt{2}}, \qquad N_{78} = -\frac{\hat{f}}{\sqrt{2}}.$$

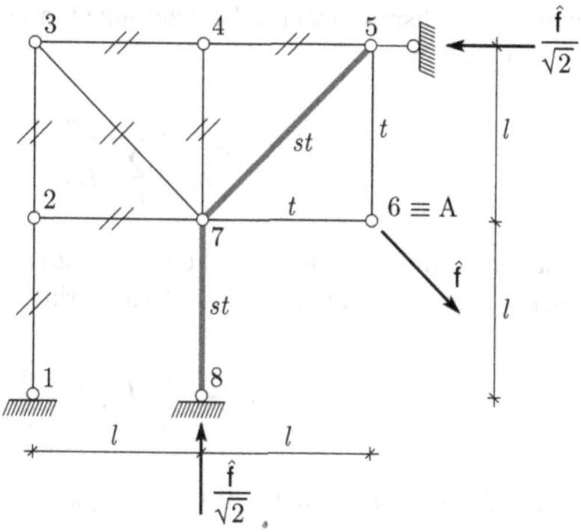

Fig. 8.56 Reactive forces and action characteristics

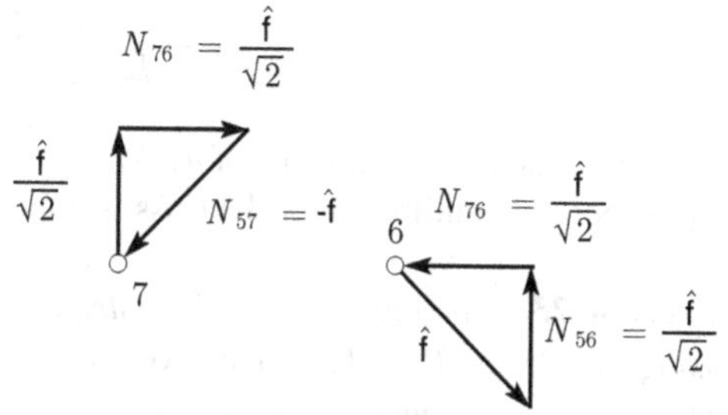

Fig. 8.57 Equilibrium of the nodes 7 and 6

In linear elasticity—the setting in which we are developing the present analysis—in the absence of prestress states, the elastic energy is

$$\mathcal{E} = \frac{1}{2}\int_{\mathcal{B}} \varepsilon \cdot \mathbb{C}\varepsilon \, d\mu = \frac{1}{2}\int_{\mathcal{B}} \sigma \cdot \varepsilon \, d\mu,$$

for $\sigma = \mathbb{C}\varepsilon$.

The last integral in the previous identity is the inner work in the body $\mathcal{B}$. Then, since the inner work is equal to the external work—a consequence of the integration in time of the equality between the external power and internal power—we can write

$$\mathcal{E} = \frac{1}{2}L^{ext};$$

this identity is commonly called **Clapeyron's theorem**.

In the specific case of the exercise, we have

$$\mathcal{E} = \frac{1}{2}\hat{f}u_A, \tag{8.72}$$

with u_A the displacement to be determined in the direction of $\hat{f}$. The elastic energy is also given by

$$\mathcal{E} = \frac{1}{2}\sum_{i=1}^{4} \frac{N_i^2}{EA} l_i + \frac{1}{2}\frac{\hat{f}^2}{2k},$$

where the summation is over all the rods with $N \neq 0$, and the last term is the elastic energy accumulated in the linear elastic spring. Then from (8.72), we get

$$u_A = \frac{\hat{f}l}{EA}\left(\sqrt{2} + \frac{3}{2}\right) + \frac{\hat{f}}{2k},$$

oriented is the direction of $\hat{f}$. If the spring is rigid ($k \to \infty$), we obtain

$$u_A = \frac{\hat{f}l}{EA}\left(\sqrt{2} + \frac{3}{2}\right),$$

and this is the minimum value of u_A. When $k \to 0$, the structure becomes labile, and $u_A \to \infty$: the equilibrium of the truss is impossible under the applied force $\hat{f}$.

Exercise 23. *Compute the elastic deflection* $z \to \mathsf{v}(z)$ *for the rod in Figure 8.58 using the Bernoulli rod scheme. Assume constant flexural stiffness* EI *and* $\hat{f} = \bar{q}l$*. Then evaluate the maximum value of* v.

Remarks and solution. The beam is isostatic. In the Bernoulli scheme, it is sufficient to consider the second-order differential equation $EI\mathsf{v}'' = -M$. The reactive forces are given by the balance equations (Fig. 8.59)

$$H_A = 0, \qquad Y_C - \hat{f} - \bar{q}l = 0, \qquad M_A - \frac{\hat{f}l}{2} + \bar{q}l\frac{l}{2} = 0,$$

which yield

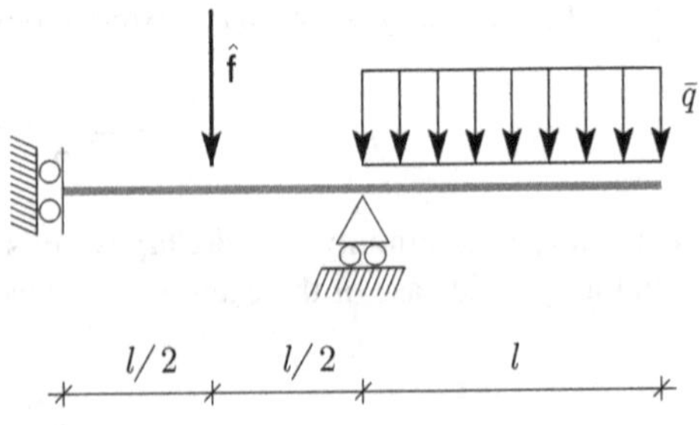

Fig. 8.58 Isostatic rod

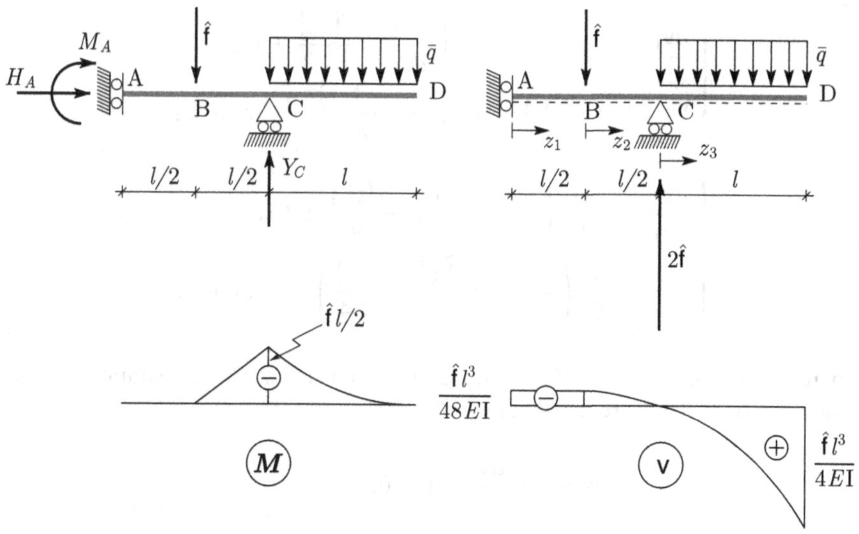

Fig. 8.59 Reaction forces, bending moment, and elastic deflection of the rod

$$H_A = 0, \qquad Y_C = 2\hat{f}, \qquad M_A = 0.$$

The expressions for the bending moments along the parts $\overline{AB}$, $\overline{BC}$, and $\overline{CD}$ (abscissas z_1, z_2, z_3; Fig. 8.59) are

$$M(z_1) = 0, \qquad M(z_2) = -\hat{f}z_2, \qquad M(z_3) = -\frac{\hat{f}l}{2} + \hat{f}z_3 - \frac{\hat{f}z_3^2}{2l}.$$

Denoting by v_1, v_2, v_3 the function v in the parts characterized by z_1, z_2, z_3, respectively, we have

$$\frac{d^2v_1}{dz_1^2} = 0,$$
$$\frac{d^2v_2}{dz_2^2} = \frac{\hat{f}z_2}{EI},$$
$$\frac{d^2v_3}{dz_3^2} = -\frac{1}{EI}\left(-\frac{\hat{f}l}{2} + \hat{f}z_3 - \frac{\hat{f}z_3^2}{2l}\right).$$

Integration gives

$$\begin{cases} \dfrac{d\mathsf{v}_1}{dz_1} = C_1, \\ \mathsf{v}_1 = C_1 z_1 + C_2 \end{cases}, \qquad \begin{cases} \dfrac{d\mathsf{v}_2}{dz_2} = \dfrac{\hat{\mathsf{f}}}{EI}\dfrac{z_2^2}{2} + C_3, \\ \mathsf{v}_2 = \dfrac{\hat{\mathsf{f}}}{EI}\dfrac{z_2^3}{6} + C_3 z_2 + C_4 \end{cases},$$

$$\begin{cases} \dfrac{d\mathsf{v}_3}{dz_3} = -\dfrac{1}{EI}\left(-\dfrac{\hat{\mathsf{f}}l}{2}z_3 + \dfrac{\hat{\mathsf{f}}}{2}z_3^2 - \dfrac{\hat{\mathsf{f}}z_3^3}{6l}\right) + C_5, \\ \mathsf{v}_3 = -\dfrac{1}{EI}\left(-\dfrac{\hat{\mathsf{f}}l}{2}\dfrac{z_3^2}{2} + \dfrac{\hat{\mathsf{f}}}{2}\dfrac{z_3^3}{3} - \dfrac{\hat{\mathsf{f}}}{6l}\dfrac{z_3^4}{4}\right) + C_5 z_3 + C_6 \end{cases},$$

where the C_k, $k = 1, \ldots, 6$, are integration constants. The constraints impose conditions on v. They are listed below:

$$\text{Point A}: \quad \frac{d\mathsf{v}_1}{dz_1}(0) = 0.$$

$$\text{Point B}: \begin{cases} \dfrac{d\mathsf{v}_1}{dz_1}\left(\dfrac{l}{2}\right) = \dfrac{d\mathsf{v}_2}{dz_2}(0), \\ \mathsf{v}_1\left(\dfrac{l}{2}\right) = \mathsf{v}_2(0). \end{cases}$$

$$\text{Point C}: \begin{cases} \mathsf{v}_2\left(\dfrac{l}{2}\right) = 0, \\ \mathsf{v}_3(0) = 0, \\ \dfrac{d\mathsf{v}_2}{dz_2}\left(\dfrac{l}{2}\right) = \dfrac{d\mathsf{v}_3}{dz_3}(0). \end{cases}$$

The first condition prevents the rotation at A; the others represent the continuity of v and the rotation at B, the absence of vertical displacement at C, the continuity of the rotations at the same point. The integration constants are

$$C_1 = 0, \quad C_2 = -\frac{\hat{\mathsf{f}}l^3}{48EI}, \quad C_3 = 0, \quad C_4 = -\frac{\hat{\mathsf{f}}l^3}{48EI}, \quad C_5 = \frac{\hat{\mathsf{f}}l^2}{8EI}, \quad C_6 = 0.$$

The elastic deflections v are then

$$\mathsf{v}_1 = -\frac{\hat{\mathsf{f}}l^3}{48EI},$$

$$\mathsf{v}_2 = \frac{\hat{\mathsf{f}}}{6EI}z_2^3 - \frac{\hat{\mathsf{f}}l^3}{48EI},$$

$$\mathsf{v}_3 = \frac{1}{EI}\left(\frac{\hat{\mathsf{f}}l}{4}z_3^2 - \frac{\hat{\mathsf{f}}}{6}z_3^3 + \frac{\hat{\mathsf{f}}}{24l}z_3^4\right) + \frac{\hat{\mathsf{f}}l^2}{8EI}z_3,$$

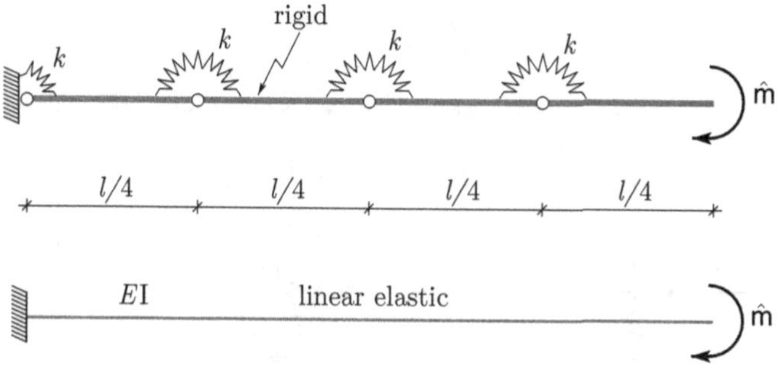

Fig. 8.60 Cantilever rods

Fig. 8.61 Isostatic structure

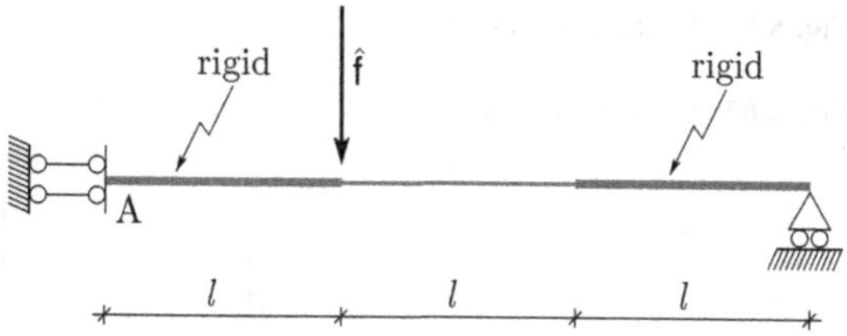

and the pertinent first derivatives represent the rotation along the rod's axis; v is portrayed in Figure 8.59. Its maximum value corresponds to the section D:

$$\mathsf{v}_{max} = \mathsf{v}_3(l) = \frac{\hat{\mathsf{f}} l^3}{4EI}.$$

Exercise 24. *Consider the two cantilevers in Figure 8.60 subjected to a concentrated couple at the free end section. Determine the value of k such that the vertical displacement of the free end is equal in the two cases. For this value of k, determine the elastic energy $\mathcal{E}$ for the two rods.*

Solution. $k = \dfrac{5EI}{l}$, $\mathcal{E} = \dfrac{2\hat{\mathsf{m}}^2 l}{5EI}$ for the first cantilever, $\mathcal{E} = \dfrac{\hat{\mathsf{m}}^2 l}{2EI}$ for the second cantilever.

Exercise 25. *The structure in Figure 8.61 consists of two rigid parts and a central portion of constant flexural stiffness EI. Find the vertical displacement of the section A. Draw the deflection $z \rightarrow \mathsf{v}(z)$ at least qualitatively.*

Solution. $\mathsf{v}_A = \dfrac{7\hat{\mathsf{f}} l^3}{3EI}$, directed downward.

Exercise 26. *The truss system in Figure 8.62 has uniform axial rigidity EA and is subjected to two concentrated forces $\hat{\mathsf{f}}$. Find the horizontal displacement w_C of the point C.*

Solution. $\mathsf{w}_C = \dfrac{\hat{\mathsf{f}}}{2k} + \dfrac{\hat{\mathsf{f}} l}{EA}\left(\dfrac{5}{2} + \sqrt{2}\right)$, directed rightward.

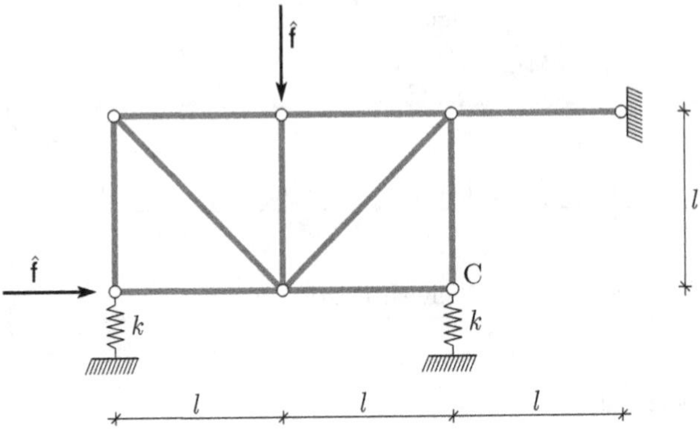

Fig. 8.62 Isostatic truss system

Fig. 8.63 Isostatic structure

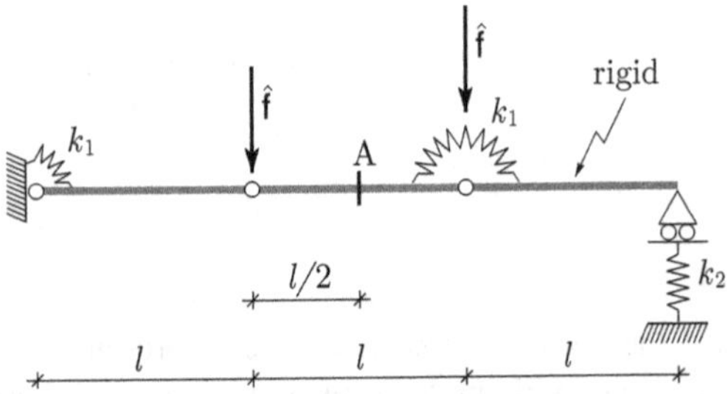

Exercise 27. *The structure in Figure 8.63 comprises three rods with infinite stiffness, connected by elastic springs. Find the vertical displacement* v_A *of the point* A. *Assume* $k_1 = k_2 l^2$.

Solution. $\mathsf{v}_A = \dfrac{11\hat{\mathrm{f}}}{8k_2}$, directed downward.

Exercise 28. *Find and draw the diagrams of the action characteristics for the structure represented in Figure 8.64. The rods have uniform sections and are made of the same material; the shear deformation is negligible. Assume* $\mathrm{EI} = EAl^2$, $\alpha\delta\theta = \frac{7\hat{\mathrm{f}}l^2}{24EI}$.

Remarks and solution. The structure has one degree of hyperstaticity. In fact, there are seven simple constraints and two rods, so

$$3K - n_v = 6 - 7 = -1 = \hat{l} - \hat{h}.$$

The structure does not admit rigid displacements ($\hat{l} = 0$); hence $\hat{h} = 1$. Also, it is symmetric and symmetrically loaded with respect to the $s - s$ axis (Fig. 8.65). The balance equations pertaining to the whole system furnish the external reactive forces, which are all equal to zero. If we eliminate the rod $\overline{\mathrm{AB}}$ and denote by X

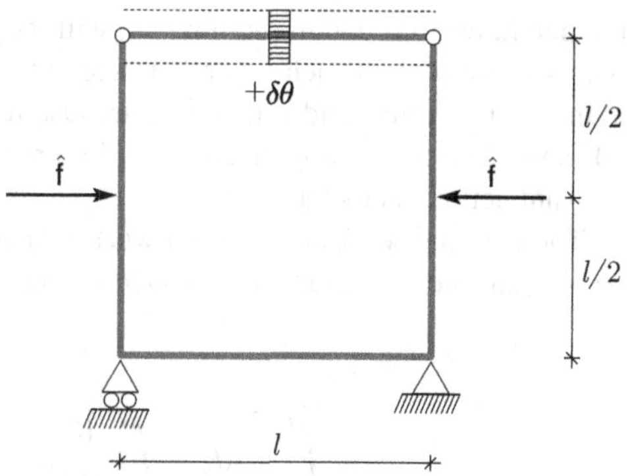

Fig. 8.64 Hyperstatic structure

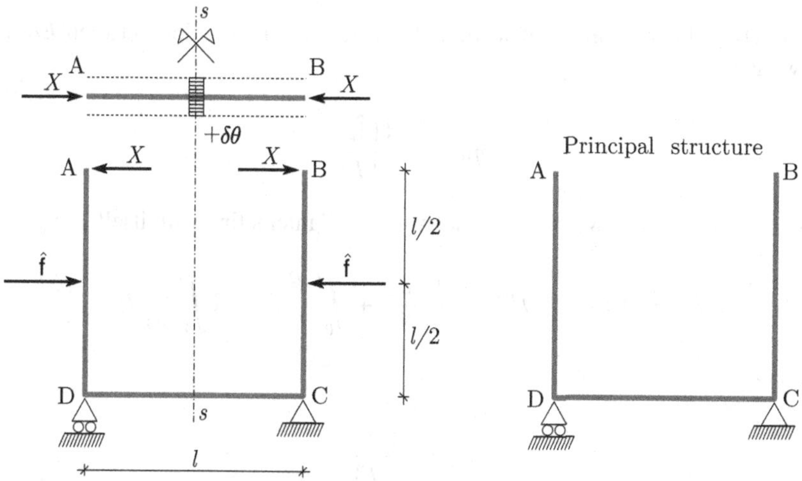

Fig. 8.65 Choice of the hyperstatic unknown and the principal structure

the axial force relating to it (there are no other action characteristics, since the rod is constrained by two hinges and is not subjected to transversal loads or applied couples), we can assume that X is a hyperstatic unknown and select the principal scheme to be the structure without the rod $\overline{\text{AB}}$. On applying the force method, we deduce the Müller–Breslau equation

$$\eta_{10} + \eta_{11} X = -\frac{Xl}{EA} + \alpha l \delta\theta.$$

Here η_{10} is the horizontal relative displacement between A and B in the 0-system (the principal system loaded by the external actions) considered with positive sign if the two points move far away from each other; η_{11} has the same meaning, but it is measured in the so-called 1-system (the principal system loaded by the force $X = 1$). The first term in the right-hand of the compatibility equation is the elongation of the pendulum $\overline{AB}$ due to the force X, with negative algebraic sign because the two points move toward each other. The second term has the same meaning but is due to the heating effect and is positive, because it describes an elongation. The systems "0" and "1" are statically determined. Figure 8.66 shows the reactive forces and the relevant action characteristics.

The external work and internal work related to the two systems "0" and "1" can be written (the abscissas are indicated in Fig. 8.66):

$$L^{ext} = 1\eta_{10},$$

$$L^{inn} = -\int_0^l \frac{\hat{f}}{EA}dz - \int_0^l \frac{\hat{f}l}{2EI}l\,dz - 2\int_0^{l/2} \frac{\hat{f}z}{EI}\left(z + \frac{l}{2}\right)dz.$$

On equating the two values of work and considering the relation between EA and EI, we get

$$\eta_{10} = -\frac{41}{24}\frac{\hat{f}l^3}{EI}.$$

In the same way, if we consider the system "1" interacting with itself, we get

$$L^{ext} = 1\eta_{11}, \qquad L^{inn} = \int_0^l \frac{dz}{EA} + \int_0^l \frac{l^2}{EI}dz + 2\int_0^l \frac{z^2}{EI}dz,$$

whence

$$\eta_{11} = \frac{8}{3}\frac{l^3}{EI}.$$

The Müller–Breslau equation becomes

$$-\frac{41}{24}\frac{\hat{f}l^3}{EI} + \frac{8}{3}\frac{l^3}{EI}X = -\frac{Xl}{EA} + \alpha l\delta\theta,$$

i.e.,

$$X = \frac{6}{11}\hat{f}.$$

The positive algebraic sign indicates that the force X has the direction assumed at the beginning, i.e., the rod $\overline{AB}$ is a strut. The diagrams of the action characteristics in the structure are represented in Figure 8.67 and were computed by direct superposition of the diagrams of the systems "0" and "1":

$$N = N_0 + \frac{6}{11}\hat{f}N_1, \qquad M = M_0 + \frac{6}{11}\hat{f}M_1, \qquad T = T_0 + \frac{6}{11}\hat{f}T_1.$$

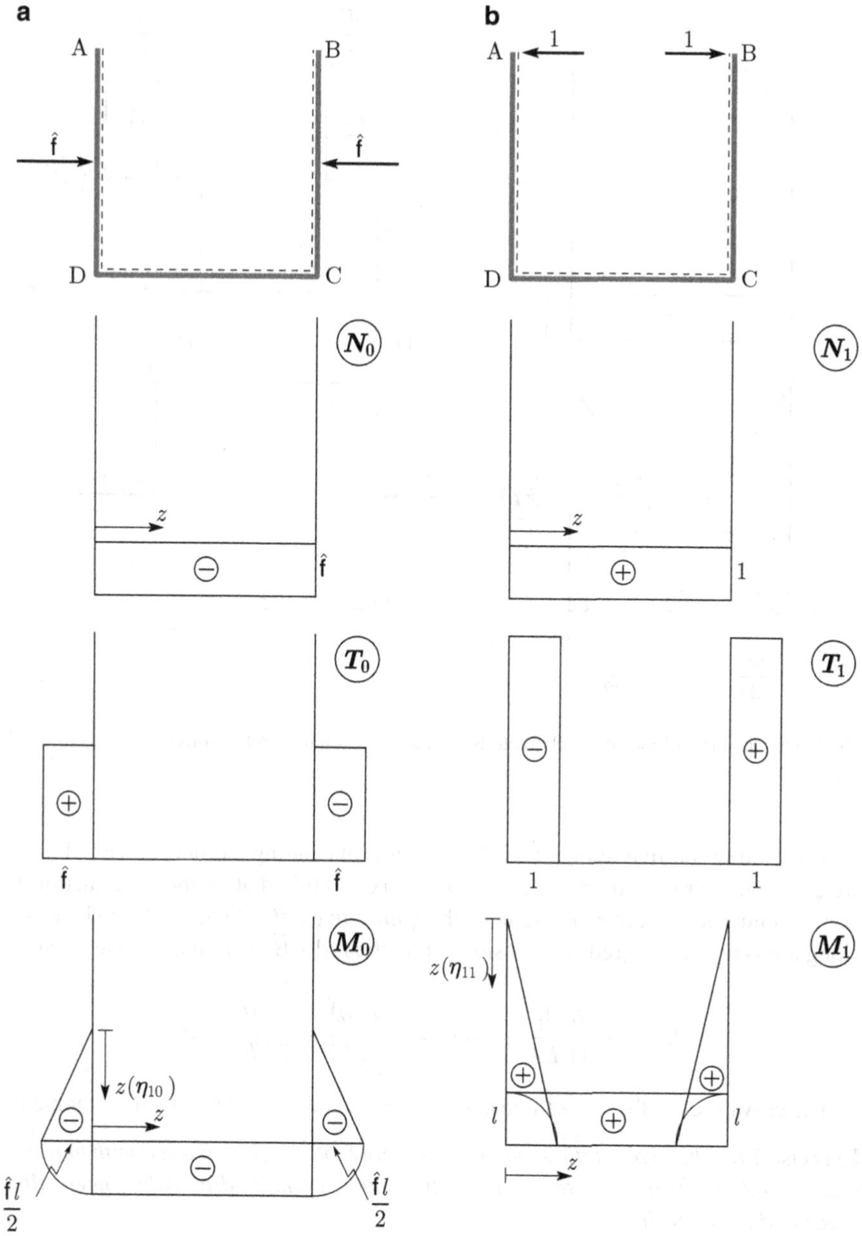

Fig. 8.66 (a) Systems "0" and (b) "1."

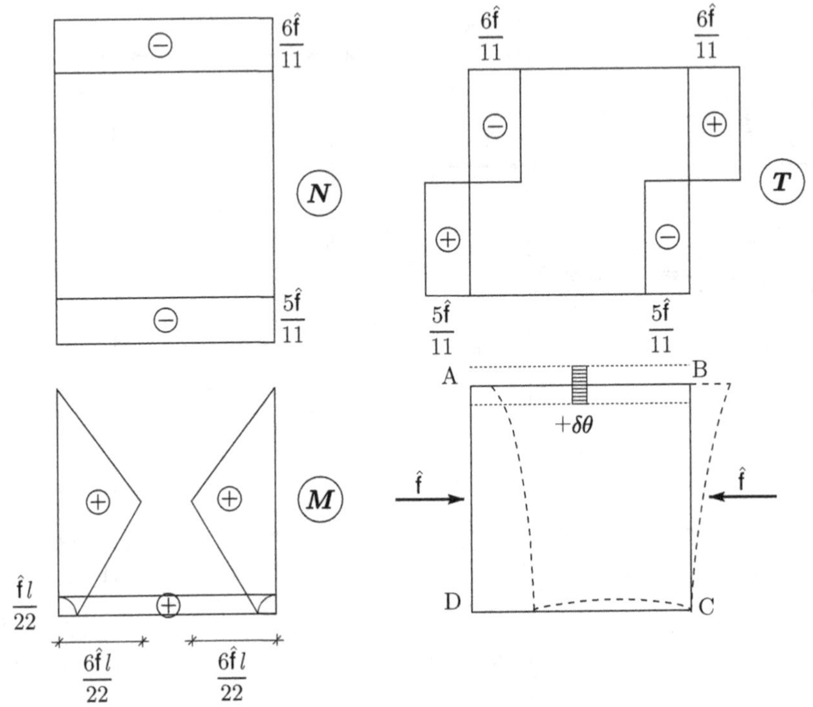

Fig. 8.67 Diagrams of the action characteristics and the qualitative deflections of the rods (dashed lines)

Since the structure is symmetric, the diagrams of bending moment and axial force are also symmetric with respect to the $s-s$ axis, while that of the shear action is skew-symmetric. Figure 8.67 includes the *qualitative deflections* of the rods. If the elongations are considered with positive sign, the rod $\overline{\text{AB}}$ shortens, and we obtain

$$\delta l_{AB} = -\frac{6}{11}\frac{\hat{f}l}{EA} + \alpha l \delta\theta = -\frac{6}{11}\frac{\hat{f}l^3}{EI} + \frac{7}{24}\frac{\hat{f}l^3}{EI} < 0.$$

The curvatures of the other three parts depend on the sign of the bending moment.

Exercise 29. *The hyperstatic structure in Figure 8.68 consists of a rod with uniform section and material. Choose a hyperstatic unknown and find its value, neglecting shear and axial strains.*

Remarks and solution. If we suppose that the springs are rigid, we have just four simple constraints (two hinges, in fact), and there is no lability; hence $\hat{h} = 1$. A possible choice of the hyperstatic unknown X is the reactive force in the horizontal spring at C. By eliminating this constraint, we get the isostatic structure in Figure 8.69-b. The relevant Müller–Breslau equation is

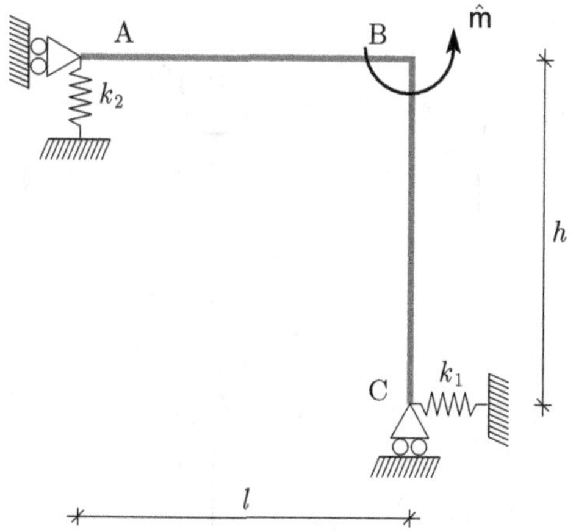

Fig. 8.68 Hyperstatic rod with elastic springs

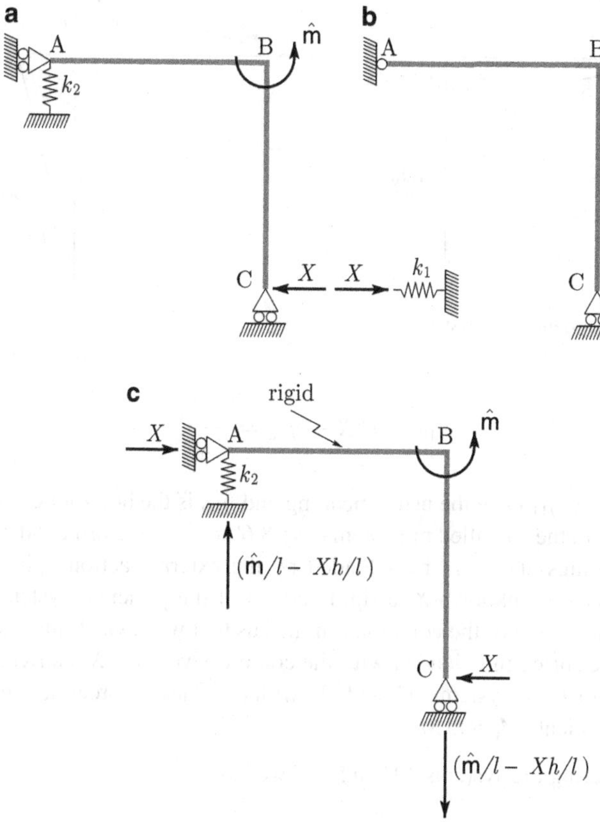

Fig. 8.69 (a) Choice of the hyperstatic unknown, (b) principal structure, (c) m-system with constraint reactions shown

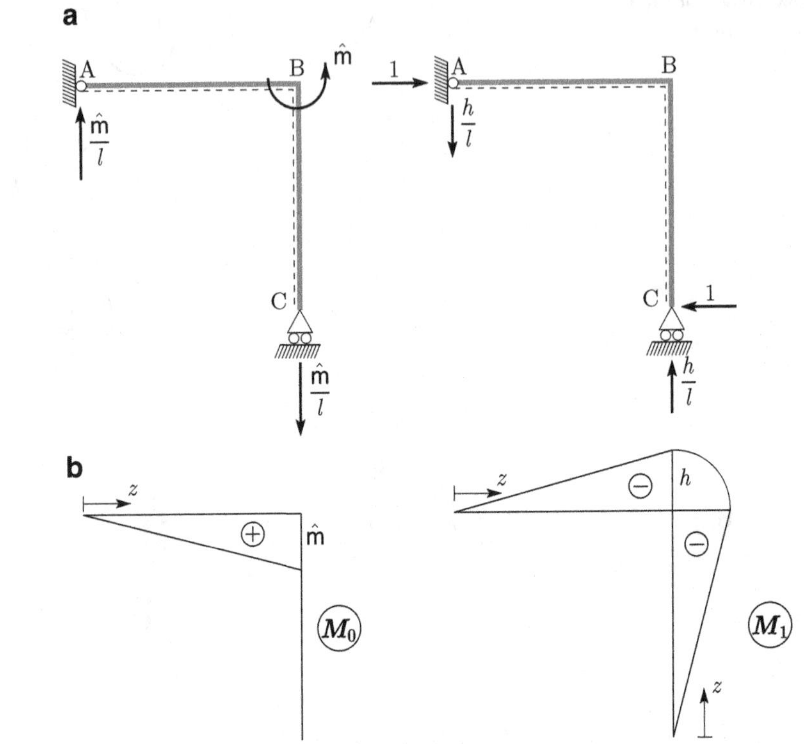

Fig. 8.70 (a) 0-system, (b) 1-system

$$\eta_{10} + \eta_{11}X + \eta_{1m} = -\frac{X}{k_1},$$

in which η_{10} and η_{11} have the usual meaning and η_{1m} is the horizontal displacement of the point C in the so-called m-system (Fig. 8.69-c), i.e., the principal system with the elastic springs and *rigid* rods loaded by the external actions (the couple $\hat{\mathsf{m}}$) and the hyperstatic unknown X, a rigid version of the principal system. The term on the right-hand side of the equation reminds us that we have eliminated a sinking constraint (the spring in C loaded with the compressive force X shortens by X/k_1). Figure 8.70 shows the systems "0" and "1" with the reactive forces and the diagrams of bending moments M_0 and M_1.

On considering the systems "0" and "1," we have

$$L^{ext} = 1\eta_{10}, \qquad L^{inn} = \int_0^l \frac{\hat{\mathsf{m}}}{l} z \left(\frac{-zh}{l}\right) \frac{dz}{EI}.$$

From $L^{ext} = L^{inn}$, we get

$$\eta_{10} = -\frac{\hat{m}hl}{3EI}.$$

In an analogous way, on considering the 1-system, we obtain

$$L^{ext} = 1\eta_{11}, \qquad L^{inn} = \int_0^l \frac{h^2z^2}{l^2EI}dz + \int_0^h \frac{z^2}{EI}dz,$$

and their equality yields

$$\eta_{11} = \frac{h^2l}{3EI} + \frac{h^3}{3EI}.$$

For calculating η_{1m}, we consider the system "m" to evaluate displacements and the system "1" for virtual action characteristics, obtaining

$$L^{ext} = 1\eta_{1m} + \left(\frac{\hat{m}}{l} - X\frac{h}{l}\right)\left(\frac{h}{l}\right)\frac{1}{k_2}, \qquad L^{inn} = 0.$$

The equality between the two work values yields

$$\eta_{1m} = -\frac{\hat{m}h}{l^2k_2} + \frac{h^2}{l^2k_2}X,$$

so that from the Müller–Breslau equation, we obtain

$$X = \frac{\dfrac{\hat{m}hl}{3EI} + \dfrac{\hat{m}h}{l^2k_2}}{\dfrac{1}{k_1} + \dfrac{h^2}{l^2k_2} + \dfrac{h^3}{3EI} + \dfrac{h^2l}{3EI}}.$$

Alternatively, without computing separately the terms of the Müller–Breslau equation, we can write directly the whole external work and inner (or internal) work on the structure, namely

$$L^{ext} = -1\frac{X}{k_1} + \left(\frac{\hat{m}}{l} - X\frac{h}{l}\right)\left(\frac{h}{l}\right)\frac{1}{k_2},$$

$$L^{inn} = \int_{st}\left(\frac{M_1(M_0 + M_1X)}{EI}\right)ds$$

$$= \frac{1}{EI}\int_0^l \left(\frac{-zh}{l}\right)\left(\frac{-zhX}{l} + \frac{z\hat{m}}{l}\right)dz + \frac{1}{EI}\int_0^h z^2X dz.$$

By equating the two values, we obtain the value of X.

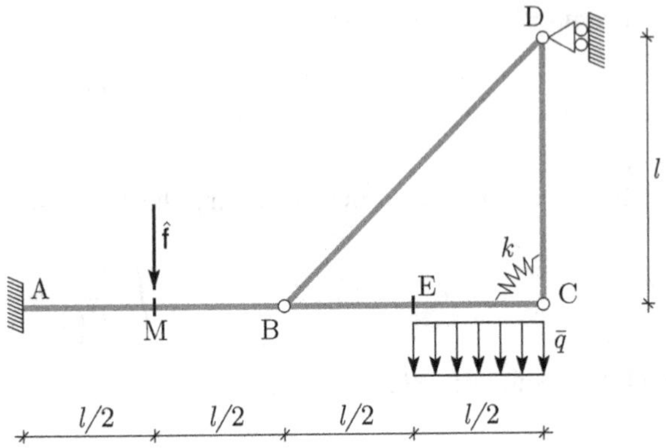

Fig. 8.71 Hyperstatic structure with an internal rotational elastic spring

Exercise 30. *The hyperstatic structure in Figure 8.71 consists of rods with uniform sections and materials. Choose a hyperstatic unknown, find its value, and draw the diagrams of the action characteristics. Neglect shear and axial strains for all the rods with the exception of the pendulum* $\overline{\mathrm{BD}}$*. Data:* $\mathrm{EI} = 2l^2 EA$, $k = \mathrm{EI}/l$, $\hat{\mathrm{f}} = \bar{q}l/2$.

Remarks and solution. If we consider the rod $\overline{\mathrm{BD}}$ to be an internal constraint, a pendulum, and the elastic spring in C to be rigid, the system comprises two rods, and there are seven simple constraints. The part $\overline{\mathrm{AMB}}$ is fixed for the joint in A; the part $\overline{\mathrm{BCD}}$ is also fixed, because it is constrained by an independent hinge (B) and carriage (D). Consequently, the structure does not admit rigid displacements, and it is one time hyperstatic ($\hat{h} = 1$).

Figure 8.72 shows a possible choice for the hyperstatic unknown X, i.e., the axial force in the pendulum $\overline{\mathrm{BD}}$, and the corresponding principal scheme. The associated Müller–Breslau equation is then

$$\eta_{10} + \eta_{11} X + \eta_{1m} = -\frac{Xl\sqrt{2}}{EA}. \tag{8.73}$$

The term on the right-hand side corresponds to the elongation of the pendulum $\overline{\mathrm{BD}}$ subjected to the axial force X. Figures 8.73 and 8.74 show the diagrams of the bending moments of the systems "0" and "1," respectively, and Table 8.4 collects the equations of these action characteristics with reference to the abscissas z in the figures.

The coefficients η_{10} and η_{11} are

$$\eta_{10} = \int_{st} \frac{M_0 M_1}{\mathrm{EI}} ds, \qquad \eta_{11} = \int_{st} \frac{M_1 M_1}{\mathrm{EI}} ds.$$

To find η_{1m}, we consider the systems "m" and "1," obtaining

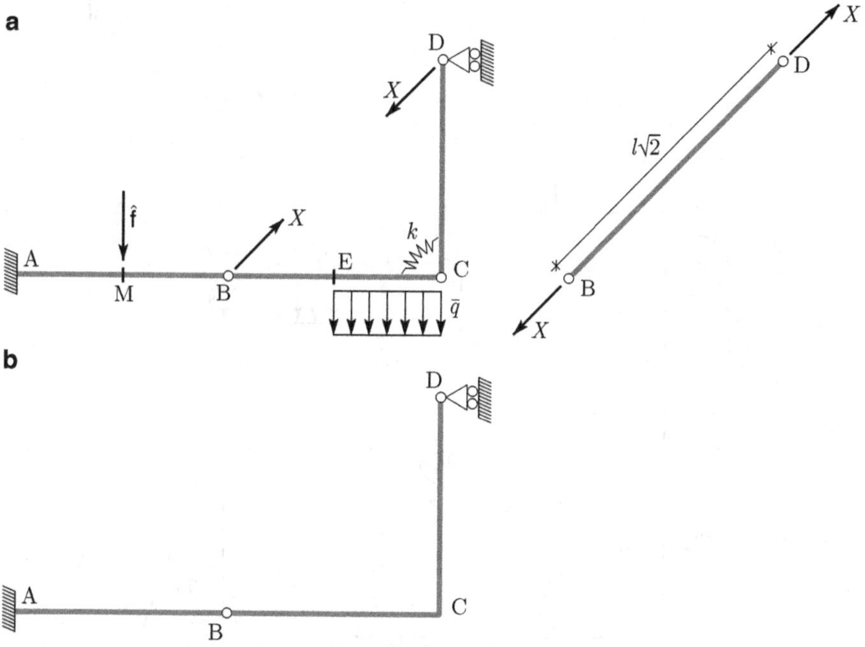

Fig. 8.72 (a) Choice of the hyperstatic unknown, (b) principal structure

$$L^{ext} = 1\eta_{1m}, \qquad L^{inn} = \frac{l}{k\sqrt{2}}\left(\frac{3\hat{\mathrm{f}}l}{4} + X\frac{l}{\sqrt{2}}\right);$$

their equality yields the identity

$$\eta_{1m} = \frac{l}{k\sqrt{2}}\left(\frac{3\hat{\mathrm{f}}l}{4} + X\frac{l}{\sqrt{2}}\right).$$

We also get

$$\eta_{10} = \frac{35\sqrt{2}l^3}{128EI}\hat{\mathrm{f}}, \qquad \eta_{11} = \frac{l^3}{3EI},$$

so that equation (8.73) gives

$$X = \left(\frac{1245\sqrt{2}}{16832} - \frac{747}{2104}\right)\hat{\mathrm{f}} \approx -0.25\hat{\mathrm{f}}.$$

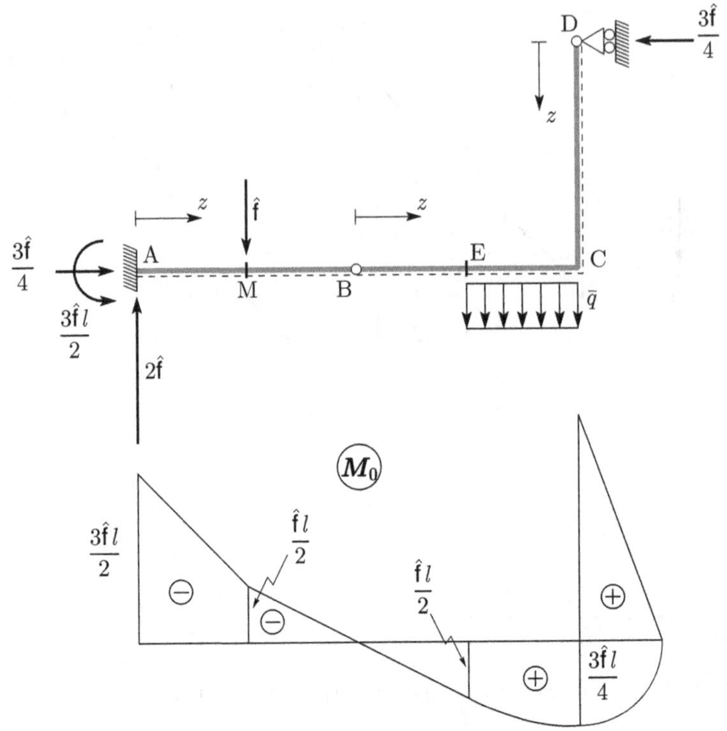

Fig. 8.73 0-system

The pendulum $\overline{BD}$ is then a strut, instead of a tie as initially assumed. Figure 8.75 shows the diagrams of N, M, and T obtained by superposition of those pertaining to the 0-system and those of the 1-system, multiplied by X.

Exercise 31. *The truss structure in Figure 8.76 consists of rods with uniform sections and materials. Find and draw the diagram of the axial force along the rods. Data:* $\alpha\delta\theta = \hat{f}/EA$.

Remarks and solution. Figure 8.77 shows the constraint reactions (they are denoted by R, N, and S). The balances of forces in the horizontal and vertical directions read

$$\frac{R}{\sqrt{2}} - \frac{S}{\sqrt{2}} + \hat{f} = 0, \qquad \frac{R}{\sqrt{2}} + N + \frac{S}{\sqrt{2}} - \hat{f} = 0.$$

We have two independent balance equations and three unknowns: the structure is one time hyperstatic (the balance of couples is always verified). We select $N = X$ as a hyperstatic unknown, so that

$$R = -\frac{X\sqrt{2}}{2}, \qquad S = \sqrt{2}\left(\hat{f} - \frac{X}{2}\right).$$

Fig. 8.74 1-system

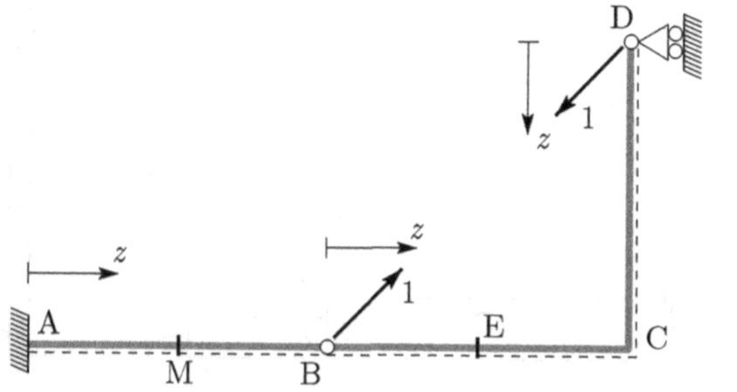

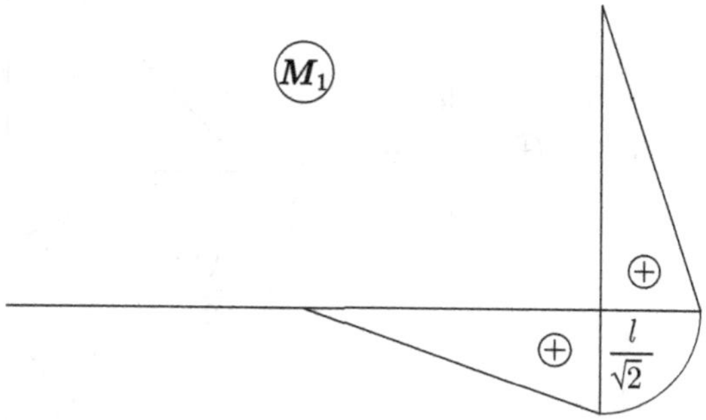

Table 8.4 Equations of bending moments in the systems "0" and "1."

Part	$M_0(z)$	$M_1(z)$
$\overline{\text{AM}},\ z \in \left[0, \dfrac{l}{2}\right]$	$-\dfrac{3}{2}\hat{f}l + 2\hat{f}z$	0
$\overline{\text{MB}},\ z \in \left[\dfrac{l}{2}, l\right]$	$\hat{f}(z-l)$	0
$\overline{\text{BE}},\ z \in \left[0, \dfrac{l}{2}\right]$	$\hat{f}z$	$\dfrac{z}{\sqrt{2}}$
$\overline{\text{EC}},\ z \in \left[\dfrac{l}{2}, l\right]$	$\dfrac{\hat{f}l}{2} - \dfrac{\bar{q}}{2}\left(z-\dfrac{l}{2}\right)^2 + \hat{f}\left(z-\dfrac{l}{2}\right)$	$\dfrac{z}{\sqrt{2}}$
$\overline{\text{DC}},\ z \in [0, l]$	$\dfrac{3}{4}\hat{f}z$	$\dfrac{z}{\sqrt{2}}$

The relevant Müller–Breslau equation for the special case treated here is

$$\eta_{10} + \eta_{11}X + \eta_{1t} = -\frac{Xl}{EA} + \alpha\delta\theta l.$$

The coefficient η_{1t} represents the contribution of the thermal variations along the two diagonal rods, whereas the coefficient $\alpha\delta\theta l$ represents the analogous contribution along the rod $\overline{\text{AB}}$. We get (Figs. 8.78 and 8.79)

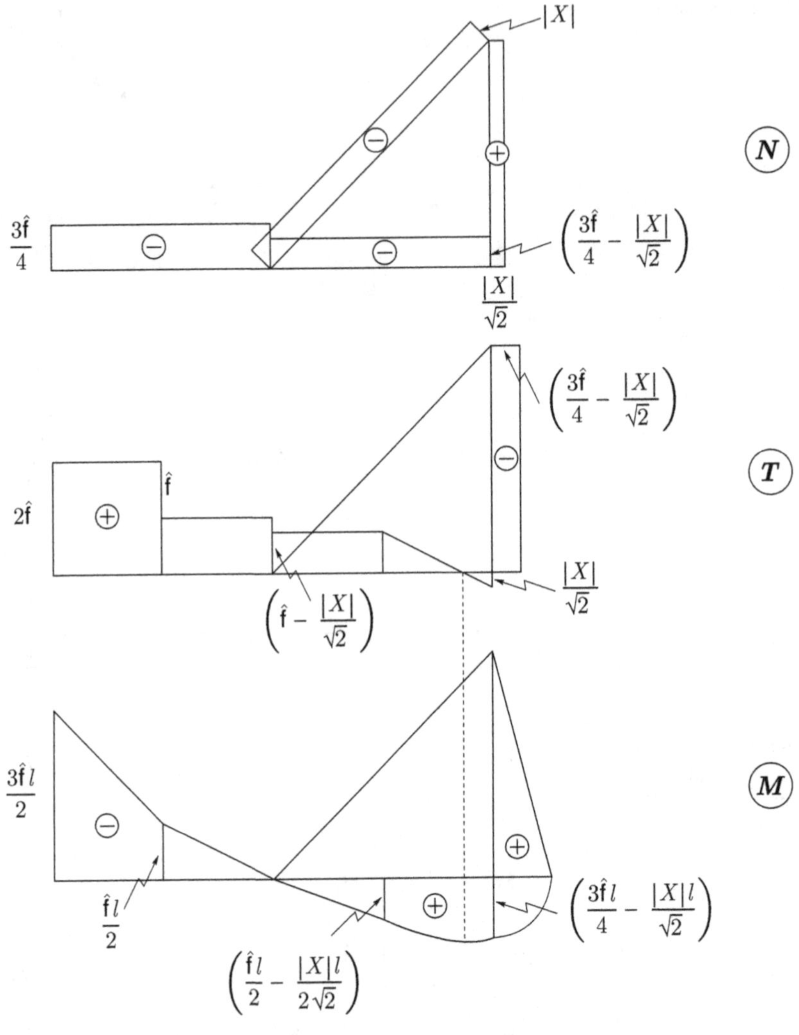

Fig. 8.75 Diagrams of the action characteristics

$$\eta_{10} = \int_0^{l\sqrt{2}} \left(-\hat{f}\sqrt{2}\right) \frac{\sqrt{2}}{2} \frac{dz}{EA} = -\frac{\hat{f}l\sqrt{2}}{EA},$$

$$\eta_{11} = 2\int_0^{l\sqrt{2}} \frac{dz}{2EA} = \frac{l\sqrt{2}}{EA},$$

$$\eta_{1t} = 2\int_0^{l\sqrt{2}} \frac{\sqrt{2}}{2} \alpha\delta\theta \, dz = 2\alpha\delta\theta l,$$

Fig. 8.76 Truss system with thermal effects

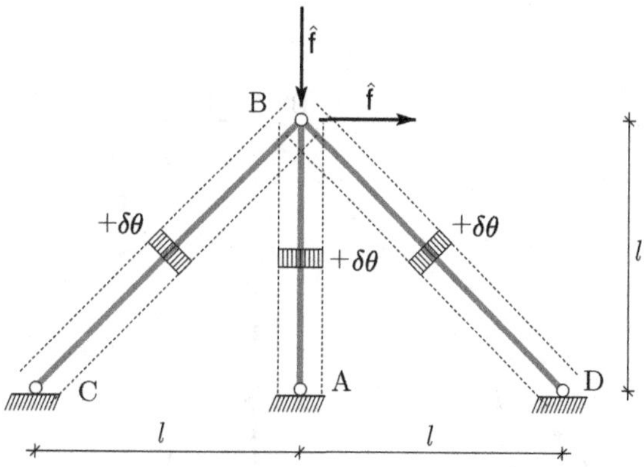

so that the hyperstatic unknown is

$$X = \hat{f}\frac{(\sqrt{2}-1)}{(\sqrt{2}+1)} \approx 0.171\hat{f}.$$

The positive sign indicates that the pendulum $\overline{\mathrm{AB}}$ is a strut. The values of S and R are

$$S = \hat{f}\frac{(3\sqrt{2}+2)}{2(\sqrt{2}+1)} \approx 1.29\hat{f}, \qquad R = \hat{f}\frac{(\sqrt{2}-2)}{2(\sqrt{2}+1)} \approx -0.121\hat{f}.$$

The rod $\overline{\mathrm{DB}}$ is a strut, while $\overline{\mathrm{CB}}$ is a tie. Figure 8.80 shows the diagram of the axial force N.

Exercise 32. *The rod in Figure 8.81 has constant section and is made of a uniform material. Determine and draw the diagrams of the action characteristics, neglecting shear strain.*

Remarks and solution. The structure is twice hyperstatic, because there are five simple constraints and the rod does not admit rigid displacements. Figure 8.82 shows two possible choices of the principal structure with the corresponding hyperstatic unknowns. For the specific analysis dealt with here, we consider the second scheme (for the continuous rods it is generally used the first scheme, instead). The relevant Müller–Breslau equations in this case are

$$\eta_{10} + \eta_{11}X_1 + \eta_{12}X_2 = \eta_{1B},$$

$$\eta_{20} + \eta_{21}X_1 + \eta_{22}X_2 = \eta_{2C},$$

where η_{1B} and η_{2C} are the vertical displacements of the sections B and C that vanish due to the translational constraints in the original structure. The systems "0," "1," and "2" are represented in Figure 8.83, where the physical meaning of

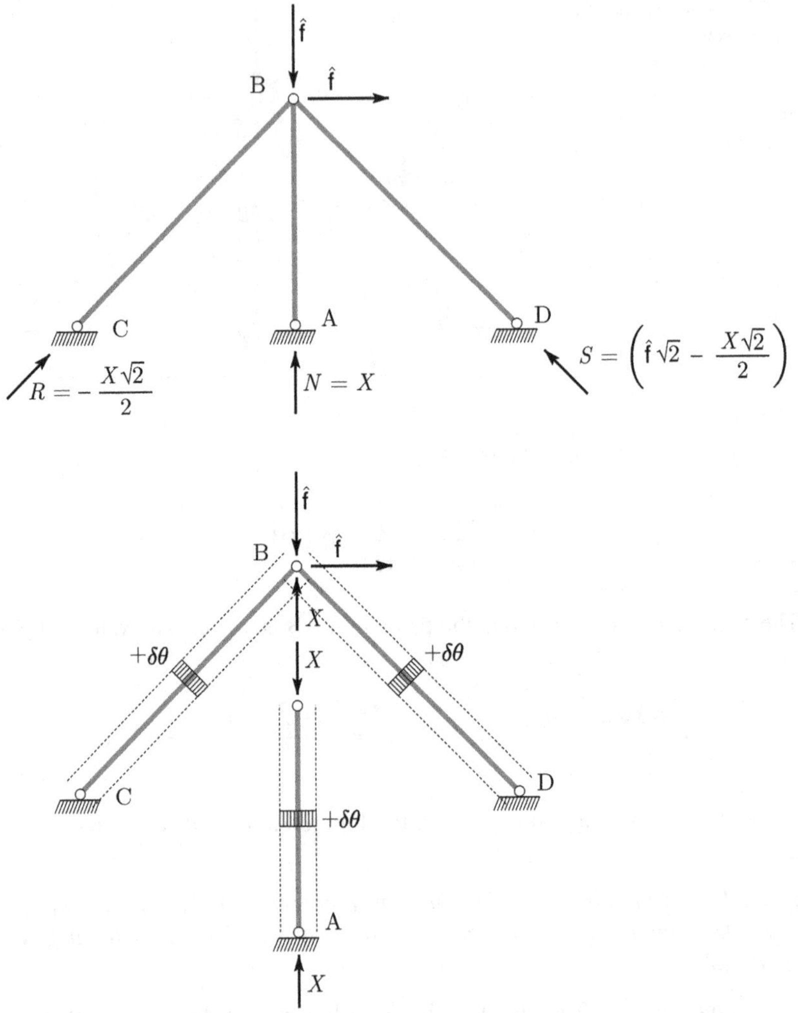

Fig. 8.77 Choice of the hyperstatic unknown

the coefficients η_{ij} appears. Using as virtual forces the action characteristics of the 1-system first, and those of the 2-system afterward, we write

$$L^{ext} = 0, \qquad L^{inn} = \int_{st} \left(\frac{M_1(M_0 + M_1X_1 + M_2X_2)}{EI} \right) ds,$$

$$L^{ext} = 0, \qquad L^{inn} = \int_{st} \left(\frac{M_2(M_0 + M_1X_1 + M_2X_2)}{EI} \right) ds,$$

where "st" indicates that the integration domain is the whole structure.

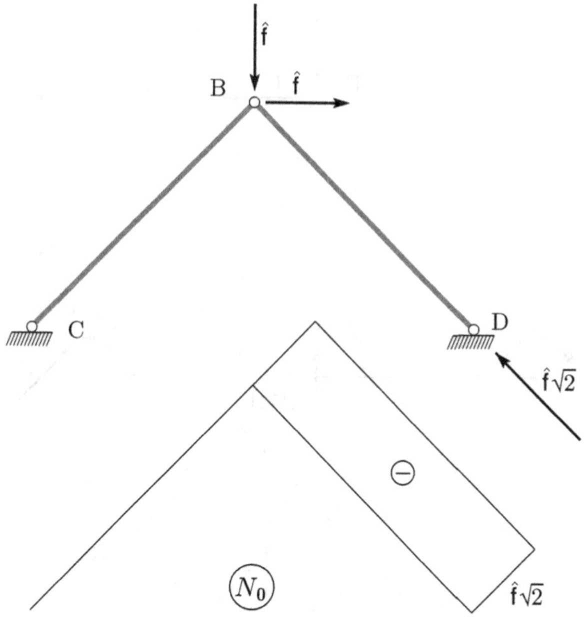

Fig. 8.78 0-system

Fig. 8.79 1-system

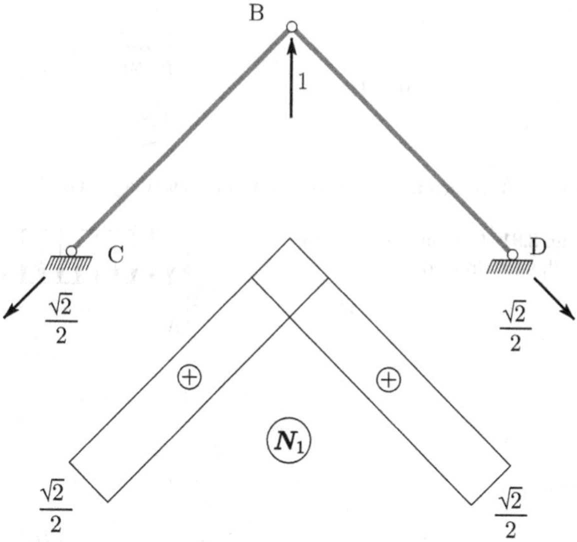

The identity between external work and inner (or internal) work leads to

$$-\frac{17\bar{q}l^4}{384EI}+\frac{l^3}{24EI}X_1+\frac{5l^3}{48EI}X_2=0,$$

$$-\frac{\bar{q}l^4}{8EI}+\frac{5l^3}{48EI}X_1+\frac{l^3}{3EI}X_2=0,$$

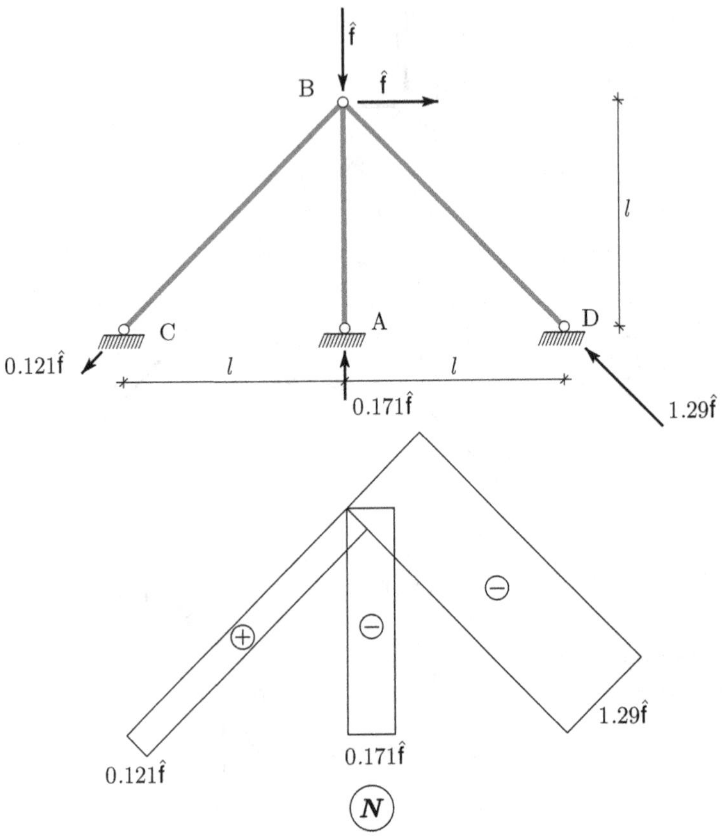

Fig. 8.80 Structure with the reaction forces and diagram of the axial force N

Fig. 8.81 Continuous rod with uniform load

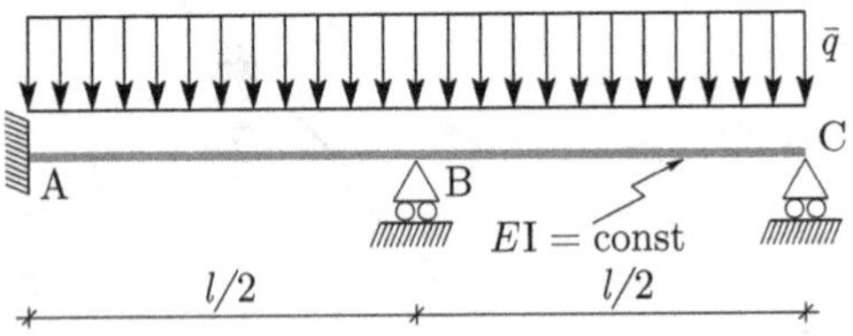

i.e.,

$$X_1 = \frac{4\bar{q}l}{7}, \qquad X_2 = \frac{11\bar{q}l}{56}.$$

The balance equations yield the reactive forces Y_A and M_A ($H_A = 0$):

$$Y_A = \frac{13\bar{q}l}{56}, \qquad M_A = \frac{\bar{q}l^2}{56}.$$

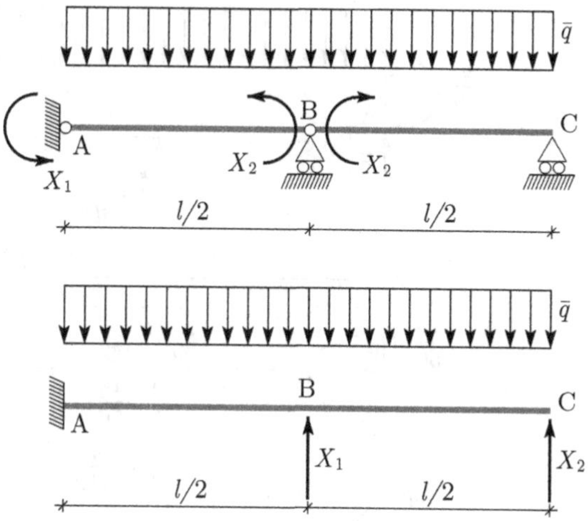

Fig. 8.82 Principal structures with the load and the hyperstatic unknowns

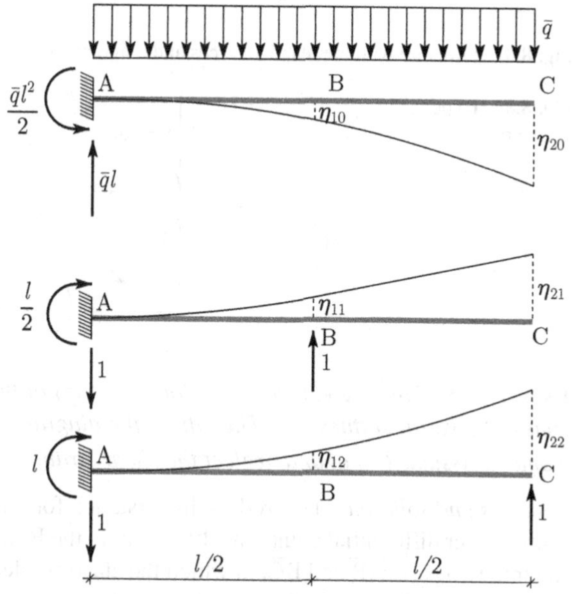

Fig. 8.83 Systems "0," "1," and "2."

Figure 8.84 shows the diagrams of the shear action T and the bending moment M (the axial force N is zero everywhere). To calculate the maximum bending moments in the two spans, it is sufficient to find the abscissas z_1 and z_2 where the shear is zero. For example, for the span $\overline{AB}$, we get

$$z_1 = \frac{13}{56}l, \qquad M(z_1) = -\frac{\bar{q}l^2}{56} + Y_A \frac{13}{56}l - \frac{\bar{q}}{2}\left(\frac{13}{56}l\right)^2 = \frac{57}{6272}\bar{q}l^2.$$

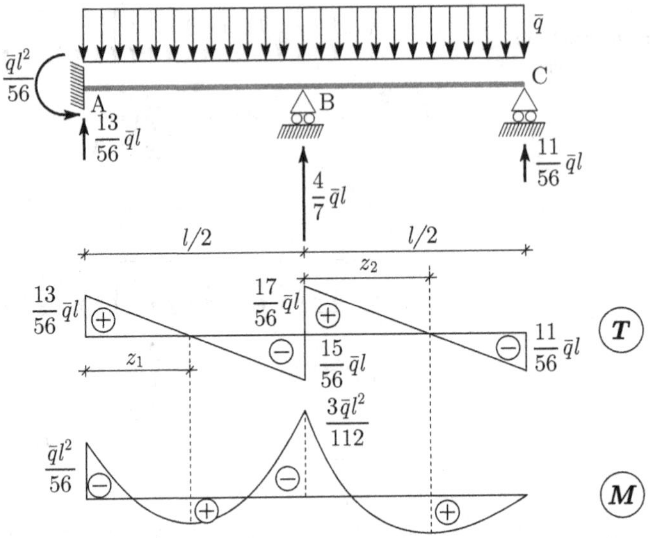

Fig. 8.84 Structure with the reactive forces and diagrams of shear force and bending moment

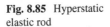

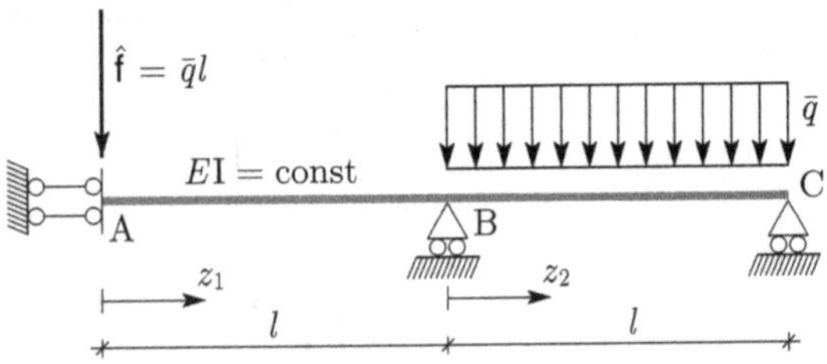

Fig. 8.85 Hyperstatic elastic rod

Exercise 33. *Find the elastic deflection $z \to v(z)$ of the rod in Figure 8.85 having constant flexural stiffness EI. Then draw the diagrams of shear force and bending moment. Assume $\hat{f} = \bar{q}l$ and neglect the shear strain.*

Remarks and solution. The rod is hyperstatic; for this reason, we consider the fourth-order differential equation $EIv^{IV} = \bar{q}$, in the Bernoulli scheme. We divide the rod into two parts $\overline{AB}$ and $\overline{BC}$, in which the elastic deflections are denoted by v_1 and v_2, and the abscissas z_1 and z_2 vary in the interval $[0, l]$. The relevant equations are

$$EI\frac{d^4v_1}{dz_1^4} = 0 , \qquad EI\frac{d^4v_2}{dz_2^4} = \bar{q} .$$

Integration is immediate. There are eight boundary conditions, two in A, two in C, and four in B. They are the kinematic and the static conditions are imposed by the constraints and by continuity in the middle section. Such conditions are

$$\text{Point A}: \begin{cases} \dfrac{d\mathsf{v}_1}{dz_1}(0) = 0, \\ EI\dfrac{d^3\mathsf{v}_1}{dz_1^3}(0) = \hat{\mathsf{f}}. \end{cases}$$

$$\text{Point B}: \begin{cases} \mathsf{v}_1(l) = 0, \\ \mathsf{v}_2(0) = 0, \\ \dfrac{d\mathsf{v}_1}{dz_1}(l) = \dfrac{d\mathsf{v}_2}{dz_2}(0), \\ \dfrac{d^2\mathsf{v}_1}{dz_1^2}(l) = \dfrac{d^2\mathsf{v}_2}{dz_2^2}(0). \end{cases}$$

$$\text{Point C}: \begin{cases} \mathsf{v}_2(l) = 0, \\ \dfrac{d^2\mathsf{v}_2}{dz_2^2}(l) = 0. \end{cases}$$

They imply

$$C_1 = \bar{q}l, \qquad C_2 = -\frac{19\bar{q}l^2}{32}, \quad C_3 = 0, \qquad C_4 = \frac{25\bar{q}l^4}{192},$$
$$D_1 = -\frac{29\bar{q}l}{32}, \quad D_2 = \frac{13\bar{q}l^2}{32}, \quad D_3 = -\frac{3\bar{q}l^3}{32}, \quad D_4 = 0.$$

Consequently, we obtain

$$\mathsf{v}_1 = \frac{\bar{q}l}{6EI}z_1^3 - \frac{19\bar{q}l^2}{64EI}z_1^2 + \frac{25\bar{q}l^4}{192EI},$$
$$\mathsf{v}_2 = \frac{\bar{q}z_2^4}{24EI} - \frac{29\bar{q}l}{192EI}z_2^3 + \frac{13\bar{q}l^2}{64EI}z_2^2 - \frac{3\bar{q}l^3}{32EI}z_2,$$
$$M_1 = -\bar{q}lz_1 + \frac{19\bar{q}l^2}{32},$$
$$M_2 = -\frac{\bar{q}z_2^2}{2} + \frac{29\bar{q}l}{32}z_2 - \frac{13\bar{q}l^2}{32},$$
$$T_1 = -\bar{q}l,$$
$$T_2 = -\bar{q}z_2 + \frac{29\bar{q}l}{32}.$$

Figure 8.86 shows the rod with the diagrams of the elastic deflection v and the action characteristics.

Exercise 34. *The structure in Figure 8.87 consists of rods with constant sections and uniform material. Find the diagrams of the action characteristics, neglecting shear strain. Further data:* $EI/EA = l^2/5$, $k = EI/l$, $\alpha\delta\theta = \bar{q}l^3/(2EI)$, $2\alpha\delta\theta/h = \bar{q}l^2/(EI)$, $\delta\vartheta_{in} = \bar{q}l^3/(4EI)$.

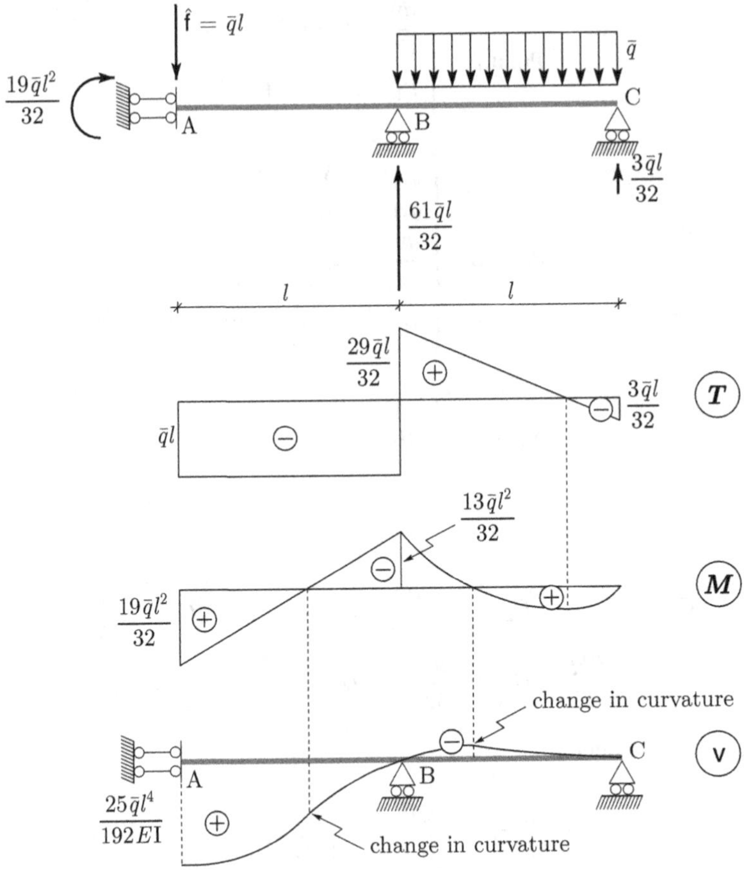

Fig. 8.86 Elastic deflection and diagrams of bending moment and shear force

Some elements of the solution: The structure is one time hyperstatic; $M_E = 15\bar{q}l^2/64$, counterclockwise.

Exercise 35. *The structure in Figure 8.88 consists of rods with constant sections and uniform material. Find the diagrams of the action characteristics, neglecting shear and axial strains. Further data:* $k = \mathrm{EI}/l$, $\delta = \bar{q}l^4/(2\mathrm{EI})$.

Some elements of the solution: The structure is one time labile, one time hyperstatic; $M_C = 21\bar{q}l^2/80$, clockwise.

Exercise 36. *The structure in Figure 8.89 consists of rods with constant sections and uniform material. Find the diagrams of the action characteristics, neglecting shear and axial strains. Datum:* $2\alpha\delta\theta/h = 2\bar{q}l^2/(3\mathrm{EI})$. *The carriages are inclined at* $45°$.

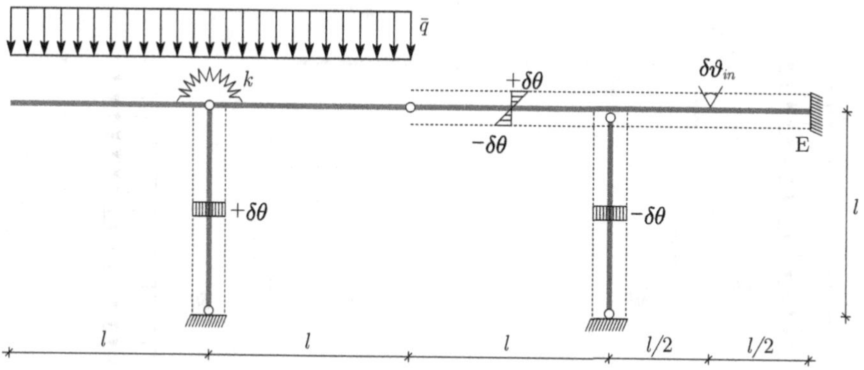

Fig. 8.87 Structure with linear thermal loads

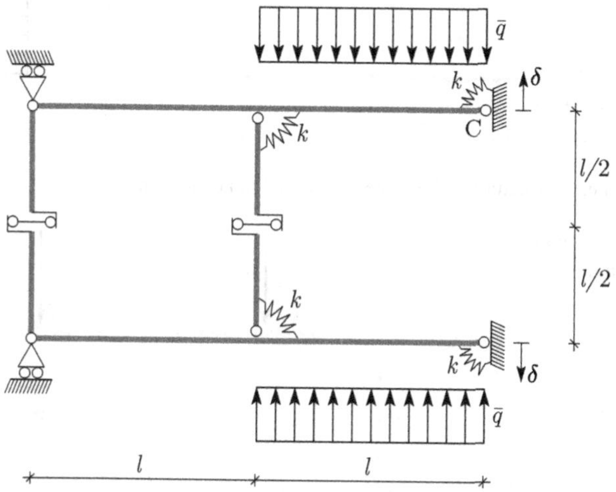

Fig. 8.88 Structure with elastic constraints

Some elements of the solution: The structure is one time labile, one time hyperstatic; $R_A = \dfrac{3\sqrt{2}}{2}\bar{q}l$.

Exercise 37. *The structure in Figure 8.90 is made of rods with constant sections and uniform material. Find the diagrams of the action characteristics, neglecting shear and axial strains.*

Some elements of the solution: The structure is one time labile, one time hyperstatic; $N_{AB} = -\dfrac{57}{80}\hat{\mathrm{f}}$.

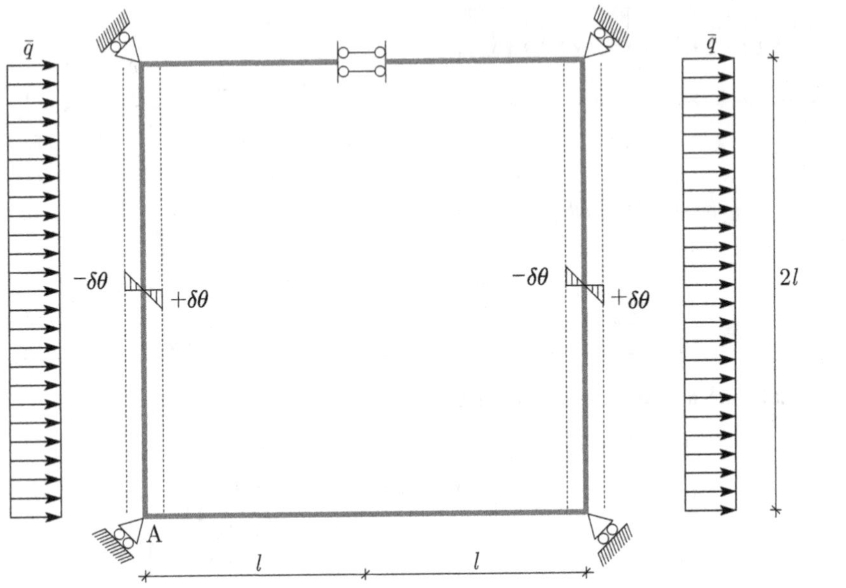

Fig. 8.89 Symmetric structure with skew-symmetric load condition

Fig. 8.90 Structure with concentrated loads

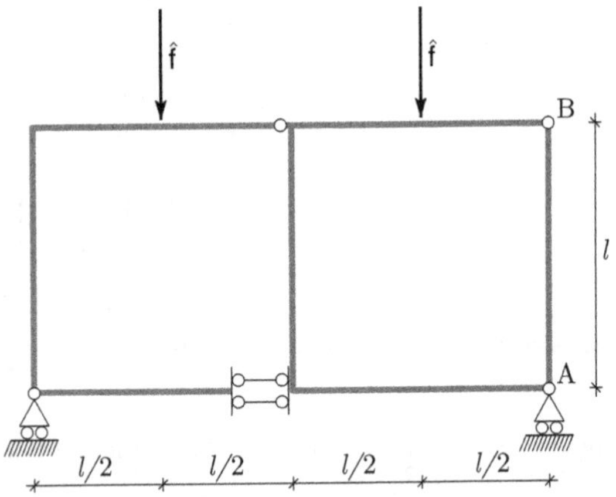

Exercise 38. *The structure in Figure 8.91 consists of a rigid rod and two deformable rods with constant sections and uniform material. Find the diagrams of the action characteristics, neglecting shear strain. Datum:* $\alpha\delta\theta = \hat{f}/(2EA)$.

Some elements of the solution: The structure is one time hyperstatic; $N_{AB} = -\dfrac{\hat{f}}{2}$.

Exercise 39. *The structure in Figure 8.92 consists of a rigid rod and other deformable rods with constant sections and uniform material. Find the diagrams*

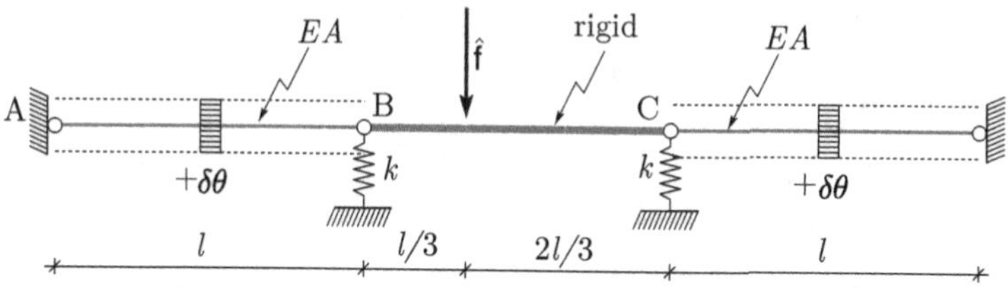

Fig. 8.91 Rod on elastic springs

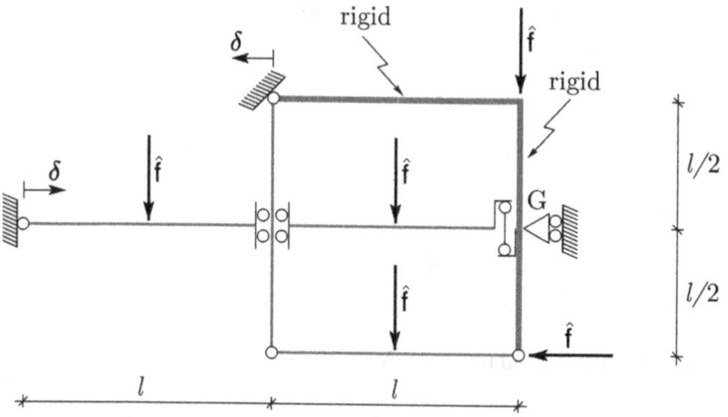

Fig. 8.92 Structure with inelastic constraint failures

of the action characteristics, neglecting shear and axial strains. Datum: $\delta = \hat{f}l^3/(12EI)$.

Some elements of the solution: The structure is one time hyperstatic; $H_G = 3\hat{f}$, directed rightward.

Exercise 40. *The structure in Figure 8.93 consists of deformable rods with constant sections and uniform material. Find the diagrams of the action characteristics, neglecting shear strain, and calculate the horizontal displacement of the point* C. *Further data:* $k = 2EA/l$, $\alpha\delta\theta = \hat{f}/(10EA)$, $EI/EA = l^2/5$.

Some elements of the solution: The structure is one time labile and one time hyperstatic; $N_{AC} = -\dfrac{3\hat{f}}{4(2+\sqrt{2})}$, $w_C = \dfrac{\hat{f}l}{EA}\dfrac{(21\sqrt{2}+16)}{20\sqrt{2}(2+\sqrt{2})}$, directed rightward.

Exercise 41. *The structure in Figure 8.94 is made of deformable rods with constant sections and uniform material. Find the diagrams of the action characteristics, neglecting shear and axial strains. Further data:* $k = EI/l$, $\hat{f} = \bar{q}l$.

Some elements of the solution: The structure is one time hyperstatic; $M_A = \dfrac{19}{64}\hat{f}l$, counterclockwise.

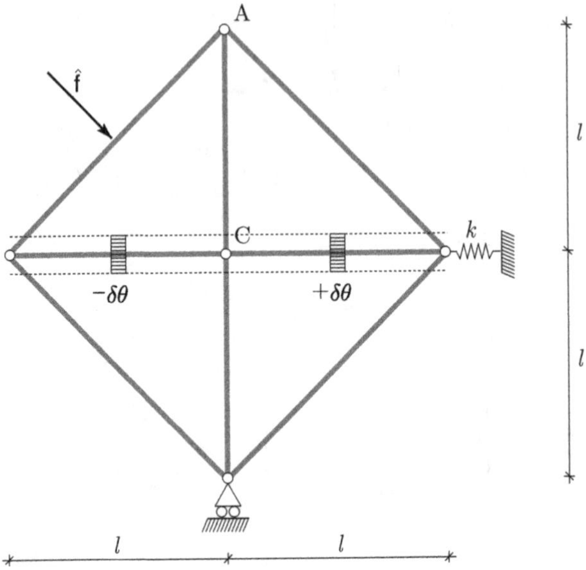

Fig. 8.93 Structure with loads and thermal effects

Fig. 8.94 Symmetric structure with skew-symmetric load condition

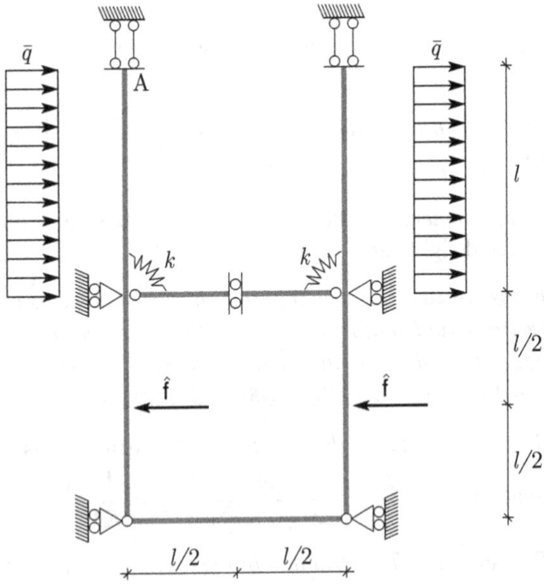

Exercise 42. *The structure in Figure 8.95 consists of deformable rods with constant sections and uniform material. Find the diagrams of the action characteristics, neglecting shear and axial strains. Further data:* $k_1 = 3EI/(8l)$, $k_2 = EI/l^3$.

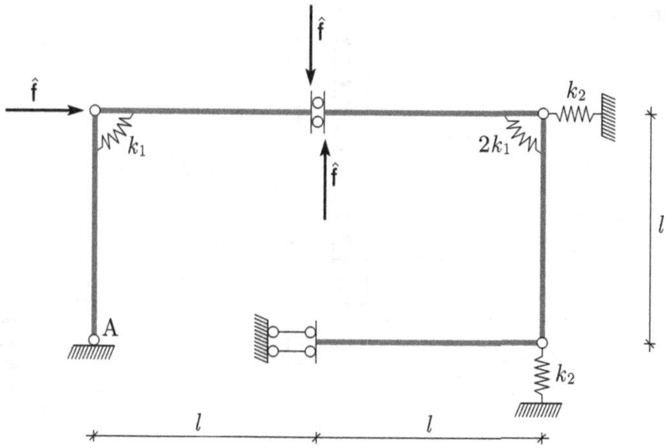

Fig. 8.95 Structure with concentrated forces and elastic constraints

Some elements of the solution: The structure is one time hyperstatic; $H_A = \frac{4}{5}\hat{f}$, directed rightward.

Exercise 43. *The structure in Figure 8.96 consists of three rigid parts and other deformable rods with constant sections and uniform material. Find the diagrams of the action characteristics, neglecting shear and axial strains. Datum:* $k = EI/l$.

Some elements of the solution: The structure is one time labile and one time hyperstatic; $M(B) = \frac{\bar{q}l^2}{6}$.

Exercise 44. *The structure in Figure 8.97 consists of four rigid rods and four deformable rods with constant sections and uniform material. Find the diagrams of the action characteristics, neglecting shear and axial strains with the exception of the two vertical pendulums with axial stiffness equal to EA. Assume* $k_1 = EAl$, $k_2 = 2EAl$, $\alpha\delta\theta = 3\hat{f}/(2EA)$.

Some elements of the solution: The structure is twice labile and one time hyperstatic; $N_{AB} = -\frac{4}{3}\hat{f}$.

Exercise 45. *The structure in Figure 8.98 consists of a rigid rod (the part* $\overline{BC}$*) and a deformable rod (the part* $\overline{AB}$*). Find the elastic deflection* $z \rightarrow v(z)$*, according to the Bernoulli model. Assume* $k_1 = EI/l$, $k_2 = EI/l^3$.

Some elements of the solution: $v_B = \frac{2\hat{f}l^3}{3EI}$, $v_C = \frac{\hat{f}l^3}{2EI}$.

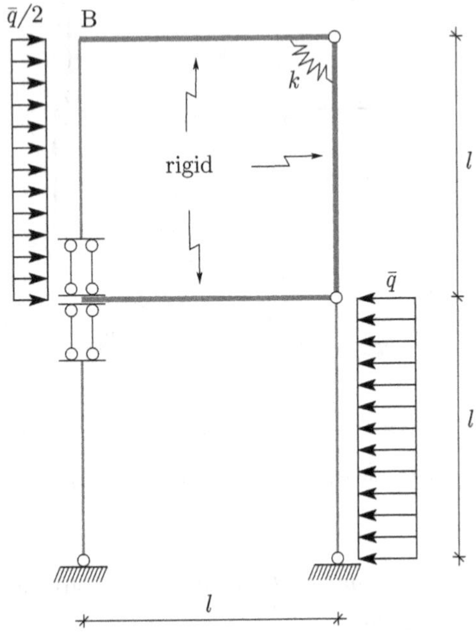

Fig. 8.96 Structure with rigid rods

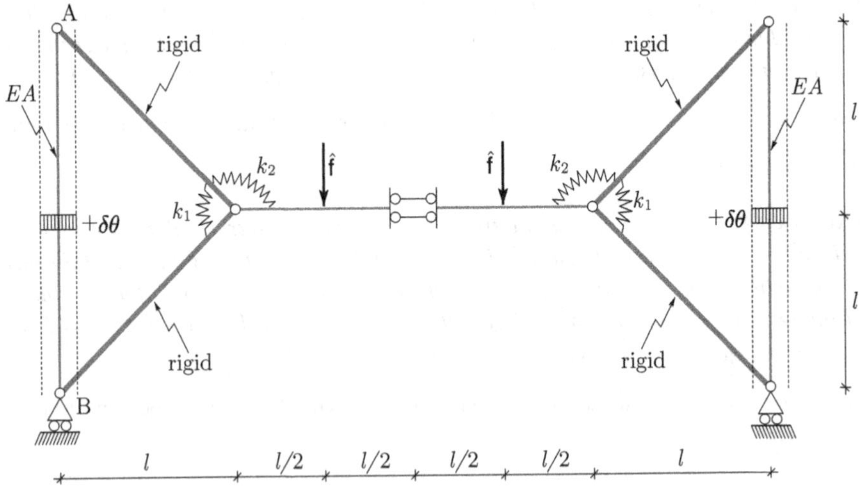

Fig. 8.97 Hyperstatic symmetric structure

Exercise 46. *The elastic rod in Figure 8.99 has constant section and consists of uniform material. Determine the maximum value of the elastic deflection* $\mathsf{v}(z)$*, according to the Bernoulli model. Assume* $\bar{q}(z) = \bar{q}_0 \sin \dfrac{2\pi z}{l}$, $\delta = \dfrac{\bar{q}_0 l^4}{20EI}$, $\vartheta_{in} = \dfrac{\bar{q}_0 l^3}{EI}$, $l = 400\,\text{cm}$, $\bar{q}_0 = 5\,\text{kN/m}$, $E = 210000\,\text{N/mm}^2$, $\text{I} = 6000\,\text{cm}^4$.

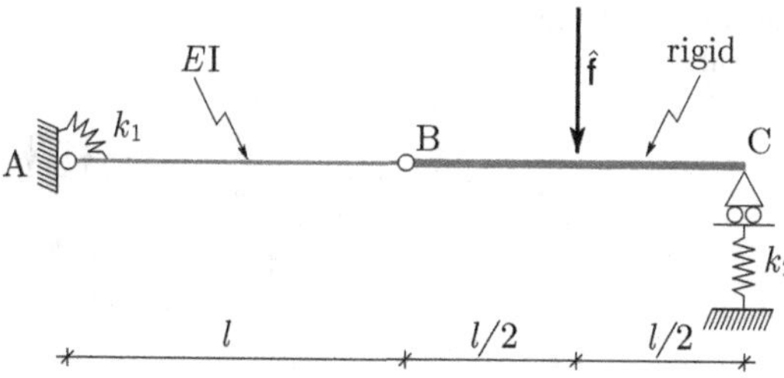

Fig. 8.98 A structure with elastic constraints

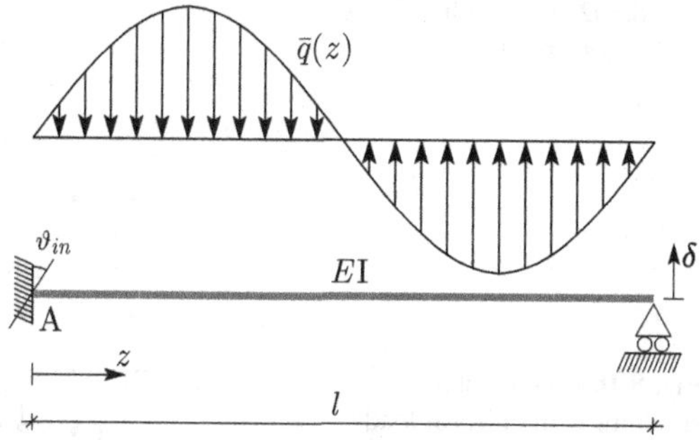

Fig. 8.99 Continuous rod with inelastic constraint failures under sinusoidal load

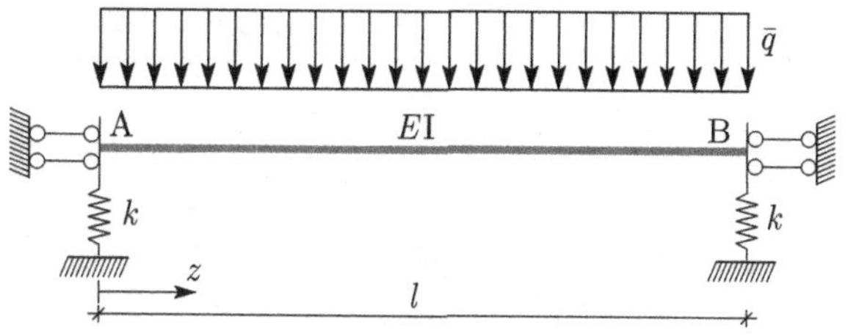

Fig. 8.100 Rod with elastic constraints

Solution. $\mathsf{v}_{max} = 1.84$ cm at $z = 157.6$ cm.

Exercise 47. *The elastic rod in Figure 8.100 has constant section and is made of a uniform material. Find the elastic deflection $z \to \mathsf{v}(z)$, according to the Bernoulli scheme, and draw the corresponding graph. Assume $k = 24\mathrm{EI}/l^3$.*

Some elements of the solution: $\mathsf{v}_A = \mathsf{v}_B = \dfrac{\bar{q}l^4}{48EI}$.

Exercise 48. *The elastic rod in Figure 8.101 has constant section and is made of a uniform material. Find the elastic deflection $z \to \mathsf{v}(z)$ and draw the corresponding graph.*

Solution. $\mathsf{v}(z) = \dfrac{\bar{q}_0 z^5}{120lEI} - \dfrac{\bar{q}_0 l^2}{48EI}z^2 + \dfrac{\bar{q}_0 l^4}{80EI}$.

Exercise 49. *The elastic rod in Figure 8.102 has constant section and is made of a uniform material. Find the elastic deflection $z \to \mathsf{v}(z)$ and its maximum value v_{max}. Assume $k = 3\mathrm{EI}/(2l^3)$.*

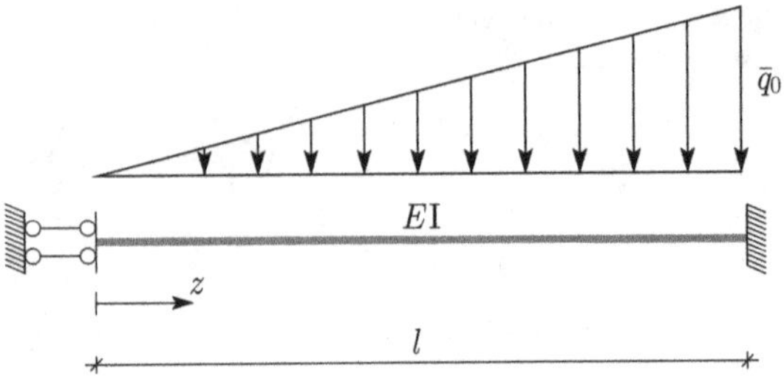

Fig. 8.101 Rod with triangular load

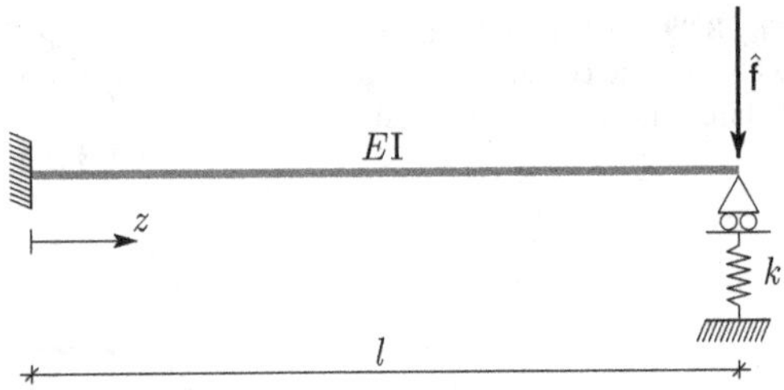

Fig. 8.102 Rod with a single elastic constraint

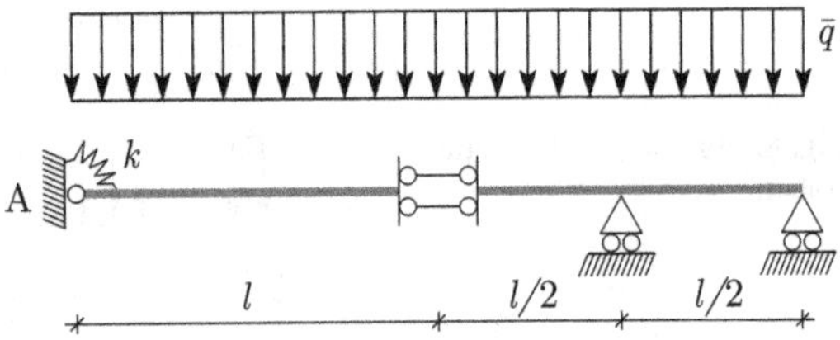

Fig. 8.103 Hyperstatic structure with uniform load

Solution. $\mathsf{v}(z) = -\dfrac{\hat{\mathsf{f}}z^3}{9EI} + \dfrac{\hat{\mathsf{f}}l}{3EI}z^2$, $\mathsf{v}_{max} = \mathsf{v}(l) = \dfrac{2\hat{\mathsf{f}}l^3}{9EI}$.

Exercise 50. *The structure in Figure 8.103 includes rods with constant section and uniform material. Find the rotation of the section* A*, neglecting the shear strain. Assume* $k = EI/l$.

Solution. The structure is one time hyperstatic; $\vartheta_A = \dfrac{121\bar{q}l^2}{512k}$, clockwise.

Exercise 51. *The structure in Figure 8.104 includes deformable rods with constant section and uniform material. Find the length variation of the pendulum* $\overline{AB}$*, neglecting shear strain. Assume* $EA = 6EI/(5l^2)$, $k_1 = EI/(4l^3)$, $k_2 = l^2 k_1$, $\alpha\delta\theta = 43\bar{q}l^3/(48EI)$.

Solution. The structure is one time hyperstatic; $\delta l_{AB} = \bar{q}l^4/(6EI)$.

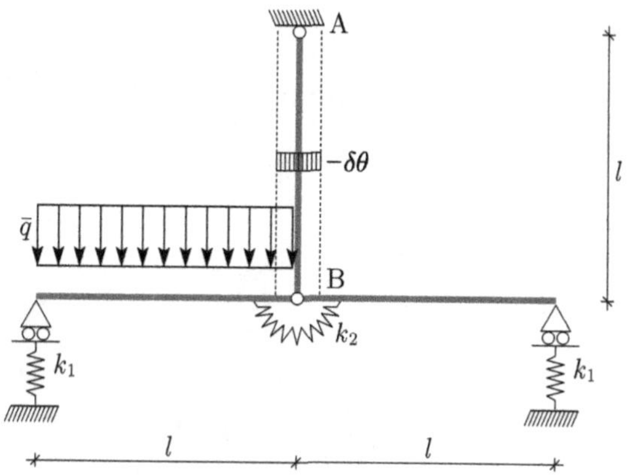

Fig. 8.104 Hyperstatic structure with uniform load and thermal variation

Chapter 9
Euler's Analysis of Critical Loads: Bifurcation Phenomena

9.1 Euler's Rod: The Critical Load

Consider a cantilever linear elastic inextensible and unshearable rod axially loaded (Fig. 9.1). In the straight configuration, the rod is at equilibrium, provided the load P is coaxial with the rod axis—exactly as depicted in Figure 9.1.

A question is whether we can have values of the load P allowing the equilibrium of deformed configurations under the assumption that the direction of the load does not change with the deformation. Consider a possible deformed shape, the one in Figure 9.1, determined by imposing a small strain. Then the constitutive relation (8.67) holds. At a generic section at z, the moment $M_P(z)$ of the external force P is given by

$$M_P(z) = -\mathsf{P}(\bar{\mathsf{v}} - \mathsf{v}(z)),$$

where $\bar{\mathsf{v}}$ is the displacement $\mathsf{v}(l)$ orthogonal to the rod's original axis. On inserting such a relation into (8.67) and putting

$$\alpha^2 = \frac{\mathsf{P}}{E\mathrm{I}}, \tag{9.1}$$

we get

$$\mathsf{v}'' + \alpha^2 \mathsf{v} - \alpha^2 \bar{\mathsf{v}} = 0, \tag{9.2}$$

which implies

$$\mathsf{v}(z) = a_1 \sin \alpha z + a_2 \cos \alpha z + \bar{\mathsf{v}},$$

with a_1 and a_2 integration constants.

P.M. Mariano, L. Galano, *Fundamentals of the Mechanics of Solids*,
DOI 10.1007/978-1-4939-3133-0_9

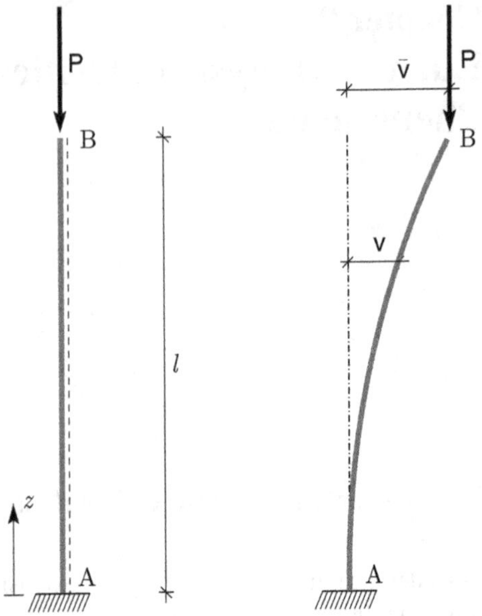

Fig. 9.1 A scheme for analyzing bifurcation phenomena in elasticity. A possible deformed shape of the cantilever rod; P does not change direction with the deformation

The boundary conditions read

$$v(0) = 0, \qquad v'(0) = 0, \qquad v(l) = \bar{v}.$$

They imply, respectively,

$$a_2 = -\bar{v}, \qquad a_1 = 0, \qquad \bar{v}(1 - \cos\alpha l) = \bar{v}.$$

The last relation requires

$$\bar{v}\cos\alpha l = 0.$$

The condition is satisfied when

$$\bar{v} = 0,$$

or

$$\cos\alpha l = 0.$$

The first condition corresponds to the undeformed rod. The second condition is satisfied when

$$\alpha l = \frac{\pi}{2} + n\pi, \qquad n = 0, 1, 2, \ldots.$$

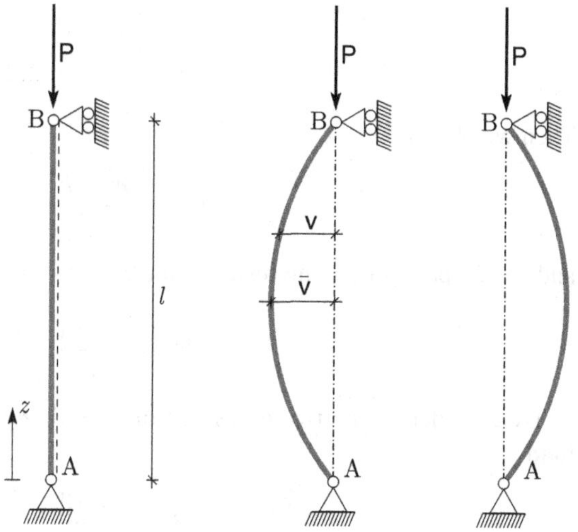

Fig. 9.2 Simply supported rod with axial load: two possible in-plane deformations

By restricting attention to the case $n = 0$, we find a critical value P_{cr} of P using the definition (9.1), namely

$$\mathsf{P}_{cr} = \frac{\pi^2 EI}{4l^2}.$$

This is a 1759 result obtained by Leonhard Euler (1707–1783).

Later (1770–1773), Joseph-Louis Lagrange (1736–1813) discussed the same problem for the supported rod, the one depicted in Figure 9.2. In this case, the question is to find a value of the load P ensuring the equilibrium of a deformed configuration like those shown in the same figure. In this case, $M_P(z) = \mathsf{P}\mathsf{v}(z)$, and equation (9.2) reduces to

$$\mathsf{v}'' + \alpha^2 \mathsf{v} = 0 \tag{9.3}$$

with boundary conditions

$$\mathsf{v}(0) = 0, \qquad \mathsf{v}(l) = 0.$$

Then equation (9.3) implies

$$\mathsf{v}(z) = a_1 \sin \alpha z + a_2 \cos \alpha z,$$

and the boundary conditions yield

$$a_2 = 0, \qquad a_1 \sin \alpha l = 0.$$

The last condition is satisfied when $a_1 = 0$, i.e., when the rod is undeformed, or

$$\sin \alpha l = 0,$$

i.e.,

$$\alpha l = n\pi, \qquad n = 1, 2, \ldots .$$

Then we obtain

$$\mathsf{P}_{cr} = \frac{n^2\pi^2 EI}{l^2},$$

and we are particularly interested in the case $n = 1$, i.e., when

$$\mathsf{P}_{cr} = \frac{\pi^2 EI}{l^2}.$$

By considering the two cases just discussed, we can write for the first **critical load**

$$\mathsf{P}_{cr} = \frac{\pi^2 EI}{l_c^2}, \tag{9.4}$$

with $l_c = 2l$ in the case of the cantilever rod and $l_c = l$ for the supported one. The validity of the formula (9.4) goes beyond the two cases above. Consider, for example, the rod in Figure 9.3 and its possible deformed shape, where the unknown reaction H_B of the support appears explicitly. In this case, the couple exerted by the external actions P and H_B on a rod section at z is given by

$$M_{PH}(z) = \mathsf{P}\mathsf{v} + H_B z;$$

it has to be balanced by $M(z) = -EI\mathsf{v}''$ to ensure equilibrium. Then we write

$$\mathsf{v}'' + \alpha^2 \mathsf{v} = -\frac{H_B}{EI} z \tag{9.5}$$

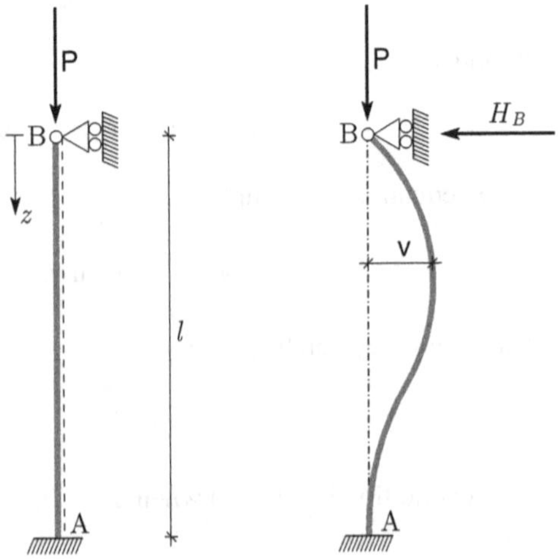

Fig. 9.3 Hyperstatic rod under axial load determining a critical state

after equating $M(z)$ with $M_{PH}(z)$. The resulting differential equation admits a solution v that is the sum of the solution v_{hom} of the homogeneous equation

$$\mathsf{v}'' + \alpha^2 \mathsf{v} = 0,$$

already discussed, and a special solution $\tilde{\mathsf{v}}$ of (9.5) given by

$$\tilde{\mathsf{v}}(z) = -\frac{H_B}{E\mathrm{I}\alpha^2} z = -\frac{H_B}{\mathsf{P}} z.$$

Then the solution of equation (9.5) is

$$\mathsf{v}(z) = a_1 \sin \alpha z + a_2 \cos \alpha z - \frac{H_B}{\mathsf{P}} z;$$

it must satisfy the boundary conditions

$$\mathsf{v}(0) = 0, \qquad \mathsf{v}'(l) = 0, \qquad \mathsf{v}(l) = 0,$$

which give respectively

$$a_2 = 0, \qquad a_1 \alpha \cos \alpha l = \frac{H_B}{\mathsf{P}}, \qquad a_1 \sin \alpha l = \frac{H_B}{\mathsf{P}} l.$$

The last two conditions imply

$$\tan \alpha l = \alpha l,$$

i.e.,

$$\alpha l \approx 4.4934$$

and

$$\mathsf{P}_{cr} \approx \frac{2.0456 \pi^2 E\mathrm{I}}{l^2}.$$

By comparison with (9.4), we obtain

$$l_c \approx 0.699\, l.$$

In all these cases, the critical load corresponds to a bifurcation of the equilibrium configuration.

9.2 An Energetic Analysis

The critical load can be derived in terms of energy. Consider the case of the supported rod and adopt the notation in Figure 9.4. When the rod deforms, the point B where P is applied undergoes a vertical displacement w given by

$$\mathsf{w} = \int_0^l (ds - dz) = \int_0^l \left(\sqrt{dz^2 + d\mathsf{v}^2} - dz\right)$$
$$= \int_0^l \left(\sqrt{1 + (\mathsf{v}'(z))^2} - 1\right) dz \approx \frac{1}{2}\int_0^l (\mathsf{v}'(z))^2\, dz.$$

Since we have assumed the rod to be inextensible and unshearable, its elastic energy depends just on bending, so that the total energy $\mathcal{E}^{tot}$ is given by

$$\mathcal{E}^{tot}(\mathsf{v}) = \frac{1}{2}\int_0^l \mathrm{EI}\,(\mathsf{v}''(z))^2\, dz - \frac{1}{2}\mathsf{P}\int_0^l (\mathsf{v}'(z))^2\, dz.$$

Consider now a smooth field $z \longmapsto h(z)$ such that $h(0) = h(l) = 0$. With $\xi \in \mathbb{R}$, we define

$$\delta_h \mathcal{E}^{tot}(\mathsf{v}) := \left.\frac{d}{d\xi}\mathcal{E}^{tot}(\mathsf{v} + \xi h)\right|_{\xi=0}$$

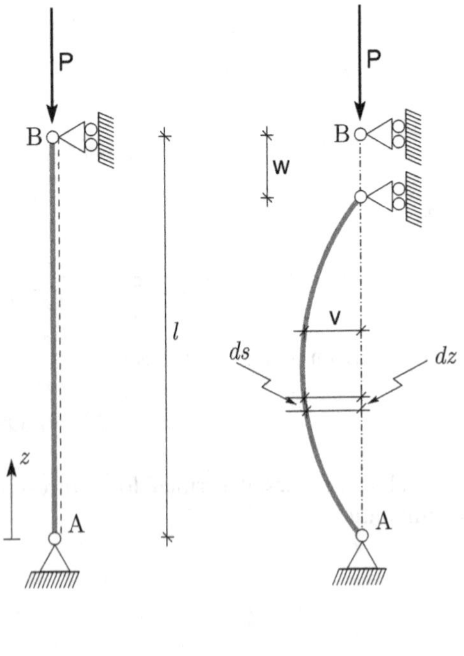

Fig. 9.4 Supported rod notation used in Section 9.2

and say (we recall) that the formal first variation of $\mathcal{E}^{tot}$ vanishes when

$$\delta_h \mathcal{E}^{tot}(\mathsf{v}) = 0 \tag{9.6}$$

for *an arbitrary* choice of h, as we have done in discussing the variational principles in linear elasticity. The vanishing of the first variation indicates critical points for the functional $\mathcal{E}^{tot}$. On computing the derivative (9.6), we get

$$\int_0^l (EI\mathsf{v}''(z)h''(z) - \mathsf{P}\mathsf{v}'(z)h'(z))dz = 0.$$

Since $\mathsf{v}(0) = \mathsf{v}(l) = 0$ and h is arbitrary, we can choose

$$\mathsf{v}(z) = h(z),$$

and we obtain

$$\mathsf{P} = \frac{\int_0^l \mathrm{EI}\left(\mathsf{v}''(z)\right)^2 dz}{\int_0^l \left(\mathsf{v}'(z)\right)^2 dz},$$

so that the critical load is given by the Rayleigh ratio

$$\mathsf{P}_{cr} = \min \frac{\int_0^l \mathrm{EI}\left(\mathsf{v}''(z)\right)^2 dz}{\int_0^l \left(\mathsf{v}'(z)\right)^2 dz}.$$

Choose

$$\mathsf{v}(z) = a \sin \frac{\pi z}{l}.$$

On inserting such an expression into the expression for P_{cr}, we obtain

$$\mathsf{P}_{cr} = \frac{\mathrm{EI}\frac{\pi^4}{l^4}a^2 \int_0^l \sin^2 \frac{\pi z}{l} dz}{\frac{\pi^2}{l^2}a^2 \int_0^l \cos^2 \frac{\pi z}{l} dz} = \frac{\pi^2 \mathrm{EI}}{l^2}.$$

9.3 Structures with Concentrated Elasticities

The analysis based on the direct balance of couples and that based on the energy can be both applied to finding the critical load of structures consisting of rigid bodies connected by springs. Examples are presented in the exercises below.

Exercise 1. *Given the stiffness k of the rotational spring connecting the two rigid bodies in Figure 9.5, find the critical load in terms of k and the length l.*

Solution. $\mathsf{P}_{cr} = \dfrac{4k}{l}$.

Exercise 2. *Find the critical load for the body in Figure 9.6 in terms of the stiffness k and the length l.*

Solution. $\mathsf{P}_{cr} = kl$.

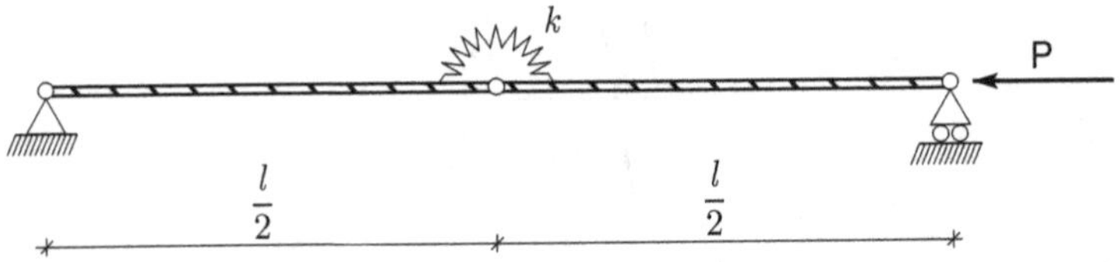

Fig. 9.5 Two rigid bodies connected by a rotational spring

Fig. 9.6 Rigid body with linear spring

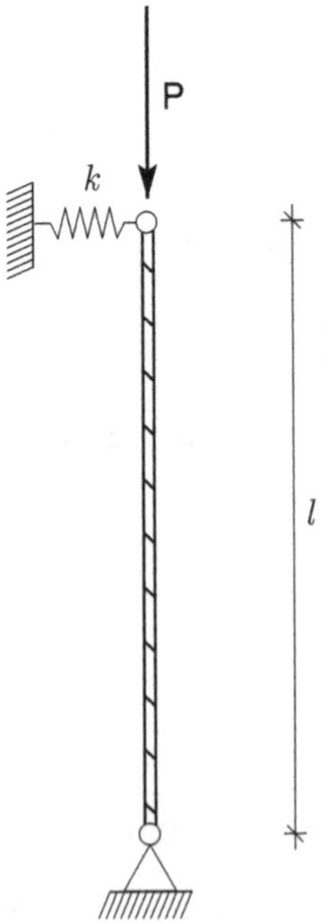

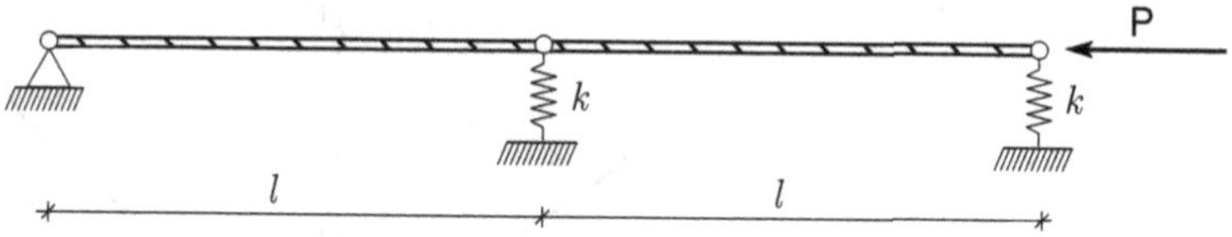

Fig. 9.7 Two rigid bodies with linear springs

Exercise 3. *Find the critical loads for the system of two rigid bodies with linear springs depicted in Figure 9.7.*

Solution. $\mathsf{P}_{1cr} = \left(\frac{3-\sqrt{5}}{2}\right) kl$, $\mathsf{P}_{2cr} = \left(\frac{3+\sqrt{5}}{2}\right) kl.$

9.4 Further Remarks

Consider the rod in Figure 9.2 and imagine that we begin applying a small load P and progressively increasing it. The previous analyses show that the rod slightly compresses first, but when P reaches P_{cr}, it buckles into one of two possible distinguished states in the plane. These states are stable, but they are not the sole ones at equilibrium. The compressed state is again possible, but it is now unstable.

With $\bar{\mathsf{v}} = \mathsf{v}(l/2)$ the displacement of the rod's center line, we can summarize what we have just described in the diagram in Figure 9.8, where the dashed line represents unstable states. We can develop a stability analysis using instead of the Bernoulli scheme, the linearized version of the elastica, as Euler himself did. The resulting diagram will be in the plane $\mathsf{P}/(E\mathrm{I})$, $\vartheta(0)$ if we consider the equation

$$E\mathrm{I}\vartheta'' + \mathsf{P}\vartheta = 0$$

with the boundary conditions

$$\vartheta'(0) = \vartheta'(1) = 0$$

for a rod of unitary length. Even in this case, the points of bifurcation are $\mathsf{P}_{cr} = E\mathrm{I}\pi^2 n^2, n = 1, 2, \ldots$.

- As we have seen, a bifurcation analysis in statics involves state variables (v or ϑ in the previous example) and one or more parameters as P, as we have already discussed. The analysis then reduces to solving an equation of type

$$f(\varsigma, \lambda) = 0, \tag{9.7}$$

where ς is the state and λ the list of parameters, so that

$$f : \Sigma \times \mathbb{R}^n \longrightarrow \mathfrak{Y},$$

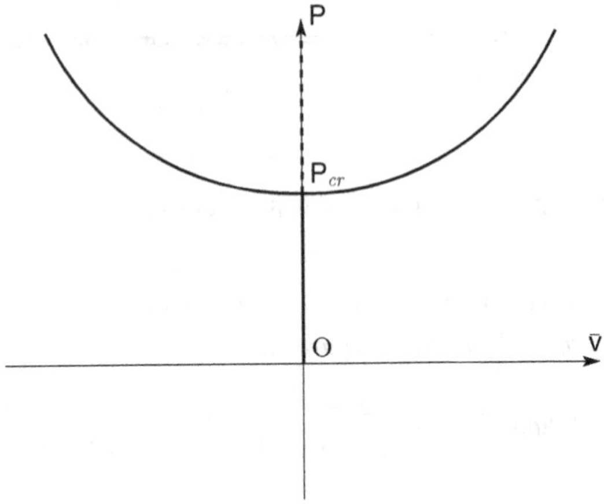

Fig. 9.8 Bifurcation diagram

where Σ is the state space, and $\mathfrak{Y}$ another space appropriate for the problem under consideration.

Once we have found solutions, we have to decide which are stable and which are unstable. The analysis may be done either in terms of the linearization of the previous equation or by testing for maxima or minima of a potential (if any) pertaining to the problem in hands. When the space of solutions is determined for the full range of the admissible states and all values of the parameters, the associated bifurcation analysis is called *global*. In contrast, when we are able to characterize the space of solutions only in a neighborhood of a given solution, the pertinent bifurcation analysis is called *local*.

- The equation (9.7) can undergo changes for small perturbations in the structure of f or the inclusion of further parameters in the list λ for modeling reasons. If the relevant bifurcation diagram is insensitive to such changes, we say that it is *structurally stable*; otherwise, it is *structurally unstable*.

  This last case is often determined in the stability of elastic structures by the presence of imperfections: the pertinent literature has had a long history since Warner Tjardus Koiter's 1945 dissertation.
- Consider the case in which a group $\mathcal{G}$ may act on both Σ and $\mathfrak{Y}$. We call the subgroup $\mathcal{G}_\varsigma$ of $\mathcal{G}$ defined by

$$\mathcal{G}_\varsigma := \left\{\tilde{h} \in \mathcal{G} | \tilde{h}\varsigma = \varsigma\right\}$$

  the *fixer (or symmetry) group* of $\varsigma \in \Sigma$, and we say that f is *covariant* when

$$f(\tilde{h}\varsigma, \lambda) = \tilde{h}f(\varsigma, \lambda).$$

Notice that f is *not necessarily* covariant with respect to the whole fixer group $\mathcal{G}_\varsigma$. Often, the trivial solution of (9.7) admits the entire group $\mathcal{G}$, while the bifurcating solution has a smaller fixer group. In these cases, we say that the bifurcation *breaks the symmetry*. Such a phenomenon can also be induced by the occurrence of imperfections that can be created during the loading program. Pertinent analyses are challenging and not easy to develop.

- All the examples discussed in the previous sections of this chapter deal with slender bodies. For them, bifurcation of static solutions can be rather easily reproduced in the laboratory. However, slenderness is not a necessary prerequisite for exhibiting bifurcation phenomena. That point was raised first in a 1930 paper by Antonio Signorini (1888–1963), who showed that the analysis of the equilibrium of a nonlinear elastic body under traction boundary conditions can have nonunique solutions even when small loads act on a natural state. Such nonuniqueness is intended in the sense that solutions are unequal with respect to a rigid change of place of bodies and loads. In many cases, such a bifurcation phenomenon in the space of equilibrated configurations depends only on the constitutive (fourth-rank) tensor of the linearized elasticity rather than the whole stored elastic energy. An interpretation by Gianfranco Capriz, dating back to the early 1970s, of the Signorini phenomenon is that some alternative solutions have a dynamic nature, rather than being genuinely static. However, in the linearization process, based on the series expansion used by Euler for the stability analysis, the inertial terms are of third rank, so that they disappear to leading order, and the relevant solutions appear to be static.
- Bifurcation analyses can be developed in dynamics. While those in statics deal with the recognition of the bifurcation of equilibrium states, in dynamics, attention is focused on qualitative phenomena such as the sudden occurrence of periodic orbits that change phase portraits.[1] Here, the analyses deal with slender bodies (beams, plates), since a dynamic bifurcation theory for three-dimensional elasticity is not yet available.

What we have presented in this chapter is just a brief overview of a wide landscape crammed with technical and conceptual difficulties, full of results but also of unexplored lands. The topic requires further extended reading, to which this chapter is just an invitation, just as this book is no more than a brief introduction to the vast field of continuum mechanics.

[1]Consider the ordinary differential equation

$$\dot{y}(t) = \mathsf{A}(\lambda)y(t)$$

with $y(t) \in \mathbb{R}^n$. The *phase portrait* is the collection of pairs $(\dot{y}, y)$ satisfying such an equation under appropriate initial conditions as λ varies.

Appendix A
Tensor Algebra and Tensor Calculus

We collect here some notions of geometry and analysis used in the previous chapters. This appendix is no more than a brief summary containing a few definitions and statements of results without proof. The conscientious reader unfamiliar with the topics collected here should take from these pages some directions for further reading in appropriate textbooks.

A.1 Vector Spaces

Definition 1. A set $\mathcal{V}$ is called a **vector space** (or a **linear space**) over a field $\mathcal{F}$ if it is endowed with two operations, $+ : \mathcal{V} \times \mathcal{V} \longrightarrow \mathcal{V}$ and $\circ : \mathcal{V} \times \mathcal{F} \longrightarrow \mathcal{V}$, called respectively *addition* and *multiplication*, such that

- addition is commutative, i.e., $v + w = w + v$ for every pair $v, w \in \mathcal{V}$,
- addition is associative, i.e., $(v+w)+z = v+(w+z)$ for every triple $v, w, z \in \mathcal{V}$,
- addition admits a unique neutral element $\mathbf{0} \in \mathcal{V}$ such that $v + \mathbf{0} = v, \forall v \in \mathcal{V}$,
- with αv denoting the product $\alpha \circ v$, for every $\alpha, \beta \in \mathcal{F}$ and $v \in \mathcal{V}$, multiplication is associative, i.e., $\alpha(\beta v) = (\alpha\beta)v$, where $\alpha\beta$ is the product in $\mathcal{F}$,
- multiplication admits a unique neutral element $1 \in \mathcal{F}$ such that $1v = v, \forall v \in \mathcal{V}$,
- multiplication is distributive with respect to addition, i.e., $\alpha(v + w) = \alpha v + \alpha w$ for every pair $v, w \in \mathcal{V}$,
- multiplication is distributive with respect to addition in $\mathcal{F}$, i.e., for every pair $\alpha, \beta \in \mathcal{F}$ and every $v \in \mathcal{V}$,

$$(\alpha + \beta)v = \alpha v + \beta v.$$

We do not recall the abstract definition of a field in the algebraic sense, because we consider here only the special case of $\mathcal{F}$ equal to the set of real numbers $\mathbb{R}$.

P.M. Mariano, L. Galano, *Fundamentals of the Mechanics of Solids*,
DOI 10.1007/978-1-4939-3133-0

Definition 2. Elements $v_1, \ldots, v_n$ of a linear space are said to be **linearly dependent** if there exist scalars $\alpha_1, \ldots, \alpha_n$ not all equal to zero such that

$$\sum_{i=1}^{n} \alpha_i v_i = 0.$$

In contrast, when the previous equation holds if and only if $\alpha_i = 0$ for every i, the elements $v_1, \ldots, v_n$ are said to be **linearly independent**.

From now on, we shall use Einstein notation, and we shall write $\alpha_i v_i$ instead of $\sum_{i=1}^{n} \alpha_i v_i$, implying summation over repeated indices.

Given $v_1, \ldots, v_n \in \mathcal{V}$, we denote by

$$\text{span}\,\{v_1, \ldots, v_n\}$$

the subset of $\mathcal{V}$ containing $v_1, \ldots, v_n$ and all their linear combinations.

Definition 3. A **basis** (or a **coordinate system**) in $\mathcal{V}$ is a set of linearly independent elements of $\mathcal{V}$, say $v_1, \ldots, v_n$, such that

$$\text{span}\,\{v_1, \ldots, v_n\} = \mathcal{V}.$$

In this case, n is the **dimension** of $\mathcal{V}$.

In particular, when $n < +\infty$, we say that $\mathcal{V}$ has finite dimension, which will be the case throughout this book. The dimension of a linear space $\mathcal{V}$ is independent of the selection of a basis for it. Every set of $m < n$ linearly independent elements of an n-dimensional linear space $\mathcal{V}$ can be extended to a basis of $\mathcal{V}$.

Let $e_1, \ldots, e_n$ be a basis of a vector space $\mathcal{V}$ over the field of real numbers, in short a real vector space. Every $v \in \mathcal{V}$ admits a representation

$$v = v^i e_i$$

with $v^i \in \mathbb{R}$ the ith component of v in that basis. In other words, v is independent of the basis, while its generic component v^i exists *with respect to* a given basis. The distinction is essential, because we want to write the laws of physics independently of specific frames of reference (coordinate systems), i.e., of the observers.

Consider the n-dimensional point space $\mathcal{E}$. We can construct a linear space $\mathcal{V}$ over it by taking the difference between points, namely $y - x$ with x and y in $\mathcal{E}$, and considering $\mathcal{V}$ the set of all possible *differences* between pairs of points in $\mathcal{E}$. Then $\mathcal{V}$ is a linear space with dimension n. We call it the **translation space** over $\mathcal{E}$. Once we select a point in $\mathcal{E}$, call it O, we can express all differences of points in reference to it, by writing

$$y - x = (y - \mathrm{O}) - (x - \mathrm{O}) = (y - \mathrm{O}) + (-1)(x - \mathrm{O}).$$

By assigning a basis in $\mathcal{V}$ consisting of elements expressed in reference to O, which becomes the origin of the coordinate system, we identify $\mathcal{V}$ with $\mathbb{R}^n$. There is, then, a hierarchy among $\mathcal{E}$, $\mathcal{V}$, and $\mathbb{R}^n$, the n-dimensional Euclidean point space, the related translational space, and finally, $\mathbb{R}^n$, determined by an increasing richness of geometric structure.

We call the elements of a given vector space (a linear space) **vectors**, although they can be far from the standard visualization of vectors as arrows in the three-dimensional ambient space or in the plane. For example, the set of $n \times m$ real matrices is a linear space, and so is the set of all polynomials with complex coefficients in a variable t, once we identify $\mathcal{F}$ with the set of complex numbers and consider addition and multiplication as the ordinary addition of two polynomials and the multiplication of a polynomial by a complex number. Several other examples of even greater complexity could be given, which are more intricate and difficult to visualize then those we have given.

Definition 4. The **scalar (inner) product** over $\mathcal{V}$ is a function

$$\langle \cdot, \cdot \rangle : \mathcal{V} \times \mathcal{V} \longrightarrow \mathbb{R}$$

that is

- commutative, i.e., $\langle v, w \rangle = \langle w, v \rangle$,
- distributive with respect to vector addition, i.e., $\langle v, w + z \rangle = \langle v, w \rangle + \langle v, z \rangle$,
- associative with respect to multiplication by elements of $\mathcal{F}$, i.e., $\alpha \langle v, w \rangle = \langle \alpha v, w \rangle = \langle v, \alpha w \rangle$, $\forall \alpha \in \mathcal{F}, v, w \in \mathcal{V}$,
- nonnegative definite, i.e., $\langle v, v \rangle \geq 0$, $\forall v \in \mathcal{V}$, and equality holds only when $v = 0$.

When $\mathcal{V}$ is endowed with a scalar product, we say that it is a **Euclidean space**.

We call a point space $\mathcal{E}$ Euclidean if its translational space is endowed with a scalar product. Let n be the dimension of $\mathcal{V}$ with respect to $\mathcal{E}$. In this case, we shall write $\mathcal{E}^n$, saying that it is an n-dimensional Euclidean point space.

Definition 5. Let $\{e_1, \ldots, e_n\}$ be a basis in a Euclidean vector space $\mathcal{V}$. We say that it is **orthonormal** if

$$\left\langle e_i, e_j \right\rangle = \delta_{ij}$$

with δ_{ij} the **Kronecker delta**:

$$\delta_{ij} = \begin{cases} 1 & \text{if} \quad i = j, \\ 0 & \text{if} \quad i \neq j. \end{cases}$$

We shall also write δ^{ij} and δ^i_j for the components of the Kronecker delta, with the same numerical values but with different meaning of the indices, as we shall make precise later.

Take two vectors $v = v^i e_i$, $w = w^j e_j$ in $\mathcal{V}$, expressed in terms of the basis $\{e_1, \ldots, e_n\}$, with n finite, which is not necessarily orthonormal. In terms of components, we write

$$\langle v, w\rangle = \langle v^i e_i, w^j e_j\rangle = v^i w^j \langle e_i, e_j\rangle = v^i g_{ij} w^j = v^i w^j g_{ij},$$

where

$$g_{ij} := \langle e_i, e_j\rangle$$

and

$$g_{ij} = g_{ji},$$

due to the commutativity of the scalar product. In particular, when the basis is orthonormal with respect to the scalar product considered, we have

$$\langle v, w\rangle = v^i \delta_{ij} w^j = v^i w_i,$$

where $w_i = \delta_{ij} w^j$ and the position of the index i has a geometric meaning that we explain below. Before going into details, however, we will say a bit about changes of frames in $\mathcal{V}$. We call g_{ij} the ijth component of the **metric tensor** over $\mathcal{V}$. The terminology recalls that g_{ij} determines in the basis $\{e_1, \ldots, e_n\}$ the **length** $|v|$ of $v \in \mathcal{V}$, defined by the square root of the scalar product of v with itself. The meaning of the word *tensor* is defined below. We compute

$$|v|^2 := \langle v, v\rangle = v^i g_{ij} v^j.$$

In particular, when $g_{ij} = \delta_{ij}$, we say in this case that the metric is *flat*, we have

$$|v|^2 = v^i \delta_{ij} v^j = v^i v_i = \sum_{i=1}^{n} (v^i)^2.$$

Assume $\mathcal{V}$ with finite dimension, as we shall do from now on. Take two bases $\{e_1, \ldots, e_n\}$ and $\{\mathrm{i}_1, \ldots, \mathrm{i}_n\}$. Assume that the latter is orthonormal. With respect to such a basis, we have

$$e_i = A_i^k \mathrm{i}_k,$$

with A_i^k appropriate proportionality factors. In turn, we have also

$$\mathrm{i}_k = \bar{A}_k^i e_i,$$

with $\bar{A}_k^i$ other factors. Consequently, we obtain

$$\mathrm{i}_k = \bar{A}^i_k e_i = \bar{A}^i_k A^l_i \mathrm{i}_l,$$

so that since $\{\mathrm{i}_1, \ldots, \mathrm{i}_n\}$ is orthonormal, we have

$$\delta_{ij} = \langle \mathrm{i}_i, \mathrm{i}_j \rangle = \langle \bar{A}^r_i A^l_r \mathrm{i}_l, \mathrm{i}_j \rangle = \bar{A}^r_i A^l_r \langle \mathrm{i}_l, \mathrm{i}_j \rangle = \bar{A}^r_i A^l_r \delta_{lj} = \bar{A}^r_i A^j_r.$$

We specialize the analysis to $\mathbb{R}^n$ to clarify further the nature of the proportionality factors. Let $z = \{z^1, \ldots, z^n\}$ be coordinates in reference to an orthonormal basis $\{\mathrm{i}_1, \ldots, \mathrm{i}_n\}$, and let $x = \{x^1, \ldots, x^n\}$ be another system of coordinates associated with a different basis, namely $\{e_1, \ldots, e_n\}$, and assume the possibility of a *coordinate change*

$$z^j \longmapsto x^i(z^j),$$

which is one-to-one and differentiable, with differentiable inverse mapping

$$x^i \longmapsto z^j(x^i).$$

Consider two smooth curves $s \longmapsto x(s)$ and $s \longmapsto \hat{x}(s)$ intersecting at a point $\bar{x}$, and compute the derivatives with respect to s at $\bar{x}$:

$$v = \frac{dx^i}{ds}(\bar{x})e_i, \qquad w = \frac{d\hat{x}^j}{ds}(\bar{x})e_j,$$

where e_i and e_j are the elements of the generic basis mentioned above as a counterpart to the orthonormal one $\{\mathrm{i}_1, \ldots, \mathrm{i}_n\}$. When we express these curves in terms of the coordinates z^i, i.e., if we consider maps $s \longmapsto z^i(x(s))$ and $s \longmapsto z^i(\hat{x}(s))$, and we compute once again v and w in the new frame, we get

$$v = \frac{dz^i(x(s))}{ds}\mathrm{i}_i = \frac{\partial z^i}{\partial x^h}\frac{dx^h}{ds}\mathrm{i}_i$$

and

$$w = \frac{dz^j(\hat{x}(s))}{ds}\mathrm{i}_j = \frac{\partial z^j}{\partial x^k}\frac{d\hat{x}^k}{ds}\mathrm{i}_j,$$

since $\dfrac{\partial z^j}{\partial \hat{x}^k} = \dfrac{\partial z^j}{\partial x^k}$ for the coordinates of the points over $x(s)$ and $\hat{x}(s)$ are always $(x^1, \ldots, x^n)$. Then we compute

$$\langle v, w \rangle = \left\langle \frac{dx^h}{ds}e_h, \frac{d\hat{x}^k}{ds}e_k \right\rangle = \frac{dx^h}{ds}\frac{d\hat{x}^k}{ds}g_{hk} = \left\langle \frac{\partial z^i}{\partial x^h}\frac{dx^h}{ds}\mathrm{i}_i, \frac{\partial z^j}{\partial x^k}\frac{d\hat{x}^k}{ds}\mathrm{i}_j \right\rangle$$
$$= \frac{\partial z^i}{\partial x^h}\frac{dx^h}{ds}\frac{\partial z^j}{\partial x^k}\frac{d\hat{x}^k}{ds}\delta_{ij},$$

i.e.,

$$g_{hk} = \frac{\partial z^i}{\partial x^h} \delta_{ij} \frac{\partial z^j}{\partial x^k}. \tag{A.1}$$

The one-to-one mapping $x \longleftrightarrow z$ allows us to consider g_{hk} the *pushforward* of δ_{ij} in the coordinate system $\{x^1, \ldots, x^n\}$. Conversely, we can write

$$\delta_{ij} = \frac{\partial x^h}{\partial z^i} g_{hk} \frac{\partial x^k}{\partial z^j}$$

(prove it as an exercise) and consider δ_{ij} the *pullback* of g_{hk} into the coordinate system $\{z^1, \ldots, z^n\}$.

Notice that here we are considering coordinate transformations in the same space. The relation (A.1) emerges by the computation of the same vectors. In Chapter 1, we use a version of the previous remarks involving two different spaces with the same dimension, each space endowed with its own independent metric. Even in this case, we can speak about pullback and pushforward of these metrics from one space to the other by means of some bijective mapping. The matrix with g_{ij} as the ijth component has nonzero determinant, due to the definition of g_{ij} introduced above. Hence, the inverse of that matrix exists and represents what we call **inverse metric** (or *the inverse of the metric*), denoted by g^{-1}, in the frame of reference defining g_{ij}. We denote by g^{ij} the generic element of g^{-1}.

Definition 6. The **cross product** of two vectors in $\mathcal{V}$ is a map $\times : \mathcal{V} \times \mathcal{V} \longrightarrow \mathcal{V}$ assigning to every pair $v, w \in \mathcal{V}$ a vector $v \times w \in \mathcal{V}$ with ith component given by

$$(v \times w)^i = \mathsf{e}^i_{jk} v^k w^j,$$

with e^i_{jk} a version of the so-called Ricci symbol e_{ijk}, defined by

$$\mathsf{e}_{ijk} = \begin{cases} 1 & \text{if } \ ijk \text{ are a even permutation of } 1, 2, 3, \\ -1 & \text{if } \ ijk \text{ are a odd permutation of } 1, 2, 3, \\ 0 & \text{elsewhere.} \end{cases}$$

The Ricci symbol is such that

$$\mathsf{e}_{ijk} \mathsf{e}_{ipq} = \delta_{jp} \delta_{kq} - \delta_{jq} \delta_{kp}, \tag{A.2}$$

with δ_{ij} the Kronecker delta. The reason for writing some indices in superscript position and others as subscripts will be clear in the subsequent section.

A.2 Dual Spaces: Covectors

Consider linear functions $f : \mathcal{V} \longrightarrow \mathbb{R}$ over a real vector space $\mathcal{V}$ with finite dimension. Linearity means that for all pairs $v, w \in \mathcal{V}$ and $\alpha, \beta \in \mathbb{R}$, we get

$$f(\alpha v + \beta w) = \alpha f(v) + \beta f(w).$$

In particular, for fixed $v = v^i e_i$, we have

$$f(v) = v^i f(e_i),$$

leaving understood, as usual, the summation over repeated indices.

If we collect in a column the values v^i, namely

$$\begin{pmatrix} v^1 \\ v^2 \\ \cdot \\ \cdot \\ \cdot \\ v^n \end{pmatrix}$$

and in a row the values $f(e_i)$, namely

$$\left(f(e_1), f(e_2), \ . \ . \ ., f(e_n)\right)$$

we write

$$f(v) = v^i f(e_i) = \left(f(e_1), f(e_2), \ . \ . \ ., f(e_n)\right) \begin{pmatrix} v^1 \\ v^2 \\ \cdot \\ \cdot \\ \cdot \\ v^n \end{pmatrix} \tag{A.3}$$

in terms of the standard row–column product. We consider

$$\left(f(e_1), f(e_2), \ . \ . \ ., f(e_n)\right)$$

the list of components of an element b of a space $\mathcal{V}^*$, with dimension n, and we write

$$f(v) = b \cdot v,$$

where the right-hand term is a product with the same meaning as the right-hand side of (A.3). We express b explicitly as

$$b = b_i e^i,$$

with e^i the ith element of the basis in $\mathcal{V}^*$, defined to be such that

$$e^i \cdot e_j = \delta^i_j,$$

where δ^i_j is the Kronecker symbol.

The traditional convention is to write in superscript position the indices of vector components (called *contravariant*) as for v^i, with reference to an element of $\mathcal{V}$ in the scheme followed so far, and in subscript position the indices of elements of $\mathcal{V}^*$ (called *covariant*), to recall the product in the identity (A.3). This convention—essentially suggested by the identity (A.3)—justifies also the position of the indices in the formulas collected in the previous section.

We call $\mathcal{V}^*$ the **dual space** of $\mathcal{V}$, and $b \in \mathcal{V}^*$ a **covector**. Given $v, w \in \mathcal{V}$, we can interpret in terms of covectors the relation expressing in coordinates the scalar product, i.e., $\langle v, w \rangle = v^i g_{ij} w^j$. In fact, the factor $v^i g_{ij}$ defines the jth component of a covector $v^\flat$, i.e.,

$$v^\flat_j = g_{ji} v^i,$$

so that we write

$$\langle v, w \rangle = gv \cdot w = v^\flat \cdot w. \tag{A.4}$$

To describe the action of g on the vector v, in the common jargon of geometry we say that we are lowering the index of v. A special, although prominent, case occurs when g_{ij} coincides with δ_{ij}, i.e., the metric is flat: the column in which we list the components of v becomes simply a row without any further numerical factor. Moreover, taking the matrix with g_{ij} as the ijth component with nonzero determinant, the inverse g^{-1} of the metric exists, and its generic ijth component is denoted by g^{ij}.

Consider a covector $c \in \mathcal{V}^*$. When we apply g^{-1} to it, we determine a vector $c^\#$ with ith component given by

$$c^{\# i} = g^{ij} c_j.$$

To describe the action of g^{-1} on the covector c, we say that we are raising the index of c. For this reason, given $c \in \mathcal{V}^*$ and $v \in \mathcal{V}$, we obtain a vector $c^\# := g^{-1} c \in \mathcal{V}$ such that

$$c \cdot v = \langle g^{-1} c, v \rangle = \langle c^\#, v \rangle. \tag{A.5}$$

The two relations (A.4) and (A.5) establish an isomorphism between $\mathcal{V}$ and its dual counterpart $\mathcal{V}^*$. We put in evidence that $\mathcal{V}^*$ is the space of linear functions over $\mathcal{V}$ by writing also $\mathrm{Hom}(\mathcal{V}, \mathbb{R})$ instead of $\mathcal{V}^*$. Such notation is particularly useful when we consider linear maps between two different spaces. We already expressed two such maps when we described the action of the metric g over a vector and the action of g^{-1} over a covector, as we specify below.

The map assigning to a vector space $\mathcal{V}$ its dual $\mathcal{V}^*$ is *involutive*, i.e., the dual of $\mathcal{V}^*$ is $\mathcal{V}$:

$$(\mathcal{V}^*)^* = \mathcal{V}.$$

A.3 Linear Maps between Generic Vector Spaces: Tensors

Take a vector space $\mathcal{V}$ over $\mathbb{R}$ (in fact, we can consider a generic field $\mathcal{F}$ instead of $\mathbb{R}$, but the restricted choice adopted here is sufficient for what is contained in the book) and consider a linear map

$$h : \mathcal{V} \longrightarrow \mathcal{V}.$$

By definition, for $v \in \mathcal{V}$, h is such that $h(v) \in \mathcal{V}$, and for $\alpha, \beta \in \mathbb{R}$ and $w \in \mathcal{V}$,

$$h(\alpha v + \beta w) = \alpha h(v) + \beta h(w).$$

Consequently, we have

$$h(v) = h(v^i e_i) = v^i h(e_i),$$

where we have adopted the summation convention over repeated indices as usual. In particular, by listing in a column the numbers v^i and denoting by A the matrix having as columns the vectors $h(e_i)$, namely

$$\left(h(e_1), h(e_2), \ . \ . \ ., h(e_n)\right),$$

we can write

$$h(v) = Av$$

by adopting the standard row–column matrix product. In fact, A is a linear operator from $\mathcal{V}$ to $\mathcal{V}$, namely $A \in \mathrm{Hom}(\mathcal{V}, \mathcal{V})$. Take the generic element e_r of the vector basis in $\mathcal{V}$. Then Ae_r is another vector in $\mathcal{V}$ with kth component in the assigned basis given by $(Ae_r)^k$. We define the krth component A_r^k of A as the real number defined by

$$A_r^k = (Ae_r)^k = e^k \cdot Ae_r.$$

Recall that in the previous formula, the indices k and r are just labels identifying the element of the basis in $\mathcal{V}$.

The matrix with generic component A_r^k is just a representation of A in the selected basis. In other words, as a linear operator, A exists in the abstract, independent of the basis that we might choose in $\mathcal{V}$, while its explicit matrix representation depends on the choice of the basis; A_r^k is characterized by a contravariant index and a covariant one. We commonly say that such an A is of type $(1, 1)$. Let $\{x^1, \ldots, x^n\}$ be the coordinates with respect to which we have calculated A_r^k. Consider a change of coordinates $\{x^1, \ldots, x^n\} \longrightarrow \{y^1, \ldots, y^n\}$, which is one-to-one and differentiable with differentiable inverse. Write $\bar{A}_r^k$ for the counterpart of A_r^k in the coordinates $\{y^1, \ldots, y^n\}$. We have

$$\bar{A}_r^k = \frac{\partial y^k}{\partial x^s} A_l^s \frac{\partial x^l}{\partial y^r}.$$

We summarize all these properties by saying that A is a 1-contravariant, 1-covariant second-rank **tensor**. Notice that in this sense, a vector is a tensor of type $(1, 0)$, while a covector is a $(0, 1)$ tensor.

In the same way, we can define other types of second-rank tensors. We can have $A \in \mathrm{Hom}(\mathcal{V}, \mathcal{V}^*)$ with components

$$A_{rk} := \langle e_r, Ae_k \rangle$$

such that, with the overbar indicating the change of variables as above, we have

$$\bar{A}_{rk} = \frac{\partial x^l}{\partial y^r} A_{ls} \frac{\partial x^s}{\partial y^k}.$$

In this case, A is of type $(0, 2)$. Moreover, we can also have $A \in \mathrm{Hom}(\mathcal{V}^*, \mathcal{V}^*)$ or $A \in \mathrm{Hom}(\mathcal{V}^*, \mathcal{V})$. In the first case, we have

$$\bar{A}_k^r = \frac{\partial x^s}{\partial y^k} A_s^l \frac{\partial y^r}{\partial x^l}.$$

In the second case, the generic component of A is

$$A^{rk} = (Ae^r) \cdot g^{-1} e^k = (Ae^r)_s \, g^{sk} e_k^j;$$

it is also such that

$$\bar{A}^{rk} = \frac{\partial y^r}{\partial x^l} A^{ls} \frac{\partial y^k}{\partial x^s}.$$

Given two such linear operators, say A and B, we define a product

$$(A, B) \longrightarrow AB$$

giving as a result a second-rank tensor. In this case, we say that we *saturate* (or *contract*) the second component of A with the first component of B, according to the row–column multiplication convention. Precisely, consider $A, B \in \mathrm{Hom}(\mathcal{V}, \mathcal{V})$. The product AB is a second-rank tensor with components

$$(AB)^i_j = A^i_r B^r_j,$$

where we consider, as usual, summation over repeated indices.

Notice that the indices saturated are respectively in covariant and contravariant positions, according to the row–column multiplication. To account for this, when we want to multiply tensors having, for example, both contravariant components, we have to lower or raise the appropriate index by the action of the metric. For example, consider $A, B \in \mathrm{Hom}(\mathcal{V}^*, \mathcal{V})$. Their product saturating one index involves the metric. To render explicit the convention above, we write sometimes AgB instead of AB, and when we adopt the more synthetic form, we mean that we are defining a second-rank tensor with components

$$(AB)^{ij} = (AgB)^{ij} = A^{ik} g_{ks} B^{sj}.$$

An analogous agreement is understood when we involve second-rank tensors with both covariant components. In this case, we have to raise one index using g^{-1}. In other words, when $A, B \in \mathrm{Hom}(\mathcal{V}, \mathcal{V}^*)$, we write

$$(AB)_{ij} = (Ag^{-1}B)_{ij} = A_{ik} g^{ks} B_{sj}.$$

All previous distinctions can be neglected when the metric is flat, i.e., when $g_{ij} = \delta_{ij}$, since no numerical factors induced by the metric appear.

In general, the product AB is noncommutative, i.e., $AB \neq BA$.

Instead of saturating just one index, we can saturate both indices, producing a scalar. In this case, we use the notation $A \cdot B$, meaning

$$A \cdot B = A_{ij} B^{ji}$$

or

$$A \cdot B = A^i_j B^j_i,$$

with the appropriate adjustments of the indices by means of the metric as in the product AB. The product $A \cdot B$ is commutative, i.e., $A \cdot B = B \cdot A$.

Consider $A \in \mathrm{Hom}(\mathcal{V}, \mathcal{V})$. We define the **transpose** of A to be the unique element of $\mathrm{Hom}(\mathcal{V}, \mathcal{V})$, denoted by A^{T}, such that

$$\langle Av, w \rangle = \langle v, A^{\mathrm{T}} w \rangle$$

for every pair $v, w \in \mathcal{V}$. We define also the **formal adjoint** of A to be the unique element of $\mathrm{Hom}(\mathcal{V}, \mathcal{V}^*)$, denoted by A^*, such that

$$b \cdot Av = (A^* b) \cdot v$$

for every $v \in \mathcal{V}$ and $b \in \mathcal{V}^*$. We observe that A^{T} and A^* also satisfy the relation

$$A^{\mathrm{T}} = g^{-1} A^* g.$$

A proof can be found in Chapter 1 in a case a bit more general in that it includes linear operators between two different vector spaces. The definitions above in fact hold also in this case. In particular, take two such spaces, namely $\mathcal{V}$ and $\mathcal{W}$, and consider $A \in \mathrm{Hom}(\mathcal{V}, \mathcal{W})$. We then have the following inclusions: $A^{\mathrm{T}} \in \mathrm{Hom}(\mathcal{W}, \mathcal{V})$, $A^* \in \mathrm{Hom}(\mathcal{W}^*, \mathcal{V}^*)$.

When $A \in \mathrm{Hom}(\mathcal{V}, \mathcal{W}^*)$ or $A \in \mathrm{Hom}(\mathcal{V}^*, \mathcal{W})$, transpose and formal adjoint coincide. They are in $\mathrm{Hom}(\mathcal{W}, \mathcal{V}^*)$ and $\mathrm{Hom}(\mathcal{W}^*, \mathcal{V})$, respectively.

The characterization just expressed of A^{T} and A^* as elements of some sets, those listed above, assumes that we apply $A \in \mathrm{Hom}(\mathcal{V}, \mathcal{W})$ to vectors of $\mathcal{V}$ *from the left*. This means that the *second* index of A saturates that of the vector components. When we want to apply A *from the right*, i.e., we want to saturate the first index of A, which means that we write cA, with c now a covector because A has the first component in contravariant position, the inclusions written above change. More specifically, if $A \in \mathrm{Hom}(\mathcal{V}, \mathcal{W})$ *when applied from the left, acting on the right* we have $A \in \mathrm{Hom}(\mathcal{W}^*, \mathcal{V}^*)$, $A^* \in \mathrm{Hom}(\mathcal{V}, \mathcal{W})$, $A^{\mathrm{T}} \in \mathrm{Hom}(\mathcal{V}^*, \mathcal{W}^*)$.

From the definition of transpose and formal adjoint, we obtain also

$$(A^{\mathrm{T}})^{\mathrm{T}} = A, \qquad (A^*)^* = A.$$

Consider, e.g., $A \in \mathrm{Hom}(\mathcal{V}, \mathcal{V})$ and define the determinant of the linear operator A as the determinant of the matrix with generic element A^i_j, representing A in some basis. In other words, for $n = \dim \mathcal{V}$, we can write

$$\det A = \sum_{i_1, \ldots, i_n = 1}^{n} \mathsf{e}_{i_1 \ldots i_n} A^{i_1}_1 \cdots A^{i_n}_n,$$

or alternatively,

$$\det A = \frac{1}{n!} \sum_{i_1, \ldots, i_n = 1}^{n} \sum_{j_1, \ldots, j_n = 1}^{n} \mathsf{e}_{i_1 \ldots i_n} \mathsf{e}_{j_1 \ldots j_n} A^{i_1}_{j_1} \cdots A^{i_n}_{j_n},$$

where $\mathsf{e}_{i_1 \ldots i_n}$ is 1 or -1, provided that $i_1, \ldots, i_n$ constitutes an even or an odd permutation of $1, 2, \ldots, n$ respectively, 0 otherwise. In particular, if $n = 3$, $\det A$ is the triple product of its vector columns, e.g.,

$$\det A = \langle A_1, A_2 \times A_3 \rangle ,$$

where A_1, A_2, and A_3 are the first column, the second column, and the third column of A. We have also the following:

- $\det(AB) = \det A \det B$,
- $\det A^{\mathrm{T}} = \det A$,
- $\det A^* = \det A$,
- with $\alpha \in \mathbb{R}$, $\det(\alpha A) = \alpha^n \det A$.

Let us denote by I the **identity** in $\mathrm{Hom}(\mathcal{V}, \mathcal{V})$, namely a second-rank tensor such that

$$AI = IA = A.$$

We define the **inverse** of A to be the unique linear operator A^{-1} such that

$$A^{-1}A = AA^{-1} = I.$$

Once we select a basis in $\mathcal{V}$, the matrix representing A^{-1} is the inverse of the matrix collecting the components of A in the same basis. We have then

$$(A^{-1})^i_j = \frac{(-1)^{i+j} \det A_{(ji)}}{\det A},$$

where $A_{(ji)}$ is the matrix obtained from the one representing A by deleting the jth row and the ith column.

The definition of determinant can be extended to tensors of type $A \in \mathrm{Hom}(\mathcal{V}, \mathcal{W})$, with $\dim \mathcal{V} = \dim \mathcal{W}$. A subclass of $\mathrm{Hom}(\mathcal{V}, \mathcal{W})$ with $\dim \mathcal{V} = \dim \mathcal{W}$ is given by invertible tensors such that

$$Q^{-1} = Q^{\mathrm{T}}.$$

We indicate the special case by writing $Q \in O(n)$, with $n = \dim \mathcal{V}$ once again, and call $O(n)$, which is endowed with a group structure, the **orthogonal group**.

With $Q \in O(n)$, $\det Q$ can be 1 or -1. In the first case, i.e., $\det Q = 1$, we say that Q belongs to the **special orthogonal group**, denoted by $SO(n)$. It is a subgroup of $O(n)$, i.e., a subset of $O(n)$ endowed with the group structure inherited from $O(n)$. Elements of $SO(n)$ represent **rotations** in the n-dimensional space $\mathcal{V}$, while those in the remaining part of $O(n)$, i.e., $O(n)\backslash SO(n)$, describe **reflections** in $\mathcal{V}$. The words "rotation" and "reflection" recall the way $Q \in O(3)$ may act on vectors in the three-dimensional Euclidean point space.

We say that a second-rank tensor A is **symmetric** if $A = A^{\mathrm{T}}$, while it is **skew-symmetric** if $A = -A^{\mathrm{T}}$. Every second-rank tensor admits the additive decomposition

$$A = \frac{1}{2}(A + A^{\mathrm{T}}) + \frac{1}{2}(A - A^{\mathrm{T}}).$$

We define

$$\mathrm{Sym}\, A := \frac{1}{2}(A + A^{\mathrm{T}}),$$

the **symmetric part** of A, and

$$\mathrm{Skw}A := \frac{1}{2}(A - A^{\mathrm{T}}),$$

the **skew-symmetric part** of A.

Consider a skew-symmetric tensor B, i.e., $B \in \mathrm{Skw}(\mathcal{V}, \mathcal{V})$. By definition, $B = -B^{\mathrm{T}}$. However, there is another property that could be used as a definition of skew-symmetry: there exists $q \in \mathcal{V}$ such that for every $v \in \mathcal{V}$, we have

$$Bv = q \times v.$$

A.4 Tensor Product

With $\mathcal{V}$, $\mathcal{W}$, and $\mathcal{U}$ real vector spaces,

$$f := \mathcal{V} \times \mathcal{W} \longrightarrow \mathcal{U}$$

is called a **bilinear mapping** if

- for every $v \in \mathcal{V}$, the map $w \longmapsto f(v, w)$ is linear;
- for every $w \in \mathcal{V}$, the map $v \longmapsto f(v, w)$ is linear.

Given two vector spaces $\mathcal{V}$ and $\mathcal{W}$ that we consider real for the purposes of this book, but that we could also take as defined over a field $\mathcal{F}$ not necessarily coinciding with $\mathbb{R}$, for $v \in \mathcal{V}$ and $w \in \mathcal{W}$, consider the map

$$(v, w) \longmapsto v \otimes w,$$

where $v \otimes w$ is what we call the **tensor product** (or the **dyad**) between v and w. Such a product is endowed with the following properties:

- for $v_1, v_2 \in \mathcal{V}$ and $w \in \mathcal{W}$,

$$(v_1 + v_2) \otimes w = v_1 \otimes w + v_2 \otimes w;$$

- for $w_1, w_2 \in \mathcal{W}$ and $v \in \mathcal{V}$,

$$v \otimes (w_1 + w_2) = v \otimes w_1 + v \otimes w_2;$$

- with $\alpha \in \mathbb{R}$ (or $\in \mathcal{F}$ in general),

$$(\alpha v) \otimes w = \alpha(v \otimes w) = v \otimes (\alpha w).$$

Here $v \otimes w$ is defined to be an element of $\mathrm{Hom}(\mathcal{W}^*, \mathcal{V})$. Precisely, for $c \in \mathcal{W}^*$, we have

$$(v \otimes w)c = (c \cdot w)v,$$

i.e., $v \otimes w$ has components $v^i w^j$.

Theorem 7. *Let $\mathcal{V}$ and $\mathcal{W}$ be finite-dimensional vector spaces over $\mathbb{R}$ (or $\mathcal{F}$). Then there exist a finite-dimensional vector space $\mathcal{T}$ over $\mathbb{R}$ (or $\mathcal{F}$) and a bilinear mapping $\mathcal{V} \times \mathcal{W} \longrightarrow \mathcal{T}$ indicated by*

$$(v, w) \longmapsto v \otimes w$$

satisfying the following properties:

- *If $\mathcal{U}$ is a vector space over $\mathbb{R}$ (or $\mathcal{F}$) and $f : \mathcal{V} \times \mathcal{W} \longrightarrow \mathcal{U}$ is a bilinear mapping, there exists a linear map*

$$\tilde{f} : \mathcal{T} \longrightarrow \mathcal{U}$$

such that for every pair (v, w), with $v \in \mathcal{V}$ and $w \in \mathcal{W}$, we have

$$f(v, w) = \tilde{f}(v \otimes w).$$

- *If $\{e_1, \ldots, e_n\}$ is a basis for $\mathcal{V}$ and $\{\tilde{e}_1, \ldots, \tilde{e}_n\}$ is a basis for $\mathcal{W}$, the elements*

$$e_i \otimes \tilde{e}_j,$$

with $i = 1, \ldots, n$, $j = 1, \ldots, n$ constitute a basis for $\mathcal{T}$.

We usually denote $\mathcal{T}$ by $\mathcal{V} \otimes \mathcal{W}$, since the dyads $e_i \otimes \tilde{e}_j$ constitute a basis within it; $\mathcal{V} \otimes \mathcal{W}$ is isomorphic to $\mathrm{Hom}(\mathcal{W}^*, \mathcal{V})$. Every $A \in \mathrm{Hom}(\mathcal{W}^*, \mathcal{V})$ admits, in fact, the representation

$$A = A^{ij} e_i \otimes \tilde{e}_j.$$

Such a representation is compatible with the definition of A^{ij}. We have, in fact,

$$A^{ij} = e^i \cdot A\tilde{e}^j = e^i \cdot A^{ij}(e_i \otimes \tilde{e}_j)\tilde{e}^j = A^{ij} e^i \cdot (e_i \otimes \tilde{e}_j)\tilde{e}^j$$
$$= A^{ij} e^i \cdot (\tilde{e}^j \cdot \tilde{e}_j) e_i = A^{ij} e^i \cdot e_i = A^{ij}.$$

By specifying $\mathcal{V}$ and $\mathcal{W}$, we have the following relations:

- If $A \in \mathrm{Hom}(\mathcal{V}, \mathcal{V})$ we have

$$A = A^i_j e_i \otimes e^j,$$

so that $\mathrm{Hom}(\mathcal{V}, \mathcal{V})$ is isomorphic to $\mathcal{V} \otimes \mathcal{V}^*$, and we write formally

$$\mathrm{Hom}(\mathcal{V}, \mathcal{V}) \simeq \mathcal{V} \otimes \mathcal{V}^*$$

to summarize the statement symbolically.
- If $A \in \mathrm{Hom}(\mathcal{V}^*, \mathcal{V}^*)$, we have

$$A = A^j_i e^i \otimes e_j,$$

so that

$$\mathrm{Hom}(\mathcal{V}^*, \mathcal{V}^*) \simeq \mathcal{V}^* \otimes \mathcal{V}.$$

- If $A \in \mathrm{Hom}(\mathcal{V}, \mathcal{V}^*)$, we have

$$A = A_{ij} e^i \otimes e^j,$$

so that

$$\mathrm{Hom}(\mathcal{V}, \mathcal{V}^*) \simeq \mathcal{V}^* \otimes \mathcal{V}^*.$$

- If $A \in \mathrm{Hom}(\mathcal{V}^*, \mathcal{V})$, we have

$$A = A^{ij} e_i \otimes e_j,$$

so that

$$\mathrm{Hom}(\mathcal{V}^*, \mathcal{V}) \simeq \mathcal{V} \otimes \mathcal{V}.$$

Consider the dyad $e_i \otimes e^j$ and imagine that the basis $\{e_1, \ldots, e_n\}$, , and consequently the dual basis $\{e^1, \ldots, e^n\}$, is rotated by $Q \in SO(n)$. Define $\bar{e}_i = Qe_i$,

$$\bar{e}_i \otimes \bar{e}^j = Qe_i \otimes e^j Q^{\mathrm{T}}.$$

Consequently, under such a rotation, $A \in \mathrm{Hom}(\mathcal{V}, \mathcal{V}^*)$ becomes

$$\bar{A} = QAQ^{\mathrm{T}},$$

so that we compute

$$\det \bar{A} = \det Q \det A \det Q^{\mathrm{T}} = \det A,$$

and we find that the determinant of a tensor is invariant under rotations of the basis. In terms of components, we have

$$\bar{A}^i_j = Q^i_r A^r_s Q^s_j.$$

In particular, we could find a rotation such that in the rotated basis, the matrix representing A has diagonal form. If A is *symmetric*, such a rotation *always exists.* The numbers on the diagonal version of A are its *eigenvalues.*

Consider $A = A_{ij} e^i \otimes e^j \in \mathrm{Sym}(\mathcal{V}, \mathcal{V}^*)$, where $\mathrm{Sym}(\mathcal{V}, \mathcal{V}^*)$ is the subset of $\mathrm{Hom}(\mathcal{V}, \mathcal{V}^*)$ containing all symmetric elements of $\mathrm{Hom}(\mathcal{V}, \mathcal{V}^*)$. Its eigenvalues are a solution of the algebraic equation

$$\det(A - \lambda I) = 0,$$

called the *characteristic equation.* By raising the first component of A, we see that the roots λ are also eigenvalues of the version $\hat{A}$ of A with components

$$\hat{A}^i_j = g^{ik} A_{kj}.$$

We obtain, in fact,

$$\det(\hat{A} - \lambda I) = \det(g^{-1}(A - \lambda I)) = \det g^{-1} \det(A - \lambda I).$$

The *eigenvectors* associated with the generic eigenvalue λ are solutions to the equation

$$A^i_k \xi^k = \lambda \xi^i.$$

The solution $\xi^1, \ldots, \xi^n$ expresses the components of the eigenvector associated with λ.

Theorem 8. *For every symmetric tensor $A \in \mathrm{Hom}(\mathcal{V}, \mathcal{V})$, with $\mathcal{V}$ a n-dimensional vector space, there exist an orthonormal basis $\{e_1, \ldots, e_n\}$ and numbers $\alpha_i \in \mathbb{R}$, $i = 1, \ldots, n$ such that*

$$A = \sum_{i=1}^{n} \alpha_i e_i \otimes e^i.$$

In the representation of A just presented, we call α_i the **eigenvalues** of A, and e_i the corresponding **eigenvectors**. For a given eigenvalue α, the set of all v such that $Av = \alpha v$ is called an **eigenspace**. If the eigenvalues are all distinct, the related eigenvectors are determined uniquely. To visualize the situation, consider the case $n = 3$. If $\alpha_1 \neq \alpha_2 \neq \alpha_3$, the eigenspaces are the lines span(e_1), span(e_3), span(e_3). If $\alpha_2 = \alpha_3$, we have $A = \alpha_1(e_1 \otimes e_1) + \alpha_2(I - e_1 \otimes e_1)$, with I the $(1, 1)$ unit tensor, so that the eigenspaces are span(e_1) and its orthogonal complement (a plane orthogonal to e_1). Finally, if $\alpha_1 = \alpha_2 = \alpha_3 = \alpha$, then A is spheric: $A = \alpha I$.

As we have already remarked, a special case of a second-rank tensor is the metric $g \in \text{Hom}(\mathcal{V}, \mathcal{V}^*)$. We derived g using the scalar product structure. Conversely, we can begin by assigning a second-rank symmetric tensor $g \in \text{Hom}(\mathcal{V}, \mathcal{V}^*)$ and defining in this way a scalar product. The metric tensor appears also in the explicit computation of the *modulus* $|A|$ of a tensor A. Consider $A = A^i_j e_i \otimes e^j$. We write

$$|A|^2 = A \cdot A^{\mathrm{T}} = A^i_j A^j_i.$$

When $A = A_{ij} e^i \otimes e^j$, we write

$$|A|^2 = A_{ij} A^{ji},$$

where $A^{ji} = g^{jk} A_{kl} g^{li}$. An analogous expression holds when $A = A^{ij} e_i \otimes e_j$.

Take $v, w \in \mathcal{V}$ and their dyad $v \otimes w \in \text{Hom}(\mathcal{V}^*, \mathcal{V})$. We define the **trace** of $v \otimes w$ to be the number

$$\text{tr}(v \otimes w) = \langle v, w \rangle = v^i g_{ij} w^j. \tag{A.6}$$

For any other $A \in \text{Hom}(\mathcal{V}^*, \mathcal{V})$, i.e., $A = A^{ij} e_i \otimes e_j$, we define its **trace** as

$$\text{tr}A = A \cdot g = A^{ij} g_{ji},$$

as a direct consequence of (A.6), because

$$\text{tr}A = \text{tr}(A^{ij} e_i \otimes e_j) = A^{ij} \text{tr}(e_i \otimes e_j) = A^{ij} \langle e_i, e_j \rangle = A^{ij} g_{ij} = A^{ij} g_{ji}.$$

Analogously, when $A \in \text{Hom}(\mathcal{V}, \mathcal{V}^*)$, i.e., $A = A_{ij} e^i \otimes e^j$, we write

$$\text{tr}A = A \cdot g^{-1} = A_{ij} g^{ji},$$

since g^{ij} is given by the scalar product of the elements of the dual basis, a product induced by the product over $\mathcal{V}$.

Finally, if $A = A^i_j e_i \otimes e^j$, we have

$$\text{tr}\, A = \text{tr}(A^i_j e_i \otimes e^j) = A^i_j \delta^j_i = A \cdot I = A^i_i,$$

since $e_i \otimes e^j = \delta^j_i$ by definition.

A.5 Higher-Rank Tensors

We can replicate the tensor product, constructing tensors with rank higher than the second. We may have, in fact, linear operators of the form

$$T = T^{i_1 \dots i_p}_{j_1 \dots j_p} e_{i_1} \otimes e_{i_2} \otimes \cdots \otimes e_{i_p} \otimes e^{j_1} \otimes e^{j_2} \otimes \cdots \otimes e^{j_q}.$$

We shall say that T is a p-covariant, q-contravariant tensor, or a tensor of rank (p, q), when under a change of coordinate induced by a diffeomorphism $x \longleftrightarrow z$, as described above, the generic component of T behaves as

$$\bar{T}^{h_1 \dots h_p}_{k_1 \dots k_q} = T^{i_1 \dots i_p}_{j_1 \dots j_q} \frac{\partial z^{h_1}}{\partial x^{i_1}} \cdots \frac{\partial z^{h_p}}{\partial x^{i_p}} \frac{\partial x^{j_1}}{\partial z^{k_1}} \cdots \frac{\partial x^{j_q}}{\partial z^{k_q}},$$

where $T^{i_1 \dots i_p}_{j_1 \dots j_q}$ is the component of the representation of T in terms of the coordinates x^r, while $\bar{T}^{h_1 \dots h_p}_{k_1 \dots k_q}$ is its counterpart in terms of the coordinates z^s. In this book, besides various types of second-rank tensors, we use just one of fourth rank, especially in the form with all covariant components, namely

$$\mathbb{C} = \mathbb{C}_{ijhk} e^i \otimes e^j \otimes e^h \otimes e^k.$$

For A a second-rank tensor, we write $(\mathbb{C}A)$ for the product contracting all the indices of A, i.e., $\mathbb{C}A$ is a second-rank tensor with ijth component given by

$$(\mathbb{C}A)_{ij} = \mathbb{C}_{ijhk} A^{kh}.$$

A.6 Some Notions in Tensor Analysis

Consider (i) a real vector space $\mathcal{V}$, (ii) a tensor space $\mathcal{T}$, e.g., $\mathcal{V} \times \mathcal{V}^*$ or something else, (iii) a domain Ω in $\mathbb{R}^n$. We discuss here only the case in which $\mathcal{V}$ and $\mathcal{T}$ have finite dimension. Take maps

$$\tilde{v} : \Omega \longrightarrow \mathcal{V} \tag{A.7}$$

and

$$\tilde{T} : \Omega \longrightarrow \mathcal{T}. \tag{A.8}$$

We call them a **vector** and a **tensor field**, respectively, and write v for $\tilde{v}(x)$ and T for $\tilde{T}(x)$. In fact, even the map (A.7) defines a tensor field, and it is of type $(1, 0)$, while if we considered a **covector field**

$$\tilde{b} : \Omega \longrightarrow \mathcal{V}^*,$$

we could always think of it as a tensor field of type (0, 1). Consider a differentiable scalar function $f : \Omega \longrightarrow \mathbb{R}$. Its differential is given by

$$df = \frac{\partial f}{\partial x^i} dx^i = \left(\frac{\partial f}{\partial x^1}, \frac{\partial f}{\partial x^2}, \ldots, \frac{\partial f}{\partial x^n} \right) \begin{pmatrix} dx^1 \\ dx^2 \\ \cdot \\ \cdot \\ \cdot \\ dx^n \end{pmatrix},$$

where the last identity considers the standard row–column multiplication, and $\{x^1, \ldots, x^n\}$ are the coordinates of a point x in Ω. We consider the list $\left(\frac{\partial f}{\partial x^1}, \ldots, \frac{\partial f}{\partial x^n} \right)$ as that of the components of a covector Df that is the **derivative** of f. We define the **gradient** of f, and denote it by ∇f, the contravariant version of Df, i.e.,

$$\nabla f := (Df)^{\#} = (Df) g^{-1},$$

with components $(\nabla f)^i = (Df)_j g^{ji}$.

A vector field $\tilde{v}$ is said to be *differentiable* at $x \in \Omega$ if there is a linear mapping $D\tilde{v}(x) \in \mathrm{Hom}(\mathbb{R}^n, \mathcal{V})$ such that with $u \in \mathbb{R}^n$,

$$\tilde{v}(x + u) = \tilde{v}(x) + (D\tilde{v}(x))u + \mathrm{o}(u)$$

as $u \to 0$. In the previous expression, $\mathrm{o}(u)$ denotes higher-order terms in the sense that when $D\tilde{v}(x)$ exists, it satisfies for $\alpha \in \mathbb{R}$ the limit

$$\lim_{\alpha \to 0} \frac{\tilde{v}(x + \alpha u) - \tilde{v}(x)}{\alpha} = (D\tilde{v}(x))u.$$

The following *rules* hold:

- 1. For $\tilde{v}$ and $\tilde{w}$ two vector fields as above,

$$D(v \cdot w) = (Dv)^* w + (Dw)^* v.$$

- 2. For f a scalar function,

$$D(fv) = v \otimes Df + fDv.$$

– 3. With $\tilde{v} : \mathbb{R}^n \longrightarrow \mathcal{V}$ and $\tilde{w} : \Omega \longrightarrow \mathbb{R}^n$, denoting by $\tilde{v} \circ \tilde{w}$ their composition ($\tilde{v}$ is a function of $\tilde{w}$), the *chain rule*

$$D(\tilde{v} \circ \tilde{w})(x) = ((D\tilde{v}) \circ \tilde{w})\, D\tilde{w}(x) = D_w \tilde{v} D\tilde{w}(x)$$

holds, with D_w denoting the derivative of $\tilde{v}$ with respect to $\tilde{w}$, evaluated at x.

Consider $\mathbb{R}^n$ and a copy $\tilde{\mathbb{R}}^n$ of it. By z^i and $\tilde{z}^i$ we denote coordinates of points in $\mathbb{R}^n$ and $\tilde{\mathbb{R}}^n$, respectively, in Cartesian frames, while x^i and $\tilde{x}^i$ are coordinates of the same points in other frames, connected to the Cartesian one by diffeomorphisms $x \longleftrightarrow z$, $\tilde{x} \longleftrightarrow \tilde{z}$. Metrics g and $\tilde{g}$ pertain to these last frames. We write v for $\tilde{v}$ as a function of the coordinates z^i, and $\bar{v}$ as a function of x^i.

Consider a vector field $\tilde{v} : \mathbb{R}^n \longrightarrow \tilde{\mathbb{R}}^n$ having derivative $Dv := D\tilde{v}(x)$ with components $\frac{\partial v^i}{\partial z^j}$ in the Cartesian frame and $\frac{\partial \bar{v}^i}{\partial x^j}$ in the other frame. We consider the relation

$$\frac{\partial v^i}{\partial z^r} = \frac{\partial \tilde{z}^i}{\partial \tilde{x}^a} \frac{\partial \bar{v}^a}{\partial x^j} \frac{\partial x^j}{\partial z^r}. \tag{A.9}$$

With the agreement that (A^i_j) is the matrix with generic component A^i_j, we compute

$$\begin{aligned} \det Dv &= \det\left(\frac{\partial v^i}{\partial z^r}\right) = \det\left(\frac{\partial \tilde{z}^i}{\partial \tilde{x}^a} \frac{\partial \bar{v}^a}{\partial x^j} \frac{\partial x^j}{\partial z^r}\right) \\ &= \det\left(\frac{\partial \tilde{z}^i}{\partial \tilde{x}^a}\right) \det\left(\frac{\partial \bar{v}^a}{\partial x^j}\right) \det\left(\frac{\partial x^j}{\partial z^r}\right). \end{aligned} \tag{A.10}$$

Moreover, since the metric in $\tilde{\mathbb{R}}^n$ pertaining to the Cartesian frame is δ_{ij}, for the metric $\tilde{g}$ pertaining to the other frame, we have

$$\tilde{g}_{ab} = \frac{\partial \tilde{z}^i}{\partial \tilde{x}^a} \delta_{ij} \frac{\partial \tilde{z}^j}{\partial \tilde{x}^b},$$

so that

$$\begin{aligned} \det(\tilde{g}_{ab}) &= \det\left(\frac{\partial \tilde{z}^i}{\partial \tilde{x}^a}\right) \det(\delta_{ij}) \det\left(\frac{\partial \tilde{z}^j}{\partial \tilde{x}^b}\right) \\ &= \det\left(\frac{\partial \tilde{z}^i}{\partial \tilde{x}^a}\right) \det\left(\frac{\partial \tilde{z}^j}{\partial \tilde{x}^b}\right) = \left(\det\left(\frac{\partial \tilde{z}^i}{\partial \tilde{x}^a}\right)\right)^2. \end{aligned}$$

For the same reason, since in $\mathbb{R}^n$,

$$\det(g^{bc}) = \det\left(\frac{\partial x^b}{\partial z^a}\right) \det(\delta^{ak}) \det\left(\frac{\partial x^c}{\partial z^k}\right),$$

we have

$$\det g^{-1} = \det(g^{bc}) = \left(\det\left(\frac{\partial x^b}{\partial z^a}\right)\right)^2.$$

Consequently, from the identity (A.10), we get

$$\det D\tilde{v}(x) = \frac{\sqrt{\det \tilde{g}}}{\sqrt{\det g}} \det\left(\frac{\partial \bar{v}^a}{\partial x^j}\right).$$

For the sake of simplicity, consider $\tilde{v}$ to be a function of $\mathbb{R}^n$ onto itself: $\tilde{v} : \mathbb{R}^n \longrightarrow \mathbb{R}^n$ and write $v := \tilde{v}(z)$ and $\bar{v} := \tilde{v}(x)$, considering different coordinates. For a one-to-one differentiable change of coordinates with differentiable inverse, $x \longleftrightarrow z$, we have, in fact,

$$v^i = \frac{\partial z^i}{\partial x^j}\bar{v}^j \qquad \text{and} \qquad \bar{v}^j = \frac{\partial x^j}{\partial z^i}v^i,$$

so that we compute

$$\begin{aligned}\frac{\partial v^i}{\partial z^k} &= \frac{\partial}{\partial z^k}\left(\frac{\partial z^i}{\partial x^j}\bar{v}^j\right) = \frac{\partial x^l}{\partial z^k}\frac{\partial}{\partial x^l}\left(\frac{\partial z^i}{\partial x^j}\bar{v}^j\right) \\ &= \frac{\partial z^i}{\partial x^j}\frac{\partial \bar{v}^j}{\partial x^l}\frac{\partial x^l}{\partial z^k} + \bar{v}^j\frac{\partial^2 z^i}{\partial x^j \partial x^l}\frac{\partial x^l}{\partial z^k} \\ &= \frac{\partial z^i}{\partial x^j}\frac{\partial \bar{v}^j}{\partial x^l}\frac{\partial x^l}{\partial z^k} + v^r\frac{\partial x^j}{\partial z^r}\frac{\partial^2 z^i}{\partial x^j \partial x^l}\frac{\partial x^l}{\partial z^k}.\end{aligned}$$

Consequently, we have

$$\frac{\partial v^i}{\partial z^k} = \frac{\partial z^i}{\partial x^j}\frac{\partial \bar{v}^j}{\partial x^l}\frac{\partial x^l}{\partial z^k},$$

i.e., the components of Dv change as those of a second-rank tensor under changes of coordinates $x \longleftrightarrow z$ as above, only when

$$\frac{\partial^2 z^i}{\partial x^j \partial x^l} = 0,$$

i.e., for changes of coordinates of the type

$$z^i = a^i_j x^j + b^i$$

with a^i_j and b^i constant; they are called *affine*. Consequently, for general changes of frame, to ensure that $D\tilde{v}(z)$ behaves tensorially, we have to define its ijth component $v^i_{,j}$, where the comma indicates the presence of a derivative with respect to the coordinate indicated by the index to the right of the comma, as

$$v^i_{,j} = \frac{\partial v^i}{\partial z^j} - v^r \frac{\partial x^k}{\partial z^r} \frac{\partial^2 z^i}{\partial x^k \partial x^l} \frac{\partial x^l}{\partial z^j}, \tag{A.11}$$

an expression that we restrict by defining

$$\Gamma^i_{kj} := -\frac{\partial x^k}{\partial z^r} \frac{\partial^2 z^i}{\partial x^k \partial x^l} \frac{\partial x^l}{\partial z^j}$$

and calling it the *Christoffel symbol*. In this sense, we may interpret $v^i_{,j}$ as a "generalized" derivative, and we call the process leading to it *covariant differentiation*. A summary of these ideas follows.

Theorem 9. *Assume that the ("generalized") derivative of a vector field $\tilde{v}$ transforms as a tensor under every change of frames, and in coordinates x^i, we have*

$$\bar{v}^i_{,j} = \frac{\partial \bar{v}^i}{\partial x^j}.$$

Then in any other coordinate system z^i, the derivative of $\tilde{v}$ has the form

$$v^i_{,j} = \frac{\partial v^i}{\partial z^j} + \Gamma^i_{kj} v^k.$$

The Christoffel symbol as defined above expresses in $\mathbb{R}^n$ what we commonly call a *Euclidean connection*. In greater generality, we define a **connection** on a manifold[1] $\mathcal{M}$ as an operation $\bar{\nabla}$ that associates to each pair of vector fields v and w on $\mathcal{M}$ a third vector field $\bar{\nabla}_w v$, called the covariant derivative of v along w such that

- (a) $\bar{\nabla}_w v$ is linear in each of w and v,
- (b) for scalar functions f we have $\bar{\nabla}_{fw} v = f \bar{\nabla}_w v$,
- (c) $\bar{\nabla}_w (fv) = f \bar{\nabla}_w f + \left(\frac{\partial f}{\partial x^i} w^i \right) v.$

[1]Consider a set $\mathcal{M}$ endowed with a topology that is Hausdorff, i.e., for every pair of distinct elements of $\mathcal{M}$, say y_1 and y_2, there exist nonintersecting open neighborhoods $\mathcal{I}(y_1)$ and $\mathcal{I}(y_2)$ containing y_1 and y_2, respectively. We say that $\mathcal{M}$ is a *topological manifold* when it is *locally Euclidean*, i.e., every $y \in \mathcal{M}$ has an open subset of $\mathcal{M}, \mathcal{U}(y)$, containing it, which is homeomorphic to an open subset $\mathcal{Y}$ of $\mathbb{R}^n$, meaning that it is possible to define a one-to-one mapping $\varphi : \mathcal{U} \longrightarrow \mathcal{Y}$ of $\mathcal{U}$ onto $\mathcal{Y}$. We call the pair $(\mathcal{U}, \varphi)$ a *coordinate chart* (or simply *chart*), and the set $\mathfrak{F} := \{(\mathcal{U}_i, \varphi_i)\}$ of charts such that $\cup_i \mathcal{U}_i = \mathcal{M}$, for $i \in I$, with I some index set, an *atlas*. In other words, $\mathfrak{F}$ determines a covering of the coordinate system over the whole of $\mathcal{M}$. In particular, we say that $\mathcal{M}$ has dimension n when all $\mathcal{U}_i$ are mapped onto sets $\mathcal{Y} \subseteq \mathbb{R}^n$ with dimension n. If for *all* $i, j \in I$, the change of coordinates between charts is of class C^k, i.e., $\varphi_i \circ \varphi_j^{-1}$ is C^k with $1 \leq k \leq +\infty$, and for every chart $(\mathcal{U}, \varphi)$ such that $\varphi \circ \varphi_i^{-1}$ and $\varphi_i \circ \varphi^{-1}$ are C^k, for all $i \in I$, we get $(\mathcal{U}, \varphi) \in \mathfrak{F}$, we say that $\mathcal{M}$ is a *differentiable manifold of class* C^k, or simply a *differentiable manifold* when $k = +\infty$.

The related Christoffel symbols with respect to coordinates x^i are defined by

$$\Gamma^i_{jk}(x)e_i = \left(\bar{\nabla}_{e_j} e_k\right)(x).$$

When the domain and the target space of the vector field considered are different, e.g., we have $\tilde{v} : \mathbb{R}^n \longrightarrow \tilde{\mathbb{R}}^n$, then the expression of $v^i_{,A}$, where the index A refers to coordinates in $\mathbb{R}^n$, while i refers to those in $\tilde{\mathbb{R}}^n$, has the same structure already determined for $v^i_{,j}$, but now in the related Christoffel symbol Γ^i_{kA} the indices i and k refer to coordinates in $\tilde{\mathbb{R}}^n$, while A refers to those in $\mathbb{R}^n$. Derive this as an exercise by following the same steps leading to the expression of $v^i_{,j}$. Along the same lines, we can compute an explicit expression for the components of the derivative of a covector field $\tilde{c} : \mathbb{R}^n \longrightarrow \mathbb{R}^{n*}$.

Theorem 10. *Assume that the ("generalized") derivative of a covector field $\tilde{c}$ transforms as a tensor under every change of frames, and in coordinates x^i, we have*

$$\bar{c}_{i,j} = \frac{\partial \bar{c}_i}{\partial x^j},$$

where the overbar recalls that $\tilde{c}$ is seen as a function of $x = \{x^1, \ldots, x^n\}$. Then in any other coordinate system z^i, the derivative of $\tilde{c}$ has the form

$$c_{i,j} = \frac{\partial c_i}{\partial x^j} - \Gamma^k_{ij} c_k.$$

Consider a tensor field $x \longmapsto T := \tilde{T}(x)$ of the type (p, q). Its derivative $D\tilde{T}(x)$, when it exists, maps T into a tensor of type $(p, q+1)$.

Theorem 11. *Assume that the ("generalized") derivative of a tensor field $\tilde{T}$ transforms as a tensor under every change of frames, and in coordinates x^i, it is expressed by*

$$\bar{T}^{i_1 \ldots i_p}_{j_1 \ldots j_q, k} = \frac{\partial \bar{T}^{i_1 \ldots i_p}_{j_1 \ldots j_q}}{\partial x^k}.$$

Then in every other coordinate system z^i, we have

$$T^{i_1 \ldots i_p}_{j_1 \ldots j_q, k} = \frac{\partial T^{i_1 \ldots i_p}_{j_1 \ldots j_q}}{\partial x^k} - T^{i_1 \ldots i_p}_{l j_2 \ldots j_q} \Gamma^l_{j_1 k} - \cdots - T^{i_1 \ldots i_p}_{j_1 \ldots j_{q-1} l} \Gamma^l_{j_q k}$$
$$+ T^{l i_2 \ldots i_p}_{j_1 \ldots j_q} \Gamma^{i_1}_{lk} + \cdots + T^{i_1 \ldots i_{p-1} l}_{j_1 \ldots j_q} \Gamma^{i_p}_{lk}.$$

The algebraic signs in the previous formula follow from the calculations leading to the expressions of $v^i_{,j}$ and $c_{i,j}$.

In the previous expression, the Christoffel symbols can vanish. In particular, we call a coordinate system such that $\Gamma^i_{jk} = 0$ *Euclidean*. In this case, we say that the connection is trivial. Such is the case in Cartesian frames, i.e., when $g_{ij} = \delta_{ij}$, called also Euclidean for this reason. The notion of metric and connection are in principle unrelated to each other. However, we can establish a correspondence between the two that is canonical in the sense that $\Gamma^i_{jk} = 0$ corresponds to $g_{ij} = \delta_{ij}$. To this end, we say that a connection Γ^k_{ij} is *compatible with a nonconstant metric* g_{ij} if

$$g_{ij,k} = \frac{\partial g_{ij}}{\partial x^k} - g_{il}\Gamma^l_{jk} - \Gamma^l_{ik}\, g_{lj} = 0.$$

Theorem 12. *Let $g(x)$ be a nondegenerate metric, meaning that $\det g \neq 0$. There exists a unique symmetric connection compatible with this metric. In any coordinate system $\{x^1, \ldots, x^n\}$, it is expressed by the Christoffel formula*

$$\Gamma^k_{ij} = \frac{1}{2} g^{kl} \left(\frac{\partial g_{lj}}{\partial x^i} + \frac{\partial g_{il}}{\partial x^j} - \frac{\partial g_{ij}}{\partial x^l} \right).$$

In particular, a constant metric is compatible with the trivial connection. On contracting the contravariant index of T with its second covariant component, we obtain

$$\Gamma^i_{ki} = \frac{1}{2} g^{il} \left(\frac{\partial g_{lk}}{\partial x^i} + \frac{\partial g_{li}}{\partial x^k} - \frac{\partial g_{ki}}{\partial x^l} \right) = \frac{1}{2} g^{il} \frac{\partial g_{il}}{\partial x^k}.$$

However, since

$$\frac{\partial \det g}{\partial x^k} = \frac{\partial \det g}{\partial g} \frac{\partial g}{\partial x^k} = (\det g) g^{-\mathrm{T}} \frac{\partial g}{\partial x^k},$$

for a second-rank tensor field $A(x)$ with $\det A \neq 0$, the derivative of its determinant with respect to A exists and is given by

$$\frac{\partial \det A}{\partial A} = (\det A) A^{-\mathrm{T}}$$

(see Chapter 3 for the proof), and we obtain

$$\frac{1}{2} g^{il} \frac{\partial g_{li}}{\partial x^k} = \frac{1}{2 \det g} \frac{\partial \det g}{\partial x^k} = \frac{\partial}{\partial x^k} \log\left(\sqrt{|\det g|} \right).$$

We define the **divergence** of a vector field $\tilde{v}$ to be the trace of its derivative. We then write

$$\operatorname{div} v = v^i_{,i} = \frac{\partial v^i}{\partial x^i} + \Gamma^i_{ki} v^k = \frac{\partial v^i}{\partial x^i} + \frac{1}{2 \det g} \frac{\partial \det g}{\partial x^k} v^k$$
$$= \frac{1}{\sqrt{|\det g|}} \frac{\partial}{\partial x^k} \left(\sqrt{|\det g|}\, v^k \right).$$

Notice that in the previous formula, the presence of the absolute value of $\det g$ appears because in the theorem giving the expression of the Christoffel formula, we have required only that $\det g \neq 0$. In Euclidean (Cartesian) coordinates, we have simply

$$\operatorname{div} v = \frac{\partial v^1}{\partial x^1} + \cdots + \frac{\partial v^n}{\partial x^n}.$$

In this case, we define the **curl** of $\tilde{v}$ to be the vector field

$$\operatorname{curl} v = (\mathsf{e}\nabla v)^{\#},$$

i.e., the vector associated with the covector with components

$$\mathsf{e}_{ijk}\left(\frac{\partial v^k}{\partial x^i}\right)^j.$$

In particular, for $n = 3$, curl v is a vector with components

$$\begin{pmatrix} \dfrac{\partial v^3}{\partial x^2} - \dfrac{\partial v^2}{\partial x^3} \\ \dfrac{\partial v^1}{\partial x^3} - \dfrac{\partial v^3}{\partial x^1} \\ \dfrac{\partial v^2}{\partial x^1} - \dfrac{\partial v^1}{\partial x^2} \end{pmatrix}.$$

Analogous definitions hold for covector fields above all in Euclidean (Cartesian) coordinates, where we can confuse them with vectors, since the metric involved is δ_{ij}. The divergence of a tensor field $x \longmapsto T := \tilde{T}(x)$ can be defined as the contraction in DT of the index of D with one of T in a way such that if T is (p, q), so that DT is $(p, q+1)$, its divergence is $(p-1, q)$ or $(p, q-1)$, depending on the choice we adopt, a choice to be specified every time, and when T is $(0, q)$, its divergence is $(0, q-1)$.

As regards the curl of a tensor, for a second-rank tensor A with components A_{ij}, denoting by $\tilde{A}$ its 1-contravariant, 1-covariant version with components $\tilde{A}^i_j$, we define $\operatorname{curl} A$ to be the $(0, 2)$ tensor with components

$$(\operatorname{curl} A)_{ij} = \mathsf{e}_{ilk}\left(\frac{\partial \tilde{A}^k_j}{\partial x}\right)^l.$$

We also write at times

$$(\operatorname{curl} A)_{ij} = \mathsf{e}_{ilk}\frac{\partial A_{lj}}{\partial x^k},$$

not distinguishing between vector and covector components, since the identification of one with the other is allowed in the Euclidean (Cartesian) ambient. In such a setting, the derivative operator D commutes, so that we have $D_j D_k = D_k D_j$. However, such a property holds just in some coordinate system, due to the presence of the Christoffel symbols. In fact, in arbitrary coordinates $\{x^1, \ldots, x^n\}$, for any differentiable n-dimensional vector field $\tilde{v}$, we compute

$$(D_k D_j - D_j D_k) v^i = \mathbf{R}^i_{lkj} v^l + \mathbf{T}^m_{kj} D_m v^i,$$

where

$$\mathbf{R}^i_{lkj} = \frac{\partial \Gamma^i_{lj}}{\partial x^k} - \frac{\partial \Gamma^i_{lk}}{\partial x^j} + \Gamma^i_{mk} \Gamma^m_{lj} - \Gamma^i_{mj} \Gamma^m_{lk}$$

is the so-called *curvature tensor* or the *Riemann tensor* and

$$\mathbf{T}^i_{jk} = \Gamma^i_{jk} - \Gamma^i_{kj}$$

is the so-called *torsion tensor*.

Consider two tensor fields $x \longmapsto T := \tilde{T}(x)$ and $x \longmapsto S := \tilde{S}(x)$, respectively, of type (p, q) and (r, s). We have

$$D(T \otimes S) = DT \otimes S + T \otimes DS,$$

where denoting by $T^{(p)}_{(q)}$ the generic component $T^{i_1 \ldots i_p}_{j_1 \ldots j_q}$, the tensor $DT \otimes S$ is intended to have components $(DT \otimes S)^{(p)(r)}_{(q)(s)k}$, with the index k referring to $\dfrac{\partial}{\partial x^k}$.

We now list some relations used often in the book in the Euclidean (Cartesian) setting. They involve scalar functions $f = \tilde{f}(x)$, vector fields $v = \tilde{v}(x)$, and second-rank tensor fields $A = \tilde{A}(x)$, all assumed to be differentiable.

– We have

$$\begin{aligned}
&\operatorname{curl} Df = 0, \\
&\operatorname{div} \operatorname{curl} v = 0, \\
&\operatorname{curl} Dv = 0, \\
&\operatorname{curl} (Dv)^{\mathrm{T}} = D \operatorname{curl} v, \\
&\operatorname{div} \operatorname{curl} A = \operatorname{curl} \operatorname{div} A^{\mathrm{T}}, \\
&(\operatorname{curl} \operatorname{curl} A)^{\mathrm{T}} = \operatorname{curl} \operatorname{curl} A^{\mathrm{T}}, \\
&\operatorname{div} (\operatorname{curl} A)^{\mathrm{T}} = 0, \\
&\operatorname{curl} (fI) = - (\operatorname{curl} (fI))^{\mathrm{T}},
\end{aligned}$$

with I the second-rank identity tensor.

– With $A \in \mathrm{Hom}(\mathcal{V}^*, \mathcal{V})$ and $v \in \mathcal{V}$, we get

$$\mathrm{div}(A^{\mathrm{T}} v) = v \cdot \mathrm{div}\, A + A \cdot Dv.$$

– For a differentiable covector field $x \longmapsto c = \tilde{c}(x)$, assume that Dc is skew-symmetric and therefore can be written

$$Dc = \frac{\partial c_i}{\partial x^j} \mathrm{Skw}(e^i \otimes e^j),$$

which we rewrite as

$$Dc = \frac{\partial c_i}{\partial x^j} e^i \wedge e^j,$$

where the wedge product extracts the skew-symmetric part of the tensor product $\otimes$. We then get

$$DDc = \frac{\partial^2 c_i}{\partial x^k \partial x^j} e^i \wedge e^j \wedge e^k = 0,$$

since $\dfrac{\partial^2}{\partial x^k \partial x^j}$ is symmetric, while $e^j \wedge e^k$ is skew-symmetric and the product $A^{sym} \cdot B^{skw}$ of a symmetric tensor A^{sym} by a skew-symmetric tensor B^{skw} always vanishes. Such a property can be written in terms of differentials of tensors with all covariant components, i.e., tensors of the type $(0, k)$, which we write here as

$$\omega = \omega_{i_1 \ldots i_k} dx^{i_1} \wedge \cdots \wedge dx^{i_k}.$$

For these we compute, in fact,

$$d^2\omega = d(d\omega) = \frac{\partial^2 \omega_{i_1 \ldots i_k}}{\partial x^p \partial x^q} dx^{i_1} \wedge \cdots \wedge dx^{i_k} \wedge dx^p \wedge dx^q = 0,$$

for A^{sym}.

– Denoting by Δ the **Laplacian operator** $\Delta = \mathrm{div}\, D$, we have

$$\mathrm{curl}\,\mathrm{curl}\, v = D\,\mathrm{div}\, v - \Delta u.$$

– If $x \longmapsto A := \tilde{A}(x)$ is symmetric, we obtain

$$\mathrm{tr}\,\mathrm{curl}\, A = 0,$$
$$\mathrm{curl}\,\mathrm{curl}\, A = -\Delta A + 2\,\mathrm{Sym} D\,\mathrm{div}\, A - DD\mathrm{tr}\, A + I(\Delta \mathrm{tr}\, A - \mathrm{div}\,\mathrm{div}\, A).$$

– We consider the integral of a vector (a tensor) field over an m-dimensional domain Ω, with boundary $\partial\Omega$ oriented by the normal n (considered as a covector) almost everywhere with respect to the "surface" measure $d\mathcal{H}^{m-1}$, as a vector (a tensor) with components given by the integral of the components of the integrand. With $d\mu$ the m-dimensional measure over Ω, we have

$$\int_{\partial\Omega} fn \, d\mathcal{H}^{m-1} = \int_{\Omega} Df \, d\mu,$$

$$\int_{\partial\Omega} v \otimes n \, d\mathcal{H}^{m-1} = \int_{\Omega} Dv \, d\mu,$$

$$\int_{\partial\Omega} v \cdot n \, d\mathcal{H}^{m-1} = \int_{\Omega} \operatorname{div} v \, d\mu,$$

$$\int_{\partial\Omega} An \, d\mathcal{H}^{m-1} = \int_{\Omega} \operatorname{div} A \, d\mu,$$

and by considering n a vector,

$$\int_{\partial\Omega} v \times n \, d\mathcal{H}^{m-1} = -\int_{\Omega} \operatorname{curl} v \, d\mu.$$

This appendix is just a rough summary of some notions in differential geometry. For further reading, one might consider the following basic treatises on the subject:

– Abraham R., Marsden J.E., Ratiu T.S. (1988), *Manifolds, Tensor Analysis, and Applications*, Springer-Verlag, Berlin;
– Dubrovin B.A., Novikov S.P., Fomenko A.T. (1990 and 1992), *Modern Geometry: Methods and Applications*, Parts I, II, III, Springer-Verlag, Berlin;
– Novikov S.P., Taimanov I.A. (2006), *Modern Geometric Structures and Fields*, American Mathematical Society, Providence, Rhode Island;
– Olver P.J. (2000), *Applications of Lie Groups to Differential Equations*, Springer-Verlag, Berlin;
– Slawianowski J. (1991), *Geometry of Phase Spaces*, PWN-Polish Scientific Publishers, Warsaw and John Wiley and Sons, Chichester;
– Warner F.W. (1983), *Foundations of Differentiable Manifolds and Lie Groups*, Springer-Verlag, Berlin.

In concluding this appendix, we mention that most of the subject matter presented in this book could be presented entirely in the Euclidean setting. However, our distinction between vectors and covectors or referential and actual metrics is not an unnecessary complication. To us, the way we have followed in presenting the subject underlines that the structure of the basic principles in the mechanics of continua does not vary when we go beyond the Euclidean setting and prepares the reader, from the first meeting with the topic, to proceed naturally toward an extended perspective—such as indicated in the analysis of material complexities, just to take one example. We are, however, conscious that some readers might have a different opinion.

Suggestions for Further Reading

We propose here a few suggestions for further reading to those who might be interested in deepening their knowledge of theoretical and applied mechanics. The list includes only books or book chapters and is not exhaustive. These books have different styles and viewpoints, and they do not all cover the same topics.

The list reflects just our actual choices, nothing more.

- Antman, S.S.: Nonlinear Problems of Elasticity. Springer-Verlag, Berlin (1995)
- Arnold, V.I.: Mathematical Methods of Classical Mechanics. Springer-Verlag, Berlin (1989)
- Baldacci, R.: Scienza delle Costruzioni, vols. I and II. UTET, Torino (1984) [in Italian]
- Bažant, Z.P., Planas, J.: Fracture and Size Effect in Concrete and Other Quasi Brittle Materials. CRC Press LLC, Baton Rouge (1998)
- Bedford, A.: Hamilton's Principle in Continuum Mechanics. Pitman Publishing Inc., London (1985)
- Berthram, A.: Elasticity and Plasticity of Large Deformations. Springer-Verlag, Berlin (2012)
- Bhattacharya, K.: Microstructure of Martensite: Why It Forms, How It Gives Rise to the Shape-Memory Effect. Oxford University Press, Oxford (2003)
- Capriz, G.: Continua with Microstructure. Springer-Verlag, Berlin (1989)
- Ciarlet, P.G.: Mathematical Elasticity, vol. I: Three-Dimensional Elasticity (1997); vol. II: Theory of Plates (2000); vol. III: Theory of Shells. North-Holland Publishing Co., Amsterdam (1988)
- Chorin, A.J., Marsden, J.E.: A Mathematical Introduction to Fluid Mechanics. Springer-Verlag, New York (2013)
- Corradi Dall'Acqua, L.: Meccanica delle Strutture, vols. 1, 2, and 3. McGraw-Hill, Milano (2010) [in Italian]
- Dafermos, C.: Hyperbolic Conservation Laws in Continuum Physics. Springer-Verlag, New York (2010)

P.M. Mariano, L. Galano, *Fundamentals of the Mechanics of Solids*,
DOI 10.1007/978-1-4939-3133-0

- Weinan, E.: Principles of Multiscale Modeling. Cambridge University Press, Cambridge (2011)
- Freund, L.B.: Dynamic Fracture Mechanics. Cambridge University Press, Cambridge (1990)
- Gallavotti, G.: Foundations of Fluid Dynamics. Springer-Verlag, Berlin (2002)
- Giaquinta, M., Hildebrandt, S.: Calculus of Variations, vol. I The Lagrangian Formalism; vol. II The Hamiltonian Formalism. Springer-Verlag, Berlin (1996)
- Giaquinta, M., Modica, G., Souček, J.: Cartesian Currents in the Calculus of Variations, vols. I and II. Springer-Verlag, Berlin (1998)
- Guckenheimer, J., Holmes, P.J.: Nonlinear Oscillations, Dynamical Systems and Bifurcations of Vector Fields. Springer-Verlag, Berlin (1986)
- Gurtin, M.E.: The linear theory of elasticity. In: Truesdell, C.A. (ed.) Handbuch der Physik, Band IVa/2, pp. 1–295. Springer-Verlag, Berlin (1973)
- Krajcinovic, D.: Damage Mechanics. North-Holland, Amsterdam (1996)
- Marsden, J.E., Hughes, T.R.J.: Mathematical Foundations of Elasticity. Prentice Hall Inc., Englewood Cliffs, NJ (1983); 1994 Dover Edition
- Marsden, J.E., Ratiu, T.: Introduction to Mechanics and Symmetry. Springer-Verlag, Berlin (1999)
- Mariano, P.M.: Elements of Multifield Theories for the Mechanics of Complex Materials. Springer-Verlag (2016)
- Mariano, P.M.: Mathematical Representation of Plasticity and Other Material Mutations. American Mathematical Society (2016)
- Maugin, G.A.: The Thermodynamics of Plasticity and Fracture. Cambridge University Press, Cambridge (1992)
- Milton, G.W.: The Theory of Composites. Cambridge University Press, Cambridge (2002)
- Müller, I.: Thermodynamics. Pittman, Boston (1985)
- Noll, W.: The Foundations of Mechanics and Thermodynamics. Springer-Verlag, Berlin (1974)
- Nemat-Nasser, S.: Plasticity. A Treatise on Finite Deformation of Heterogeneous Inelastic Materials. Cambridge University Press, Cambridge (2004)
- Ogden, R.W.: Non-linear Elastic Deformations. Ellis Horwood Ltd., Chichester (1984)
- Šilhavý, M.: The Mechanics and Thermodynamics of Continuous Media. Springer-Verlag, Berlin (1997)
- Simo, J.C., Hughes, T.R.J.: Computational Inelasticity. Springer-Verlag, Berlin (1998)
- Truesdell, C.A.: Six Lectures on Modern Natural Philosophy. Springer-Verlag, Berlin (1966)
- Truesdell, C.A.: A First Course in Rational Continuum Mechanics. Academic Press, New York (1977)
- Truesdell, C.A., Muncaster, R.G.: Fundamentals of Maxwell's Kinetic Theory of a Simple Monoatomic Gas. Academic Press, New York (1980)
- Truesdell, C.A., Noll, W.: The non-linear field theories of mechanics. In: Handbuch der Physik, Band III/3, pp. 1–602. Springer-Verlag, Berlin (1965)

- Truesdell, C.A., Rajagopal, K.R.: An Introduction to the Mechanics of Fluids. Birkhäuser, Boston (2005)
- Truesdell, C.A., Toupin, R.A.: Classical field theories of mechanics. In: Handbuch der Physik, Band III/1, pp. 226–793. Springer-Verlag, Berlin (1960)
- Villaggio, P.: Qualitative Methods in Elasticity. Noordhoff Int. Publ. Co., Leyden (1977)
- Villaggio, P.: Mathematical Models of Elastic Structures. Cambridge University Press, Cambridge (2005)

Index

P.M. Mariano, L. Galano, *Fundamentals of the Mechanics of Solids*,
DOI 10.1007/978-1-4939-3133-0

Printed in the United States
By Bookmasters

Printed in the United States
By Bookmasters